Transdisciplinary Challenges in Landscape Ecology and Restoration Ecology

Landscape Series

Volume 7

Series Editors:

Aims & Scope:

Springer's innovative Landscape Series is committed to publishing high-quality manuscripts that approach the concept of landscape from a broad range of perspectives. Encouraging contributions on theory development, as well as more applied studies, the series attracts outstanding research from the natural and social sciences, and from the humanities and the arts. It also provides a leading forum for publications from interdisciplinary and transdisciplinary teams.

Drawing on, and synthesising, this integrative approach the Springer Landscape Series aims to add new and innovative insights into the multidimensional nature of landscapes. Landscapes provide homes and livelihoods to diverse peoples; they house historic – and prehistoric – artefacts; and they comprise complex physical, chemical and biological systems. They are also shaped and governed by human societies who base their existence on the use of the natural resources; people enjoy the aesthetic qualities and recreational facilities of landscapes, and people design new landscapes.

As interested in identifying best practice as it is in progressing landscape theory, the Landscape Series particularly welcomes problem-solving approaches and contributions to landscape management and planning. The ultimate goal is to facilitate both the application of landscape research to practice, and the feedback from practice into research.

Transdisciplinary Challenges in Landscape Ecology and Restoration Ecology - An Anthology

with Forewords by E. Laszlo and M. Antrop and Epilogue by E. Allen

by

ZEV NAVEH

Faculty of Civil and Environmental Engineering,
Technion, Israel Institute of Technology, Haifa, Israel

Springer

A C.I.P. Catalogue record for this book is available from the Library of Congress.

ISBN-10 1-4020-4420-8 (HB)
ISBN-13 978-1-4020-4420-5 (HB)
ISBN-10 1-4020-4422-4 (e-book)
ISBN-13 978-1-4020-4422-9 (e-book)

Published by Springer,
P.O. Box 17, 3300 AA Dordrecht, The Netherlands.

www.springer.com

Cover photograph by Barbel Tress and Gunther Tress

Printed on acid-free paper

Transdisciplinary Challenges in Landscape Ecology and Restoration Ecology - An Anthology

with Forewords by E. Laszlo and M. Antrop and Epilogue by E. Allen

by

ZEV NAVEH

Faculty of Civil and Environmental Engineering, Technion, Israel Institute of Technology, Haifa, Israel

Springer

A C.I.P. Catalogue record for this book is available from the Library of Congress.

ISBN-10 1-4020-4420-8 (HB)
ISBN-13 978-1-4020-4420-5 (HB)
ISBN-10 1-4020-4422-4 (e-book)
ISBN-13 978-1-4020-4422-9 (e-book)

Published by Springer,
P.O. Box 17, 3300 AA Dordrecht, The Netherlands.

www.springer.com

Cover photograph by Barbel Tress and Gunther Tress

Printed on acid-free paper

ACKNOWLEDGEMENTS

I acknowledge with gratitude the kind help of Dr Yohay Carmel, Prof. Emeritus Arthur S. Lieberman and Sonia Danziger during the composition of this anthology. My warmest thanks are also extended to Ms Ria Kanters, the publishing assistant of Springer and the editors of the Landscape Series Prof. Henri Decamps, Gunther and Baerbel Tress. I appreciate especially the encouragement I enjoyed throughout, and the readiness of Dr. Gunther and Baerbel Tress to help me in completing this book. I acknowledge gratefully the forewords contributed by Prof. E Laszlo and by M. Antrop, and the epilogue by E. Allen.

I am grateful to all those many scholars and students, landscape and restoration scientists, ecologists and geographers, planners and conservationists, whose interest in my work was a source of inspiration and pride for me. I also thank the publishers of the journals and books, for granting me permission to republish my studies in chapters 1–13.

I dedicate this book to
my dear wife Ziona,
whose love and devotion, patience
and encouragement have been
the strongest pillars of my life and work
for 55 years.

CONTENTS

Acknowledgements ix
Foreword by the Series Editors xi
Preface by Marc Antrop xiii
Foreword by Ervin Laszlo xvii

Introduction – A Review of the Chapters Content with some Autobiographical Comments 1

A. MEDITERRANEAN ISSUES

1 *Naveh Z.* (1967) Mediterranean ecosystems and vegetation types in California and Israel Ecology 48: 445–459. 35

2 *Naveh Z., R.H. Whittaker* (1979) Structural and floristic diversity of shrublands and woodlands in northern Israel and other Mediterranean areas Vegetation 4:171–190. 61

3 *Naveh Z.* (1990) Fire in the Mediterranean–a landscape ecological perspective. In: Goldammer, J.F., Jenkins M.J. (eds.) Fire in Ecosystems Dynamics. Proceedings of the Third International Symposium in Freiburg, FRG, May 1989. SPB Academic Publishing by, The Hague, the Netherlands, pp. 1–20. 95

4 *Naveh Z.* (1994) From Biodiversity to ecodiversity: a landscape-ecology approach to conservation and restoration. Restoration Ecology 2:180–189. 117

5 *Naveh Z.* (1995) Conservation, restoration, and research priorities for Mediterranean uplands threatened by global climate change. In: Moreno J, Oechel WE (eds.) Global Change and Mediterranean–Type Ecosystems Ecological Studies Vol. 117. Springer, New York, pp. 482–508. 135

6 *Naveh Z.* (1998) From biodiversity to ecodiversity–holistic conservation of the biological and cultural diversity of Mediterranean landscapes. In: Rundel P., Montenegro G., Jaksic F. M.(eds.) Landscape Disturbance and Biodiversity in Mediterranean–Type Ecosystems. Ecological Studies Vol. 136 Springer, Berlin pp. 23–50. 159

7 *Naveh Z.* (2005) The Role of Fire in the Evolution of the Mediterranean Cultural Landscape in the pleistocene and Early Holocene with special reference to Mt. Carmel. (Based on: The evolution of the cultural landscape in Israel as affected by fire, grazing, and human activities. Z. Naveh and Y. Carmel. In: Wasser S.P. (ed.) Papers in Honour of Eviatar Nevo. Evolutionary Theory and Processes: Modern Horizons. Kluwer Academic Publishers, Dodrecht pp. 337–409). 187

B. GLOBAL ISSUES

8 *Naveh Z.* (1966) The development of Tanzania Masailand. A sociological and ecological challenge. African Soils: 499–518. 217

9 *Naveh Z.* (1978) The role of landscape ecology in development Environmental Conservation 5: 57–63. 235

10 *Naveh Z.* (1998) Culture and landscape conservation: A landscape-ecological Perspective. In: Gopal B.P., Pathak P., Sayena K.G. (eds.) Ecology Today: An Anthology of Contemporary Ecological Research International Scientific Publications, New Delhi, pp.19–48. 249

11 *Naveh Z.* (1998) Ecological and cultural landscape restoration and the cultural evolution towards a post-industrial symbiosis between human society and nature. Restoration Ecology 6:135–143. 281

12 *Naveh Z.* (2001) Ten major premises for a holistic conception of multifunctional landscapes. Landscape and Urban Planning 57: 269–284. 297

13 *Naveh Z.* (2004) Multifunctional, self-organizing biosphere landscapes and the future of our Total Human Ecosystem. World Futures 60: 469–503. 319

C. RECAPITULATION

14 *Naveh Z.* (2005) The transformation of landscape ecology and restoration ecology into transdisciplinary sciences of holistic landscape study, management and planning, conservation and restoration. 353

EPILOGUE: A Lifetime of Leadership in the Fields of Landscape Ecology, Restoration Ecology, and Ecological Complexity- by Edith Allen, University of California, Riverside, California. 403

Author Index 407

Subject Index 415

FOREWORD BY THE SERIES EDITORS

With the Springer Landscape Series we want to provide a much-needed forum for dealing with the complexity of landscape types that occur, and are studied, globally. It is crucial that the series highlights the richness of this diversity – both in the landscapes themselves and in the approaches used in their study. Moreover, while the multiplicity of relevant academic disciplines and approaches is characteristic of landscape research, we also aim to provide a place where the synthesis and integration of different knowledge cultures is common practice.

Transdisciplinary Challenges for Landscape Ecology and Restoration Ecology is the seventh volume of the series. It is an anthology of global and Mediterranean issues that assembles the author's work of over 40 years of publications in the field of landscape ecology and restoration ecology. Professor Zev Naveh dedicated his work to the study of landscapes around the world and has been one of the pioneers in introducing holistic concepts into landscape studies demanding collaboration across disciplinary boundaries. The fourteen chapters in this book give unique evidence of the various stages of development and progress in landscape ecology and restoration ecology, which were and are influenced by the author until now. The anthology illustrates the increasing importance of landscape ecology in coping with societal and ecological challenges. It is the author's restless attempt to offer landscapes as a conceptual basis to approach recent environmental, social and economic challenges and problems in our world, understood as the Total Human Ecosystem. The chapters propose an academic development in the study of landscapes, which ultimately will lead to a holistic landscape science, including natural sciences, social sciences, humanities and the arts.

We are proud to publish this anthology in our series and recommend the book to students and all academics interested in understanding the development and need for landscape science. The anthology is an impressing record of landscape ecology and restoration ecology and will be an important source of inspiration for the current and future generation of academics involved in the study of landscapes.

Toulouse and Aberdeen, January 2006
Henri Décamps
Bärbel Tress
Gunther Tress

PREFACE

by Marc Antrop

When Zev Naveh asked me to write a foreword to his anthology, I felt honoured and glad. Honoured because it proved me that he appreciated the many discussions we used to have; he as an ecologist and I as a geographer, who met each other in the field of landscape ecology. I was also glad that he was able to finalize his idea of an anthology which he had been thinking about for a long time. In my opinion it is an important work. Many people know – as citations prove – Naveh from the book *Landscape Ecology – Theory and Applications* which he published with Arthur Lieberman for the first time in 1984. At present it continues to be a milestone in the revival of landscape ecology in the second half of the 20th century. Many things have changed since then though. Not only the landscape, but the whole of the environment and society have changed drastically. Adhering to his own holistic approach to the landscape as part of the Total Human Ecosystem (THE), Naveh broadened his work accordingly. He has become even more convinced of the importance of holistic landscape research and its application in societal matters. As such, he has become a promoter of transdisciplinary discipline with the landscape as an integrating concept, important for the future ecological, socio-economical and cultural development. He has stressed this in many publications and many enthusiastic lectures all over the world. It is this personal transition towards transdisciplinary landscape studies in a more global context Naveh has wanted to express clearly in this anthology. Using a selection of previous work, put in a logical order and completed with new comments and a recapitulation. Not looking back, but stressing the important issues and also pointing to the future and the enormous responsibility scientists have in what he calls the revolution of sustainability. This makes this anthology special and a valuable narrative as well.

Naveh started as a vegetation scientist working in Mediterranean areas. The first part of this book deals with his experiences there. Rapidly, he broadened his 'perceptual and conceptual "looking glass" from ecosystems to landscapes', as he writes in the introduction. He became a landscape ecologist focussing on conservation and restoration ecology. As a geographer I was struck by the similarities between Naveh's perception and conception of Mediterranean landscapes and the ones of the first geographers describing these landscapes such as Vidal de la Blache and Philippson. As a young geographer, I also started my career in the Mediterranean and I am convinced that my holistic perspective on landscape was mainly forged by these landscapes too. The great variation of the natural conditions and the ancient history of human use of the land resulted in landscapes where the "interaction between natural processes and human activities", as defined in the recent European Landscape Convention, is obvious. Mediterranean landscapes are also severely threatened by mismanagement and I agree with Naveh that there is little optimism for the future. New transdisciplinary approaches based on sound landscape ecological principles are urgently needed here, but, as Naveh points out, landscape ecologists still play a minor role in decision making. Consequently, landscape ecologists cannot be pure scientists, but they have to transmit and explain their work in a broader sense, to educate decision makers, to make the general public aware of the severe problems and possible solutions.

The second part of the book consists of a selection of theoretical, methodological and practical issues in a global perspective. Here Naveh focuses on his Total Human Ecosystem (THE) and its implementation in a transdisciplinary landscape science in the perspective of sustainable development. In his introductory comments, he stresses the applicability of this theoretical paradigm and the selected papers offer practical examples. Even when familiar with the selected publications, I found it very worthwhile re-reading them after reading Naveh's comments on the subject. Clearly, this anthology is not just a collection of previous work, but a reflection of a scholar on his past work looking at it with greater wisdom and concern for the future development. This second part deals as much with spiritualism, culture, socio-economics as ecology. Their integration forms the basis for "the transdisciplinary process for the overall improvement in the quality of living creatures and their environment", Naveh writes. Also, the focus is no longer on natural and pastoral landscapes, but is broadened to all kind of landscapes, also to what Naveh calls the techno-landscapes. It becomes clear that he sees landscapes as the tangible meeting points of nature and mind in the context of a holistic THE. When he uses the term 'landscape ecology' it has a much broader meaning than just a kind of ecology or a specific discipline. He speaks of a transdisciplinary landscape science; biology-ecology and culture always go hand in hand when it comes to diversity, conservation and restoration. The real significance of ecological research is given by its contribution within the THE as part of the Grand Synthesis of Laszlo, who inspired Naveh a lot.

The third part of the book consists of one chapter devoted on this gradual transformation of landscape ecology and restoration ecology into transdisciplinary approach of sciences to holistic landscape studies with applications in management, planning, conservation and restoration. It is an extensive recapitulation of Naveh's thinking and evolution through his career, and also a new step in consolidating his thoughts on a universal philosophical theory by Laszlo. It is certainly not an end, but as Naveh puts it hopefully a new cornerstone in the further development of a unified transdisciplinary landscape science.

The anthology combines the theoretical and methodological development of landscape ecology and the personal narrative of an eminent scientist and scholar in his attempt to make his science more useful and meaningful for mankind. For me, his message to the scientific community is clearly stated in Chapter 10: *scientists will have to integrate scientific knowledge with ecological wisdom and ecological ethics, helping us to learn from past experience, comprehending the present and envisaging the future of our cultural landscapes.* It reminds me also to one of his famous quotations in one of his challenging lectures for young PhD students in landscape ecology: "The things that count, can not be counted."

Marc Antrop
University of Ghent,
Belgium
January 2006

FOREWORD

by Ervin Laszlo

It is a genuine pleasure and distinct privilege to write this introductory note to Zev Naveh's comprehensive summary of his life-work, an opus of truly epoch-making dimensions and significance.

Writing this Preface is a personal pleasure, because it is seldom that a theoretician can discover that his ideas have fallen on fertile grounds not only on the level of theory, but also on the level of practice – or, as a first step, as practice-oriented applied theory. This, however, is what I have found in Naveh's writings. I am delighted that his THE (Total Human Ecosystem) model not only uses the holistic thinking I have long championed as a general framework, but finds use for such specific theoretical concepts as my general-evolution theoretical model of bifurcation-based sequential evolution, and even more remarkably, for the Integral Theory of Everything I have outlined recently, rooted in the concept of a universal information field, the Akashic or A-field.

Writing this Preface is even more importantly a privilege, and also a moral and intellectual duty, for the importance of Naveh's work is not limited to the choice of theoretical model. It is not an exaggeration to claim that Naveh's contribution to scientific advance in the field of landscape ecology, and beyond that to the grasp of the interrelated processes that frame our life on this planet, are entirely fundamental. Naveh brings to the domain of ecology a systemic understanding that, to – an outsider surprisingly – has been lacking in this field of inquiry, although it is reputed to be generally systemically oriented. He points out that the study and management, and conservation and restoration of landscapes in every part of the world require a holistic approach, transcending the frontiers of the natural sciences into the social sciences, and even the humanities and the arts. This is a fundamental insight, and a radical innovation for the dominant paradigm in ecology.

Naveh discusses the urgent need for a transdisciplinary shift in both theory and practice in all environmental fields, especially in landscape ecology, restoration ecology, and landscape studies in general. This shift is the essence of the urgently needed environmental revolution that could lead to a post-industrial symbiosis of humans and nature, as expressed in the interaction of humans and landscapes. In the essays that comprise this volume, the warrant for and the practical utility of this transdisciplinary shift is convincingly demonstrated through the use of concepts developed in the systems sciences, in cybernetics and complexity theory, and through the general evolutionary and integrated worldview conceptions elaborated by the present writer.

As Naveh shows – and as it is also widely recognized – most ecologies in the world are presently on a degenerative path. They need not only to be maintained in their current state, but restored to a more balanced, healthy condition. In this regard, Naveh notes, consideration needs to be given to the restoration not only of vegetation patterns, but of all the processes that ensure a sustainably healthy and attractive landscape. This calls for a holistic approach that encompasses the maintenance and restoration of the homeorhetic flow-equilibrium between biodiversity and ecological and human cultural heterogeneity in regard to the patterns of human land-use. Ultimately, the requirement is for a holistic land-use policy. This, alas, is still far from reality, in the scientific as well as in the political domain. Yet Naveh's point, that the true role of ecology in an integrated socio-economic, ecological and cultural development is to lead to an overall improvement in the quality of living creatures and the environment, is good common sense.

The point this volume makes is one that few people outside the technical disciplinary domain of ecology may be acquainted with. It is that the reductionist and mechanistic worldview in science and society has left a deep and unwelcome mark on the paradigm that dominates work in the field of landscape ecology. This has to be overcome, for it encourages a false separation of human social and cultural activity from the biological processes that hallmark the environment in which that activity takes place. The most insidious consequence of this separation is the often voiced view that environmental problems are problems of nature and not of society, and thus they concern biologists, chemists, and landscape engineers and not politicians, business managers, and civil society. This view ignores the multiple strands of interaction between society and the environment that makes it not just misleading, but factually false, to speak of nature as a system of biological organisms without considering the human beings and their cultures and technologies that interact with the system.

The view that society and nature are distinct and separate is a logical but pernicious consequence of the dynamics that drive society progressively further from integration with the rhythms and processes of nature. As other living species, our distant forebears were well integrated into their natural environment. This level of integration was progressively impaired in the course of evolution. Around 40 million years ago the family of primates split off from the then existing species of mammals, and about 9.2 million years ago the primate family split into two groups, of which one, the pongids, stayed with the arboreal life, and several others (such as *Gigantopithecus* and *Sivapithecus*) later became extinct. The group of pongids evolved into the modern apes: chimpanzees, gorillas, orangutans, and gibbons. Out of this family evolved a subgroup of terrestrially based bipedalists: the hominids.

About 2.5 million years ago, the early hominid *Australopithecines* split into different branches. Many became later extinct (for example, *boisei* and *robustus),* and the surviving branch evolved into *habilis* and *erectus,* and ultimately into *sapiens.* Some 40 000 years ago, sapients appeared in Europe, probably co-inhabiting the continent with *Homo neanderthalis.* The latter disappeared around 30 000 years ago, and since then *Homo sapiens sapiens* has been the sole survivor of the hominid branch.

With *Homo sapiens sapiens,* integration with the environment became entrusted to culture rather to genes: human evolution shifted from the biological to the sociocultural domain. For the past thirty to fifty thousand years it has been the sociocultural organization of groups of individuals that has mutated rather than their gene pool. A series of sociocultural mutations led to the increasing detachment of the processes and dynamics of the societies formed by humans from their ecologies. This detachment culminated with the appearance of modern industrial societies. At this point human societies appeared, and still appear, as systems with *in sui generis* laws and processes. The environment is relegated mainly to the role of source and sink: to the role of supplier of natural resources and habitable space, and sink of wastes.

Industrial societies possess a dynamic that is out of sync with the natural rhythms of their environment, but they are not, for all that, separate systems. Society and the environment are properly viewed as systems with a dynamic of their own, but not as systems of such autonomy that their modeling can ignore the fact that they are elements in a larger common system. Naveh is correct in insisting that landscapes must be regarded as organic parts of the Total Human Ecosystem Gestalt; as meeting points between dynamic processes of nature and the workings of human mind and culture.

The links of industrial societies to the biosphere have intensified, and in many areas transformed into relations of critical dependence. Human societies have always depended on clean air, water, fertile soils, and the availability of sufficient quantities and qualities of air and biological produce, but industrial societies are critically dependent on a vast range of minerals and energy sources in addition. This has prompted some social scientists to include the environment in their modeling of the pertinent flows and processes; for example, economics is now viewing nature as a subsystem of the economy. This indicates that the separation of the natural and the human domains is being narrowed, but it also shows that the new conceptualizations are still inadequate. It is more correct to speak of the economy (and indeed, of human society as a whole) as a subsystem of nature, rather than of nature as a subsystem of the economy.

The total dynamics of the human/cultural/technological/biological system – the socio-biosphere, or in Naveh's terminology, the Total Human Ecosystem – constitutes the basis of our life. Recovering our integration with it, and restoring its damaged integrity is a precondition of our continued survival. We should be grateful to courageous and insightful ecologists such as Zev Naveh for pointing to the fallacy of the fragmented reductionist approach, and insisting that, both in the interest of scientific accuracy and of our life and future, applying the holistic paradigm of complex, interacting and information-imbued systems is of urgent and crucial importance.

If any reader may still entertain any doubt about the scientific validity and practical utility of the holistic approach in landscape ecology, he or she has only to read this volume. It marks a milestone in the vital advance of integrated evolutionary systems thinking in the empirical sciences.

Ervin Laszlo
Tuscany, Italy
January 2006

IN MEMORIAM OF FRANCESCO DI CASTRI (1930–2005)

Just before the final submission of this anthology to Springer, the sad news reached me that Prof. Francesco Di Castri died. As I stated in the introduction to Chapter 1, I had the great fortune to be invited by him to participate in an international symposium on the comparison of Mediterranean climate types in Chile and California at Valdivia in 1971, conducted under his brilliant leadership. Since then we created warm personal contacts. He was without doubt the most influential ecologist who had deep roots in the Mediterranean and I regarded him highly as a fine human and great scientist and concerned ecologist. I would therefore not like to close this book without devoting a few sentences to his memory.

UNESCO, in which he played a leading role, has devoted in their website a full account of his numerous contributions as an eminent scientist, deeply involved in environmental issues from the local to the global levels, summarized here as follows: "UNESCO is saddened to announce the death of Francesco di Castri, eminent ecologist, first director of the MAB Program (1971–1984), later Coordinator of UNESCO environmental Programs. Francesco di Castri played a creative and leading role in many international activities on the environment, at both non-governmental and governmental levels. He was a four–language polyglot, estimated by those in government and those in science, in North and South, East and West, a prolific writer. Authoring and coauthoring some 700 scientific and popular articles, monographs and book chapters. He was a noble, complex and highly erudite signor in the tradition of his native city (Venice), a skilled and subtle debater, with a fine sense of irony and metaphor, an exceptional motivator of those around him, a man of imagination and foresight, of courage and rigour. He left his mark. He will be missed and remembered by many."

Among his numerous achievements, probably the most significant and of lasting relevance were his contributions as the founding director of the Division of Ecological Sciences of UNESCO when he shaped and directed what came to be considered one of UNESCO's principal contributions in promoting international cooperation on environmental issues within the framework of the MAB program. Together with Michael Battise he nurtured and developed the Biosphere Reserve concept. He furthered its worldwide implementation as the transdisciplinary basis, bridging the natural and social sciences for the rational use and conservation of resources of the biosphere, and for the improvement of the relationships between man and the environment. In the same vein, he initiated also the World Cultural Heritage Convention, what became one of the most successful legal instruments for the conservation of the world-wide natural and cultural heritage. What I most admired was his realistic, but at the same time optimistic forward-looking view of ecology (as well as landscape ecology) as sciences, rich in opportunities, instilling hope, rather than despair, and rejecting dogma and rhetoric, instead of action and involvement. We were specially pleased and honoured when Francesco responded readily to contribute a short and very positive introduction to the Spanish edition of the Naveh- Lieberman Springer book on landscape ecology in 2001. I could not think of anybody else who could serve as such a fine model for a Mediterranean ecologist as Francesco di Castri.

INTRODUCTION – A REVIEW OF THE CHAPTERS CONTENT WITH SOME AUTOBIOGRAPHICAL COMMENTS

The author

1. A SHORT PERSONAL INTRODUCTION

I was born in 1919 in Amsterdam, and grew up in Germany. Escaping the Nazi regime, I immigrated in 1935 to Israel (then Palestine) together with other members of the Zionist Pathfinder youth movement to a Kibbutz (collective settlement) in the Jezreel Valley. In 1938 we founded our own Kibbutz "Mazuba" in the foothills of the Western Galilee. Here, among other jobs, I helped to reclaim the rocky slopes for cultivation by uprooting the dense shrub cover, removing the rock outcrops and using these for the construction of terraces, similar to those built already more than 5000 years ago by the Israelite tribes, settling in the hills of the Galilee. My main occupation was to work as shepherd of goats and sheep and finally as cattle herder and breeder for milk and meat production. This marked the beginning of my attachment to the rocky Mediterranean hill and mountain landscape, and its rich natural and cultural assets, resulting from a very long human history of coevolution with these landscapes and their utilization for better or worse. To these Mediterranean landscapes I have devoted most of my professional and scientific career, first as a pasture research scientist after I received a M Sc degree in Agronomy from the Hebrew University, Jerusalem in 1950, at the Governmental Experimental Station, and from 1965 at the Lowdermilk Faculty of Agricultural Engineering of the Technion, the Israel Institute of

The author with pasture specialists from the Israel Soil Conservation Authority (ISCA) inspecting the striking natural regeneration of perennial grasses after temporary removal of shrub competition by controlled burning, chiefly. Note in the background Carob trees, protected from fire. Mazuba 1950, Photo, courtesy I.S.C.A.

pre-European California Native American land use. All these scientists and scholars opened new vistas for me and had great influence on my work and thinking.

The same holds true for the far-sighted holistic ecologist Frank Egler, whom I met for the first time (and several times later on) at his famous Eaton Forest estate in Connecticut on the way back to the East Coast before returning home to Israel. I adopted his "Total Human Ecosystem" (THE) concept, integrating humans with their total environment, by which humans are treated as co-evolutionary partners at a higher ecological hierarchic level above the natural ecosystem. At the University of Georgia at Athens, Georgia, where I presented a guest lecture on my work in California, I made the acquaintance of Prof. E.P. Odum, the "father" of ecosystem ecology. Although I did not agree with his rather deterministic succession to climax paradigm, his personality and his holistic ecosystem concept made a great impression on me. I introduced his important innovative textbook on the fundamentals of ecology in Israel, and used it extensively while teaching ecology courses at the Tel Aviv University. These were the first ecology courses ever conducted in Israel which enabled students specializing in botany or in zoology to participate jointly. I used Odum's ecosystem concept as the common ecological foundation, bridging the then existing gaps in the ecological education between botanists and zoologists. Later on at the Technion, selected chapters of the third edition of his book (Odum 1971) became obligatory reading for my agricultural and environmental engineering and architecture graduate students. At Brooklyn College, I met Prof. R.H. Whittaker and realized that I was dealing with a giant in vegetation ecology.

In conclusion, these two years in the U.S.A. were formative for me as a young ecologist at the beginning of his professional career and I am most grateful to the N.A.S – N.R.C. for affording me this unique opportunity.

In Chapter 1, I examined both Mediterranean-climate type regions in through integrative comparisons of climatic, topographic, edaphic and botanical features, as well as other landscape characteristics with special attention to the impact of human land uses. About ten years later this was followed by a comprehensive comparison of Mediterranean-climate type ecosystems in Chile and California, conducted under the leadership of Prof. F. Di Castri, whose results were first presented at an international symposium at Valdivia, Chile in 1971, together with a comprehensive collection of lectures on the origin and structure of Mediterranean-type ecosystems (Di Castri and Mooney 1973). I had the great fortune to participate in these groundbreaking discussions and to create warm personal contacts with Prof. Francesco di Castri, who became the most prominent and influentional Mediterranean ecologist. This conference opened the way for closer interactions between ecologists dealing with the comparison of such Mediterranean-type ecosystems and landscapes on all continents and lead to the formation of "ISOMED", the International Society for the Study of Mediterranean Ecology, of which I had the honour to be a distinguished member.

I concluded that in order to conduct an overall landscape comparison, climate alone is insufficient to ensure a resemblance in the structure and physiognomy of vegetation if edaphic and biotic factors are divergent, even if their human uses and reaction to burning are rather similar. For example, this is the case of the grazed and browsed marquis and garigue shrub vegetation of *Quercus* and *Pistacia* on Terra Rossa and Rendzina soils in Israel and in other Mediterranean countries, as opposed to the chamise (*Adenostoma*) chaparral on very poor non-calcic, lithosols in California.

However, I found the closest overall resemblance between the oak savannas of Central Coastal California, between Santa Barbara and Carmel Valley, dominated by *Quercus douglasii* H.&A. and by *Quercus itaburensis* (Decne) Boiss in the Lower Galilee. I defined such landscapes as *ecologically equivalent*, indicating that resemblance in climatic conditions correlates to parent rock, topography and land use. In these, even great part of the herbaceous vegetation is taxonomically synonymous and some of the dominating woody plants are ecological and taxonomical "vicariants". That means they have close physiognomic features in both Mediterranean-climate type regions. I assumed that the most successful invaders and colonizers from the Mediterranean, such as *Avena, Bromus, Erodium* and *Medicago* were pre-adapted to the combined agro-pastoral Spanish land use, which the latter introduced to California from similar field-grasslands. In fact, according to Stebbins (personal communication 1959) on the few calcareous soils of this region in California, the replacement of indigenous plants by Mediterranean invaders was greatest.

It is of interest to note that by using more advanced statistical methods for the climatic comparison between California and Israel, my young colleague, Y. Carmel, who replaced me as teacher and researcher of ecology at the Technion, and with whom I co-authored a study on the evolution of Mt. Carmel cultural landscapes (see Chapter 7), reached similar conclusions, 30 years later. He found also the closest climatic resemblance existing between the Mediterranean Zone in northern Israel and the Central Coastal Region and choose the Carmel Valley, dominated by *Q. Douglasii,* as his main study area in California, which was also one of my chief study areas.

CHAPTER 2 – Naveh Z. and R.H. Whittaker (1979). STRUCTURAL AND FLORISTIC DIVERSITY OF SHRUBLANDS AND WOODLANDS IN NORTHERN ISRAEL AND OTHER MEDITERRANEAN AREAS (Vegetatio 4:171–190.)

This study was carried out in collaboration with Prof. R.H. Whittaker whom I met again at Cornell University. Since our first encounter he had founded the most advanced school of quantitative vegetation science in North America. I had the opportunity to get closer acquainted with his work and especially with his innovative, quantitative approach to biological diversity. I was very fortunate to gain his collaboration for a joint study of structural and floristic diversity in Israel and other Mediterranean areas. We applied the method that Whittaker himself had developed, used also for his precious studies in other mediterranean climate regions in California, Chile, Australia and South Africa.

As part of the first Mediterranean biodiversity study (and one of the first studies, supported by the Israel–USA Research Foundation), we also included some of the results of these studies, as well as of my research in Northern California. We also developed the system of using stratified vegetation cover and richness as parameters for structural diversity. Despite of the great importance of these parameters for determining the ecological niche diversity for understory plants and for birds, it is largely neglected in biodiversity studies. Our investigation of vascular plant diversity in Israel was combined with a study of animal species diversity lead by Prof. M. Warburg.

Our Israeli study provided convincing proof that richness and diversity of herbaceous species was dependent on the kind and intensity of defoliation pressures in shrublands and on grazing intensities in open Tabor oak woodlands. Biodiversity was highest under moderate grazing during the main winter rain growth season. This was later on corroborated by the findings of Noy-Meir and Kaplan (1991) on a much larger scale. In addition, we found that diversity increased along a declining moisture gradient, with highest woody and herb species richness in open *Pistacia lentiscus* shrubland on the xeric borders of the Mediterranean climate zone on Mt. Gilboa (described already in the Bible as a dry place, "lacking dew and rain"). Here we found shrubs and herbs with one of the highest alpha diversity in the world. Mt. Gilboa also ranked highest in animal species diversity. As mentioned in Chapter 5, the ecotype of *P. lentiscus* on Mt. Gilboa proved itself as the most drought resistant of all those we tested in Northern Israel, and we can assume that this is true not only for *P. lentiscus* in Israel, but also for all other sclerophyll shrubs throughout the Mediterranean.

Comparing the diversity of the different Mediterranean climate types, we clearly distinguished between the much older Gondwanan heath-like plant communities in Australia and South Africa, and the younger Pleistocene Californian chaparral and woodlands, the Chilean matorral and the Old World Mediterranean shrublands and woodlands. The former are well adapted to very old, nutrient-poor soils, which are lacking in annual species, but are extremely rich in woody species. In this context we also discussed the evolutionary implications of the increasing duration of human disturbances in California, Chile and the Mediterranean, mentioned also in Chapters 2 and 7.

CHAPTER 3 – FIRE IN THE MEDITERRANEAN – A LANDSCAPES ECOLOGICAL PERSPECTIVE. In: Goldammer, J. F., Jenkins M. J. (Eds) Fire in Ecosystems Dynamics. Proceedings of the Third International Symposium in Freiburg, FRG, May 1989. SPB Academic Publishing, the Hague, the Netherlands, pp. 1–20.

This chapter is based on the outcomes of our fire ecology research until 1975. I had commenced this research already in 1951, as part of my studies on brush range improvement of mixed Marquis shrub communities in the Western Galilee, while working as pasture and range research scientist at the Neve Yaar Governmental Experimental Station. It served at the same time also as a part of my PhD thesis at the Hebrew University, Jerusalem.

In this study I followed the example of California, using controlled, prescribed burning of the brush canopy as a favourable seedbed for perennial grasses in the ashes, but simultaneously also saving the scattered *Quercus calliprinus* oaks and *Ceratonia siliqua* carob trees from the fire. The object was to create a more or less stable and productive perennial grass – tree savanna. However, we tried in vain to prevent the vigorous resprouting of the shrub canopy by applying brush killers 2–4D and 2–4–5T from airplanes or by selective spot spraying of the resprouting of undesirable woody plants in the first years after the fire. This was one of the first applications of herbicides in Israel and as mentioned below I did not have at that time any misgivings about herbicides spraying.

In addition to my smaller-scale intensive experimental areas, the large-scale controlled burning, reseeding and spraying trials were carried out on the non-tillable rocky shrub hill pastures of Kibbutz (collective settlement) Mazuba in close cooperation with the Soil Conservation Department of the Ministry of Agriculture and "Chim Avir", the first commercial company in Israel, spraying herbicides by airplanes. As a former member of Kibbutz Mazuba, in charge of the cattle herd, I was well acquainted with these shrublands and used the Kibbutz as my "home base" for these studies. At the same time I could also study the effect of widespread wildfires in adjacent marquis shrubland on the post-fire regeneration of the herbaceous and woody vegetation grazed by cattle and goats under traditional grazing regimes.

However, in spite of the successful establishment and early high production of some of the reseeded grasses, and especially *Oryzopsis (Piptatherum) milicea* and *Phalaris tuberosa*, these highly palatable grasses were very sensitive to heavy grazing in the spring and were soon crowded out by the vigorously regenerating shrubs. Therefore, the long range results were disappointing, both from a practical, economic and ecological point of view. However, this taught me a valuable lesson: you cannot fight nature by futile attempts to turn such resilient shrub communities into grass pastures. In my thesis in 1958, I still cited recent studies which claimed that 2–4D and 2–4–5T had no detrimental effects on cattle. When I found out about the detrimental effects of these herbicides, I became one of the strongest opponents of the use of these herbicides. Our landscape ecological and restoration studies, carried out since 1965 at the Faculty of Agricultural Engineering of the Technion, also convinced me that the evergreen sclerophyllous shrub, *Pistacia lentiscus* which we previously attempted to eliminate as a low-palatable and heavy competitor for grasses and trees was a most valuable soil protector and builder, with high photosynthetic and water efficiency. Combining low palatability with high drought resistance and limestone tolerance, this vigorously resprouting after fire and cutting, deep-rooted shrub maximizes its overall survival potentials. Therefore, we also used it as a bio-engineering and ecological model plant for rehabilitation and slope stabilization (Yogev and Naveh 1986) and other restoration studies (Naveh 1988) (See also Chapters 4 and 6).

Thanks to these earlier studies, I became aware of the significance of fire in the Mediterranean uplands – and fire ecology studies both in the Western Galilee and on Mt. Carmel became an important part of my research programme.

Here is the proper place to mention how much I was inspired in my fire ecological studies in Israel by Prof. Biswell's fire ecology research in California. His untiring efforts to introduce controlled burning to prevent destructive hot wildfires, and at the same time to show its beneficial effects on wildlife and browsing game in forests and shrublands and in nature reserves and parks, faced fierce opposition by most of his colleagues of the School of Forestry. He even was not allowed to use the University's experimental forest for his studies and had to carry out his experiments on privately owned forest land in Northern California. This turned out to be a unique opportunity to prove how right he was: one of the most destructive wildfires, raging in the Ponderosa pine forests of Northern California could be stopped right on the border of the forest in which Biswell had cleaned the entire brush understory by controlled burning. Thus, in spite of all obstacles he finally made his point on a much larger scale and in a real situation, and not in experimental plots of the University forest which he could never have simulated on experimental plots at the University forest. He showed convincingly that the "Smoky Bear" syndrome could be replaced by more efficient ways to prevent hot and uncontrollable wildfires, namely by the judicious application of fire and fuel management, especially in national Parks and Reserves. He served for me as an example to present a similar message in Israel and in the Mediterranean Basin to replace the wholesale condemnation of fire in Mediterranean uplands by a more rational approach, realizing that controlled and prescribed fire has advantages.

In this International Symposium on Fire in Ecosystem Dynamics at Freiburg I discussed the role of fire from a landscape ecological perspective, emphasizing the most relevant aspects for the study and application of fire as a tool in integrated and dynamic conservation management for Mediterranean uplands.

This chapter also summarizes our studies on the effect of burning on soil fertility and stability, showing that fire serves as an important link in the recycling of nutrients to the soil. In the first years after the fire, this has striking effects on the herbaceous fire followers, and especially on the rise in dry matter and seed production of grasses. If this was also the case in the Late Pleistocene, it could have had far-reaching implications for the role of fire in the domestication of grasses, such as *Triticum dicoccoides* and *Hordeum spontaneum* (see Chapter 7). In the following years also deeper-rooted woody plants could take advantage of this.

Like in the earlier studies in the Western Galilee we similarly could not detect on Mt. Carmel any evidence of run-off or soil erosion, after hot wildfires in dense Marquis and mixed oak and pine forests. However, this does not mean that under heavy grazing or other post-fire disturbances, such as the removal of dead wood, no erosion could occur. Indeed the greatest damage can be inflicted upon the vegetation and soil by goat and cattle grazing in the first winter and spring after fire. I therefore warned against jumping into too sweeping general conclusions on the effect of fire from single, short-term studies and superficial observations. I further pointed out that the study of fire should be treated in a holistic way as part of multifactorial landscape functions. In such a function, a clear distinction has to be made between controlling landscape state factors, such as land use, soil parent material, relief, climatic fluxes and organisms, and their depending post-fire variables, such as soil, vegetation, human-made artefacts.

I hypothesized that these fire-swept Mediterranean upland landscapes could be considered perturbation dependent dissipative structures, maintained and stabilized only by permanent energy/matter and entropy exchange with the environment. I illustrated this with the help of Prigogine's dissipative function under different fire perturbation regimes. The determination of optimum fire perturbation regimes requires systematic long-term studies on all landscape scales. In such optimum regimes, intervals between fires should neither be too frequent nor occurring with too long intervals, so that the capacity of constant self-organization and stabilization can be maintained for the conservation and enhancement of biological diversity and other desirable functions. For this purpose, new directions in research and land management are necessary, which aim not only at the suppression of destructive wildfires, but also at the beneficial application of fire and fuel management. To my great satisfaction the results of these studies could serve as guidelines for the extensive fire ecology research, following two wildfires on Mt. Carmel and included prescribed burning in pine forests.

This chapter is the first in this anthology reflecting the broadening of my perceptional and conceptional scientific "looking glass" from ecosystems to landscapes. This paradigm shift towards landscape ecology was influenced to a great deal by my visits to Central Europe, where I became acquainted with LE and created personal, long-lasting warm contacts with some of its outstanding representatives in Germany and the Netherlands. Their broad holistic approach and impressive, problem-solving oriented research enriched my thinking and inspired the future direction of my work. Among these distinguished landscape ecologists were Prof. Wolfgang Haber from the Technical University of Muenchen in Weihenstephan, Prof. Karl-Friedrich Schreiber from the University of Muenster. I was introduced to them by their former teacher and mentor, Prof. H. Ellenberg, my gracious and impressive host at the University of Goettingen. He was the most influential plant ecologist of Central Europe, broadening its scope by initiating integrated ecosystems studies, and especially the Solling project as one of the most comprehensive multidisciplinary International Biological Program in Europe. As reflected in Naveh and Lieberman (1994), his ecosystem concept inspired the distinction between natural biosphere and artificial technosphere ecosystems and landscapes. Also in the Netherlands I made the acquaintance of two of the leading landscape ecologists, Prof. Edie Van der Maarel from the University of Nijmegen and Prof. Isaac Zonneveld from the important ITC institute in Enschede, who was chosen the first president of the International Association of Landscape Ecology (IALE), founded in 1984. These developments paved the way for our first edition of the above-mentioned textbook on LE, published in 1984 by Springer, containing also a detailed description of the development of LE in Central Europe (Naveh and Lieberman 1994). My collaborator in this book and dear friend, Prof. Arthur Lieberman from Cornell University also introduced me to the emergence of this science in North American as a most important and active branch of IALE.

CHAPTER 4 – FROM BIODIVERSITY TO ECODIVERSITY: A LANDSCAPE–ECOLOGICAL APPROACH TO CONSERVATION AND RESTORATION (*Restoration Ecology* 2:180–189, 1994)

My first paper in my first newly founded journal "Restoration Ecology" – I attempted to present RE in the framework of a holistic landscape conception of whole landscape restoration. I considered such restoration as part of the urgently needed environmental revolution, which should lead to a post-industrial symbiosis of humans and nature, as expressed through human–landscape interaction. This conception was the outcome of more than 20 years research and its application in practical restoration and biological protection projects on denuded slopes and roadsides. This included the rehabilitation of an abandoned limestone quarry (the very first carried out in any semi-arid Mediterranean country (see also Chapter 9)). The first inspiration for such a landscape restoration concept I gained in my above-mentioned visit to Germany, and especially from the impressive, large scale, whole landscape restoration work, carried out along the Mosel and the Ruhr rivers, and from the open coal mine rehabilitation in the greatest industrial center of Germany in the Ruhr district. This was carried out in the 1960s as a major effort to improve the quality of life and to prevent many of the highly qualified professional labourers and engineers immigrating to more attractive regions, especially to the rapidly developing industrial areas in Bavaria.

I was also very fortunate to have the opportunity of later learning from the experience and expertise of restoration research and practice in many other countries and conditions where I enjoyed the kind hospitality of my guides. Here I can mention only very briefly some of these. Thus, for instance, I gained much inspiration for our restoration research, and for what I introduced as "vegetation engineering" to Israel, from the closer acquaintance with such grand bio-engineering projects along the Brenner Pass highway and from similar, smaller-scale work in the Swiss and Austrian Alps.

I benefited very much from my personal contacts with Prof. A.D. Bradshaw from the University of Liverpool. He was one of the first ecologists to devote himself to the foundation of a practical restoration science and its application to the reclamation of industrial dereliction and abandoned copper mines. He published together with M.J. Chadwick the first groundbreaking book on restoration (Bradshaw and Chadwick 1980).

Spending part of my sabbatical year, 1978, in Australia and presenting a series of guest lectures, in which I introduced LE which at that time was not known in any English-speaking country, I also learned much about the Australian restoration work. My excellent guide was C.V. Malcolm from the Agricultural Department of South-West Australia at Perth, one of the first ecological experts, choosing restoration not only as a profession, but also as a mission. Thanks to him. I had the opportunity to see much of the reclamation work of the salt affected Yarrah Eucalyptus forest slopes in the Wheat Belt in South Western Australia, and the use of halophytic *Atriplex* species as pasture shrubs, wherever the degradation had reached the point of no return. Among those shrubs from Australia, I tested in Israel, the limestone tolerant *A. nummularia* was the most promising, palatable and productive fodder shrub. It competed successfully under heavy goat grazing pressure with *Sarcopoterium spinosa*, dominating the dwarfshrub "Batha" communities in poor calcareous Rendzina soils and re-invading these reclaimed sites.

In the USA I had the great fortune to create close and lasting ties with Prof. Edith B. Allen, who invited me to an excellent symposium on advanced restoration research in arid and semi-arid land (Allen 1988). Prof. Allen is a leading restoration ecologist in

California, who, as the chief editor of "Restoration Ecology" until 2005, turned this journal from an essentially American into an important international source of information. It has dealt successfully with the rapidly advancing restoration research as a high-quality scientific and professional communication tool. Her extensive research in collaboration with her husband, Prof. Michael F. Allen – a prominent micro-biologist and ecologist – on the use of vesicular–arbuscular mycorrhisas for the rehabilitation of heavily disturbed land opened new vistas for restoration ecology. Thanks to these contacts I became an active member of the Society for Restoration Ecology and learned much about the ongoing restoration work in the grasslands and chaparral of California. At the 2001 European IALE conference we co-authored a lecture on the win–win relationships between RE and LE (Naveh and Allen 2001). This was our final conclusion from our restoration work which we could draw from our research in such diverse conditions as Israel, California and Tropical Mexico.

Thanks to this world-wide experience. I could present a broad picture of the application of the principles of ecological engineering as a science and art to my agricultural engineering, landscape architecture and regional planning students, which were attending the course on restoration ecology. This course was conducted in the Ecological Garden of the Technion, which I established in 1982, near our faculty, to serve as a major research tool for the introduction and testing of drought resistant, low maintenance plants, as well as for teaching and demonstration. Several graduate students carried out their research projects in the garden. The Eco-garden has become over time the most attractive site of the Technion campus, a place of nature solitude and beauty for students and the staff. I am very happy that it serves now, after the fusion of our Agricultural Engineering Faculty with the Faculty of Civil and Environmental Engineering as a major center in Israel for advanced, outdoor environmental education.

As described in detail elsewhere (Naveh 1988), we developed our own version of restoration and rehabilitation of degraded slopes in Mediterranean uplands. Where we can no longer rely on the natural regenerating process, our main concern is to achieve rapid and efficient revegetation and stabilization by establishing semi-natural, multistructural, stable and attractive plant communities, resembling, as far as possible the less disturbed counterparts in similar sites. For this purpose, we simulated a process of condensed natural succession in two steps: 1) the establishment of rapidly spreading and low resource demanding, indigenous "pioneer" grasses, legumes and low cover shrubs with dense foliage canopies for rapid soil protection and erosion control, and 2) the simultaneous planting of slower growing, taller and more demanding, but persistent and deep-rooting shrubs and trees with maximum soil ameliorating effects. These restoration strategies were also tested in the Ecological Garden and on disturbed slopes, adjacent to the new buildings on the Technion campus, in addition to larger-scale restoration projects, carried out in cooperation with the governmental Soil Conservation Authorities.

In this chapter, I made a strong point that for whole landscape restoration not only vegetation patterns should be restored but also all those processes ensuring sustainable healthy and attractive landscapes. This requires the maintenance and restoration of the homeorhetic flow equilibrium between biodiversity, ecological and cultural heterogeneity, as influenced by human land uses. To achieve this goal it is essential to gain full knowledge and comprehension of the evolution and history of the regional landscape.

As an example of the close interaction between natural and cultural processes I used in this chapter a joint study of ancient abandoned and disintegrating terraces in the southern Judean Hills in Israel (Naveh and Dan 1973). We expanded our investigations from the smallest landscape units or "ecotopes" of single terraces to the regional landscape system, realizing that this consists of catenas of contrasting slopes in ecological and pedological features and their resulting different land use histories.

Based on the results of our biodiversity studies, I presented a model of the landscape shaping forces of the biological-ecological-cultural triangle. Following Whittaker's subdivision of the biological diversity indices (see also Chapter 2), I suggested the transformation from his species indices into holistic, spatial ecodiversity indices and cited an example for the classification of small biotopes, used for a Danish landscape-ecology study in agricultural fields.

Finally I pointed out the development of holistic landscape conservation and restoration strategies in cultural landscapes, and the endorsement by IUCN of the proposal of the Working Group on Green Books for Landscape Conservation and Restoration, urging the preparation of a world-wide Red List of highly valuable, endangered landscapes. If this were implemented, it could serve as an excellent guide for a world-wide strategy of landscape restoration and a major contribution to such a post-industrial human-nature symbiosis.

CHAPTER 5 – CONSERVATION, RESTORATION AND RESEARCH PRIORITIES FOR MEDITERRANEAN UPLANDS THREATENED BY GLOBAL CLIMATE CHANGE. In: Moreno J, Oechel WC (eds) Global Change and Mediterranean-Type Ecosystems (*Ecological Studies* Vol. 117. Springer, New York, pp. 482–508, 1995)

This essay was originally presented as a lecture in an international post-graduate course. To demonstrate the implications of climate change for Mediterranean uplands and threatened by increasing climatic variability and uncertainty, I presented the results of a six years' study on the effects of inter- and intra-seasonal variability in rainfall and winter temperatures on herbaceous plants. This was part of a major natural pasture grazing and improvement trial, at my former workplace at the Neve Yaar Experimental Station in typical open Tabor oak woodland, including both a very dry and a very wet year. Anticipating (rightly as we realize now, 12 years later) further aggravations through the disruptive changes in rainfall and temperature patterns, and in extreme climatic events in the Eastern Mediterranean and Israel, I outlined urgently needed new strategies for research and management. These should provide better tools for decision making for conservation and restoration strategies, based on an entirely new, holistic land use policy. However, as will be shown in the next chapter, we are still very far from such a policy.

CHAPTER 6 – FROM BIODIVERSITY TO ECODIVERSITY – HOLISTIC CONSERVATION OF THE BIOLOGICAL AND CULTURAL DIVERSITY OF MEDITERRANEAN LANDSCAPES. In: Rundel P., Montenegro G., Jaksic F.M.(eds) Landscape Disturbance and Biodiversity in Mediterranean-Type Ecosystems (*Ecological Studies* Vol. 136 Springer, Berlin pp. 23–50, 1998).

In a keynote lecture at the Sixth International Conference of Mediterranean-type Ecosystem Scientists (MEDECOS) in Chile, I delineated prevailing, alarming trends and future prospects of Mediterranean uplands in the light of the global environmental crisis, of growing population pressures on natural resources, especially air, water and soil pollution, and global climate changes. The resulting exponential speed and extension of "neo-technological" and traditional landscape degradation and desertification, threaten the unique co-evolutionary biological and cultural richness of Mediterranean uplands and their intrinsic and instrumental values.

As an innovative tool for the public and decision makers, and to drive home the vital need for preserving the natural and cultural wealth of these landscapes for further generations, I reported on the outcome of a EU-sponsored pilot project of the above-mentioned IUCN Working Group for the preparation of "Green Books for Landscape Conservation" in Western Crete.

I was able to initiate this project, thanks to my close connection with Prof. V. Papanastis from the Aristotle University at Thessalonica, an outstanding ecologist and range specialist with a broad transdisciplinary outlook. I accepted his suggestion using this region of Crete as the best example of the severe problems facing the conservation and restoration of the unique biological and cultural assets of Greece and other Mediterranean countries. This, in spite of the fact that the most scenic and touristicially attractive mountain landscapes are protected as nature reserves and have been declared as UNESCO Cultural Heritage, but their management is suffering from misconceived "total non-interference", and from too heavy visitor pressure, and from neglect of their most valuable cultural remnants.

In addition, I outlined some major landscape ecological principles, pointing to the basic differences between bio- and techno-landscapes. I reported on our findings on holistic dynamic conservation management of human-perturbation dependent semi-natural and cultural biosphere landscapes, and stressed the importance of their homeorhetic flow equilibrium for ensuring highest attainable total landscape ecodiversity. For this purpose I used the Portuguese montados – oak savannas as example.

Presently, in spite of the growing public awareness and many attempts, supported by MAP and the Blue Plan to slow down the exponential process of environmental degradation of the Mediterranean Sea and the Coast, there is not much room for optimism on their future. In this respect the Mediterranean region is following the global trends of the early twenty-first century, moving ever closer to decline, desertification and collapse.

This is especially the case in the Iberian Peninsula, hit in summer 2005 by the severest drought, causing desiccation, drying up wells and lowering river levels, and by the worst raging wildfires which destroyed 160,000 hectares of forest, woodland and shrubland. At the same time, the tourist boom at the Mediterranean Sea shores is even greater than was predicted, already reaching 364 millions visitors each year, and making up to 30 percent of total world tourism! Thus, in Spain alone, 180,000 houses, devoted to tourism, have been constructed along the coast in 2004. Until 2025, an additional 136 millions visitors are predicted. A quarter of a million ships cross the Mediterranean each year, and 40 percent of the 46000 km of coastal shores are built up. Till 2025 these will grow by 4000 km more and 20 million people more will crowd these shores. One hundred and sixty additional power stations will be needed to accommodate the exponentially growing local population and tourists. All this put tremendous pressures both on seascapes and landscapes, but the sad fact is that the richest countries are interested in economic development on account of their Coastal and Sea-rich biological and cultural diversity, and the poor nations claim that their resources are not sufficient to deal with the pollution. Only about two-thirds of the 601 coastal cities have any installations to clear their waste disposal to the Mediterranean Sea. (The Po river alone adds each year 270 000 tons of nitrogen to the Sea!) But in some countries, like Israel, almost all polluted waste disposal to the Sea has been stopped, thanks to the efforts to implement the Barcelona Declaration obliging each country who joined it to prevent all sea pollution.

As a result, in the Mediterranean 104 marine species, including big mammals and reptiles, are endangered, and the rapid despoliation of many of the most precious coastal landscapes is continuing, even in Spain, that until recently has made great efforts to avoid this menace. Here, also the pressures of African "environmental refugees", driven by hunger and misery are becoming almost unbearable, in spite of the inhuman treatment suffered by those succeeding to reach the Mediterranean Coast. At the same time, the rate of population growth in the Moslem countries of the Southern and Eastern Mediterranean Basin is the highest in the world.

Unfortunately, in most Mediterranean countries conservation-oriented scientists, including landscape ecologists, play only a minor role in the Academic World, and they have very little influence on land use decision making. The only hope comes from the steadily growing public pressures and these could be further enhanced by intensified involvement of concerned scientist and by broader and more efficient public environmental education, making more use of the major mass media, namely television.

CHAPTER 7 – THE ECOLOGICAL AND CULTURAL SIGNIFICANCE OF FIRE IN THE EVOLUTION OF MEDITERRANEAN CULTURAL LANDSCAPES WITH SPECIAL REFERENCE TO MT. CARMEL. (Based on: The evolution of the cultural landscape in Israel as affected by fire, grazing and human activities. Z. Naveh and Y. Carmel. pp. 337–409 in: Wasser S.P. (ed.) *Papers in Honour of Eviatar Nevo. Evolutionary Theory and Processes: Modern Horizons.* Kluwer Academic Publishers, Dodrecht. (2004).

In this condensed version of the above paper, the clock has been turned far back to the study of the emerging of the human-modified, semi-natural cultural landscapes in the Pleistocene, coinciding with the biological and cultural evolution of the Paleolithic gatherer–hunters. The original paper dealt with paleontological and archeological findings on Mt. Carmel, and our studies on fire ecology, as well as the comparative dynamics of ecological equivalent Mediterranean landscapes in Israel and California. These are considered in the light of the theories of self-organization of living systems and landscapes and of non-linear, general evolution. We assumed that this was a co-evolutionary process between humans and their landscapes, progressing in several distinct bifurcations, in which natural and human-set fires played an increasingly important role. This co-evolution and the intentional use of fire as a vegetation management tool to increase fodder for game and food for humans, presumably reached its peak in the Late Pleistocene, used by the pre-agricultural Carmel Natufians.

This chapter focuses on our central hypothesis, namely, the role of fire as a major driving force in this co-evolutionary process. The above, original paper was preceded by my earlier contribution to the study of Prof. A. Ronen on the Sefunim cave prehistory on Mt. Carmel (Naveh 1984). Since then, this fire hypothesis has been corroborated thanks to more recent archeological findings on Mt. Carmel, and by new insights on the role of fire in plant evolution through selection of adaptive traits in woody and herbaceous plants in Mediterranean landscapes in the Pleistocene. These adaptive traits developed most probably together with those against all other environmental stresses, especially drought and foraging, and were intensified throughout the long history of human use of fire in this region.

The striking ecological and ethnographical similarity of the Epipaleolithic Natufians of Mt. Carmel and the pre-European Indians of the Central California Coast in the sustainable proto- and semi-agricultural food-collecting, hunting and fishing economies and their intensive fire vegetation management give even more weight to our assumptions of the important role of fire. However, whereas the Natufians lived more than 10,000 years ago at the threshold of the Neolithic agricultural evolution, only about 300 years have passed since these pre-agricultural Indian tribes had their first contact with the Spanish missioners. Californian landscapes missed, therefore, the long period of agro-pastoral land uses and the great resilience acquired by Mediterranean landscapes during this period.

These assumptions of an active co-evolutionary role of the Pleistocene foragers in shaping their landscapes were contrary to the generally accepted view of a passive human role until the Neolithic agricultural revolution in the early Holocene. Contrary to our claims on the evolutionary and ecological significance of fire, it was regarded as an entirely detrimental and destructive factor in Mediterranean landscapes, and therefore to be prevented at all costs.

The role of fire utilized by Late Paleolithic hunters and food gatherers in proto-agricultural economies and therefore in the emergence of agriculture in the Late Pleistocene in the Levant and the early Holocene in Europe, has also been completely ignored by many archeologists. This is reflected, for instance, in the earlier, as well in the most recent studies by Bar-Josef – a world-renowned expert on the Natufian and other cultures of the Late Paleolithic foragers and Neolithic early agriculturist in the Levant (Bar Yosef 1998 and 2001; Bar-Yosef and Meadow 1955). As explained already in this chapter, although wildfires and human-set fires could not leave any direct archeological evidence, there should be little doubt that such wildfires, caused by volcanic eruptions or by lightening occurred in the Pleistocene. We should also not reject the possibility that early human uses of fire for heating and cooking, leaving many traces of hearths in caves of Mt. Carmel, made the Paleolithic hunters and food gatherers aware of its benefits for their hunting and gathering activities, and applied this knowledge.

This does not mean that fire played a similar important role for the emergence of agriculture everywhere in the Levant as it did on Mt. Carmel. However here, and in other more humid and chiefly mountainous sites, with dense woody cover, fire was most probably the only efficient tool to create enough niches for herbaceous plants and especially for the cereal grasses which were domesticated.

Bar-Yosef (1998, 159) described the Mediterranean Levant "an oak-dominated parkland and woodland that provided the highest biomass of food exploitable by humans", without distinguishing between these, and more xeric sites, which could not support such dense tree cover. They supported, therefore, much more open and grassy plant communities, like the Tabor oak woodland and the semi-arid shrub-grasslands on the ecotones of the Mediterranean zone in Israel. But even in the very wide-spread parklands on lower elevations and slightly warmer and drier conditions, such as the coastal foothills *Ceratonia-Pistacia lentiscus* communities in Israel, their scattered tree cover has a dense shrub-understory. As I mentioned earlier in this chapter, without distinct climatic desiccation or without widespread fires, there are very few naturally bare and open sites that allow dense continuous grass stands and colonization of typical rock dweller rodents.

Finally I would like to comment, that archeologists and anthropologists concern themselves almost exclusively with questions of the "when, where and why" of the Neolithic revolution in the Levant, but not with the "how", with which I have dealt in this chapter. In my opinion this problem is of greatest scientific and practical significance in the context of the biodiversity evolution that has become a major focus in evolutionary biology studies, driven by the accelerating realization of extensive biodiversity extinction. Therefore, it should also be a major concern for landscape and restoration ecologists and for all conservationists and should be approached with a broad transdisciplinary view.

Part two is a collection of selected studies and essays, devoted to global theoretical, methodological and practical issues, as reflected in the consolidation of the THE paradigm and its transdisciplinary implementation in general, and specifically in landscape research. It should be realized that in such a collection of studies, published in different journals and books, designated for readers with different backgrounds and interests, it is very difficult to avoid some repetitions in the presentation of premises and principles. Therefore, in each chapter they are viewed from different angles and perspectives. At the same time, as these chapters are arranged in chronological order, these studies also reflect the development and consolidation of my THE paradigm and its implementation in the theory and practice of a transdisciplinary landscape science.

CHAPTER 8 – THE DEVELOPMENT OF TANZANIA MASAILAND. A SOCIOLOGICAL AND ECOLOGICAL CHALLENGE (*African Soils*: 499–518, 1966)

I decided to open the global part of the anthology with my early Masailand development study, because for the formulation of the THE paradigm, my work in pasture research and development in northern Tanzania from 1962 to 1965 was most relevant. To this challenging mission I was seconded by the Department of International Cooperation of the Israel Ministry of Foreign Affairs in my capacity as range and pasture scientist in the governmental Neve Yaar Experimental Station. It was part of the extensive technical and scientific aid program to developing countries, and especially to those gaining independence in Africa.

I realized very soon that in order to prevent further ecological and cultural degradation of Masailand and its people, we have to reconcile the development of the Masai with the need for conserving and restoring the unique and expansive landscapes of Masailand and its biological and cultural assets. The world-wide implications of this problem have not lost their actuality. In fact, all recommendations I made on development more than 30 years ago, mentioned in Chapter 9, are still relevant today.

For the preparation of the actual development project proposal I was very fortunate to gain the collaboration of David Branagan, an open-minded and nonconformist British veterinarian, who had worked for many years with the Masai. I convinced him that for our purpose, comprehensive and innovative solutions of integrated ecological and socio-economical development and conservation (which now we would call "sustainable development") are required, based on a holistic and transdisciplinary THE concept. My research in Masailand convinced me that in such a closed, very short and fragile food chain of the Nilo-Hamitic pastoral Masai tribe the productivity of the dry savanna grazing land cannot be separated from the productivity of the Masai people and their livestock. Their present drought-starvation cycle could only be broken by a holistic approach, embracing all links of this food chain, starting from the soil by raising its low nitrogen status by N-fixing legumes. This is the major limiting factor for the rise of the productivity and the quality of the pasture forage, vital for higher milk production, on which the Masai survival depends. We realized that this was the most important and highly valued staple food and cash commodity of the Masai. Together with wildlife and tourism, these are the greatest assets of Masailand, from which the Masai and non-governmental or private enterprises should benefit. The improvement of pasture, livestock husbandry and range management must be combined with the conservation and improvement of the socio-economic and cultural status of the Masai themselves, and their total physical and biotic environment – namely the Total Masai ecosystem. This will require a radical transformation of their disrupted subsistence economy into a sophisticated, modern diversified livestock and wildlife economy, ensuring flexibility in pasture utilization and allowing conservation of wildlife and its habitats for direct revenues of the Masai in marginal pasture land, where modern livestock ranching is neither feasible nor desirable. As a first step, we proposed a combined research, development and demonstration pilot scheme in one of the first Masai cooperatives, in which I already carried out the first pasture improvement trials and found the most promising legume-grass mixture. This proposal was accepted by FAO and the Tanzania government, but because of political considerations it has lost interest in furthering the modernization of such a Masai pastoral economy. Together with the discontinuation of the Israeli aid program this combined Masai-livestock-wildlife project was abandoned.

This approach to the Masai and the development of Tanzania Masailand differed very much from that of the British pasture and livestock experts, who tried in vain to convince the Masai that their only hope for survival is to reduce their livestock, confine it to fenced pastures for meat – and not milk production and its conversion into a cash economy, similar to the beef cattle ranches of the white settlers.

Already at this early stage of independence of African nations, the mistakes in developments became very apparent. Unfortunately, most – if not all – attempts to transform traditional, well-adapted so-called "primitive" pastoral and other subsistence economies into "modern" – that means settled ones – in Africa and elsewhere have failed. As I explained in detail elsewhere (Naveh 1989), most of those in charge of these development schemes lacked the understanding and appreciation of the importance of the inbuilt resilience and the true nature of the mutual adaptation and co-evolution of these pastoralists or foragers, their livestock, pasture plants and wild animals, and the physical and biotic features of their landscapes. In East Africa and in the Sahel, most negative feedbacks of protection and regulation functions have been removed by the provision of supplemented food and water, but all attempts to replace these by new cultural regulative feedbacks of rational resource management and population control have failed. This resulted in vicious run-away feedbacks of landscape deterioration and population explosions. One of the major handicaps is the communication gaps between the scientists and the experts from different disciplines on one hand, and between them and the local population on the other hand. I have experienced this and the disregard of the ecological wisdom of these so-called "primitive" pastoralists and farmers during the preparation of the Masai development scheme.

Fortunately, I could learn much about the Masai and their life thanks to our close contacts with the Catholic missionaries, Father Hillman and Father Donovan, who had intimate knowledge of the people after living and working in Masailand for many years. They established a school near the mission as a second home for the Masai children, but did not try to convert their parents to Catholism. These missionaries had deep empathy and understanding for the Masai and supported them in their desperate efforts not to lose their grazing land and their semi-nomadic pastoral way of life, together with their identity and culture. Through their influence, we succeeded in earning the trust of the Masai, especially after my wife taught them how to prepare cheese from their milk surplus. Near the school with the assistance of my devoted technicians from the Chagga (Mt. Kilimanjaro) tribe, we established improved, but unfenced grass-legume pastures for their milking cows. Of great value in this respect was our acquaintance with the bright Masai youngster from

Longido I mentioned in this chapter. He was one of the best students in the agricultural college that was attached to our Experimental Station at Tengeru. Like most educated Masai, he knew perfect English. We "adopted" him and he spent many hours in our home and with me in the field and I got from him first-hand knowledge about the Masai, their Zebu cattle, their pasture management and learned the main pasture grasses, and their Masai names.

I also benefited very much from my acquaintance with several Australian pasture researchers and specialists and their important work in the semi-arid tropical pastures of North Australia, who visited Masailand to collect promising legumes for similar grass/legume mixtures which I tried to introduce to these improved pastures in our experimental plots.

More than 20 years later, most of our assumptions on the ecological wisdom of the pastoral and animal husbandry strategies of the Masai were confirmed in elaborate field, laboratory and computer studies (Western and Finch 1986; Krummel et al. 1986). These studies showed convincingly that Masai reliance on great numbers of milk producing cows is their main safeguard for survival in severe drought periods. Western and Finch (1986) concluded that a pastoralist's immediate survival and ultimate evolutionary success emerges from the tight link he makes between livestock production and social influence by big families, achieved through large herds for bridal wealth. As chief conservation officer in the Amboseli National Park in Kenya, Western implemented in the 1980s what we proposed already in the 1960s for the Masai in Tanzania, namely, that the Masai living there in symbiosis with the wildlife should obtain their full share of revenues from tourism (Western 1997).

Our suggestions for the development of Masailand to reconcile socio-economic advancement and conservation of biological and cultural diversity by linking economic development with nature conservation have also become the major premises of the IUCN World Conservation Movement Strategy globally. These principles were also adopted later on as the major strategies for sustainable development. However, their interpretation and implementation in developing countries are open to heavy criticism (Naveh 1997). Foreign aid to Africa was not only wasted by corrupt political leaders, it was insufficient to prevent the present catastrophic situation of many African countries. It is highly doubtful that the most recent decisions taken by the eight wealthy nations at their meeting in Scotland in June 2005 to extend more help will be able to change this situation.

CHAPTER 9 – THE ROLE OF LANDSCAPE ECOLOGY IN DEVELOPMENT (*Environmental Conservation* 5:57–63 1978)

This essay was introduced by citing the major imperatives for development which I presented in a dialogue on development by engineers and architects (Naveh 1970). For these, as in the previous chapter, the THE paradigm served as the core concept for integrated socio-economic, ecological and cultural development, implemented as a transdisciplinary process for the overall improvement in the quality of living creatures and their environment. I claimed that the true role of ecology in this process can only be appreciated if we broaden our view from narrow economic benefits, measured by the increase of material goods values, into an overall quest for the advancement of spiritual, cultural and intellectual values enabling the fulfillment of human and natural potentials.

These definitions come close to the contemporary holistic concept of *sustainable development* we applied in a multinational EU - sponsored project for modelling regional sustainable development in the information society (EU-Project MOSES. 2000). (see also Chapter 13).

Dealing not only with natural and semi-natural landscapes but also with cultural-rural and urban industrial landscapes and their inputs of fossil energy, artefacts, and cultural information and control, LE could play an important role in this process. For the research on conservation management and restoration of degraded Mediterranean uplands, carried out until 1978, I used a classification model of landscape units, ordinated according to the kind and amount of material/energy and information inputs and the gradual modification, conversion and replacement of the biosphere ecotopes. The management and restoration strategies for Mediterranean uplands and their multi-beneficial land uses and environmental functions were summarized by a flow diagram. This indicated that highest overall benefit for nature conservation, wildlife and recreation amenities, protection of environmental quality, as well as for livestock, forestry production and water yield can be expected by multi-purpose reafforestation and revegetation.

Although I did not refer specifically to "transdisciplinarity", I applied an essentially transdisciplinary approach to the challenges of LE for development, and for the replacement of the relative values of these parameters by actual quantitative ecological, socio-ecological and economic parameters.

CHAPTER 10 – CULTURE AND LANDSCAPE CONSERVATION: A LANDSCAPE ECOLOGICAL PERSPECTIVE. In: Gopal B.P., Pathak P, Saxena K.G. (eds) Ecology Today: An Anthology of Contemporary Ecological Research. International Scientific Publications, New Delhi, pp.19–48. (1998).

In my contribution to this collection of essays on contemporary ecological research, I attempted to present a coherent description of my views on culture and landscape conservation and restoration, based on a transdisciplinary systems approach. In this, landscapes are regarded as a concrete Gestalt system of our THE and as such, they become the tangible meeting points of nature and mind.

This essay updates and broadens our cultural conservation concepts dealing with the theoretical and practical aspects of LE (Naveh and Lieberman 1994). Its theoretical part of such a holistic landscape-ecological approach to nature and culture includes the following:

1. The place of landscapes in the ecological hierarchy, exemplified by a functional ordination model from the natural biosphere pole to the artificial technosphere pole;

2. The transdisciplinary nature of biosphere landscapes, illustrated by the dimensional approach of Victor Frankl, the renown psychotherapist and father of the "Third Viennese Psychology School" of Logo therapy;
3. Higher orders of landscapes beyond the Cartesian mechanistic orders, based on the theory of implicate and generative order by the prominent theoretical physicians and science philosophers, Bohm and Peat;
4. The use of fuzzy logic as an innovative tool to deal with vague and qualitative values in the study of cultural landscapes;
5. And finally, the cultural dimensions of landscape evolution and perception, presenting relevant definitions for culture, and reviewing some of the different approaches to our perceptions of human-environment interaction and landscapes by philosophers and cultural geographers. To these interactions between landscapes and culture, a special essay was devoted (Naveh 1995).

Of special importance for LE and RE is the evolutionary interpretation of culture by Laszlo, the renowned system scientist and philosopher, whom I cited extensively in many chapters of this anthology. He views cultural evolution as an integral part of the biological and social evolution within the "Grand Synthesis" of cosmos and of life on Earth, and its major steps as progressively higher organization levels through bifurcations.

The methodological part is devoted to the conservation and restoration of total (ecological + cultural) landscape ecodiversity and its intrinsic and instrumental values. This is to be achieved by dynamic landscape management for the conservation and restoration of biosphere landscapes and their ecodiversity. For this purpose I emphasized the need for a balanced ecocentric approach and for better communication tools, and the importance of the conservation and restoration of ecological processes. In a special subchapter, I dealt with the homeorhetic flow equilibrium of perturbation-dependent semi-natural landscapes.

In the concluding part, I provided some encouraging examples of cultural landscape conservation and restoration, as one important step toward the cultural and environmental revolution. I maintained that to ensure its success, *scientists will have to integrate scientific knowledge with ecological wisdom and ecological ethics, helping us to learn from past experience, comprehending the present and envisaging the future of our cultural landscapes.*

CHAPTER 11 – ECOLOGICAL AND CULTURAL LANDSCAPE RESTORATION AND THE CULTURAL EVOLUTION TOWARDS A POST-INDUSTRIAL SYMBIOSIS OF HUMAN SOCIETY AND NATURE (*Restoration Ecology* 6:135–143, 1998)

Whereas in my earlier publication in "Restoration Ecology" (see Chapter 4), I focused chiefly on Mediterranean landscapes, here I broadened the scope and discussed RE from a global perspective of ecological and cultural restoration. This was done in the context of the cultural evolution of human society, of which landscape restoration is an integral part. Introducing the notion of culture requires a transdisciplinary approach. Not only the biological-ecological aspects, but also those of human ecology have to be taken into consideration in restoration of open and built-up landscape units (or ecotopes) of the Total Human Ecosystem. In the ordination model of the functional classification of these ecotopes I pointed to a unique, promising *regenerative ecotope* as an outstanding example of both agricultural landscape rehabilitation, and ecological and cultural restoration. In this, not only the natural regenerative capacity of cultivated land is restored, but also the basic cyclic flows of energy, water and nutrients of natural biosphere landscapes, driven by the conversion of solar energy. Designed by the late, visionary landscape architect and planner John Lyle, it was realized at the California Polytechnical State University, Pomona, in an interdisciplinary teaching, research and demonstration project in regenerative technologies, open for graduate students of all faculties. In fact, my visit to this impressive center, and meeting Prof. Lyle, and the teachers and students of the project inspired me to publish this essay.

We were very fortunate that we could invite Prof. Lyle, together with other prominent landscape and restoration ecologists, to a symposium on ecological and cultural restoration in 1996, organized jointly with my colleague and close collaborator Dr. S. Burmil from the Faculty of Architecture and Town Planning. It was followed by field-workshops at "Neoth Kedumim" – the biblical landscape restoration (see Naveh 1990 for a detailed description), established and headed by Dr. Nogah Hareuveni, and at the Ecological Garden of the Technion, mentioned above. In addition to the lively discussions on the meaning of ecological and cultural restoration (see also Allen and Naveh 1996 in this chapter!), these field workshops provided the opportunity to obtain very useful feedback on our restoration work. All the participants expressed their admiration about the unique multifunctional ecological, cultural, educational and historical restoration project of "Neoth Kedumim" in the rocky and desolate Judean Hills, South of Jerusalem, which revived the glory of the natural vegetation and animals, and the reconstructed terraces of vineyards, olive and fruit orchards with their sophisticated irrigation network as in Classical Times. They also praised our attempts to teach our students how to combine natural beauty with engineering utility and how to restore a small biosphere island of the ancient Carmel forest in the heart of the Technion campus.

CHAPTER 12 – TEN MAJOR PREMISES FOR A HOLISTIC CONCEPTION OF MULTIFUNCTIONAL LANDSCAPES. Landscape and Urban Planning 57: 269–284. Special Issue. (2001).

Introducing a workshop on the theoretical foundations of multifunctional landscape I postulated ten major premises for a holistic conception of multifunctional landscapes. These present a condensed version of the theory of multifunctional landscapes as the concrete, self-transcendent and self-organizing mixed natural and cultural Gestalt system of our THE. Treated with a *biperspective systems view,* functioning simultaneously as material, biogeophysical systems and as mental, cognitive systems, these landscapes become a multifunctional, tangible bridge between nature and mind.

I showed how these concepts could be applied by a goal-oriented and mission-driven landscape researcher for the overarching transdisciplinary vision of a sustainable future. Serving in the dual position of experts in their own field and helping to integrate innovative, future-oriented study and action, these could fulfill an important role in bridging the still wide gaps between all biological and human ecological aspects related to land use.

This symposium, at which this and other workshops were carried out, was a very well prepared and organized scientific meeting of landscape ecologists at Roskilde University – the active research and teaching center for holistic, interdisciplinary and transdisciplinary-oriented LE. Organized by J. Brandt and G. and B. Tress, its productive dialogues in smaller working party groups facilitated the interaction, which is so often missing at these meetings. It culminated in joint recommendations regarding the research of multifunctional landscapes, and in addition led to the special issue of "Landscape and Urban Planning" edited by Tress and Tress (2001) on the workshop in which this chapter served as the opening lecture, and also in the publication of the proceedings of the symposium meeting in three volumes (Brandt and Vejre 2004).

CHAPTER 13 – MULTIFUNCTIONAL, SELF-ORGANIZING BIOSPHERE LANDSCAPES AND THE FUTURE OF OUR TOTAL HUMAN ECOSYSTEM (*World Futures* 60: 469–503, 2004)

In an invited essay, I outlined the vital multiple functions of solar powered self-creating and regenerating natural and cultural biosphere landscapes for the biological evolution and sustainable future of organic life, and for human physical and mental health. This essay addressed a very different reader circle from all my previous publications; therefore, I had to provide a rather detailed introduction to the major premises of LE, in the light of the transdisciplinary challenges of the emerging information age. For this purpose, I presented landscapes as tangible multifunctional Gestalt systems of our THE, adopting and further developing new insights into self-organization and evolutionary processes in non-equilibrium systems and their auto- and cross-catalytic cycles driving them. I showed that these biosphere landscapes and their multiple functions are endangered by the exponential growth and waste products of urban-industrial technosphere and agro-industrial landscapes. Adopting Laszlo's contentions on the urgently needed, all-embracing sustainable revolution, driven by humans who are conscious for their responsibility as stewards of the fragile web of life, I pointed to the important role of landscape ecologists and restorationists in the evolutionary process. As manifested in a multinational project for modelling regional sustainable development, these can help to convert the antagonistic, destructive relations between biosphere and technosphere landscapes into mutually supportive, symbiotic relations, fostered by restorative cultural and economic cross-catalytic networks in our THE.

CHAPTER 14 – THE TRANSFORMATION OF LANDSCAPE ECOLOGY AND RESTORATION ECOLOGY INTO TRANSDISCIPLINARY SCIENCES OF HOLISTIC LANDSCAPE STUDY, MANAGEMENT AND PLANNING, CONSERVATION AND RESTORATION (2005).

In this chapter, especially prepared as a recapitulation of this anthology, I discuss the need for a transdisciplinary transformation of both LE and RE sciences and practices to cope with the severe ecological, cultural and socio-economic global crisis and to contribute to the sustainable revolution. Landscape and restoration ecologists, together with all other concerned environmental researchers, professionals and practitioners can contribute to this goal through undergoing a radical shift in thinking and acting, and by abandoning the deeply ingrained reductionistic and mechanistic world view on which conventional nature sciences were based. In the previous chapters, I have mentioned earlier the need for such a transdisciplinary transformation and paradigm shift, to overcome the syndrome of fragmentized scientific disciplines. However, in this chapter, I present a much more in-depth discussion of all aspects of transdisciplinarity relevant for this transformation.

The concluding subchapters are devoted to the most significant breakthrough in the ongoing holistic and transdisciplinary revolution, achieved most recently by Laszlo in his genuine formulation of the *Integral Theory of Everything.* This is based on the revolutionary discovery that at the roots of reality there is an interconnecting, information-conserving and conveying cosmic field. Laszlo calls it "the Akashic or A" Field, after the Sanskrit word for ether. This field is not limited to the physical world but informs all living things – the entire web of life – including our consciousness. The discovery of the underlying mechanism for both the physical-geospheric and the cognitive mental and spiritual noospheric space spheres of the THE, opens new vistas for the comprehension interactions of human society and its landscapes. Integrating the findings at the cutting edge of many innovative scientific fields from cosmology to quantum physics, and from evolutionary biology to consciousness research, Laszlo infers that all the elements of a system, from atoms to ecological system up to galaxies have a coherence of "quasi-timing together", and are so correlated that what happens to one of these parts, also happens to other parts of the system. I claim that for landscapes and our Total Human Ecosystem, integrating humans and their biosphere and technosphere landscapes this unity of correlation and coherence could have far-reaching implications for the nature-humans-culture triangle. Finally, I present the evolutionary and historical relations of this triangle in a hierarchical model of the THE as part of Laszlo's *"Informed Universe"* and its underlying A field.

The new insights gained for these reciprocal relations require a revision of our holistic and transdisciplinary landscape concepts, in the light of Laszlo's Integral

Theory of Everything. This will reinforce our contentions that healthy and attractive landscapes also have favourable effects on our mental and physical well-being, which until now were hard to explain by conventional scientific methodologies. Landscape restoration projects may be the most convincing examples of these positive reciprocal relations, thanks to their active human intervention in the conservation and rehabilitation of nature and culture.

Gaining the cooperation of environmental-oriented quantum physicists and biologists for these studies will be one of the most demanding transdisciplinary challenges for the future of both TLS: Landscape and restoration ecology.

I am closing this introduction by some additional autobiographical comments, summarizing my road from the rocky pastures of the foothills of the Western Galilee in Mazuba to the vision of a transdisciplinary, unified landscape metascience. In the introduction to the first part of this anthology I have already mentioned the first important stations after I finished my studies as agronomist and ecologist at the Hebrew University in Jerusalem and started my professional career in range and pasture research at the Neve Yaar Experimental Station of the Ministry of Agriculture. The years 1958–1960 at the University of California in Berkeley were formative for me in becoming an open-minded systems ecologist.

The second most significant station along this road was my assignment in Tanzania from 1962 to 1965 for pasture research and development from the cultivated slopes of snow-capped Mt. Kilimanjaro to the semi-arid Masailand acacia savanna. Here, I became aware of the significance of a holistic approach to the people and their land as expressed in the Total Human Ecosystem and applied it to the development of Masailand.

After returning home and starting my final professional position as teacher and researcher at the Technion and as I described above, thanks to my encounter with European landscape ecology I arrived at the third important station on my road to being a holistic, problem-solving landscape ecologist.

During sabbaticals at U.C. Berkeley, at the State University of San Diego, and Rutgers University, I could renew and further broaden my contacts in the USA, and now also in the field of restoration ecology, thanks to Edie Allen who worked at that time at the State University of California, San Diego.

A further significant station on my road to a transdisciplinary world view and its projection on landscapes was my acquaintance with the renowned systems scientist, educator and philosopher Erich Jantsch and his writings. I had the great fortune to meet him, during my sabbatical in 1979 at U.C. Berkeley, while attending his brilliant seminar lectures on the evolution of the self-organizing universe, representing the most advanced, transdisciplinary systems theory. In our talks after his lectures he introduced me to his revolutionary ideas of self-transcendence in evolving human systems which he had presented in a lucid form in "Evolution and Consciousness" (Jantsch and Waddington 1976). He gave me this book, claiming that it was one of the first truly transdisciplinary books dealing with human systems in totality. Among other important contributions, it contains also essays by Holling (1976) and by Prigogine (1976). I made him aware of our THE paradigm and the holistic conception of landscapes, which in my opinion expressed a more comprehensive concept and fitting better his transdisciplinary approach than ecosystems. His seminar lectures were published as his seminal contribution to the emerging transdisciplinary scientific revolution – but it became unfortunately his last book (Jantsch 1980).

As reflected in the chapters of the second part, in the consolidation of our conceptional foundations I reached far beyond ecology and related disciplines, relying on some of the most important contributors to the scientific revolution, namely Laszlo, Bohm and Frankel.

Of great significance for my work in more recent years were my close connections with Prof. Wolf Grossman, from the Center for Environmental Research UFZ Leipzig/Halle, Germany. Thanks to his visionary forward-looking approach to the emerging new information- and network-based economy, and his intimate knowledge of the most advanced simulation models and their practical implementation, he opened new vistas for me. I was especially impressed by his innovative transdisciplinary approach to "viable systems management", to which I referred in the concluding chapter. I accepted, therefore, readily his invitation to join him together with my Israeli collaborators, Dr. Shmuel Burmil and Dr. Didi Kaplan, in the EU "MOSES project" for regional development, as landscape experts. Under his able leadership this multinational interdisciplinary research project yielded valuable results which I mentioned in Chapter 13. We used the THE paradigm as the basis for our transdisciplinary simulation model and presented a joint lecture on this project at the European Congress of European Ecological Economy Society at the 3rd International Conference of the European Society for Ecological Economics, Vienna 2000 (Grossmann and Naveh 2000). Our lecture was chosen to open the proceedings of the conference.

It is very gratifying for me, that the Ecological Garden of the Technion is providing the opportunity to continue such field-oriented environmental education, in combination with the ecological teaching program of systems and landscape ecology by my able young colleague Dr. Yohay Carmel, who replaced me after my retirement.

Throughout the years I participated actively in most European and International IALE meetings, and some of the excellent annual symposiums of the US branch of IUCL. In this way I had the opportunity to meet the younger generation of landscape ecologists and to follow with great interest their work, forming closer bonds with some of its outstanding representatives and enjoying these mutually beneficial relations.

I realized also with great satisfaction that the holistic landscape conception that we introduced with our Springer landscape ecology book (Naveh and Lieberman 1994) gained more and more followers and – as reported in a special symposium on "Holistic LE in Action" (Palang et al. 2000) – was applied in practice all over the world. As described in Chapter 14, in recent years we also witness a growing trend towards transdisciplinary and RE, and I hope that this anthology will be a catalyst for a further, more far-reaching shift towards a unified transdisciplinary landscape science for which Laszlo's Integral Theory of Everything will serve as a solid cornerstone.

REFERENCES

Allen EB (ed.) (1988) *The Reconstruction of disturbed Arid Lands. An Ecological Approach.* American Association for the Advancement of Sciences. Selected Symposium 109. Westview Press, Inc. Boulder, Colorado.

Allen AB and Naveh Z (1996) The limits to ecological and cultural restoration in mediterranean-type landscapes. In: Steinberger Y (ed.) *Preservation of Our World in the Wake of Change.* Proceedings Sixth International Conference of Israel Society of Ecology and Environment. Quality Sciences. (ISEEQS) Ed. Vol. VI B., pp 775–778.

Bakshi T and Naveh Z (eds) (1980) *Environmental Education Principles, Methods, and Applications.* Plenum Press, New York.

Bar-Yosef O (1998) The Natufian culture in the Levant. Threshold to the origins of agriculture. *Evolutionary Anthropology* **6:**159–177.

Bar-Yosef O (2001) The world around Cyprus: from Epi – Paleolithic foragers to the collapse of the PPNB civilizations. In: Saiwy S (ed.) *The Earliest Prehistory of Cyprus.* ASOR, Boston, pp 129–164.

Bar-Yosef O and Meadow RH (1995) The Origins of Agriculture in the Near East. In Price TD and Gebauer AB (eds) *Last Hunters-First Farmers. New Perspectives on the Prehistoric Transition to Agriculture.* School of American Research Press Santa FE, New Mexico, pp 19–94.

Biswell H (1989) Prescribed Burning in California wildlands. University of California Press, Berkeley.

Bradshaw AD (1994) The need for good science-beware of straw men. Some answers for comments by Eric Higgs. *The Restoration of Land.*

Bradshaw AD and Chadwick MJ (1980) The Restoration of Land. Blackwell Scientific Publication, Oxford.

Brandt J andVejre H (2004) *Multifunctional Landscapes.* WIT Press Southampton, Boston.

Di Castri F and Mooney HA (eds) 1973. *Mediterranean Type Ecosystems Origin and Structure.* Springer-Verlag Berlin, Heidelberg, New York.

Grossmann WD and Naveh Z (2000) Transdisciplinary challenges for regional sustainable development toward the post-industrial information society. *3rd International Conference of the European Society for Ecological Economics. 2000, Vienna. pp 1–17.*

Holling CE (1976) Resilience and stability of Ecosystems. In: Jantsch E and Waddington CH (eds) HH *Evolution and Consciousness Human Systems in Transition* Addison-Wesley Publishing Company Reading, Massachusetts, pp 73–92.

Jantsch E (1980) *The Self-organizing Universe. Scientific and Human Implications of the Emerging Paradigm of Evolution.* Pergamon Press, Oxford.

Jantsch E and Waddington CH (eds) (1996). *Evolution and Consciousness Human Systems in Transition.* Addison-Wesley Publishing Company Reading, Massachusetts.

Krummel JR, O'Neill, RV and Mankin JB (1986) Regional environmental simulation of African cattle herding societies. *Human Ecology* **14:**117–130.

Naveh Z (1968) The determination of agro-ecological potential in the drier tropics with special references to East Africa. *Journal of Tropical Ecology* **3:**601–611.

Naveh Z (1970) Ecology and development. *Second World Congress of Engineers and Architects: Dialogue in Development,* Jerusalem, December 14–23.

Naveh Z (1984) The vegetation of the Carmel and Nahal Sefunim and the evolution, of the cultural landscape. In: Ronen A (ed.) *Sefunim Prehistoric Sites in Mount Carmel Israel.* pp 23–63. B.A.R. International Series 230. Oxford.

Naveh Z (1988) Multifactorial reconstruction of semiarid Mediterranean landscapes for multipurpose land use. In: Allen EB (ed.) *The Reconstruction of disturbed Arid Lands. An Ecological Approach.* American Association for the Advancement of Sciences. Selected Symposium 109. Westview Press, Inc. Boulder, Colorado, pp 234–256.

Naveh Z (1989) The challenges of desert landscape ecology as a transdisciplinary problem-solving oriented science. *J. Arid Environment* **17:**245–253.

Naveh Z (1990) Neoth Kedumim – restoration of biblical landscapes in Israel. *Restoration Management Notes* **7:**9–12.

Naveh Z (1996) Introduction to concepts and principles of ecological and cultural landscape restoration. In: Steinberger Y (ed.) *Preservation of Our World in the Wake of Change.* Proceedings Sixth International Conference of Israel Society of Ecology and Environment. Quality Sciences. (ISEEQS) Ed. Vol. VI B., pp 761–766.

Naveh Z (1995) Interactions of landscapes and cultures. *Landscape and Urban Planning* **32:**43–54.

Naveh Z (1997) A critical appraisal of sustainable development. Introduction to workshop on sustainable land use. In: *The First European Dialogue Conference on: Science for a Sustainable Society-Integrating Natural and Social Sciences,* October 1997. Department of Environment, Technology and Social Science, Roskilde University, Denmark, pp 26–29.

Naveh Z (1998) The Ecological Dimensions of Regional Sustainable Development. In: Gabriel I and Narodoslawsky M (eds) *Regions-Cornerstones for Sustainable Development.* International Workshop Series on Sustainable Regional Development. Austrian Environment Network Research. Graz, 1998, pp 33–44.

Naveh Z (2000) What is holistic landscape ecology? A conceptual introduction. Special issue, *Landscape and Urban Planning* **50:**7–26.

Naveh Z (2005a) Towards a transdisciplinary landscape science. In: Wiens JA and Moss M (eds.) *Issues and Perspectives in Landscape Ecology.* Cambridge University Press, Cambridge, pp 346–354

Naveh Z (2005b) Epilogue: Toward a transdisciplinary science of ecological and cultural landscape restoration. *Restoration Ecology* **13:**1–7.

Naveh Z and Dan J (1973) The Human Degradation of Mediterranean Landscapes. In: Di Castri F and Mooney HA (eds) *Mediterranean Type Ecosystems Origin and Structure.* Springer-Verlag Berlin. Heidelberg. New York, pp 373–390.

Naveh Z and Lieberman AS (1984) *Landscape Ecology Theory and Applications.* First Edition. Springer-Verlag New York, USA.

Naveh Z and Lieberman AS (1994) *Landscape Ecology Theory and Applications.* Second Edition. Springer-Verlag New York, USA.

Naveh Z and Allen EB (2001) Restoration Ecology and Landscape Ecology – A win win relationship. Plenary lecture at the IALE European Conference 2001: Developments of European Landscapes, Stockholm University and University of Tartu. *Publicationes Instituti Geographici Universitatis Tartunsis* 92., pp 38–41.

Nevo E (2004) Evolution of genome dynamics under ecological stress. *Dynamical Genetics 1–27.*

Noy-Meir E and Kaplan D (1991) *The effect of grazing on the herbaceous Mediterranean vegetation and its implications on the management of nature reserves.* Interim Report to the Nature Conservation Authorities, Jerusalem (in Hebrew).

Odum E (1971) *Fundamentals of Ecology.* Third Edition. WB Saunders Company, Philadelphia.

Palang H, Mander U and Naveh Z (2000) Holistic Landscape Ecology in Action. Editorial introduction to Special Edition. *Landscape and Urban Planning* **50:**1–6.

Prigogine I (1976) Order through fluctuation. In: Jantsch E and Waddington CH (eds) *Evolution and Consciousness Human Systems in Transition* Addison-Wesley Publishing Company Reading, Massachusetts, pp 93–132.

Tress B and Tress G (eds) (2001) Bridging Human and Natural Sciences in Landscape Research. *Special Issue Landscape and Urban Planning 57.*

Warburg M (1977) Animal species diversity along environmental gradients in the Mediterranean landscape of Israel. *Research Report 450. Binational Israel-American Science Foundation. Technion, Israel Institute of Technology,* Haifa.

Western D and Finch V (1986) Cattle and pastoralism: Survival and production in arid lands. *Human Ecology* **14:**77–94.

Western D (1997) In the Dust of Kilimanjaro. Washington DC. Covelo, California: *Island Press/ Shearwater Books.*

Yogev A and Naveh Z (1986) The effect of fertilizing and mulching on the establishment of Pistacia lentiscus in limestone debris. In: *Proceedings, Third International conference of the Israel Society for Ecology and Environmental Quality Sciences. Jerusalem,* 1–4 June 1986, pp 403–412.

A. MEDITERRANEAN ISSUES

View of Mt. Carmel and the Coastal Plain. Photo courtesy S. Danziger.

CHAPTER 1

NAVEH Z. (1967)

MEDITERRANEAN ECOSYSTEMS AND VEGETATION TYPES IN CALIFORNIA AND ISRAEL

ECOLOGY 48: 445–459.

Dense *Q. calliprinos* and *Pistacia lentiscus* maquis on Terra Rosa soil in Upper Galilee in Israel.

Abstract A comparison of climate, soils, vegetation, and biotic history of mediterranean ecosystems revealed closest ecological equivalence between the blue oak grassland in California and the vallonea oak grassland in Israel, both mediterranean oak savannas in which overall environmental resemblance seems closest. Ecological amplitude and syndynamics in most other comparable types are similar, but the chamise chaparral on non-calcic brown upland and lithosols differs from its maqui counterpart by its one-layered structure and by its adverse influence on the ecosystem. Annual grasslands in California foothills are comparable to seral batha dwarfshrub and grassland in Israel as human-induced degradation stages of chaparral and maqui. For such ecological comparisons a holistic ecosystem approach seems more suitable than any preconceived phytoclimatic and climax concepts

INTRODUCTION

Comparisons of mediterranean vegetation types are as old as ecology itself. Griesebach (1872) used resemblance in physiognomy of the dominating sclerophyll plants in different countries with a mediterranean climate type as one of his main examples for the dependence of life form on climate. Schimper (1898) was the first to give this epharmony a physiological interpretation and to stress its adaptive value for climate with prolonged summer drought. In a more recent comparison Ruebel (1930) stated that the dominance of durilignosa communities, such as maqui and garigue, characterizes all regions with mild winters and long, dry summers, classified by Koeppen (1923) as Csb climates.

Using a different approach of biological spectra of life forms, Raunkiaer (1934) called the mediterranean climate type a therophyte climate because of the high percentage of this life form in the Mediterranean floras. Cain (1950) confirmed this also for other countries with a mediterranean climate type, including California. He proposed that further study should take into account the quantitative role of the various species and the successional stage of the vegetation.

My object is to compare the mediterranean vegetation types of California and Israel from a dynamic and holistic viewpoint. This necessitates not only a thorough comparison of climatic conditions and their ecological significance, but also of edaphic and physiographic features, as well as the effect of human and other biotic modifications in molding the evolution and aggregation of vegetation types and communities in both countries. Ecological equivalence of these vegetation types will be considered not only in phytogeographic and physiognomic, but also in dynamic and synthetic features.

THE COMPARATIVE MEDITERRANEAN ENVIRONMENT OF CALIFORNIA AND ISRAEL

Phytogeographic Subdivision of California and Israel

Israel, despite its limited area (one-twentieth that of California), is distinguished by a large ecological diversity of climate and physiography, with its marked gradients of rainfall, humidity, temperature, and resulting evapotranspiration from north to south and from the coastal plains to the mountain ranges and inner valleys. It has much in common with the north-south and west-east gradients and climatical patterns in California—many times condensed and concentrated. This highly contrasting environment inspired Eig, Zohary, and their co-workers in their intensive phytogeographic and phytosociologic studies, summarized recently by Zohary (1962).

The only comparable study in California was carried out by Cooper (1922) in the broad sclerophyll vegetation. A more detailed subdivision of California into biotic provinces, vegetation types, and plant communities was published by Munz and Keck (1959). In this subdivision the California biotic province—west of the Sierra Nevada and the southern mountains with the interior valleys and foothills in Central and North California and the southern coastal area and coastal ranges, south of the San Francisco Bay—has the most comparable type of mediterranean climate. It includes Cooper's broad sclerophyll region (1922), with a total annual precipitation of 15–30 inches, less than 15% of which occurs between May 1 and October 31, and with the most moderate temperatures in California. This area corresponds roughly with Eig and Zohary's mediterranean phytogeographic territory in northern and central Israel, between the 15- and 30-inch isohyets.

The California province is bordered on the north by the cooler and moister Oregonian province in the north coastal region, and by the Sierran province of the great montane area with its higher elevations on the east. Similarly more humid regions with more mesic vegetation types can be found north of Israel, in the higher Syrian and Lebanon mountains. Like the mediterranean territory in Israel, the California province is bordered also by more arid zones: the southern desert which has much affinity to the Saharo-Sindian territory in Israel, and the semi-arid Nevadan (Great Basin) province, which differs, however, from the semi-arid Irano-Turanian territory in Israel in many respects. It has a much more severe winter climate and is geographically and floristically isolated from the California province by the Sierra mountains. The Irano-Turanian territory, on the other hand, is actually a transition zone, and is in close contact with the mediterranean region. This facilitated the invasion of more arid elements into depleted habitats in the mediterranean region.

The Comparative Mediterranean Climate in California and Israel

The apparent similarity of mediterranean type climates in California and Israel has led to several attempts at climatological comparisons. The most comprehensive attempt at a bioclimatical map of the Mediterranean zone and its homologues in the world was published recently by the Unesco Arid Zone Research team (1963). According to this bioclimatical classification, the mediterranean phytogeographic territory in Israel is subdivided into a wetter accentuated thermomediterranean part in the northern coast area and the Upper Galilee, and a drier xerothermomediterranean part, which merges with the attenuated subdesert climate. In California the same accentuated thermomediterranean climate extends from Monterey Bay over the coastal highlands as far as San Bernardino throughout the whole plain, north of Sacramento and between the coastal ranges and the Sierra Nevada, and on the western slopes of the southern stretches of the Sierra Nevada. The wetter parts of the central valley, the heights around Los Angeles, and the coastal strip of Lower California are mapped xerothermomediterranean. The wettest part of the California province, in the San Francisco Bay region, is attenuated thermomediterranean, like the whole northern coastal ranges and the coastal region. This corresponds to the wetter parts of Lebanon, north of Tripoli, and the Lebanon highlands.

According to these climatological classifications several homological mediterranean bioclimates have been outlined in California and Israel. However, the possibilities of establishing ecological homoclimes, based on broad characteristics of precipitation, temperature, and evaporation and expressed as single indices, should not be exaggerated. Their limitations can be demonstrated by a closer study of comparative patterns of temperatures and their diurnal amplitude and of rainfall distribution and its reliability during the growth season.

That these factors are of greatest significance for the actual plant climate in a mediterranean environment was shown for the herbaceous pasture vegetation in Australia by Trumble (1952), in California by Bentley and Talbot (1951), and in Israel by Naveh (1957).

Comparative temperature situations.—For this purpose we have chosen three stations in California and Israel, typical for the mediterranean type climate and vegetation and all mapped within the accentuated thermomediterranean bioclimatic region. Mt. Canaan in the Upper Galilee is the coldest station in Israel and has about the same average temperature as Auberry, a typical station of the Sierra foothills, but at the latter the January minima are considerably lower and the daily range is more than twice as high (Table 1). Haifa, which is typical of the mild and moderate winter temperatures that prevail in the coastal region of Israel, has about the same January average as Santa Barbara in the southern Coastal Range. However, here the diurnal range is twice as large and the minimum as low as in the cooler stations in Israel. Berkeley in the San Francisco Bay has a pronounced mild maritime climate, which is typical of the central coast, the central coastal valleys, and the lower elevations of the central coastal ranges. However, its winter climate is readily comparable, both in its absolute values and diurnal amplitude, only to the cooler stations of the Galilee and Judean hills in Israel. Thus, the winter temperature regime in comparable stations is more severe in California and even where the daily means are similar, the diurnal amplitude is much greater in California.

During the warm and dry season the Pacific Ocean exerts a much more pronounced cooling effect than the Mediterranean Sea. As a result, at Berkeley and similar stations under the direct coastal influence, summer temperatures are lower and humidity and cloudiness are much higher. Therefore soil moisture and growth conditions for plants with summer activity are quite different in both environments. This is clearly reflected in the coastal fog-belt redwoods and perennial grasslands,

Table 1. January temperature regime at representative stations with a mediterranean type climate in California and Israel

	Mean temperatures (F°)			
Station and region	Daily mean	Minimum	Maximum	Dailyrange
California				
Auberry-Sierra Nevada foothills	43.1	31.4	55.0	23.6
Berkeley-central coast area	48.0	41.6	54.4	13.2
Santa Barbara-Southern Coastal Range	53.4	42.7	64.4	21.7
Israel				
Mt. Canaan-Upper Galilee	45.0	40.0	50.0	10.0
Tabor-Lower Galilee	54.0	46.0	61.0	15.0
Haifa-northern coast area	53.1	50.0	58.8	8.8

which cannot be considered mediterranean vegetation types and are actually included in the Oregonian province. These differences in winter and summer temperature regimes are not well represented in homoclimatical maps.

Comparative rainfall distribution and its variability.—Comparative rainfall patterns cannot be expressed by annual averages alone and are also poorly represented on homoclimatic maps. Not only the variability in amount of annual rainfall should be considered, but also the patterns of the monthly rainfall distribution and its variability. In California (Varney 1925) and in Israel (Katznelson 1956) the relative variability of rainfall (V_r), as measured by the mean deviation from the longtime average, within the rainfall belt of 400–800 mm is 20–30% of this average. With the smallest variability in California, the Sierra foothills are closest to the Mediterranean hill region in Israel, whereas in the Coastal Range of California and especially in their southern parts, the variability is well over 30%. The same is true also for the interseasonal variability (V_s), which is measured by the average variation between two consecutive years over a long period. This is about 30% of the average rainfall in both countries. However, in the southern Coastal Range it is more than 40%.

Monthly rainfall distribution will be considered in Auberry in the Sierra foothills and Nazareth in Central Lower Galilee, since they represent about the same annual average and variability. At the California station the same amount of total annual rain to initiate germination and plant growth may be expected 1 month earlier and the last rains in spring and early summer 1–2 months later than in Israel (Table 2). That is, Israel has the typically eastern mediterranean winter-spring rainfall pattern, in which 90% of the total annual precipitation may be expected between November and March. This contrasts to the autumn-winter-spring and early summer rainfall pattern in California, in which less than 80% may occur within this period. However, this does not necessarily mean that in general growth conditions are more favorable and the growth period longer in California. If the same amount of rainfall is spread over a longer period, reliability of a continuous moisture supply during the growth period may be lower than in a more cencentrated rainfall situation. This is the case, in fact, when the variability of the monthly rainfall pattern is compared. This variability is presented in Table 2 not only as V_r, but also as the coefficient of variation (V_c) and as the interquartile deviation, expressed as the percentage of the median (V_q) (Katznelson 1956). Any station with a lower amount in a given month has a higher V_c and V_r value for that month. However, V_q is independent of the arithmetical averages and standard deviations and reveals the distinctly higher variability of the monthly distribution for the main rain and growth period in Auberry.

Therefore the longer rain season with the same total amount of rainfall in California may lead to a longer period of unreliability in moisture regime and to a greater probability of moisture stress, especially during the critical period of establishment and early growth. Together with the more severe temperature regime during this period, a more erratic and unfavorable plant climate is apparent. As a result, the production and composition of plant communities which are dominated by plants germinating and developing chiefly in the winter months may be even more variable than in Israel under the same total annual rainfall.

Table 2. Variability of monthly rainfall distribution at Auberry, California, and Nazareth, Israel[a]

	$\bar{p}$ (inches)		$\%T_p$		$V_r\%$		$V_c\%$		$V_q\%$	
Month	Auberry	Nazareth	Auberry	Nazareth	Auberry	Nazareth	Auberry	Nazareth	Auberry	Nazareth
August	.02		.1		50					
September	.28	.02	1.2	.1	79	167	203	333	467	
October	1.24	.46	5.1	1.8	59	113	85	177	100	200
November	2.35	2.83	9.6	11.2	64	77	88	102	87	74
December	4.37	5.48	17.9	21.7	57	57	69	60	62	43
January	3.99	6.75	16.4	26.8	55	40	85	46	62	38
February	4.87	5.81	20.0	23.1	54	46	67	53	69	49
March	4.22	2.50	17.3	9.9	55	60	69	86	61	52
April	2.05	1.10	8.4	4.4	100	83	94	116	108	109
May	.79	.25	3.4	1.0	73	123	96	96	184	
June	.11		.5		72					
July	.03		.1		100					
Total	24.40	25.20	100	100	22	22	26	28	18	18

$\bar{p}$ = perennial average precipitation $= \frac{1}{n}\Sigma Pi$

$\%T_p$ = % of total annual precipitation

$V_r\%$ = (average) relative variability $= \dfrac{p\frac{-1}{n} = \Sigma(Pi - p)}{100\,\bar{p}}$

V_c = coefficient of variation $= 100\dfrac{\sqrt{\frac{1}{n}\Sigma(Pi-p)^2}}{\bar{p}}$

V_q = interquartile variability $= 100\dfrac{q_3 q_1}{2q_2}$

(q_1 = lower quartile; q_2 = median; q_3 = upper quartile)

[a]Averages for the years 1920–21 to 1951–52. Data and statistical calculations for Nazareth by courtesy of the Government Meteorological Service, Israel.

Topography and Soils

Only the calcareous Rendzina soils, like the Zacca, Linne, and Nacimiento series in the central and south coastal ranges, are comparable in their more favorable structure, chemical, and physical features to the typical Red Mediterranean soils of north and central Israel and especially to the Pale Rendzina soils. Most other non-cultivated soils within the California province are coarse structured, neutral or acid non-calcic Brown Upland soils and lithosols, inferior in fertility, water-holding capacities, and grazing resistance. If California had experienced long-time and intensive land use similar to that of Israel and other Mediterranean countries, most of the better sites, such as swales and bottom valleys, rolling grasslands, and woodlands would have been converted into cultivated fields many centuries ago.

In comparable climatic conditions, the main limiting edaphic and topographic factors in Israel are probably shallowness and amount of rock outcrops and stones of these non-arable eroded and residual wildland soils. In California, on the other hand, probably steepness of the slopes, lower fertility, and higher erodability are of greater ecological significance. Closest resemblance can be found between the deeper Rendzina soils of foothills and plateaux in the Lower Galilee in Israel and the calcareous soils of rolling hills in the Coastal Ranges of California.

Biotic History

The differences in structure and dynamics of the mediterranean vegetation types in both countries can be appreciated fully only after taking into account their very different biotic history.

The natural vegetation of Israel has been modified since Mesolithic and early Neolithic times (seventh to sixth millenium B.C.) by intensive coppicing, bush clearing, burning, heavy grazing and browsing of domestic livestock, and by cultivation (Whyte 1961). This caused the degradation of dense climax forest and maqui formations into scrub and annual grassland, accompanied by erosion and depletion of the soil and the habitat as a whole. However, one should not ignore the possibility that the period of active human interference has been long enough to allow evolution of biotypes better adapted to these disturbed conditions and that selection pressure has favored the survival of species and ecotypes with highest resistance to fire, browsing, and grazing. As patch-cultivated fields were grazed before seeding and after harvest, the transfer of seeds to and from adjacent untillable roughlands was facilitated, and ideal conditions were created for introgression between biotypes and species and spontaneous hybridization of wild indigenous grasses and cultivated plants on field edges. The latter was proved by Zohary (1960) in the case of cultivated barley (*Hordeum vulgare*) and *H. spontaneum*. At the same time, the degradation of the mediterranean ecosystems facilitated the invasion of more xeric elements from adjacent more arid regions which replaced those mediterranean plants which succumbed to the pressure of man and its livestock. In such conditions very few if any remnants of the original pre-human vegetation could be left, and probably a new equilibrium has been reached in these modified ecosystems between the degraded physical habitat and those living components which could adapt themselves to these semi-natural conditions. Since the foundation of the State of Israel, however, this equilibrium has been disturbed again by the introduction of mechanized methods of land preparation and hill cultivation, large scale afforestation, range reseeding and improvements, water and road developments, intensive seasonal grazing of large beef cattle and milk sheep herds in the vicinity of hill settlements, and above all, the rapid process of urbanization.

In California similar agricultural, pastoral, and urbanic disturbance of the original climax vegetation began only with the arrival of the Spanish missionaries and European settlers, about two hundred years ago. Here, therefore, the process of active modification of vegetation types and hybridization is very evident, as shown in the case of *Salvia* and *Adenostoma* species by Anderson (1956). One of its other striking symptoms is the invasion of weedy and aggressive plants, chiefly annuals from the Mediterranean region. These plants evolved in similar conditions of integrated human and pastoral disturbance (described above) and are therefore pre-adapted to their new, modified mediterranean habitat.

ECOLOGICALLY EQUIVALENT VEGETATION TYPES IN CALIFORNIA AND ISRAEL

Tree and Shrub Formation

Affinities in amplitude and physiognomy.—In his comprehensive study Cooper (1922) considered the shrub-dominated chaparral to be the real ecological equivalent of Mediterranean maquis. Its center of distribution is in the southern Coastal Ranges, particularly in those locations where rainfall occurs mainly during the winter and where winters are mildest. These chaparral communities are bordered on their mesic side by broad-leaved sclerophyll forest (Cooper 1922) or mixed evergreen forest (Munz and Keck 1959). Similarly, in northern Israel and even more in Lebanon and Syria, in higher elevations and in more favorable and sheltered sites with regard to slope, exposure, insolation, and soil, the shrub-dominated maqui and garigue are replaced by tree-dominated forest and maqui.

In addition to their similar ecological amplitude, these tree and shrub formations have many ecological vicariants (Table 3). It is evident that the typical chaparral is comparable with the shrub maqui and garigue formations. (Garigue is the French term for sclerophyllous shrubs about 1 m tall.) Here, as in the chaparral, sclerophyll shrubs with flat and thickly cutinized and mostly evergreen leaves attain dominance and here the California scrub oaks, *Q. dumosa* and (on serpentine) *Q. durata,* have their closest physiognomic and taxonomic Mediterranean counterparts in the dwarfed and shrubby variant of *Quercus calliprinos.* The higher and more tree-like variety of the latter, on the other hand, is typical of the more mesic forests and maquis and more comparable to the true live oaks such as *Q. agrifolia.*

Differences in influence on habitat and in structure.—Some of the most widespread and important species, such as *Pistacia lentiscus* in Israel and *Adenostoma fasciculatum* (Chemise) in California, have no counterparts. The former is a sclerophyllous,

Table 3. Some ecological and taxonomical "vicariants" of mediterranean vegetation types in California and Israel—trees, shrubs, and climbers—chiefly in forests, maqui, and chaparral

Israel	California
Quercus infectoria Oliv.	*Q. Kelloggii* Newb.
Quercus calliprinos Webb.	*Q. agrifolia* Nee.
" " "	*Q. dumosa Nutt.*
" " "	*Q. durata Jeps.*
Quercus ithaburensis (Decne) Boiss.	*Q. Douglasii* H. & A.
Pinus halepensis Mill.	*P. Sabiniana* Dougl.
" " "	*P. Coulteri* D. Don
Arbutus Andrachne L.	*A. Menziesii* Pursh.
Acer syriacum Boiss. et Gaill.	*A. macrophyllum* Pursh.
Alnus orientalis Decne	*A. rhombi folia* Nutt.
Laurus nobilis L.	*Umbellularia californica* (H. & A.) Nutt.
Crategus Azarolus L.	*C. Douglasii* Lindl.
Cercis Siliquastrum L.	*C. occidentalis* Torr. exGray.
Styrax officinalis L.	*S. Officinalis* L. *var California* (Torr.) Rehd.
Rhamnus palaestina Boiss.	*R. crocea* Nutt. in T. & G.
Rubus sangiineus Friv.	*R. vitifolius* Cham & Schlecht.
Smilax aspera L.	*S. californica* (A.DC.) Gray
Aristolochia altissima Desf.	*A. californica* Torr.
Dwarf shrubs—chiefly in batha in Israel and coastal sage in California	
	S. apiana Jep.
Salvia triloba L.	*S. leucophylla* Greene
Salvia graveolens Vahl.	*S. mellifera* Greene
Satureja Thymbra L.	*S. mimuloides* (Benth.) Briq.
Helianthemum lavandulifolium Mill.	*H. scoparium* Nutt.
Stachys palaestina L.	*S. bullata* Benth.

evergreen shrub and has some physiognomic resemblance to sprouting *Ceanothus* species, but seems highly superior to any of the chaparral shrubs in its vitality and its vigorous regeneration and rock penetration capacities. In addition, the accumulation of large amounts of litter and humus beneath the densely spreading and prostrate shrubs (Fig. 1) make it an outstanding soil protector even under recurrent fire and heavy browsing conditions (Naveh 1960). This is in striking contrast to the minute and, under certain circumstances, even adverse influence of chamise and presumably also other species, like *Arctostaphylos Manzanita,* on their habitat. Chamise has small needle-like leaves and produces only small amounts of mor-like humus, acidic in reaction, which is slow in decomposition, nitrifies poorly, but has a large fungal population. It is highly inflammable and regenerates vigorously after fire, both from resprouting crown stumps and from readily germinating seeds. In this respect it behaves like *Calicotome villosa,* the most common garigue shrub, and like most dwarf shrubs of batha communities. (Batha is a biblical term, applied first by Eig (1927) to dwarf shrub, perennial grass and herb associations, not exceeding 50 cm in height).

In contrast to the many-layered, heterogeneous structure of the mediterranean shrub communities, the mature chamise chaparral is mostly single layered and

Fig. 1. Pistacia lentiscus on edge of brushburn on dark Rendzina in Western Galilee, Israel. A dense layer of ash, litter, and charred humus is evident in the foreground.

Fig. 2. Adenostoma fasciculatum (chamise) separated from annual grassland by a wide, natural, bare strip on a shallow and steep non-calcic Brown Upland lithosol, Mount Diablo, California.

almost monotonously uniform, devoid of any subshrub and grass understory. On steeper slopes, on poor, non-calcic stony soils and shallow lithosols, and on sandy soils, there is even a conspicuous bare strip, up to 3 m wide, separating the chamise stand from adjacent annual grassland (Fig. 2).

Phytotoxic influences of chamise and other shrubs and their ecological significances.—Combined field and laboratory studies (Naveh 1961, Landers 1962) revealed that leaves, flowers, and their litter, as well as roots of chamise plants, contain water-soluble and heat-labile saponins and unsaturated lactones, accumulating in the soil and acting as germination and growth inhibitors on grasses. This may explain the total absence of any grass understory and the bare strip in soils with inadequate physico-chemical and biological properties for rapid absorption, decomposition, and inactivation of these phytotoxic agents. Conceivably, they could also interfere with normal decomposition, humification, and nitrification. This condition would explain the lack of structure, severe erosion, and low nitrogen availability under certain old but dense chamise stands and the striking depletion of nitrogen from soils in a lysimeter at San Dimas in Southern California in which chamise was grown for 20 years (Zinke 1959).

A similar absence of therophytes under garigue with *Erica arborea* in southern France was attributed by Guyot (1951), Deleuil (1951), and their co-workers to phytotoxic root excretions. Indications of germination- and growth-inhibiting effects of leaves and litter of maqui trees and shrubs were shown also in Israel by Yardeni and Evenari (1952), Naveh (1960), and Dinoor (1962). However, in the calcareous soils of Israel and other Mediterranean countries, as well as in other soils with high pH, large amounts of free cations, high absorption capacity, micro-biological activity, and fertility, these phytotoxic leachates can be effectively absorbed and/or decomposed. This is apparently the case also for chamise on calcareous soils and on serpentine in California, as well as on the deeper and more fertile clays on which chamise can sometimes be found. In these conditions these shrub communities have the same multi-layered structure as their Mediterranean counterparts, and lack also the sharp brush-grass borderlines described above (Fig. 3).

Fig. 3. A mediterranean shrub formation, the *Quercus calliprinos–Pistacia Palaestina* or maqui on Terra Rossa soil in Upper Galilee, Israel. The maqui has a multi-layered structure and lacks a sharp brush-grass border.

More recently, Muller and his co-workers showed that a similar absence of a grass understory and a sharp grass-brush borderline in certain situations in coastal sagescrub of southern California can also be explained by the phytotoxic and germination-preventing activity of the shrubs. But in this case these are volatile excretions from the leaves of aromatic shrubs, like *Artemisia californica* and *Salvia* species containing terpens and phenolic compounds, which are apparently absorbed and concentrated in the upper soil layers during the summer, especially in calcareous soils like the Zacca series, and thereby inhibiting germination after the first rains of fall and winter (Muller, 1966 and *personal communication*).

Ecological importance may be attributed, therefore, to such phytotoxic agents in special chemical, physical, and topographical soil conditions which favor the accumulation and action of these kolines, and where their accumulation is not prevented by frequent defoliation and fire.

Dwarf Shrub Formations

In both countries the shrub formations are bordered on their xeric sides by dwarf shrub formations—the coastal sage scrub in the coast region of central and south California, and the batha on the ecotones of the Mediterranean and Irano-Turanian territories (Fig. 4). These dwarf shrub communities not only have the same ecological position as the most xeric mediterranean woody vegetation types, but are more similar in physiognomy and structure than most other shrub formations of the chaparral and maqui. In both countries they are dominated by aromatic and xerophytic dwarf shrubs, many of them from the same families (Labiatae, Compositae, Rosacae) and some of the most characteristic species even from the same genera (Table 3).

However, physiognomic resemblance should not blind us to the differences in floristic composition of these communities, due to their different geology and biotic history. Not only are the most important and abundant species, such as *Poterium spinosum* in Israel and *Artemisia californica, Eriogonum fasciculatum,* and *Encelia farinosa* in

Fig. 4. Coastal sage scrub of *Artemesia californica* in central coastal California. This dwarf shrub formation is equivalent to the *Poterium spinosum* batha in Western Galilee.

California, without ecological vicariants, but also their relative ecological amplitude and distribution seems to be quite different. Whereas the California coastal sage is geographically restricted to the coastal area and to the mesas and lower foothills of south California, the batha dwarf shrub communities are widespread all over the mediterranean range in Israel. This position is occupied by annual grasslands in central and north California, which, like batha in Israel, are in close contact with the shrub communities and sometimes form an intricate mosaic-like pattern with them.

More detailed phytosociological work in California may separate the coastal sage into *Artemisia californica* associations in the coastal region and *Eriogonum fasciculatum* and *Salvia mellifera* associations in southern and Lower California. The former could be more readily compared with dwarf shrub associations on coastal shore Kurkarhills and the latter to those on the semi-arid ecotones of the Mediterranean territory.

Woodland–grassland and the mediterranean oak savanna

The main woodland type within the Mediterranean territory in Israel is the deciduous vallonea oak (*Quercus ithaburense*) parkland of the Lower Galilee foothills which, according to Eig (1933), covered large areas in central Palestine before the First World War and has counterparts in other East Mediterranean countries.

Of the different woodland types in California, probably the blue oak (*Quercus douglasii*) of the inner coastal ranges and Sierra foothills are most similar in ecological amplitude and physiognomic appearance to the vallonea oak in Israel. Both oaks are deciduous, broadleaved, with stout trunks and rounded crown, 5–20 m high, and may be considered the most xeric oak species in the mediterranean climate type, occurring on the drier side of denser shrub and maqui communities. This is especially true for their open woodland–grassland or savanna variants on the rolling hills with calcareous soils, surrounded by intensively cultivated fields and plantations in Lower Galilee and the coastal valleys between Carmel and Santa Barbara in California (Fig. 5). There is great similarity in climatic, edaphic, and topographic conditions, and the herbaceous cover of the blue oak–grassland shows greater resemblance than any other grass cover in California to the more rapid and vigorous winter-growing, denser and more productive vallonea oak grasslands in Israel. According to Stebbins (1959, *personal communication*) the replacement of indigenous California plants by Mediterranean invaders was greatest on these calcareous soils. These oak–annual grassland types in both countries therefore could be called truly the mediterranean oak savanna.

Annual grasslands

Mediterranean grasslands in Israel are either derived from above-mentioned oak woodland or from maqui and shrub communities. On the ecotones with the Irano-Turanian territory there are also hemicryptophytic batha and steppe grasslands, which have no California counterparts. But those annual grasslands in the Mediterranean hill region, which are degradation stages of adjacent brush—chiefly patch-cultivated and later on abandoned fields—are closest in resemblance and in botanical composition to the annual grasslands in northern and central California. Apparently it was from similar

Fig. 5. Mediterranean oak savanna. *Quercus douglasii* in the Sierra Nevada foothills, California, similar to *Quercus ithaburense* in the Lower Galilee foothills, Israel.

field–grasslands in European Mediterranean countries, and chiefly in Spain, that the most successful invaders in California were derived, including species of *Avena, Bromus, Erodium,* and *Medicago hispida* (Table 4).

COMPARATIVE SYNGENETICS AND SYNDYNAMICS OF EQUIVALENT VEGETATION TYPES IN CALIFORNIA AND ISRAEL

In Israel, as well as in other Mediterranean countries, herbaceous plant communities, dwarf shrub, and lower shrub are regarded as distinct successional stages leading to stable tree and dense shrub communities of maqui and forest climax formations (Zohary 1962). However, in California the syngenetical relationships of such communities with the chaparral are still a matter of controversy and even of confusion. Since direct evidence of primary successions is insufficient in both countries and there is no objective way of comparing the different theories and hypotheses, attention will be directed chiefly to the comparison of secondary successions as caused by human and other biotic interferences in these vegetation types.

Reaction to Fire and Grazing

Secondary succession after fire is an unsatisfactory term for the rapid vegetative and reproductive regeneration of the same individuals, even with a shift in generation through germination of certain species already present before the fire. This process is obscured in general by a short interlude of herbaceous—and chiefly grass—domination, and its dynamics are very similar in California (Cooper 1922, Sampson 1944, Horton and Kraebel 1955, Sweeney 1956) and in Israel (Naveh 1960).

Table 4. Some common plants in Israel grasslands, fields, waste places, and roadsides which are naturalized in California, their phytogeographic origin, and observations on their occurrence in the two countries

Scientific name[a]	Phytogeographic origin[a]	Israel	California
Gramineae			
Aegilops ovata L.	M	In heavily grazed ranges	
Avena barbata Pott.	M-IT(EB)	Abundant	Abundant
Avena sterilis L.		Closely related to *Avena fatua* L. in California; very abundant in lightly grazed range	
Avena fatua L.			Closely related to *Avena sterilis* L. in Israel; very abundant in lightly grazed ranges
Brachypodium dystachium L. Roem et Schult.	M-IT-(SD)	In heavily grazed ranges	Rapidly spreading
Bromus rigens L.	M-IT(EB) = *B. rigidus* Roth.		Very abundant
Bromus tectorum L.	M-IT-SS(EB)		Replacing perennial grasses in semi-arid California,Nevada, and Idaho
Bromus mollis L.	EB-M-IT		Most abundant grass in California ranges
Elymus Caput Medusae L.	M-IT(EB) M(EB, IT)		Very troublesome and spreading rapidly from north to south
Hordeum murinum L.	EB-M-IT = *H. leporinum* Link.	Chiefly on roadsides	Abundant on depleted ranges
Lolium rigidum Gaud.	M-IT(EB)	Closely related to *Lolium strictum* in California	
Lolium strictum Presl.			Closely related to *Lolium rigidum* in Israel
Lolium perenne L.	EB-M-IT		Escaped from cultivation
Phalaris paradoxa L.	M-IT(SD)	Very abundant, chiefly in fields	Very abundant, chiefly in fields
Poa bulbosa L.	EB-M-IT	Most xeric perennial grass in med. grasslands; abundant	
Dactylis glomerata = .(subsp. *Judaica*)	EB-M-IT		More mesic subspecies from Europe, escaped from cultivation
Schismus basbatus (L.) Thell.	IT-SS(M)	In low rainfall areas	In low rainfall areas
Schismus arabicus Nees.	IT-SS(M)	In low rainfall areas	In low rainfall areas

(*Continued*)

Table 4. Continued

Scientific name[a]	Phytogeographic origin[a]	Israel	California
Compositae			
Centaurea iberica Trev.	EM-IT	Chiefly in fields and roadsides	Chiefly in fields and roadsides
Chrysanthemum coronarium L.	M	Abundant, chiefly in waste places	
Scolymus hispanicus L.	M	Abundant in abandoned fields	
Senecio vulgaris L.	EB-M-IT	Abundant	Abundant
Silybium Marianum (L.) Gaert.	M-IT	Abundant	Abundant
Xanthium spumosum L.	B-Tr	Chiefly in waste places	Chiefly in waste places
Umbelliferae			
Ammi majus L.	M-IT(SD)	Chiefly in fields and waste places	Chiefly in fields and waste places
Ammi Visnaga (L.) Lam.	M(IT	Chiefly in fields and	Chiefly in fields and
Scandix Pecten Veneris L.	M-IT(EB)	waste places	waste places
Erodium Botrys (Cav.) Bertol	M	On sandy soils	Very abundant in grass-lands
Erodium cicutarium (L.) L'Her.	EB-M-IT	In roadsides, waste places	Abundant in grassland
Erodium moschatum (L.) L'Her.	EB-M-IT	In roadsides, waste places	Abundant in grassland
Geranium molle L.	EB-M-IT	Chiefly in shaded places	Chiefly in shaded places
Cruciferae			
Brassica nigra (L.) Koch	M-IT(EB, SD)	Chiefly in fields and roadsides	Chiefly in fields and road-sides
Raphanus Raphanistrum L.	M(EB, IT)	Chiefly in fields	Chiefly in fields
Sysimbrum orientale L.	M-IT(EB)	Chiefly in fields and roadsides	Chiefly in fields and road-sides
Sysimbrum officinale (L.) Scop.	EB-M-IT	Chiefly in fields and roadsides	Chiefly in fields and road-sides
Leguminosae			
Medicago hispida Gaert.	M-IT(EB, SD)	Chiefly in abandoned fields; less in grasslands, used widely for reseeding in Isreal as "California bur clover"	Most abundant and important legume in grasslands
Medicago minima (L.) Bartal.	M-IT(EB, SD)	Abundant	Rare
Medicago orbicularis (L.) Bart.	M-IT(SD)	Abundant	Rare
Melilotus indicus (L.) All.	M-IT(Tr)	On wet places	On wet places
Melilotus albus Med.	EB-M-IT	On wet places	On wet places

(*Continued*)

Table 4. Continued

Scientific name[a]	Phytogeographic origin[a]	Israel	California
Vicia dasycarpa Ten.	M(EB)	Chiefly in fields, used widely for reseeding in Israel as California lana vetch	
Diverse families			
Chenopodium album L.	B Tr	Chiefly waste places, fields	Chiefly waste places, fields
Chenopodium murale L.	B Tr	Roadsides and waste places	
Chenopodium rubrum L.	B Tr	Chiefly waste places	Chiefly waste places
Chenopodium vulvaria L.	B Tr	Chiefly waste places	Chiefly waste places
Galium tricorne Stockes	M-IT(EB)	Abundant, chiefly in fields	Abundant, chiefly in fields
Lavatera cretica L.	M		
Malva sylvestris L.	EB-M-IT		Garden escape
Papaver Rhoeas L.	M-IT(EB)	Abundant, chiefly in fields	Garden escape
Plantago lanceolata L.	EB-M-IT	On wet places	On wet places
Reseda alba L.	M-IT	Chiefly in waste places	Chiefly in waste places

[a]Nomenclature and phytogeographic origin according to Eig, Zohari, and Feinbrun (1948).

M-Mediterannean	IT-Irano Turanic	Tr-Tropic
NM-North Mediterranean	SS-Saharo Sindic	EB-Eurosibiric-North American
EM-East Mediterranean	SD-Sudano Decanic	B-Borealic (Northern)
		( )-region or regions invaded secondarily

The similar response to fire of both the individual plant and the community as a whole is one of the most striking common features of these mediterranean shrub communities. Their resistance to recurrent fires and even pyrophytism seems to be no less significant than their well-known resistance to prolonged summer drought, and may be closely related to the latter.

In the Mediterranean countries fire is still considered a wholly condemnable, destructive element, and not as an integral part of the brush ecosystem itself. On the other hand, its role in the evolution of the different chaparral species was stressed over 30 years ago by Jepson (1930). It probably plays a similar role in maqui and garigue species. Here, fire is also playing a comparable role in the maintenance of the brush ecosystem, by reverting the woody plants into a more rejuvenated and vigorous state, and by mobilizing nutrients tied up in the dead and very slowly decomposing lignin-rich wood. As first suggested by Went, Juhren, and Juhren (1951), fire may also destroy phytotoxic and antibiotic agents in the litter and upper soil layer and thereby cause a desirable change in the soil micropopulation. This, and the temporary opening of favorable niches, rich in light, moisture, and water-soluble nutrients from the brush ashes, apparently encourages the phenomenal but short-lived increase of herbaceous oppressed remnants and new invaders.

If the fire is followed by heavy grazing and browsing, as practiced in Mediterranean countries, a shift in domination to the least palatable shrub species, like *Salvia* and *Cistus,* may occur (Braun-Blanquet 1951, Naveh 1960). This can also be observed in California on burned and heavily browsed chaparral (Biswell 1956 and *personal communication*). Continuous and heavy browsing may cause the opening-up of dense brush thicket into scattered islands of stunted shrubs and annual grasses. This happened in the Mediterranean hill region in Israel in the vicinity of Arabic villages with large goat populations, and can also be found in a few spots in the coastal ranges in California which have a high, year-round deer population (Fig. 6). In California the understory consists of annual grasses and in Israel of dwarf shrubs, chiefly *Poterium spinosum.*

In general, fire has been more important in California, whereas in Israel browsing and cutting for fuel took over part of the same defoliation function and had a more pronounced effect on the evolution of these shrub species and communities. According to Cooper (1922) chaparral in northern and central California is degraded directly into oak–grassland. But similarity in secondary successions seems to be more pronounced in southern California, where it appears that coastal sage communities play a similar role to that of batha in Israel. This is very evident in the inner valleys of the coastal ranges in the Carmel and Santa Barbara areas.

Chamise behaves as a stable dominant on the drier and poorer sites, but as a fire-induced invader in more favorable conditions, where it will be replaced in time by more mesic species after the elimination of fire. A similar dual position is

Fig. 6. Adenostoma fasciculatum chaparral opened by year-round deer browsing in northern coastal ranges, California.

occupied by *Calicotome villosa,* the most important garigue shrub. Like chamise, it has low mineral requirements and is a most aggressive fire-follower, in which it is aided by a double mechanism of regeneration by fire-stimulated germination and root-crown sprouting.

Syngenetical Relations between Annual Grasslands and Shrub Formations

If the annual grasslands in central and northern California occupy a similar seral position to the chaparral as seral coastal communities in southern California, then they are also comparable to the degraded anthropogenic batha communities in Israel. According to Cooper (1922) the California brush was pushed back from its original wide occurrence and dominance by frequent burning, hand grubbing, and ploughing on the more open and productive sites, to the steeper and less accessible slopes in the coastal ranges. Even today there is still first-hand evidence from ranchers who burned and cleared dense chamise brush on the slopes of Mount Diablo. These recent annual grasslands do not differ in any respect from all other annual grasslands in the coastal ranges. The eradication of the brush could be achieved in different ways, but, as pointed out by Biswell (1956), its re-establishment in a dense herbaceous cover, which is grazed and sometimes burned in a mediterranean environment, is improbable. The same is true also for the annual grassland islands and patches within the brush region in Israel. Here, since the foundation of the State of Israel and the release of heavy grazing pressure from many areas, it is evident that secondary succession towards brush seems to be much more a process of vegetative re-encroachment by resprouting and rebranching of already existent, stunted, and suppressed shrubs and trees, than actual ecesis of new seedlings. Even *Poterium spinosum,* the most aggressive dwarf shrub pioneer, cannot crowd out dense stands of wild oats and other grasses (Litav, Kupernick, and Orshan 1962), and its invasion powers are restricted chiefly to poor edaphic conditions and to abandoned fields surrounded by batha with large amounts of viable seeds, germinating readily after fire (Naveh 1960).

The present stability and absence of secondary successions from these annual grasslands does not mean that they are not syngenetically related to the adjacent shrub communities. It merely shows the decisive influences of initial floristic composition and present biotic plant community management (Egler 1954).

No conclusive answers can be given on syngenetical relations of the different shrub, dwarf shrub, woody, and herbaceous formations. In both countries they are still a matter of speculation and generalization based on local circumstantial proof, and not on long-range systematic observations and experimental-causative study. However, Cooper's statements on the seral position of the coastal sage communities in South California are very much in accord with Eig and Zohary's conceptions of the seral, batha degradation stages of the maqui. That chaparral in northern and central California could have degraded into wooded and open grassland without any intermediate dwarf shrub stage may be explained by the above-described one-layered structure of these communities, lacking a well-developed subshrub layer. Therefore the newly opened niches could be filled immediately by annual pioneer species, as well as by perennial grass fire followers. Among the latter *Stipa pulchra* is most conspicuous. This apparently led Clements (1934) to believe that annual grasslands in California should be considered a disclimax of the pre-white man California bunchgrass prairie. Biswell (1956) showed that the railroad rights-of-way, which served Clements as proof of his *Stipa pulchra* relicts, were burned annually during the period of his observations. Possibly the frequent burning by Indians and protection from grazing may have favored this and other bunchgrasses in the past as well. In Israel, if brush burns are followed by protection from grazing, xerophytic bunchgrasses (among them *Stipa bromoides*) may increase (Naveh 1960). The occurrence of steppe-like grasslands in the more arid Great Valley, and prairie-like grasslands in the more humid and cooler coastal fogbelt, should certainly not be confused with the occurrence of certain perennial grasses as a result of a specific response to local biotic conditions and modifications. The probability of a climatic bunchgrass climax in the foot-hills and lower elevations of the coastal ranges and the Sierra Nevada in a typically mediterranean type climate seems very low.

The Dynamics of Invasion and Degradation of Mediterranean Annual Grasslands

Whether the annual grasslands of California originated from frequently burned, hand-cleared, and/or ploughed-off brushland, or from degraded perennial bunchgrass prairies, at present they are composed of a dense and highly plastic cover of annual grasses, legumes, and forbs, dominated by alien species which were unintentionally introduced in the last 200 years, chiefly from the Mediterranean region and Europe. According to Robbins (1940), more than 400 alien species can be found in these grasslands. This process of uncontrolled modification is still continuing.

In my opinion this process of invasion and degradation has as bases (1) the open unstable nature of annual grasslands; (2) the pattern of closely integrated agricultural and pastoral land use, and resulting biotic modifications in Mediterranean countries, which was repeated in California by the Spaniards and first settlers; and (3) the genecological potentials of Mediterranean, chiefly self-fertilized annuals, which evolved under intensive land use and could therefore successfully invade, colonize, and dominate their new habitats.

As mediterranean grasslands are composed chiefly of winter-germinating and developing plants, their fate is determined year after year by a new set of highly variable and unpredictable moisture, temperature, and light conditions after the first months and during the cooler winter months. This period of germination and early establishment may be even longer in California, because of the greater unreliability in early rainfall distribution and more extreme temperature regime. Therefore, California annual grasslands may be considered an even more unstable and modifiable ecosystem than their counterparts in Israel.

After the elimination of the original perennial and presumably woody vegetation, ground spaces not occupied in the beginning of the rainy season by living plants can be regarded as new habitat, open to invasion of pioneering species. In general, those plants which are first to germinate and emerge may have the best chances to succeed. Annuals from the Mediterranean region were among the most successful early pioneers and invaders. They could invade temporary habitats around human dwellings and waste places, but at the same time they could also establish themselves as weeds in fields and annual pasture plants in the cleared bushfields and untillable, heavily grazed range-lands. Under grazing pressure they could compete successfully with the indigenous flora and spread further.

Most of these plants, like *Avena, Bromus,* and *Hordeum* species and *Medicago hispida,* have the genecological advantage of being self-fertilized. As shown by Stebbins (1957) those few, well-adapted individuals which colonized first available niches could quickly build up large, homogeneous, and highly reproductive populations, but through occasional outbreeding still be able to produce occasional bursts of genetic variability. In this way they could occupy a wide range of diverse habitats and, as shown in the case of *Bromus mollis* by Knowles (1944), undergo ecotypic differentiation. However, with increasing livestock populations on the ranges, the early introductions of highly valuable pasture plants were replaced by less and less palatable and more and more undesirable species, which are better adapted to the lower, pastorally depleted stages of mediterranean grasslands, and since the beginning of this century, when stocking of California ranges reached its peak, they became very troublesome (Burcham 1957).

A similar process of degradation, which occurred in California ranges during a period of 150 years, may take place in Israel in any overgrazed pasture in a couple of years. The same highly valuable grasses or their taxonomic relatives (Table 4) are crowded out by other grasses which have more ephemeral and xeric characters along with greater seed production. They also have lower fertility requirements. A good example of this group is *Brachypodium distachium.* With further depletion, totally unpalatable forbs and thistles, which are already present, may take over.

One may expect that this process will be much more disastrous to the California ranges. They lack resistance capacities and recovery potentials acquired by their counterparts in Israel in the course of a long biotic history as grazing grounds. Such functions of stabilization and protection are provided in the Mediterranean region by certain perennial and annual plants, chiefly grasses and legumes, like *Hordeum bulbosum, Dactylis glomerata* subsp. *Judaica, Trifolium purpureum, Trifolium campestre,* and many others, which have not only high endurance to grazing pressure and drought, but may also respond vigorously to range improvement, such as deferred grazing and fertilization (Naveh 1957). Although of lower productivity when compared with cultivated forage plants under field conditions, these plants are highly superior in the difficult environment of the open ranges in Israel, and this may also be true for California. The higher drought resistance and endurance of the Israeli *Dactylis glomerata* subsp. *Judaica,* as compared with the much more productive subspecies *lusitanica,* was shown experimentally by McKell, Perrier, and Stebbins (1960). *Trifolium hirtum* was introduced from Turkey to California and is used successfully as a pioneer plant for reseeded pastures.

RESEMBLANCES TO VEGETATION TYPES OUTSIDE THE MEDITERRANEAN RANGE

Before concluding our comparison it is worthwhile to mention other sclerophyll vegetation types with physiognomic resemblance to maqui, but with different ecological positions and amplitudes outside the mediterranean climate type. These cannot be considered as true ecological equivalents of the California and Israeli vegetation types.

The actual distribution of chaparral shrubs reaches far beyond their mediterranean range in North California and South Oregon. In the conifer forests of the Sierran biotic province, chaparral shrubs are widespread both as understory of the pine forest and, after its destruction by fire and logging, as the dominant life form, with manzanita (*Arctostaphylos* spp.) as the most conspicuous genus. In contrast to the climax chaparral within the California province, these communities are considered successional and, as Cooper (1922) maintains, "Chaparral has transgressed its normal climatic limits along its mesophytic border through its invasion of the forest, fire being the causative agent."

Other broad sclerophyll formations, dominated chiefly by scrub oaks and called by Clements petran chaparral and by Cooper the deciduous thicket type, are abundant all over the Rocky Mountains in very different semi-arid, summer rain type climate. The landscape of some of these open oak savannas, grazed by cattle, sheep, and goats on rocky and chiefly calcareous soils in south-western Texas, has much resemblance to the above-mentioned, heavily grazed and browsed brush and oak savannas in the Karstic Galilee mountains of Israel. However, here the grass components are entirely different; they are summer-growing, perennial grasses of the short grass prairie and the desert plains grasslands.

RECAPITULATION AND CONCLUSION

The closer the resemblance in the overall environment, the closer the ecological equivalence and homology appears. This is the case in the mediterranean oak savanna types, where resemblance in bioclimatic conditions is correlated with resemblance in parent rock, topography, and land use, and where even a great part of the herbaceous vegetation is taxonomically synonymous. On the other hand, climate alone is not sufficient to ensure resemblance in structure and physiognomy, if edaphic and biotic factors are divergent as in the case of the chamise chaparral on non-calcic, poor lithosols in California and the grazed and browsed maqui and garigue of *Quercus* and *Pistacia* on Terra Rossa and Rendzina soils in Israel.

Although sclerophyll vegetation types are used as classical examples of phytoclimatic epharmony in a mediterranean type climate, it is apparent that such vegetation types may occur also as pseudo-equivalent types in a summer rain climate like that of southern Arizona and Texas. For deep-rooted, woody plants with an almost year-round growth rhythm, the total amount of rainfall and its storage in the soil are of greater ecological significance than the monthly rain pattern.

Our study has substantiated the warnings of Sauer (1950) and Beadle (1951) of the fallacy of too broad climatic-ecological generalizations, in which bioclimate is expressed by mean annual rainfall and monthly temperatures or by a single bioclimatic index. On the other hand, there is great similarity in ecological amplitude and in the syndynamics of all these mediterranean vegetation types. There is even a striking analogy in the reaction of woody and herbaceous plants and communities to biotic influences in general and especially to fire and grazing.

It appears that the results of uncontrolled modifications of these mediterranean ecosystems are more disastrous in California, because of more severe climatic, edaphic, and physiographic limitations and the inherent inferiority of the California flora in its resistance capacities and soil-protection and formation potentials.

These considerations may have some practical implications. The significance for economic range management and improvement of the highly modifiable nature of these dynamic grassland ecosystems and their quick response to agronomic modifications was recognized in both countries. It is therefore not astonishing that a similar, agronomistic approach of manipulation of soil-plant animal complex (Love and Williams 1956, Love 1961) and plant community management (Naveh 1955) was developed independently in California and Israel. Departing from traditional reliance in range management on climax and succession theories, this integrated approach also makes use of principles and concepts developed in Europe and the British Commonwealth for intensive pasture management.

Part of this program in California is the planned introduction and breeding of mediterranean pasture plants for arable lands. Israel and other Mediterranean countries are benefiting already from this improved plant material from California, as well as from mediterranean climate type regions of Australia. However, it will of course be much more difficult to achieve a planned replacement of low value California indigenous and exotic species on untillable wildlands with those above-mentioned perennial grasses, annual legumes, and shrubs which are highly superior in their grazing resistance and ability to stabilize soil. It might be of great economic importance for the highly inflammable and erodable chaparral watersheds of southern California if ways could be found for reseeding these plants in combination with prescribed burning of small units, raising of soil fertility, and counteracting phytotoxicity by fertilization and liming. It becomes more evident that in their present stage these brush ecosystems cannot maintain themselves without fire, and the longer we try to keep it out, the greater the fire and erosion hazard because of accumulation of dead material and phytotoxic kolines, and the more costly the prevention of wildfires and the more disastrous their results.

For pasture plant introductions, Israel and other Mediterranean countries, on the other hand, should not rely only on strain and cultivars developed in California for local purposes, but make use of new biotypes and mutations which arose spontaneously on the open ranges in California in similar conditions to their original habitats in the Mediterranean regions.

ACKNOWLEDGEMENTS

This study was carried out under the World Wide Visiting Research Scientist Program, sponsored by the National Academy of Science, National Research Council, Washington, D.C., at the University of California, Berkeley, Calif. I am indebted to many members of the staff of the School of Forestry, Departments of Botany, Soils, and Agronomy of this university, and especially to Drs. H. H. Biswell, A. M. Schultz, A. W. Sampson, H. G. Baker, and G. L. Stebbins for sharing their knowledge of the vegetation and ecology of California. Thanks are also due to Mr. J. Katznelson from the Meteorological Service, Israel, for climatological data and advice and to Drs. H. G. Baker, D. Zohary, M. Litav, and J. Tothill for their valuable comments on the manuscript.

REFERENCES

Anderson E (1956) Man as maker of new plants and new plant communities. In: Thomas WL (ed.) *Man's role in changing the face of the earth.* University of Chicago Press, Chicago, I11, pp 763–777.

Beadle NCW (1951) The misuse of climate as an indicator of vegetation and soils. *Ecology* **32:**343–345.

Bentley JR, Talbot M (1951) Efficient use of annual plants on cattle ranges in California foothills. U.S. Dept. Agr. Circ. 870. 52 p.

Biswell HH (1956) Ecology of California grasslands. J. Range Manage **9:**19–24.

Braun-Blanquet J (1951) Pflanzensoziologie, 2nd edn. Springer, Wien, 631 p.

Burcham L T (1957) California range land. California Div. Forestry, Dept. Nat. Resources. Sacramento, Calif., 261 p.

Cain SA (1950) Lifeforms and phytoclimate. Bot. Rev. **16:**1–33.

Clements FE (1934) The relict method in dynamic ecology. J. Ecol. **22:**39–68.

Cooper WS (1922) The broad sclerophyll vegetation of California. Carnegie Inst. Wash., Publ. 319. 124 p.

Deleuil G (1951) Origine des substances toxiques du sol des associations sans therophytes due Rosmariono-Ericion. Compt. Rend. Acad. Sci. Paris **232:**2038–2039.

Dinoor A (1962) The effect of *Quercus ithaburensis* on its herbaceous associates. M.Sc. Thesis, Hebrew University, Jerusalem, 214 p.

Egler FL (1954) Vegetation science concepts I. Initial floristic composition a factor in old field vegetation development. Vegetatio **4:**412–417.

Eig A (1927) On the vegetation of Palestine. Inst. Agr. and Nat. Hist. Agr. Exp. Sta. Bull. Tel-Aviv, **7:**1–88.

——— (1933) A historical-phytosociological essay on Palestinian forests of *Quercus aegilops* L. ssp. *ithaburensis* (Desc.) in past and present. Beih. Bot. Centralbl. **51,** Abt. II:225–272.

Eig A, Zohary M, Feinbrun N (1948) Analytical flora of Palestine. Palestine J. Botany. Jerusalem, 504 p.

Grisebach A (1872) Die Vegetation der Erde nachihrer klimatischen Anordnung. Engelmann, Leipzig, 603 p.

Guyot L (1951) Sur un aspect du determinisme biologique de l'evolution floristique de quelques groupements vegetaux. Compt. Rend. Som. Se. Soc. Biogeogr. **293:**3–14.

Horton JS, Kraebel HR (1955) Development of vegetation after fire in the chamise chaparral of southern California. Ecology **36:**244–262.

Jepson WC (1930) The role of fire in relation to the differentiation of species in the chaparral. Proc. Int. Bot. Cong. **5:**114–116.

Katznelson J (1956) The variability of rainfall in Palestine and the statistical methods of its measurements. Israel Meteorol. Serv. Ser. E4. 32 p. (Hebrew).

Knowles PF (1944) Interspecific hybridization of *Bromus mollis.* Genetics **29:**128–140.

Koeppen W (1923) Die Klimate der Erde. Walter de Grunde, Berlin.

Landers RG (1962) The influence of chamise, *Adenostoma fasciculatum,* on vegetation and soil along chamise grassland boundaries. Ph.D. Thesis, University of California, Berkeley, Calif., 258 p.
Litav M, Kupernick G, Orshan G (1962) The role of competition as a factor in determining the distribution of dwarf shrub communities in the Mediterranean territory of Israel. J. Ecol. **51**:467–480.
Love RM (1961) The range–natural plant communities or modified ecosystems? J. Brit. Grassl. Soc. **16**:89–99.
Love RM, Williams WA (1956) Range land development by manipulation of the soil-plant-animal complex in the difficult environment of a mediterranean type climate. Int. Grassl. Congr. **7**:48.
McKell CM, Perrier EP, Stebbins GL (1960) Responses of two sub-species of orchardgrass (*Dactylis glomerata* subsp. *Lusitanica* and *judaica*) to increasing soil moisture stress. Ecology **41**:772–778.
Muller CH (1966) The role of chemical inhibition (allelopathy) in vegetational composition. Bull. Torrey Bot. Club **93**:332–351.
Munz PA, Keck DD (1959) A California flora. University of California Press, Berkeley & Los Angeles, 1681 p.
Naveh Z (1955) Some aspects of range improvement in a mediterranean environment. J. Range Manage. **8**:265–270.
——— (1957) Report on range research. FAO working party on mediterranean pasture and fodder development. 5th meeting, Tel-Aviv (Mimeo.).
——— (1960) Agro-ecological aspects of brush range improvement in the maqui belt of Israel. Ph.D. Thesis, Hebrew University, Jerusalem, 212 p.
——— (1961) Toxic effects of *Adenostoma fasciculatum* (chamise) in the California chaparral. Proc. 4th Congr. Scient. Soc. Rehovot, April 5, 1961:74 and unpublished data.
Raunkiaer C (1934) The lifeforms of plants and statistical plant geography. Clarendon Press, 632 p.
Robbins WW (1940) Alien plants growing without cultivation in California. Calif. Agr. Exp. Sta. Bull. 637:128 p.
Ruebel E (1930) Pflanzengesellschaften der Erden. Hans Huber, Bern-Berlin, 464 p.
Sampson AW (1944) Plant succession in burned chaparral lands in northern California. Calif. Agr. Exp. Sta. Bull. 635:144 p.
Sauer K (1950) Grassland, climax, fire and man. J. Range Manage. **3**:16–21.
Schimper AFW (1898) Pflanzengeographie auf physiologischer Grundlage. G. Fischer, Jena. 877 p.
Stebbins GL (1957) Self fertilization and population variability in plants. Amer. Natur. **91**:337–354.
Sweeney JR (1956) Responses of vegetation to fire. Univ. of Calif. Publ. Bot. **28**:143–250.
Trumble HC (1952) Grassland agronomy in Australia. Advances in Agronomy. Academic Press, New York and London, **4**:1–65.
UNESCO Arid Zone Research Team (1963) A bioclimatological map of the Mediterranean zone and its homologues. UNESCO, Paris.
Varney BM (1925) Seasonal precipitation in California and its variability. Monthly Weather Rep. **53**:48–163, 208–218.
Went F, Juhren G, Juhren MC (1951) Fire and biotic factors effecting germination. Ecology **33**:351–363.
Whyte RO (1961) Evolution of land use in South Eastern Asia. Arid Zone Research UNESCO, Paris, **16**:57–118.
Yardeni D, Evenari M (1952) The germination inhibiting and phytotoxic effects of certain leaves and leaf extracts. Phyton **2**:11–17.
Zinke P (1959) Nitrogen status of lysimeter soils under different plant cover at the St. Dimas Forest Exp. Sta. (Mimeo.).
Zohary D (1960) Studies on the origin of cultivated barley. Bull. Res. Counc. Israel, Sect. D, **9B**:21–42.
Zohary M (1962) Plant life of Palestine. Ronald Press, New York, 262 p.

CHAPTER 2

NAVEH Z., R.H. WHITTAKER (1979)

STRUCTURAL AND FLORISTIC DIVERSITY OF SHRUB-LANDS AND WOODLANDS IN NORTHERN ISRAEL AND OTHER MEDITERRANEAN AREAS* VEGETATION 4: 171–190.

Multi-layered shrub-grass community in abandoned terrace. Note the tall *Q. calliprinos* shrub, growing along terrace wall on Mt. Carmel in Israel.

---

*Nomenclature follows Zohary et al. (1948).

This study, sponsored by the U.S.-Israel Binational Science Foundation (No. 450), was carried out by Z. Naveh as principal investigator with R.H. Whittaker as American collaborator. We thank Mr. A. Mann,

Keywords Biotic diversity, Grazing pressure, Israel, Mediterranean vegetation, Shrublands, Species diversity, Stratal diversity, Vegetation disturbance, Vegetation samples, Woodlands

INTRODUCTION

Problems of species diversity in relation to evolution, stability, and maturity of natural ecosystems are central issues in ecology. Diversity is also of vital importance for conservation, as natural communities are increasingly threatened by industrial and urban expansion and forest clearing. We discuss here problems of diversity on the untillable uplands of the Mediterranean region of Israel, under the combined impact of traditional pastoral and agricultural exploitation (Naveh & Dan 1973). We summarize results on alpha and stratal diversity of Mediterranean shrublands and woodlands in northern Israel, as related to community structure and niche differentiation, as affected by environmental gradients of climate and grazing intensity, and as part of a joint botanical and zoological study of these ecosystems. We also compare diversity and structure of these communities with mediterranean communities on other continents.

STUDY SITES IN ISRAEL

This study was carried out in mountainous northern Israel in eastern Mediterranean sub-humid to semiarid climates on rocky, shallow, untillable terra rossa and brown rendzina soils. The vegetation ranges from dense, impenetrable one-layered maquis shrub thickets to more open and patchy, multilayered or mosaic-like tree-shrub-grass communities and from denser woodlands to open savanna-like parklands and treeless, derived grasslands. Larger- and smaller-scale complexity in this vegetation results from: (a) steep climatic gradients, (b) heterogeneity in soil depth, rockiness, and stoniness, as well as slope degree and exposure, (c) within-stand contrasts in soil from one point to another, (d) marked year-to-year and seasonal fluctuations in the herbaceous strata, and as a now most significant factor (e) diverse histories of disturbance, degradation, and recovery.

In order to cope with these problems, representative sites were chosen so that one major factor – climate or disturbance – would differ in sites compared while other factors were the same or varied much less. Sites were, so far as possible, on even land – hilltops or very gentle slopes. In each site a rectangular plot of one-tenth ha was marked for intensive botanical and zoological studies. Plots were taken along a climatic gradient, ranging from the sub-humid higher rainfall and the cooler Upper Galilee and Carmel Mountains and the slightly drier and lower Lower Galilee

S. Burmil, Mrs. Chaim, and Mrs. A. Kleen for botanical field work and statistical computations, Mr. D. Feigin and S. Ben Ezrah for technical assistance, Mr. S. Asherow for identification of young plants, and the Nature Reserve Authorities, the Neve Yaar Experimental Station, the Agricultural School Kfar Hanoar Hadati, and Kibbutzim Allonim and Allone Abba for allowing us to use their land for this study. The work by R.H. Whittaker was supported by the U.S. National Science Foundation, Australian National University, and Canberra Botanic Gardens; and we thank all the collaborators in this work.
[1] *Faculty of Agricultural Engineering, Technion – Israel Institute of Technology, Haifa, Israel*[**]
[2] *Section of Ecology and Systematics, Cornell University, Ithaca, New York 14850, USA*

foothills to the semiarid Mt. Gilboa, and along a biogradient of disturbance by man ranging from protection in nature reserves to moderately and heavily grazed and disturbed plots.

1. *Mt. Meron – Upper Galilee*
   Here one plot was chosen (ME) as the most mesic site, in the completely protected Mt. Meron Nature Reserve at an elevation of about 1000 m on a brown rendzina soil on hard limestone of Cenomanian origin. Mean annual precipitation is around 900 mm. The stand represents the closed, undisturbed *Quercus calliprinos* maquis with a dense woody canopy that occurs in the higher rainfall region of north Israel and other east Mediterranean countries and has been described by Zohary (1962) and others.
2. *Mt. Carmel – near Haifa*
   Here three plots were established, all at about 400 m elevation, with annual rainfall between 650 and 750 mm.
   a. Muhraqa (MU)
      This site is on a typical terra rossa on hard limestone of Cenomanian origin, about 750 mm mean annual precipitation. Total winter rainfall was 802 mm in 1974–75, 552 mm in 1975–76, and 720 mm in 1976–77. *Quercus calliprinos* forms a dense, uniform, almost impenetrable maquis with *Laurus nobilis* and few other species. The stand has not been disturbed for more than 30 years, although the area has been subjected to goat grazing, wood cutting, and coppicing since the opening of a road on the northern crest of Mt. Carmel.
   b. Forty Oaks Grove (FO)
      This site, adjacent to the tall oak grove in the Carmel Nature Reserve not far from the Haifa University, is fenced and protected, but in the past was grazed and coppiced. The soil is a dark rendzina on a complex of soft and hard limestones. In contrast to MU, this site and BO are much more open maquis stands with lower dominance of *Quercus calliprinos* and higher cover of *Pistacia lentiscus* and other shrubs. Total winter rainfall was 604 mm in 1974–75, 848 mm in 1975–76, and 726 mm in 1976–77.
   c. Bet Oren (BO)
      This site, in the vicinity of Kibbutz Bet Oren, has a mosaic of denser shrub canopy (*Quercus calliprinos* and *Pistacia lentiscus* codominant) with open, grassy patches and intermediate dwarfshrubs. The soil is a terra rossa on dolomite; like FO it has been heavily disturbed in the past but is now grazed only lightly by cattle and goats. Both stands would be regarded by phytosociologists as degraded successional stages of the maquis climax, with the climax represented by sites MU and ME. Winter rainfall was 808 mm in 1974–75, 649 mm in 1975–76, and 872 mm in 1976–77.
3. *Lower Galilee*
   Three plots were established in this region in 1974 and and three additional plots in 1976 and 1977 with different grazing intensities. All were located at elevations of 100–140 m on dark rendzina soils of Eocenic origin with a Nari (caliche) crust and with mean annual rainfall of 550–600 mm. The following sites are all more or

less open woodlands of the deciduous Tabor oak *(Quercus ithaburensis).* Such stands, which occupied large areas in central Palestine before the First World War, were first described by Eig (1933) as the only true woodlands in Mediterranean Israel. According to Zohary (1962) the Tabor oak woodlands (*Quercus ithaburensis* alliance) belong to a large group of broad-leaved, deciduous woodlands that are common in the east Mediterranean countries, notably Turkey, Iran, and Iraq, and reach in Israel their southern limits of distribution.

a. Allonim (AL)

This site, not far from Kibbutz Allonim, is part of the fenced pastures with scattered Tabor oak trees and a dense herbaceous understory, representing an open phase of the '*Quercus ithaburense-Styrax officinalis typicum*' community (Aloni 1972). It is rotationally grazed by beef cattle 3–5 times during the year, chiefly between December and May. Winter rainfall was 609 mm in 1974–75, 538 mm in 1975–76, and 713 mm in 1976–77.

b. Allone Abba (AA)

This stand, on the hill of the Allone Abba settlement, is grazed lightly, chiefly in spring, by sheep and by cattle. In addition to somewhat higher cover of Tabor oak it has more shrubs and is closer to the '*Quercus ithaburense-Styrax officinalis calliprinosum*' variant of Aloni (1972). Winter rainfall was 595 mm in 1974–75, 637 mm in 1975–76, and 713 mm in 1976–77.

c. Neve Yaar (NY)

This fenced site near the entrance of Neve Yaar Experimental Station was added in 1976 to represent a Tabor oak woodland under very light grazing (only occasionally by cattle). It is similar to AL in its scattered Tabor oak cover and rainfall.

d. Kfar Hanoar Hadati (KN)

Vegetation of this site, on a gentle south slope of the agricultural school near Kfar Chassidim, comes closest to the open phase of the Quercus ithaburense-Styrax officinalis typicum (Aloni 1972). Here, two adjacent plots were established in 1977: one inside the fence of the school and completely protected in the last three years (KN−) and the other (KN+) outside the fence and grazed heavily yearround by Arab goats and cattle. Winter rainfall was 640 mm in 1974–75, 844 mm in 975–76, and 684 mm in 1976–77.

e. Bosmat Tivon (BT)

This site, located in the vicinity of the Bedouin village Bosmat and very heavily grazed and browsed year-round by goats, sheep, and cattle, differs from all other plots in the Lower Galilee in its mosaic-like vegetation of dense shrub patches dominated by *Pistacia lentiscus,* alternating with grassy patches, described as '*Quercus ithaburense-Styrax officinalis lentiscetosum*' variant by Aloni (1972). Winter rainfall was 682 mm in 1974–75, 562 mm in 1975–76, and 567 mm in 1976–77.

4. *Mt. Gilboa (GI)*

This most xeric site is on the central part of Mt. Gilboa at 450 m elevation, close to Kibbutz Maaleh Hagilboa, with a dark rendzina soil on Nari limestone of Cenomanian-Turone origin. The vegetation is a low, open mixed shrub and grass stand; *Pistacia lentiscus* is the major shrub with widely scattered *Ceratonia siliqua* trees. The stand was heavily grazed until 1967, but since then only lightly by *Gazella gazella* and occasionally by goats. In phytosociological terminology it is a very degraded, xeric mixed batha and garigue stage of the *Ceratonieto-Pistacietum lentisii.* Mean annual precipitation is about 400 mm; winter rainfall was 387 mm in 1974–75, 462 mm in 1975–76, and 482 mm in 1976–77.

COMPARATIVE DATA FROM MEDITERRANEAN AREAS

Comparative data from mediterranean areas on different continents (using standard 1–1000 m^2 diversity samples in all cases except the IBP studies of Mooney et al. 1977) were compiled as follows:

1. *Israel*
   a. Woodlands, 3 samples, Allonim and Allone Abba as described, plus a sample from Mt. Carmel.
   b. Disturbed, open shrublands, 3 samples; Forty Oaks Grove, Bet Oren, and Mt. Gilboa as described.
   c. Maquis, 3 samples including Mt. Meron and Murhaqa as described, plus one other sample on Mt. Carmel.
   d. Batha, 1 sample from Mt. Carmel, level, *Sarcopoterium spinosum* dominant, data of Z. Naveh and R.H. Whittaker.
2. *California*
   a. Woodlands, 6 open oak woodlands near Santa Barbara, protected to moderate grazing, data of A. Shmida, L.S. Olsvig, and R.H. Whittaker, April 1978.
   b. Woodlands, 7 open oak woodlands near Carmel, currently protected (but subject to past grazing by live-stock and present effects of deer and rodents), data of Z. Naveh and J.R. Griffin, April 1979.
   c. Woodlands, 5 open oak woodlands near Carmel, light to heavy grazing, data of J.R. Griffin and Z. Naveh, April 1979.
   d. Chaparral, 5 stands of mature *Adenostoma fasciculatum* chamise, San Jacinto Mts., data of A. Shmida, L.S. Olsvig, and R.H. Whittaker, April 1978.
   e. Chaparral, 5 stands of mature *Ceanothus megacarpus* chaparral, Santa Ynez Mts., data of A. Shmida, W.H. Schlesinger, and R.H. Whittaker, April 1978.
   f. Chaparral, 1 0.1 ha sample, combining 10, 100 m^2 plots, with *Adenostoma fasciculatum* most important, Echo Valley, IBP sample of Mooney et al. (1977).
   g. Coastal sage, IBP sample, *Salvia mellifera* most important, Camp Pendleton.
3. *Chile*
   a. Interior matorral (compares with California chaparral), IBP sample, Fundo Santa Laura.
   b. Coastal shrubhland (compares with coastal sage), IBP sample, Papudo.
4. *Australia*
   a. Southwest Australian heath, 7 samples from various soils on level sites near Perth, data of R.H. Whittaker, W.M. MacArthur, and J.S. Beard, December 1975.

b. Mallee communities, 6 samples representing different phases in New South Wales, data of R.H. Whittaker, W.A. Niering, and M.O. Crisp, November 1975.

5. *South Africa*

a. Fynbos, 9 samples representing lowland and mountain (Groenlandberg) sites in the Cape Province, data of H.C. Taylor, F.J. Kruger, and R.H. Whittaker, October 1975.

METHODS

Species Richness and Importance

For determination of species richness we recorded all vascular plant species within the 0.1 ha (20 × 50 m) plots throughout the year (additional woody species in a surrounding 1 ha plot were noted). Direct visual estimation of plant coverage was used to give relative importance values for species. Percentage of rock and total plant cover, and percentage of ground cover of all perennial species, were estimated in 40 subplots of 5 × 5 m within each study plot. These subplots were also used for the determination of stratified species richness, described below. Herbaceous plant presence and covers were also recorded in 10, 1 × 1 m squares in all study plots. In spring 1975 these squares were chosen at random; in spring 1976 and 1977 they were located along a transect in the centers of the plots and were expanded to two 2 × 5, one 10 × 10, and the full 20 × 50 m quadrats to give species counts for 1, 10, 100, and 1000 m^2 (Whittaker et al. 1979). These values can be fitted by the semilog form $S = b + d \log A$, in which S is species number and A is area in m^2. The coefficients b (mean number of plant species in a 1–m^2 plot) and d (rate of increase in species number with increasing plot area) are then diversity expressions that may be compared with S for 1000 m^2 (Whittaker 1972, Whittaker et al. 1979). Other details of the sample and analysis techniques are given by Naveh & Whittaker (1979).

Stratified Cover and Richness

Vegetation structure and foliage height diversity are important for animal diversity, especially for birds (MacArthur 1965, Karr & Roth 1971, Willson 1974) and insects (Murdoch et al. 1972, Morris 1973), and were of concern for the animal diversity part of this study. We estimated cover percentage and counted stratal species richness (in 0.1 ha) for five strata. From the stratal covers a foliage height diversity in the sense of MacArthur (1964) can be calculated: FHD $= -\Sigma p_i \log p_i$, in which p_i is a stratal cover as a decimal fraction. Stratified plant species richness was determined for the following height strata and growth-forms.

1. High tree layer, mainly single-stemmed trees casting shade, 5–10 m tall
2. Tall shrub layer, 1.5–5 m
3. Mid-shrub layer, 0.5–1.5 m
4. Subshrub layer, below 0.5 m

5. Herbaceous understory, subdivided for other purposes into: a. Perennial grasses b. Perennial forbs (except geophytes) c. Geophytes d. Annual grasses e. Annual legumes f. Annual forbs, non-legumes

Dominance Concentration and Equitability

Two measures were used to express the relative importance value distribution among species (and strata) in each site: relative concentration of dominance was expressed as the Simpson index,

$$C = \sum_{i=1}^{S} \mathrm{p}_i^2$$

and relative equitability or evenness of the importance values of all species as the Shannon-Wiener index,

$$H' = -\sum_{i=1}^{S} p_i \log p_i$$

(p_i is the relative importance of species, or stratum, *i,* measured as a decimal fraction of total plant cover for the site). Whereas the Simpson index is strongly affected by the importance of the most abundant one or two species, the Shannon-Wiener index gives more weight to the intermediate species and the full importance-value sequence (Whittaker 1972, 1977a). For convenient comparison of trends of these indices with those for *S,* the reciprocal of the Simpson index and the exponentiated form of the Shannon-Wiener index were used (Hill 1973, Peet 1974).

In addition to overall dominance concentration (C_{veg}, for which p_i is the relative cover of a species, summed for all species in the sample), relative dominance among the woody species only (C_{wood}), and relative stratal concentration) C_{str}, in which p_i is the relative cover of a stratum as a decimal fraction of cover for all stata) were determined. Coefficients of correlation were calculated between *S* and other measures, and between strata and growth-forms. As a first survey of the diversity of a mediterranean landscape, the means of these measures and their coefficients of variation are presented (Table 1).

Results

Site comparisons

Comparisons of these measures are summarized in Table 1, Fig. 1, and Fig. 2. Differences, sometimes striking, in community structure and in alpha and stratal diversity appear between the following pairs of samples:

1. Closed *Quercus calliprinos* maquis shrubland: Mt. Meron and Muhraqa (ME and MV) These undisturbed sites occupy the mesic extreme of the climatic gradient. Their dense, almost impenetrable tall-shrub canopies are dominated by *Quercus calliprinos*

Table 1. Stratified and overall plant species diversity in shrubland and woodland sites of 1000 m^2 in northern Israel in 1975–1976

Growth-forms and diversity measures*	Woody species species number							Perennial herbs species number				Annual herbs species number				Total herb species	Total species, 1000 m^2				
	HT	TS	MS	SS	Cl	STR†	Total	Ge	PGr	PFo	Total	AGr	ALe	AFo	Total			C_{veg}	C_{wood}	C_{str}	H′
Mt. Meron	1	4	14	8	6	33	32	5	6	8	19	2	5	9	16	35	57	.495	.394	.344	.805
Muhraqa	–	7	5	4	5	21	13	3	2	3	8	–	–	–	–	8	21	.408	.600	.574	.581
40 Oaks Grove	1	6	14	16	5	42	24	13	6	20	39	7	24	25	56	95	119	.177	.270	.190	1.352
Bet Oren	1	8	13	8	14	44	25	14	7	13	34	7	22	30	59	93	118	.166	.277	.221	1.371
Mt. Gilboa	1	–	7	20	6	34	32	11	9	26	46	11	27	63	101	147	179	.112	.437	.160	.385
Bosmat Tivon	–	5	5	10	8	28	19	14	5	3	22	7	14	22	43	65	84	.336	.569	.282	.904
Allonim	2	2	6	5	4	19	15	8	3	12	23	16	28	53	97	120	135	.029	.449	.253	1.748
Allone Abba	1	2	6	8	8	25	25	11	4	15	30	14	22	46	82	112	137	.031	.536	.119	1.634
MEAN	.75	4.25	8.75	9.87	7	30.7	21.8	9.87	5.25	12.5	27.62	8	17.75	31	56.75	84.3	106.25	.219	.454	.268	1.248
COEFFICIENT OF VARIATION	94	70	47	55	45	30	28	42	43	64	44	69	59	70	64	54	47	79	24	53	35

* Growth-form codes and diversity measures: HT = high tree; TS, MS, and SS = tall, middle, and low shrub; Cl = climber; Ge = geophyte, PGr = perennial grass; PFo = perennial forb; AGr = annual grass; ALe = annual legume; AFo = annual forb (excluding legumes). *C* is the Simpson index of dominance concentration, using relative cover for all plant species (C_{veg}), relative cover of species within the woody strata only (C_{wood}, and relative cover of five strata as wholes (see Methods); *H′* is the Shannon-Wiener index for relative cover of all plant species, $\log_{10}$.

† Includes species present in more than one stratum.

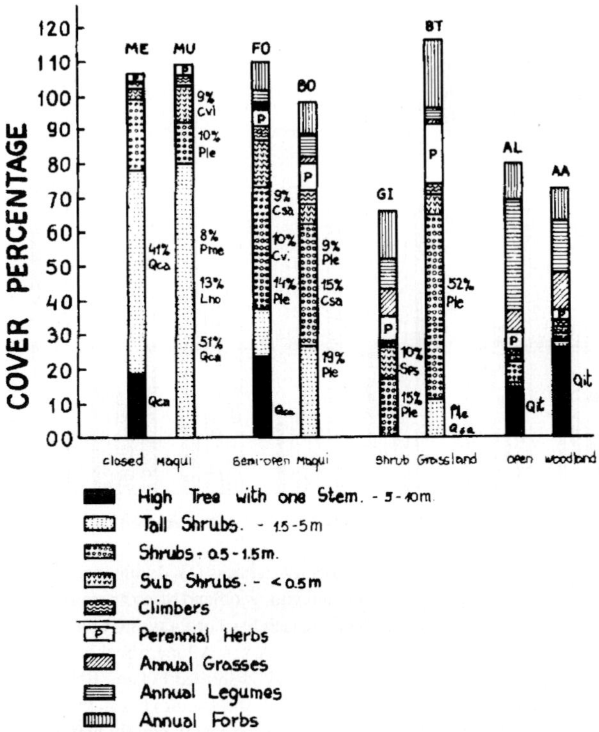

Fig. 1. Plant cover by strata and growth-forms in shrubland and woodland samples, northern Israel, 1975–1976. For sample initials see text or Fig. 2. Major species as abbreviated: *Quercus calliprinos, Q. ithaburensis, Pistacia lentiscus, Sarcopoterium spinosum, Cistus salvifolia, C. villosa, Laurus nobilis, Phillyrea media.*

with relatively few subordinate shrubs. In the most monotonous Muhraqa community, the lowest values for total and stratified woody S (21 and 13) and for total S (21) and for H' (0.581) were reached, and conversely, the highest values for woody C (0.600) and stratified C (0.574), followed by slightly higher values in the Mt. Meron site, which had the highest C (0.495) of all sites. Both have a very sparse herbaceous understory of a few shade-tolerant perennial herbs and also on Mt. Meron a few therophytes, which colonized chiefly around larger rock outcrops and in a grassy patch. Thus these dense maquis, which could be considered climax, had the lowest over-all stratal and alpha diversity.

2. Semi-open, mixed maquis shrubland: Forty Oaks Grove and Bet Oren (FO and BO) These sites are slightly less mesic, but they differ much in their more complex structure, consequent on cutting, coppicing, grazing, etc. In the case of Forty Oaks, the pruning and protecting of the 'holy oak trees' in the distant past has led to an overstory of *Quercus calliprinos* trees huge for their species. Woody cover is high but well partitioned between the different strata and species,

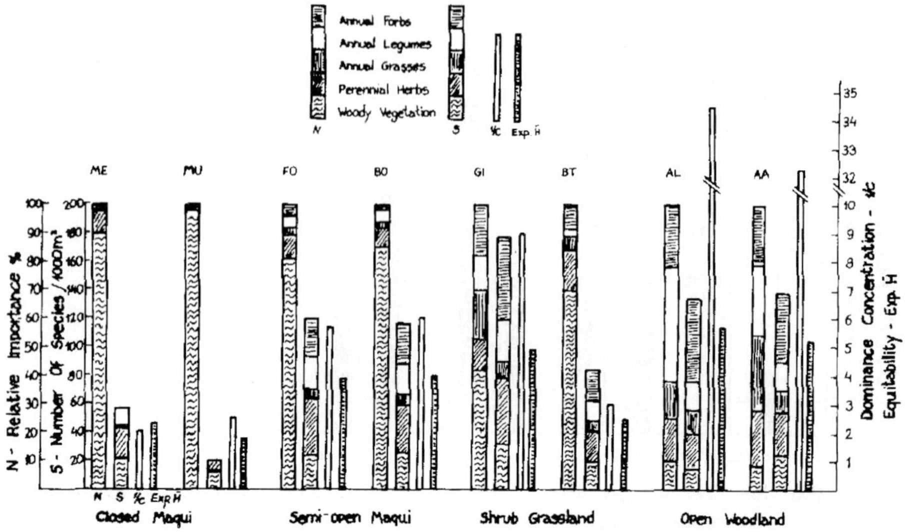

Fig. 2. Relative importance (cover) of vascular plant groups (*N*), species richness (*S*), dominance concentration reciprocal (1/*C*), and equitability (exp *H′*) of shrubland and woodland of northern Israel, 1975–1976. Study sites (1000 m^2): ME = Mt. Meron; MU = Muhraqa; FO = Forty Oaks Grove; BO = Bet Oren; GI = Mt. Gilboa; BT = Bosmat Tivon; AL = Allonim; AA = Allone Abba. Note that protected ME and MU have chiefly woody vegetation, but lowest diversity; disturbed FO and BO have also chiefly woody vegetation but much higher diversity; disturbed GI with shrub-grass has high diversity, but heavily disturbed BT low diversity; moderately grazed AL and AA with chiefly herbaceous vegetation have highest diversity.

leading to lowest woody *C* values (0.270). The rich herbaceous understory contains a wealth of flowering geophytes, as well as hemicryptophytes and therophytes in openings, as expressed in total herb *S* values of 93 and 95 and high *H′* values.

In Fig. 3 dominance-diversity curves for contrasting community types are presented. The closed maquis 'climax' community shows a very steep slope, in which the dominant *Quercus calliprinos* monopolizes a large share of resources, followed by another steep slope for subordinate woody species and a more gradual one for the few herbs. In the richer Forty Oaks community, a group of shrubs form a rather steep slope suggesting a geometric series, followed by another, more gradual slope for the densely packed minor woody species and for all herb species. Not too much inference can be drawn from these curves (May 1975, Whittaker 1975), but they suggest much finer division of understory resources and niches among species in the more open communities.

3. Open *Pistacia lentiscus* shrublands: Mt. Gilboa and Bosmat Tivon (GI and BT)
These sites occupy different positions along the climatic gradient, with Gilboa more xeric. But both have a poorly stratified shrub-grass and herb structure in which *Pistacia lentiscus* dominates. Both were exposed to heavy human

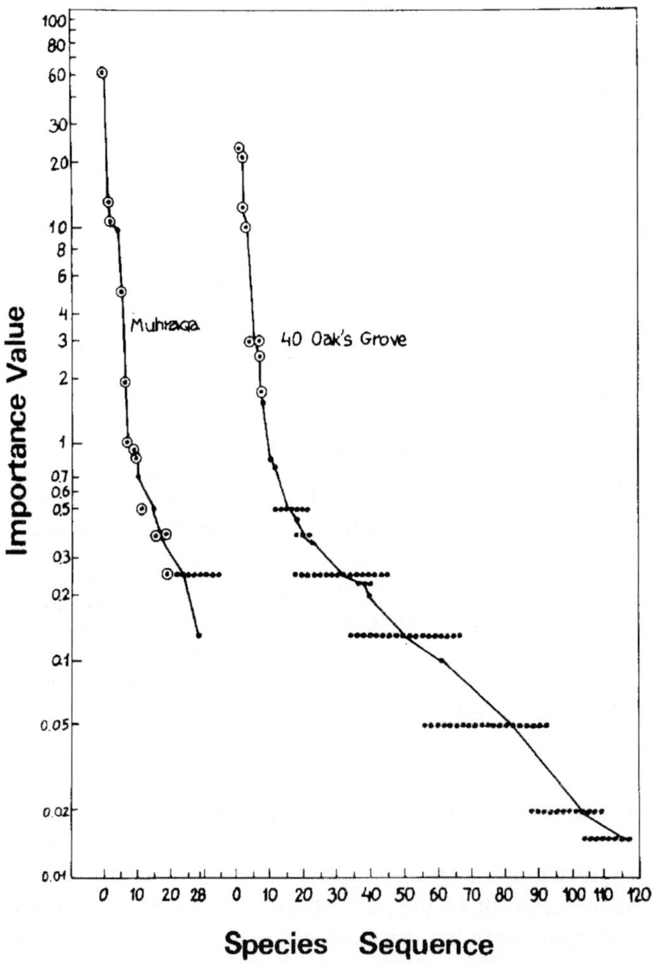

Fig. 3. Dominance-diversity curves for all vascular plant species in 1000 m^2 samples of closed and semi-open maquis shrubland on Mt. Carmel, 1975–1976. Species percentages of mean cover are used as the importance values in a logarithmic scale on the ordinate, and species numbers in the sequence from the most important to the least important on the abscissa. The curves suggest geometric series with different slopes for different strata, but in the richer community these are more complex and the lower slope is much more gradual. Dots are herb species and circled dots are woody species.

pressure, which is still continued in Bosmat Tivon and in phytosociological terminology they would be regarded as 'degraded' maquis and garigue, or even batha (in the case of Gilboa). The Bosmat Tivon site has a much higher woody cover and is distinguished by a mosaic structure of alternating dense shrub and grass patches (Whittaker & Naveh 1979), with geophytes and perennial grasses occupying the ecotones between these shrub and grass 'islands.' However, here, in spite of the larger extent, the woody canopy is much lower in stratified and

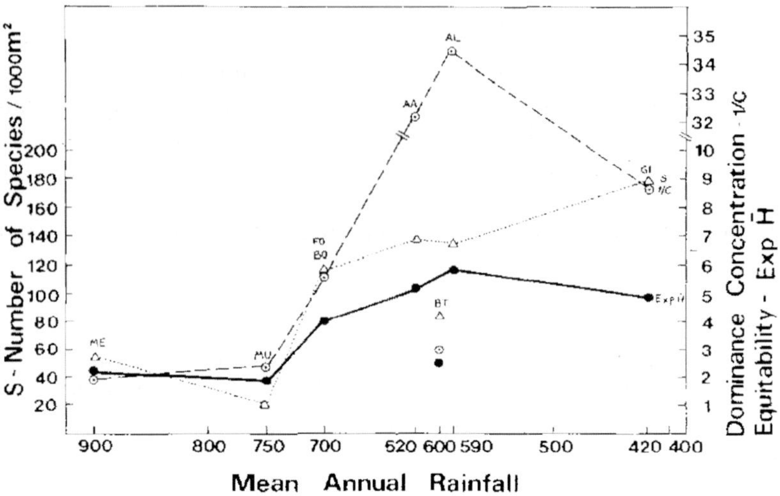

Fig. 4. Diversity trends in relation to mean annual rainfall, shrublands and woodlands in northern Israel, 1975–1976: vascular plant species per 1000 m^2 (*S*, triangles and dotted line), dominance concentration reciprocal (1/*C*, circled dots and broken line), and equitability (exp *H'*, solid dots and line); study sites as in Fig. 2. Diversities are higher in drier climates, but overgrazed BT is less diverse than AL and AA.

total *S* and has, after Muhraqa, the highest woody *C* and a relatively poor herb cover of less palatable and grazing-tolerant species. Gilboa, on the other hand, has the greatest number of small shrubs and dwarfshrubs (20) and of total woody *S* (32) of all sites, and its herb cover, in spite of the great area occupied by rocks, has by far the greatest number of perennial herbs (46), especially legumes and forbs, reaching thereby the highest total *S* of all sites (179). This is well reflected in the low stratal and over-all *C,* and the high *H'* value. This site can be regarded as a meeting point of the sub-humid Samaria Mountains (as their isolated outpost), merging with the drier, semiarid Mt. Gilboa proper and influenced also by the lower altitude semi-desert Jordan plains. This phytogeographic situation may contribute together with disturbance to the high vascular plant (and animal) diversity.

4. Open Tabor oak woodland-grassland: Allonim and Allone Abba (AL and AA)
These sites, together with Bosmat Tivon, are somewhat drier than the Carmel sites. They differ from all others by their two-layered structure of tall tree and low grass-herb strata, with shrubs and dwarfshrubs limited to rock edges and crevices. Allonim is representative of fenced and rotationally grazed pastures; Allone Abba was only lightly and irregularly grazed. Allone Abba has a higher woody *S* and higher importance values for perennial herbs and thistles; Allonim has higher *S* for grasses, legumes, and forbs, with lower cover of grasses and higher cover of legumes. Altogether, both sites had (after Gilboa) the highest total *S* and very high *H'* values (1.748 and 1.634), but by far the lowest *C* values (0.29 and .031).

Diversity Pattern and Climatic Trends

The means and coefficients of variation for diversity measures in sites, shown in Table 1, give a first approximation of the diversity of mediterranean shrublands and woodlands in northern Israel. Of the high mean total *S,* woody plants make up only a fifth and perennial herbs a fourth, therophytes more than half. This points to the great herbaceous and especially therophytic floristic richness of Mediterranean landscapes, but also to their low richness in woody species (cf. Raunkiaer 1934 and Cain 1950, on other Mediterranean areas). On the average the woody cover is about two-thirds and the total vegetation cover is close to 100% in spite of 16% rock and stones. The mean woody *C* of 0.454 points to the strong dominance of a few woody species – mainly *Quercus calliprinos* or *Pistacia lentiscus.* If batha communities had been included, their strong dominance by *Sarcopoterium spinosum* would probably have raised this value even more. The wide variation in most measures, including stratified and over-all *C,* is indicative of the complex and diverse structures of these vegetation types. Because of the rich and diverse herbaceous vegetation, the over-all mean value of *C* (0.219) is rather low and that of H' (1.248) rather high.

The histograms in Fig. 2 permit comparison of the sites by importances of major plant groups and alpha diversity measures; and their relation to mean rainfall is shown in Fig. 4. The most humid sites have the highest shrub importances and lowest diversities, and the most xeric sites have the lowest shrub importances and highest diversities. The over-all trend is one of increasing diversity with increasing drought as shown by *S* in Fig. 4 and by other diversity measures. The changes in $1/C$ and exp H' are not monotonic along the rainfall gradient, but (a) these, for the influence of structure on them, may be less effective in showing diversity trends than *S,* and

Table 2. Diversity correlations. Regression coefficients, coefficients of correlation (r) for species numbers (S) in 0.1 ha against structural measures – relative dominance (cover) concentration for all species (C_{veg}), relative species equitability (H', for species covers), and relative stratal concentration (C_{str}, using covers of 5 strata). See also Methods. Regressions are in the forms $y = a_1 + a_2x$, and $y = a_3e^{a4x}$

Independent, x	Dependent, y	a_1	a_2	r	S/n - p
C_{veg}	S (woody)	−0.0013	0.8336	0.78*	0.0139
C_{veg}	S (perennial herbs)	−0.339	1.996	0.77**	0.0142
C_{veg}	S (annuals)	0.5095	−0.6369	−0.81*	0.0123
H'	S (woody)	1.888	−2.382	−0.63*	0.0418
H'	S (perennial herbs)	2.687	−5.147	−0.81**	0.0748
H'	S (annuals)	0.465	1.717	0.89*	0.0468
		a_3	a_4	r	S/n-p
C_{str}	S (woody)	0.132	2.38	0.91**	0.0032
C_{str}	S (perennial herbs)	0.306	7.30	0.74**	0.0097
C_{str}	S (annuals)	0.569	−1.85	−0.87**	0.0039

* $p < 0.01$, ** $p < 0.001$.

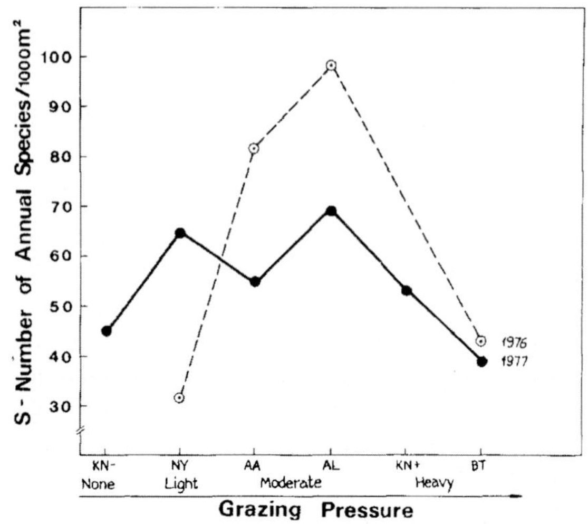

Fig. 5. Annual species richness of open woodland and shrubland in 1976 and 1977 in relation to grazing pressure. Pasture plots: KN − = Kfar Hanoar Hadati, 3 years protection; NY = Neve Yaar, very lightly grazed in 1976 and lightly grazed in 1977; AA = Allone Abba, moderately grazed; AL = Allonim, rotationally grazed; KN + = Kfar Hanoar Hadati, heavily grazed; BT = Bosmat Tivon, very heavily grazed. (Spring sampling shortly after grazing reduced *S* in AA and AL in 1977, and more intensive grazing increased *S* in NY in 1977.)

(b) $1/C$ and exp H' indicate, at least, lower dominance concentration and higher equitability in the drier sites.

Diversity trends may also be expressed as correlations of species numbers with structural measures (Table 2). Both relative concentration of dominance (cover) for species, C_{veg}, and relative concentration of stratal cover, C_{str}, are positively correlated with richness in woody plants and perennial herbs, but negatively correlated with richness in annual plants. The regressions for C_{str} were stronger in the exponential form, $y = a_3 e_4^a x$, than the linear form $y = a_1 + a_2 x$. Equitability, measured as H' using relative cover decimals for all species, was strongly correlated with annual plant richness, but negatively correlated with woody and perennial plant richness.

These results are consistent with the suppression of undergrowth diversity – notably of small, light-demanding annuals–by shrub dominance. One notes, however, that the perennial herbs are correlated with the woody plants rather than with the annual herbs. Some of the taller hemicryptophytes, and especially bunchgrasses such as *Oryzopsis miliacea* and *Stipa bromoides,* are more shadetolerant than most therophytes and occur as subordinates even in dense shrub patches or on their edges. When growing in these conditions and having high cover under protection from grazing, they tend like the shrubs to reduce therophyte and therefore total diversity. One notes further that these shrub-associated perennial grasses can behave not as hemicryptophytes but as chamaephyte semishrubs; they can resume growth from intercalary meristems along their stems

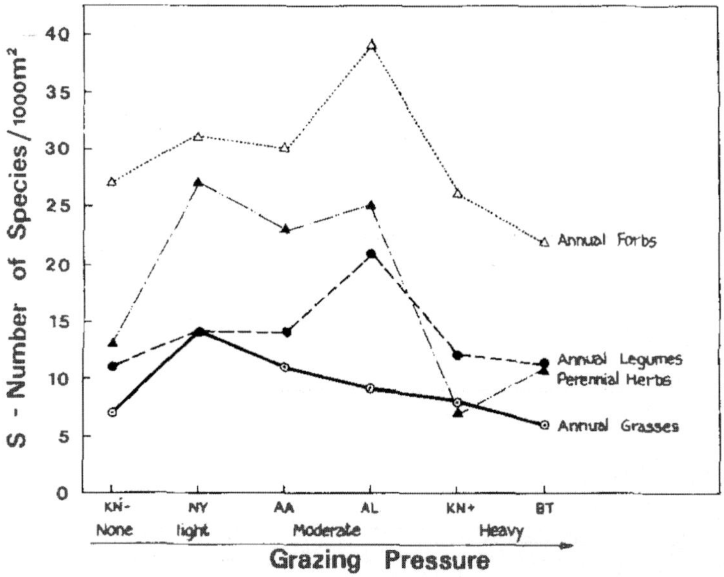

Fig. 6. Species richness of major herb groups of open woodlands and shrublands in relation to grazing pressure, 1977. Pasture plots as in Fig. 5.

after summer dormancy or burning, before the rains in fall, and reduce their leaf area in late spring and remain in a semi-green state for several months afterwards. In all these respects they differ from true hemicryptophytic perennial grasses like *Hordeum bulbosum* and *Dactylis glomerata,* which are more closely associated with the herophytes and have a strictly winter-and-spring growth rhythm. Over-all diversity is high in these communities – in some cases extremely high – when dominance of both the woody strata and the perennial herbs is limited by drought, by grazing or other disturbance, or by some combination of these.

The correlations with precipitation in the present study are not consistent. Both open maquis communities on Carmel, in spite of their higher moisture index and in spite of having high shrub importance, have also relatively high diversity, whereas the Bosmat Tivon community, closer to the xeric pole of the gradient, has also high shrub importance but low alpha diversity. The additional factor with a major influence on structure and diversity is human disturbance, including cutting, burning, coppicing, thinning, pruning, browsing, grazing, and trampling. We seek in the next section to deal with grazing effects.

The Effect of Grazing Intensity on Open Oak Woodland

In 1976 and 1977 several plots were added in the Lower Galilee to study in more detail the effect of different grazing intensities. The changes in annual species richness in the different sites in 1976 and 1977 along a gradient of increasing grazing

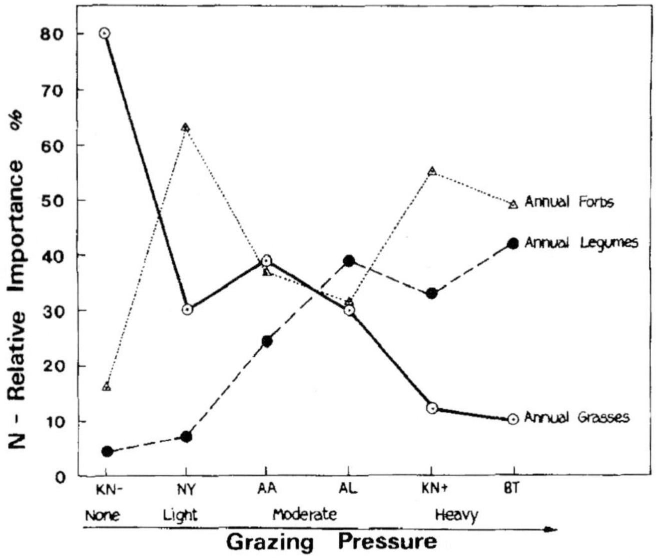

Fig. 7. Relative cover importance of annual grasses, legumes, and forbs of open woodlands and shrublands in relation to grazing pressure, 1977. Pasture plots as in Fig. 5.

pressure are shown in Fig. 5. In 1976 the KN plots were not yet included, but *S* showed a distinct trend of increase from a low (32 species) in the lightly grazed NY pasture to a high (97) in the rotationally grazed AL pasture and again a sharp decrease to a low (43) in the heavily grazed BT site. In 1977 both the protected and very heavily grazed pastures had lowest *S* and AL highest; but because in this year spring sampling in AA and AL was carried out after grazing, and in NY grazing intensity was slightly increased, the contrasts in S along the gradient were reduced.

Better insight into the dynamics of floristic changes and species richness under different grazing regimes can be gained by observing major botanical groups (Fig. 6). Lowest values for all these groups are confined to the extremes of the grazing gradient. However, distinct trends of increasing *S* up to a peak at AL are most conspicuous in the legumes, rising from 11 to 21 species and again decreasing to 11, and for forbs, rising from 27 to 39 and again decreasing to 28. Perennial herbs have distinctly higher *S* values in the lightly and moderately grazed pastures, as compared with the rested and heavily grazed ones, but in the annual grasses there is no clear trend.

However, if the relative importance values (per cent of total cover), instead of species numbers from these groups are considered, the picture is quite different, especially regarding grasses (Fig. 7). KN- has by far the highest grass relative cover (80 %), and, as can be seen in Table 3 and Fig. 8, the bulk of the grass cover (and biomass) is contributed by *Avena sterilis.* In the light and moderately grazed pastures the grass relative cover is 30–40 %, but with increasing pressure only 10 %. Striking contrasts in the two KN pastures on opposite sides of the fence are shown in Table 3. In the grazed KN+, the grass dominance of the rested pasture

Table 3. Comparison of species richness (S), relative cover per cent (N), and frequency per cent in 1-m^2 quadrats (F) of the major botanical groups and dominant species in adjacent protected and heavily grazed Tabor oak woodland (Kfar Hanoar Hadati, spring 1977). *C* is the Simpson index and *H′* the Shannon-Wiener index ($\log_{10}$), calculated for relative covers of all plant species

Grazing pressure	KN− (Protected)					KN+ (Grazed)				
Sample size	10 × 1 m^2			1000 m^2		10 × 1 m^2			1000 m^2	
	S	N	F	S	N	S	N	F	S	N
WOODY PLANTS										
Medium shrubs	–	–	–	7	8	–	–	–	3	4
PERENNIAL HERBS										
Geophytes	–	–	–	3	1.3	–	–	–	2	1
Perennial grasses	2	0.5		3	5	2	6		2	11
Dominant species:										
Hordeum bulbosum		0.5	–		4		4.1	70		9
Stipa bromoides		0.1	–		0.1		2.2	–		2
Perennial forbs	1	–		7	3.5	2	1		3	2
Dominant species:										
Echinops blanch.		–	–		0.4		0.4	20		2
ANNUAL HERBS										
Annual grasses	4	80.0		7	64	8	11		8	11
Dominant species:										
Avena ster.		80.0	100		62		–	–		1
Bromus alop.		–	–		–		5	100		4
Bromus scop.		–	–		–		2	100		2
Brachypodium dist.		–	–		–		2	100		2
Annual legumes	9	3.9		11	5	11	31		12	22
Dominant species:										
Trifolium pilul.		0.2	30		0.4		13	100		9
Trifolium purpureum		0.1	10		0.4		1	60		1
Trifolium xerocephalum		0.7	60		0.4		13	100		9
Annual forbs	21	15.6		27	14	13	51		26	49
Dominant species:										
Artedia squamata		–	–		–		4	80		9
Chaetosciadium t.		5.8	90		3		–	–		–
Crepis palaestina		1.0	60		0.4		3	100		27
TOTAL	37	100		65	100	36	100		56	100
C		0.63			0.378		0.143			0.106
1/C		1.59			2.64		2.32			9.43
H′		1.469			0.953		1.085			1.221
exp H′		1.6			2.59		2.96			3.39

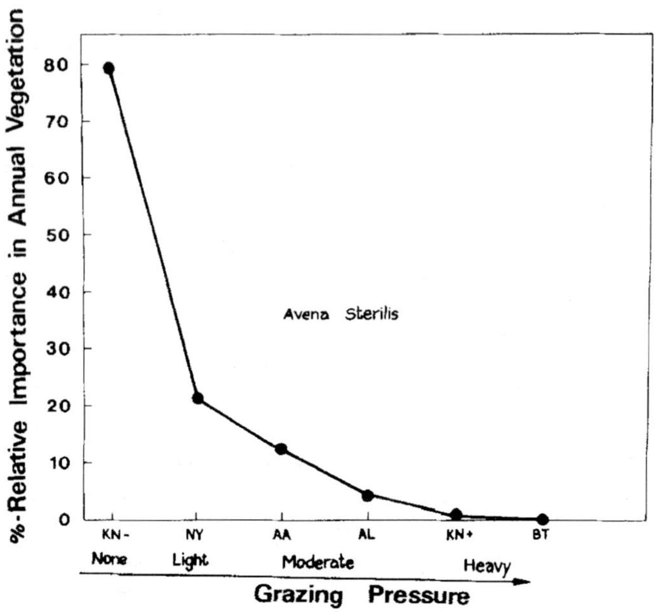

Fig. 8. Relative cover importance of *Avena sterils* in annual vegetation of open woodlands and shrublands in relation to grazing presure, 1977. Pasture plots as in Fig. 5.

KN− (chiefly *Avena sterilis*) is replaced by dominance of forbs, chiefly by the low palatable *Crepis palaestina* and the unpalatable *Arthedia squamata.* In KN+ (Table 3) and other more heavily grazed pastures, the remaining grasses are chiefly short-seasoned, low palatable *Bromus* species, and *Brachypodium distachyum.* Also perennial grass cover is higher with *Hordeum bulbosum* most abundant. As has been shown by Naveh (1955), this important pasture grass is not eaten readily in the reproductive stage from middle March until drying off in May, and can therefore produce seeds even under heavy grazing pressure. The annual grass *Avena sterilis,* on the other hand, is palatable throughout this critical stage and is thus prevented from seed production by defoliation. This species is therefore a very sensitive indicator of grazing intensity, as shown in Fig. 8.

As can be seen in Figs. 6 and 7, legumes respond differently from grasses and other forbs: In the rested pasture legumes are only 4% of relative cover, and slightly more in the lightly grazed NY pasture, but they increase with grazing pressure to reach 42% in BT. These differences in legume importance as related to grazing intensity are also shown in Table 3. In the heavily grazed KN pastures two low, short-seasoned, and low-palatable clovers – *Trifolium pilulare* and *T. xerocephalum* – contribute most of the production. However, in the moderately grazed pastures, more palatable and also more productive clovers such as *Trifolium purpureum, T. clypeatum,* and *T. campestre* are much more abundant. Most legumes, and especially those maturing earlier, are prolific seed producers, even under defoliation

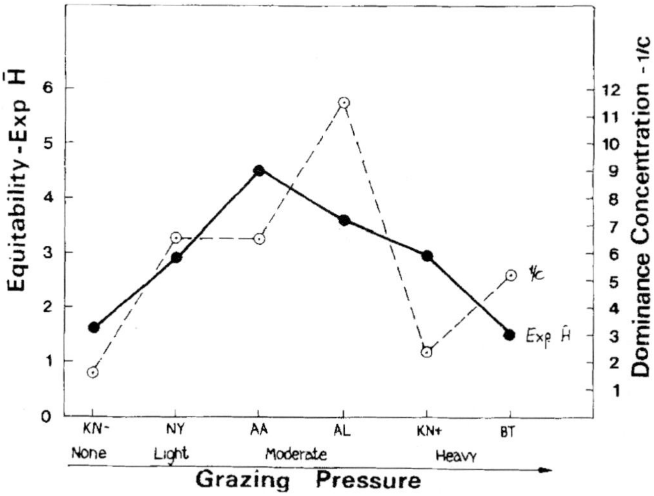

Fig. 9. Dominance concentration reciprocal ($1/C$) and equitability (exp H') of vascular plant species of open woodlands and shrublands in relation to grazing pressure, 1977. Pasture plots as in Fig. 5.

pressure. They are therefore less affected by grazing at this stage and show greater endurance under heavy grazing, and can even benefit from the removal of their taller grass competitors. These latter germinate and develop earlier in winter and, if not defoliated, can smother most minor species, including many legumes and forbs. Timing and intensity of grazing throughout the growth and reproductive stage therefore can be used to manipulate the botanical composition (Naveh & Kinsky 1975).

The changes in C and H' values along the grazing biocline (Fig. 9) show a similar trend in species richness: $1/C$ rises from a very low value of 1.6 (caused by the the almost exclusive dominance of *Avena sterilis*) to a high peak of 11.5 in AL and again a sharp drop with increasing grazing pressure in KN+ and a slight rise in BT. The latter may be related to the fact that under heavy grazing throughout the year, with the exception of the unpalatable *Anthemis pseudocotula,* not a single species reached importance values of 10%. The exp H' curve, however, shows almost similar low values at both extreme poles and a distinct peak at AA.

Dominance-diversity curves for the protected and heavily grazed KN pastures, as well as the rotationally grazed AL pasture in 1976 and 1977 are presented in Fig. 10. In KN- a great part of the relative cover is taken by *Avena sterilis,* allowing few subordinates with more than 1% importance values and very low covers for the remaining, minor species. In KN+ several codominant legumes and forbs occur, all low, or unpalatable and grazing tolerant. These curves are quite similar to those presented by McNaughton (1968) for annual mediterranean grassland in California, rested for 5 years and dominated by annual grasses from the Mediterranean region such as *Avena fatua* and *Bromus* species. But, in spite of the similar structure, the rested pasture with lowest S in Israel is richer than the California ones. The AL pasture is richer, with less dominance and higher equitability – no species reached

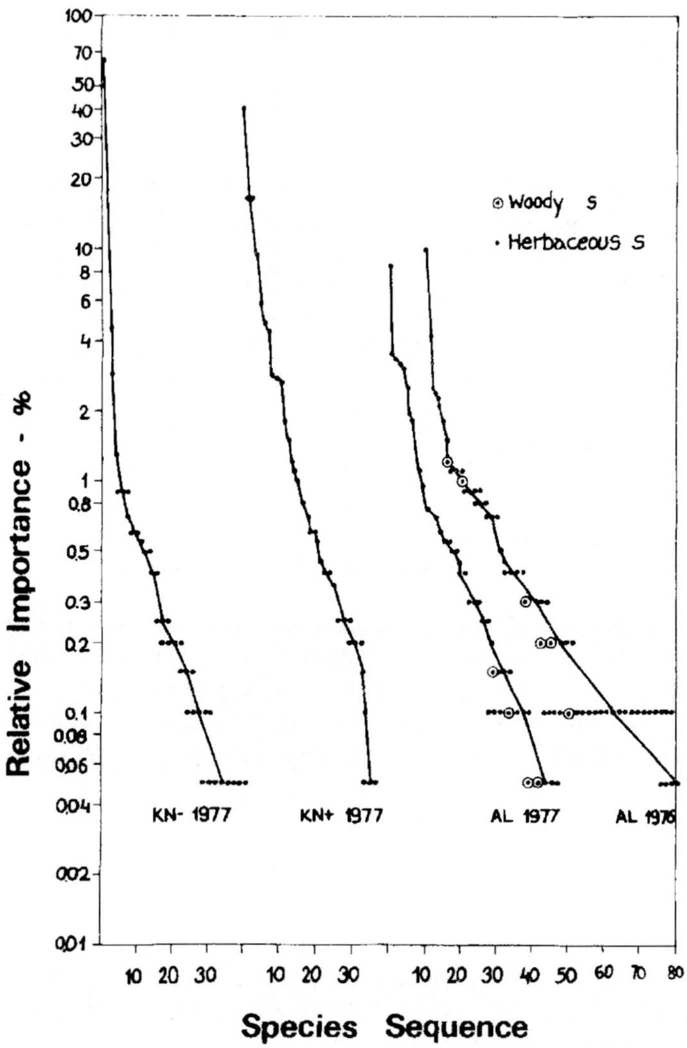

Fig. 10. Dominance-diversity curves for all vascular plant species in ten 1 m^2 quadrats of open Tabor oak woodland under different grazing regimes, spring 1976 and 1977. Explanation as in Fig. 3. Pasture plots: KN−, 3 years protection from grazing; KN+, heavily grazed; AL, rotationally grazed. Importance value slopes are steep in both the protected and heavily grazed pastures and less steep with intermediate, rotational grazing; at AL grazing was somewhat lighter in 1976 than in 1977.

10% relative cover. In the 1976 curve alpha diversity was even higher, and a greater number of minor species had 0.5–0.05 values.

The results on grazing effects in Tabor oak woodlands are consistent with those on paired communities along the climatic gradient, but indicate more clearly: (a) the two-slope relation of diversity to grazing, with minima at lowest and highest grazing pressures, and (b) the complex interplay among plant groups in response to grazing that underlies this two-slope pattern.

Correlation of Plant and Animal Communities

Zoological studies were carried out simultaneously with this study (Warburg 1977, Warburg et al. 1978). Except for certain groups that are adapted to shaded maquis habitats, animal richness, diversity, and abundance followed trends similar to those for plants. Thus, in several animal groups, the more xeric Gilboa grassland-shrubland reached highest animal *S* and the dense, mesic maquis shrublands, lowest. Gilboa ranked highest in reptile *S* (14), followed by Allonim (12) and Forty Oaks Grove (9), and both dense maquis plots were lowest (4). Also in rodent *S,* Gilboa ranked highest (7), with only 4 at Mt. Meron and the other maquis plots. Surprisingly, the open woodland sites were very poor in reptiles. In bird *S,* the high tree cover of the open woodlands stimulated the high species richness (40), but also here Gilboa ranked second with 31, as compared to only 6 at Mt. Meron. Also in isopods, the only invertebrate group studied in detail, Gilboa had highest *S* (13) and the closed maquis plots lowest (5 and 7).

With regard to animal abundance, as measured by the number of specimens collected per visit, Gilboa was highest for reptiles, spiders, scorpions, and grasshoppers. Bird densities were highest at Allonim, together with Forty Oaks Grove and Gilboa, and lowest in both dense maquis sites. Allonim also had the highest abundance of isopods and the dense maquis plots lowest. Broadly speaking, bird abundance and diversity is correlated with structural complexity of the vegetation here as elsewhere (McArthur 1965, Karr & Roth 1971, Willson 1974, Cody 1975), whereas abundance and density of ground-living vertebrates and most invertebrate groups are correlated with herb stratum diversity and increase to the drier and more open sites.

DISCUSSION

Three points that emerge from this study are: (a) the diversity response to the moisture gradient, (b) the increase of diversity with disturbance, and (c) the great structural and floristic diversity of Mediterranean shrublands and related communities. We consider the broader application of these points.

Diversity Along the Moisture Gradient

The increase in diversity with decreasing rainfall contradicts the idea that diversity should be correlated with productivity (Connell & Orias 1964) but is consistent with American studies. McNaughton (1968) noted an inverse relation of production and diversity. Comparisons (Whittaker & Niering 1965, 1975, del Moral 1972, Peet 1978) and regression analyses (Glenn-Lewin 1977) have shown higher diversities in open woodlands than in forests with closed canopies. Presumably the woodland, with its mosaic of light and partial shade, relatively abundant light for the undergrowth, and less dense occupation of the soil by tree roots than in a forest, offers a greater range of resources that may be divided among plant niches as a basis of understory diversity. In the Santa Catalina Mountains, Arizona, as in the present study, the high diversity of woodlands was part of a broader trend of increasing diversity with decreasing precipitation from closed and productive forests to open and less productive woodlands, grasslands, and semideserts (followed by decreased diversity from semidesert to desert) (Whittaker & Niering 1975).

Diversity and Disturbance

Effects of grazing on plant species diversity have been discussed by Harper (1969) and Grime (1973, 1979), Zeevalking & Fresco (1977), and Lubchenco (1978) have illustrated two-slope diversity responses to grazing. Let grazing itself be considered a 'biogradient' of increasing intensity of a biological effect – defoliation of plants by animals. There is then a certain optimum grazing pressure for maximum species diversity. (This optimum need not correspond to that at which maximum productivity or yield occurs, and it need not correspond between different communities.) Below this optimum grazing pressure strong dominance by woody plants or taller, aggressive grasses reduces diversity; above this optimum severe grazing reduces diversity by limiting the number of species able to survive to a few most aggressive and unpalatable ones.

Observations on grazing on American range lands and in Israel suggest a paradox (Whittaker 1977b): In the United States plant species diversity normally decreases progressively with increasing grazing pressure (though diversity is less with no

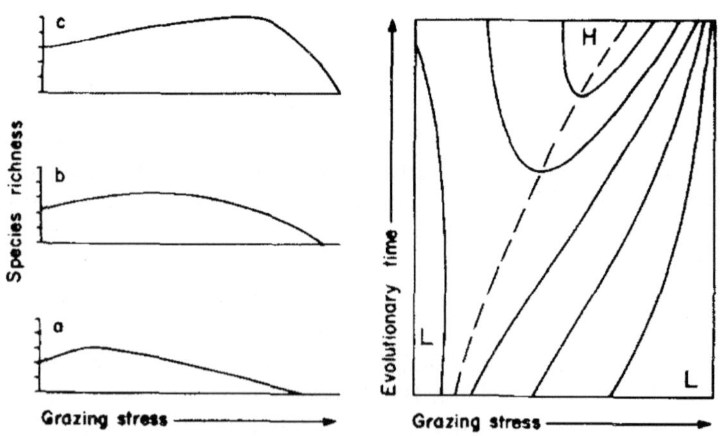

Fig. 11. Evolution of a flora under grazing: hypothetical curves and a trend surface for the relationship of plant species diversity to grazing pressure. Curve *a* represents the response of natural vegetation with maximum diversity under light grazing by native animals; curves *b* and *c* show the accumulation through time of grazing-adapted species to produce an altered, species-rich vegetation in which diversity is highest with relatively heavy grazing.

grazing than with light grazing). In Israel diversity increases over a wide range of increasing grazing pressures, and maximum diversities are at grazing pressures that would be considered severe or extreme in the United States. The apparent contrast may be resolved by an evolutionary interpretation. A community evolved in relation to light, natural grazing before the effects of man would probably show decrease in diversity with complete release from grazing; but over a wide range of increasing pressure from domestic stock as a new stress would show progressive reduction in diversity (Fig. 11a). Through extended time under grazing – some centuries, say – species adapted to grazing would evolve in or reach the area. As these species accumulated the communities under moderate and heavy grazing would be enriched by them, as shown in Fig. 11b. Through millenia of heavy grazing species adapted to this would accumulate, producing a pattern of diversity response in which the community is richest under severe (but not the most severe) grazing stress (Fig. 11c). This, we judge, is the pattern in Israel. The three curves can be combined into a response surface (cf. Auclair & Goff 1971) in which the diversity response to grazing through evolutionary time is outlined by contours. The ridge connecting peak diversities is oblique because of the faster accumulation of disturbance-adapted than of climax-adapted species.

The speed of this evolutionary process is a point of interest. We judge that before occupation by man the Levant had, like California before the Spanish, a flora that included annuals and was adapted to maquis and woodland fire cycles and to grazing by native ungulates. Israel and the Levant have been subject to fairly dense human occupation for probably 8 000–10 000 years (Naveh & Dan 1973). Archeological and paleontological evidence from Israel, France, and Greece, however, can trace hunter-gatherer economies and the human use of fire back into the middle Pleistocene, more than half a million years ago (Naveh 1975, 1977). The richness of Levantine communities is then the product of accelerating evolution in response to progressively intensifying human stress from the Pleistocene to historic times. Rapid introduction of species (and evolution of grazing-adapted ecotypes) is implied, especially for the last 8 000–10 000 years. Many of the species can be traced into the great, arid and semiarid Irano-Turanian heartland of Asia (Zohary 1962). Some of the species of arid lands, particularly annuals adapted to rapid, low growth and fruiting in full sunlight during a short rainy season, are to a degree preadapted to survival by such growth in open, overgrazed pastures. The richness of these pastures then represents the accumulation of grazing-adapted species and ecotypes, brought from the more arid parts of the Near East and from Asia by trade and armies passing through the geographic crossroads that Israel represents.

The high diversities observed are the combined result of a long evolutionary history under constant stress by drought, fire, and grazing, in a warm-temperate climate, and the small-scale heterogeneity of the mountainous habitats. In these rocky soils microrelief and soil depth are variable, and rock outcrops, tree cover, litter cover, shelter, and shade create a complex mosaic of microsites. The constant grazing pressure shifts advantage to smaller plants that can divide the soil microsites more finely; the effect of such division is clearest in the many legume species in the grazed sites.

Further directions of niche difference among the plant species are provided by: (a) adaptations to both seasonal time and year-to-year variation in climate, (b) relative palatability and tolerance of the various livestock species – cattle, sheep, horses, donkeys, goats, and camels, and (c) pollination and seed dispersal and behavior. (The latter includes some notable adaptions, among them heterocarpy, seed polymorphism in legumes and composites, and trypanocarpy, the drilling of seeds into the soil by hygroscopic awns in *Erodium* and a number of grasses.)

The consequence is the occurrence of some of the highest alpha (within-stand) diversity in the world. It is, however, our impression that beta diversity (species turnover between different topographic positions and soils in the same area) and delta diversity (extent of species turnover with change from one geographic area to another) are relatively low. We cannot document differences in beta and delta diversity, but note the widespread occurrence of many species of the rich annual flora on different topographies and soils and over a wide geographic area surrounding the Mediterranean Sea. Some of these widespread plants have also been most successful invading other mediterranean communities, particularly disturbed ones, in California and Chile (Naveh 1967). The relative consistency of the annual floras in different Mediterranean countries seems in striking contrast to the great geographic differentiation (delta diversity) within the Cape flora (Good 1974, Taylor 1978).

Despite heavy grazing, severe overgrazing of Levantine pastures was generally prevented until recently, because the numbers of livestock that could be supported during the critical period of low food availability in early winter were not sufficient to overgraze pastures during the spring flush of growth and seed-setting. The pastures and open shrublands thus were parts of a man-modified landscape that was relatively stable, and much richer in species than the natural landscape that had preceded it. This relative equilibrium is now disturbed in opposite directions. In some pastures the carrying of stock through the winter by supplementary feedings permits real overgrazing during the spring growth, with consequent impoverishment in plant species as on the far-right slope of Fig. 11c.

Where, in contrast, grazing is removed and the woody canopy returns in maquis, or the perennial herb cover increases in woodland, there is a drastic reduction in species richness in the 'climax community, toward the left of Fig. 11 c. In these communities it is by no means the case that maximum diversity is in the undisturbed or climax condition as suggested by Margalef (1968) and Odum (1969, cf. Loucks 1970, Auclair & Goff 1971, Bazzaz 1975, Pickett 1976). In Israel development from the open and disturbed shrublands to closed-canopy maquis can produce a fourfold reduction in species diversity (Table 4). Management of grazing, with intensity somewhere near the optimum in relation to which these communities have evolved, is needed to maintain the floristic richness and the notable flowering beauty of these landscapes. In the shrublands fire or cutting also is needed to maintain their open structure. Full protection, which disturbs the present semi-natural equilibrium and reduces the diversity and attractiveness of these landscapes, should not here be the objective of conservation (cf. Naveh 1971, van der Maarel 1975).

Table 4. Comparative diversity measures for mediterranean communities

	M^1	Species number/0.1 ha mn	σ	min	max	Species /1 m^2	Coefficients[3] b	d	c^4_{ssp}	c^4_{str}
ISRAEL										
Woodlands, grazed	3	136.7		135	138	22	24.2	37.5	0.07	0.36
Open shrubland	3	138.7		118	179	25	24.5	38.1	0.15	0.19
Closed shrubland (maquis)	3	35		21	57	4	4.3	10.1	0.34	0.64
Batha, Mt. Carmel	1	65				14	15.9	16.9	0.46	0.73
CALIFORNIA										
Woodlands, Santa Barbara	6	47	6.2	35	53	10.7	9.1	12.1	0.23	0.60
Woodlands, Carmel, protected	6	58	9.9	40	68	13.8	12.0	13.9	0.25	0.67
Woodlands, Carmel, grazed	5	64	14.9	50	89	15.6	12.2	15.7	0.16	0.68
Chaparral—Adenostoma	5	24	8.1	16	34	7.3	8.0	5.6	0.58	0.77
Chaparral—Ceanothus	5	32	2.9	27	34	3.9	3.1	9.2	0.56	0.96
CALIFORNIA—IBP data										
Chaparral[2]	–	44							0.35	
Coastal sage[2]	–	65							0.34	
CHILE—IBP data										
Interior mattoral[2]	–	108							0.15	
Coastal[2]	–	109							0.11	
AUSTRALIA										
Heath, southwest Australia	7	65	10.5	46	82	13.3	14.1	17.1	0.21	0.67
Mallee, New South Wales	6	49	10.4	39	68	6.1	3.8	14.5	0.30	0.52
SOUTH AFRICA										
Fynbos, Cape Province	10	75	22.7	52	128	16.7	17.3	19.0	0.22	0.55

[1] M = number of samples averaged, standard 1–1000 m^2 samples except IBP data.
[2] Data from Mooney et al. (1977). Numbers of species are totals in ten separate 100 m^2 plots; C_{ssp} computed for means of woody species coverages for four directions of exposure.
[3] From least squares regressions, $S = b + d \log_{10} A$; S = vascular plant species numbers rooted in plots of areas (A) of 1 m^2 (mean of 10), 10 m^2 (mean of 2), 100, and 1000 m^2.

[4] Simpson index, $C = \sum p_i^2$, using relative covers of plant species for C_{spp}, and relative cover of strata for C_{str}.

Comparative Mediterranean Diversities

Data now available permit preliminary comparison of plant alpha diversities in mediterranean communities on different continents. These communities, with emphasis on their convergent adaptations, have been the subject of recent research under the International Biological Programme (Castri & Mooney 1973, Mooney 1977, Cody & Mooney 1978).

In Tables 4 and 5 we have compiled mean diversity measures and growth-form compositions for sets of samples (see *Study sites*) representing plant communities in the five or six mediterranean areas of the world. There are dangers for such comparisons: Each of the mediterranean areas includes a wide range of climates, soils, and communities. Each type of community varies in diversity from one stand to another, and for herbs from one year to another. Comparisons can easily be biased by the particular stands chosen, or the particular types of communities chosen. We hope these dangers may be reduced by the use of sets of samples, and by reasonable care in the types of communities compared with one another.

In California, as in Israel, woodlands have higher diversity and lower dominance concentration than closed shrublands; maquis and the two chaparral samples are of comparable, low diversities (Naveh 1967 and Table 4). The contrast is primarily the greater richness in herbs, and especially annuals, in the woodlands (Table 5). Comparing Israel and California (woodland vs. woodland, and maquis vs. chaparral) there are marked differences in growth-form representation. Israel is much richer in climbers, geophytes, and woodland legumes; almost all growth-forms are richer in Israel, however, except annual herbs in the dense maquis. Some of the California samples were taken in the spring of 1978, a year with exceptionally high rainfall; and the representation of annuals in these samples is higher than in more average years.

The IBP samples from Chile and California (Parsons & Moldenke 1975, Mooney [a] 1977) are not strictly comparable, for they are based on combining 10 smaller (100 m^2) plots rather than single quadrats of 1000 m^2. Species numbers should be higher in these combined plots, as is the case for woody plants in the IBP chaparral compared with the other chaparral samples. The smaller number of annuals in the IBP chaparral presumably represents a less favorable spring; it is consistent with the well-known poverty of herbaceous species consequent in part on allelopathy in mature stands of *Adenostoma* (Naveh 1967, Muller et al. 1968, McPherson & Muller 1969, Christensen & Muller 1975). Comparing California with Chile, the latter is richer in all growth-form categories and has higher total diversity and lower concentration of dominance. The greater richness of Chile in woody plants may be a consequence of long-term evolutionary history, but the greater richness in herbs is a consequence of a longer history of more severe human disturbance that has opened the canopies of the Chilean shrub communities (Aschmann & Bahre 1977).

Two other southern hemisphere shrublands are the South African Cape scrub or fynbos, and the Australian heaths or kwongan. [The Australian 'heaths' include a wide range of communities, some true heathlands (dominated by *Epacridaceae*) and others that are not heathlands. For the extensive dry shrublands of the Western

Table 5. Comparative growth-form representation in mediterranean communities[1]

					Perennial herbs[2]			Annual herbs[3]			
	Tree and shrub	Sub-shrub	Climber	Total	Forb	Grami-noid	Geo-phyte	Forb, nonleg.	Grami-noid	Legume	Total herbs
ISRAEL											
Woodlands	6.7	6.0	4.7	14.4	12.0	2.7	8.0	54.3	15.0	27.3	119.3
Open shrublands	12.3	7.7	7.0	27.0	19.7	7.3	12.7	39.3	8.3	24.3	111.7
Closed shrublands (maquis)	8.7	3.3	5.3	17.3	5.0	3.3	3.7	3.0	0.7	1.7	17.3
Batha	3	7	1	11	7	3	11	19	1	13	54
CALIFORNIA											
Woodlands, Santa Barbara	4.0	1.8	.3	6.2	8.8	3.2	2.5	17.2	5.5	3.5	40.7
Woodlands, Carmel, protected	2.7	0.3	1.2	4.2	9.0	4.7	2.5	23.2	6.8	7.5	53.7
Woodlands, Carmel, grazed	1.4	0.8	1.2	3.4	8.6	3.4	2.4	28.0	9.4	8.4	60.2
Chaparral—Adenostoma	4.6	2.4	.2	7.2	2.4	1.0	2.0	8.6	1.6	1.0	16.6
Chaparral—Ceanothus	8.0	3.2	.4	11.6	3.8	1.2	.6	9.6	3.6	1.6	20.4
CALIFORNIA—IBP data											
Chaparral[4]	18	7		25	(8)		5	(6)			19
Coastal sage[4]	11	12		23	(13)		2	(27)			42
CHILE—IBP data											
Interior, mattoral[4]	26	15		41	(22)		6	(39)			67
Coastal[4]	13	17		30	(19)		10	(50)			79
AUSTRALIA											
Heath, southwest Australia	26.3	10.4	0.7	37.4	5.4	15.1[5]	1.1	5.9	.1	0	27.7
Mallee, New South Wales	12.0	3.5	1.2	16.7	7.5	6.0	1.0	15.9	2.0	0.3	32.7
SOUTH AFRICA											
Fynbos, Cape Province	24.0	10.7	0.6	35.3	7.1	19.4[5]	11.7	1.1	0.4	0	39.8

[1] Numbers of samples are the same as in Table 3; values (except IBP data) are numbers of species in 0.1 ha plots.
[2] Forbs exclude geophytes as well as graminoids (grasses and grass-like plants) under perennial herbs.
[3] Forbs exclude legumes as well as graminoids under annual herbs.
[4] Based on percentages in Mooney et al. (1977: Table 5.1). (Herbs given only as perennials and annuals.)
[5] Perennial graminoids in Australia and South Africa include restioids: mean species numbers of 2.3 in the heath, 0 in the mallee, and 7.6 in the fynbos.

Australian sandplain vegetation Beard (1976) suggests use of the native word '*kwongan*' as a term for this formation, in parallel with such regional terms as 'fynbos' and 'mallee'.] The fynbos and kwongan are far richer in woody species than any other of the mediterranean communities in Tables 4 and 5, and richer in total species than any except the Israeli woodlands. (The fynbos and kwongan are richer than the Chilean samples, despite the high species numbers in the combined samples for the latter.) Our samples of fynbos and Australian heath or kwongan are not particularly rich as such (cf. Taylor 1972, Werger et al. 1972, Westman 1975, Lamont et al. 1977). The South African samples include several of mountain fynbos, less rich than some of the lowland communities; one fynbos, less rich than some of the lowland communities; one fynbos sample with 128 species (at Jonkershoek, near Stellenbosch) is well within the high-diversity range of the Israeli woodlands, despite its lack of annuals. In the Australian heath and the fynbos, as in the Israeli woodlands, high diversity for 1000 m^2 is consequent on both high small-plot richness (S/m^2 and b) and steep slopes of species increase with area (d). Werger (1972) gives three other fynbos samples with mean values of 10.7 and 18.8 for b and d. Taylor (1972, 1977) notes that fynbos species counts are normally 50 to more than 120 per 0.1 ha, and that one count gave 121 species in 100 m^2.

The mallee and heath of south Australia are often referred to as a sixth mediterranean area, along with the Old World Mediterranean, coastal California and Chile, the Cape Province, and southwestern Australia (Aschmann 1973, Miller et al. 1977). It is doubtful that the mallee should be considered part of the mediterranean grouping, although mallee communities without heaths intergrade, through mallee with ericoid undergrowth, with some of the Australian heath communities. The mallee communities, which extend far inland, may be more nearly comparable with the pygmy conifer woodlands of the United States in climate and diversity (cf. Harner & Harper 1976). The mallee samples (see also Whittaker et al. 1979) are not particularly rich; they are comparable with the California woodlands sampled, though richer in woody plants and less rich in herbs.

Setting the mallee aside, there is striking contrast in growth-form representation between the fynbos and Australian heath, on the one hand, and the remaining samples. Cowling & Campbell (1980) demonstrate the marked divergences between South African communities, and the Californian and Chilean ones sampled by Parsons & Moldenke (1975). Annual plants, so conspicuous in the other communities, are absent from many fynbos and Australian heath communities. The latter have, in contrast, diverse representation of leptophyll shrubs (with minute sclerophyll leaves, less than 25 mm^2) together with the restioids that are not easily placed in any northern-hemisphere growth-form or life-form, and evergreen geophytes that are virtually unknown in the northern hemisphere. Like the other mediterranean communities the fynbos and kwongan are sclerophyll; but with their leptophyll dominants they are sclerophyllous in a different way, and in other respects their structures are divergent from those of the remaining communities. Extreme nutrient poverty of soils is a major adaptive problem and reason for nonconvergence of the fynbos (Cowling & Campbell 1979) and Australian heathlands with other mediterranean areas.

It may then be judged that there are two divergent groups of mediterranean communities. One group comprises the true Mediterranean, California, and Chile; this group includes broad-sclerophyll shrublands of relatively recent evolutionary history (Raven 1973) and of low species diversities, except as affected by human disturbance that enriches them in herbs. The three areas seem to form a sequence of increasing length and intensity of human disturbance and increasing consequent species diversity, from California through Chile to the Mediterranean as represented in Israel. Chile is probably part of this group because of a history similar to that of California (Solbrig et al. 1977). We infer that after South America had been part of Gondwana, much of the ancient flora was destroyed in southern South America by mountain uplift, glacial period climatic change, and rejuvenation of soils comparable to that in California. The present shrublands and other communities are products of more recent evolution. The Cape flora, in contrast, probably had its Cretaceous and earlier Tertiary development in what is now central Africa, and moved southward to the Cape, where this flora already rich in genera was further enriched in species by biotic pulsation (Whittaker 1977a) – speciation during the expansion and contraction of species areas produced by Pleistocene climatic fluctuations in a landscape of mountain ranges dispersed in lowlands (Levyns 1964, Zinderen Bakker 1976, Axelrod & Raven 1978, Taylor 1978).

The basis for the richness of the southwest Australian communities is less easily stated (Burbidge 1960, Hopper 1979) but may involve Pleistocene biotic pulsation of a flora rich in genera on the mosaic of parent materials, and valleys and uplands, of a larger geographic area of lower topographic relief than that of the Cape Province. The evolutionary connection between fynbos and kwongan is ancient, probably Cretaceous, and is expressed at higher taxonomic levels (with most genera, and many subfamilies and families different in the two areas). Derivation of both from heathlike Gondwanan shrublands of sterile soils seeems likely. In both Cape Province and southwest Australia rich floras of Gondwanan derivation persist and are adapted to the very old and nutrient-poor soils of these landscapes. These communities have had long histories in climates with less drastic fluctuation than those of other areas; and they have accumulated their extraordinary richness in species through speciation (partly in response to more moderate climatic fluctuations) through at least Pleistocene and probably a longer time.

Much has been written about convergence among mediterranean communities. These communities share temperate climates with maritime influence (but other differences), sclerophylly (which occurs more widely, and may be as much an adaptation to nutrient poverty as to climate in the fynbos and heathlands), and fire cycles (characteristic, in one way or another, of most communities between moist forest and semidesert). All five or six areas include ranges of shrub communities from those with heaths (*Ericaceae,* or *Epacridaceae* in Australia, or *Empetraceae* in Chile) dominant or well represented, primarily on acid soils, to other shrubland types without heaths. Among them, however, the Gondwanan vs. the three more recent mediterranean plant communities are profoundly different in structure, soil and nutrient relationships, and diversity. The richest temperate communities in the world are apparently the stable, climax Gondwanan shrublands, and the chronically disturbed and grazed Mediterranean pastures. These represent virtually polar types of species diversity (cf. Grassle & Sanders 1973).

The convergences within and between the two sets of mediterranean communities are clearly of interest (Cody & Mooney 1978). Between the Gondwanan and the more recent mediterranean communities, however, the divergences in diversity and other characteristics seem more significant and more interesting than the convergences.

SUMMARY

Between 1974 and 1978 structure and diversity of shrublands and woodlands of northern Israel were studied along climatic and human-disturbance gradients using 0.1 ha vegetation samples. Diversity increased along the moisture gradient, with highest woody and herb species richness in open *Pistacia* shrubland on the xeric border of the Mediterranean region, and highest equitability and lowest dominance concentration in sub-humid, moderately grazed, open oak woodlands. Semi-open disturbed shrublands were rich in herbs and had much higher structural, plant species, and animal species diversities than the closed, mature, 'climax' maquis. Diversity showed a two-slope response to grazing with highest species numbers in heavily (but not the most severely) grazed woodlands and shrublands. These communities have some of the highest plant alpha diversities in the world; the richness of their floras (especially in annual plants) is the product of relatively rapid evolution under stress by drought, fire, grazing, and cutting.

Comparative data on diversity and growth-form composition are compiled for mediterranean communities: Israeli shrublands and woodlands, California chaparral and woodlands, Chilean matorral, South African fynbos, and Australian heath and mallee. Communities of three of these areas are of more recent (primarily Pleistocene) development and share some similarities; these three form a sequence (California, Chile, and Old World Mediterranean) of increasing length of human disturbance and consequent species diversity. The southwest Australian heath or kwongan and the South African fynbos are, in contrast, derived from ancient Gondwanan heath like communities and are adapted to very old, nutrient-poor soils. The Gondwanan communities are quite different in growth-form structure and soil and nutrient relationships from communities of the three more recent mediterranean areas; the Gondwanan communities are almost lacking in annual species and are exceedingly rich in woody species. The richest temperate plant communities known – grazed Mediterranean pastures vs. fynbos and Australian heath – are in almost polar contrast in their growth-form structures and the bases of their species diversities.

REFERENCES

Aloni R (1972) The vegetation of the Lower Galil. M. Sc. Thesis, Hebrew University, Jerusalem (Hebrew), 128 pp.

Aschmann H (1973) Distribution and peculiarity of mediterranean ecosystems. In: di Castri F, Mooney HA (eds) Mediterranean Type Ecosystems. Ecol. Stud. 7:11–19.

Aschmann H, Bahre C (1977) Man's impact on the wild landscape. In: Mooney HA (ed.) Convergent Evolution in Chile and California, US/IBP Synth. Ser. 5:73–84.
Auclair AN, Goff FG (1971) Diversity relations of upland forests in the western Great Lakes area. Amer. Nat. 105:499–528.
Axelrod DI, Raven PH (1978) Late Cretaceous and Tertiary vegetation history of Africa. In: Werger MJA (ed.) Bio-geography and Ecology of Southern Africa. Junk, The Hague, pp. 77–130.
Bazzaz FA (1975) Plant species diversity in old-field successional ecosystems in southern Illinois. Ecology 56:485–488.
Beard JS (1976) An indigenous term for the western Australian sandplain and its vegetation. J.R. Soc. W.Austr. 59:55–57.
Burbidge NT (1960) The phytogeography of the Australian region. Austr. J. Bot. 8:75–209.
Cain SA (1950) Life-forms and phytoclimate. Bot. Rev. 16:1–33.
Castri F di, Mooney HA (eds) (1973) Mediterranean Type Ecosystems: Origin and Structure. Ecol. Stud. Springer, New York, 7:1–405.
Christensen NL, Muller CH (1975) Effects of fire on factors controlling plant growth in Adenostoma chaparral. Ecol. Monogr. 45:29–55.
Cody ML (1975) Towards a theory of continental species diversities: bird distributions over mediterranean habitat gradients. In: Cody ML, Diamond JM (eds). Ecology and Evolution of Communities, Harvard Univ. Press, Cambridge, pp. 214–257
Cody ML, Mooney HA (1978) Convergence versus non-convergence in mediterranean-climate ecosystems. Ann. Rev. Ecol. Syst. 9:265–321.
Connell JH, Orias E (1964) The ecological regulation of species diversity. Amer. Nat. 98:399–414.
Cowling R, Campbell B (1980) Convergence in vegetation structure in the mediterranean communities of California, Chile, and South Africa. Vegetatio (in press).
Eig A (1933) A historical-phytosociological essay on Palestinian forests of Quercus aegilops L. ssp. ithaburensis (Desc.) in past and present. Beih, Bot. Contr. 51, Abt. II:225–272.
Glenn-Lewin DC (1977) Species diversity in North American temperate forests. Vegetatio 33:153–162.
Good R (1974) The Geography of the Flowering Plants, 4th edn. Longmans, London.
Grassle JF, Sanders HL (1973) Life histories and the role of disturbance. Deep-Sea Res. 20:643–659.
Grime JP (1973) Control of species density in herbaceous vegetation. J. Environ. Mangt. 1:151–167.
Grime JP (1979) Plant strategies and vegetation processes. Wiley, New York, 222 pp.
Harner RF, Harper KT (1976) The role of area, heterogeneity, and favorability in plant species diversity of pinyon-juniper ecosystems. Ecology 57:1254–1263.
Harper JL (1969) The role of predation in vegetational diversity. In: Woodwell GM (ed.). Diversity and stability in ecological systems, Brookhaven Symp. Biol. 22:48–62.
Hill MO (1973) Diversity and evenness: a unifying notation and its consequences. Ecology 54:427–432.
Hopper SD (1979) Biogeographical aspects of speciation in the southwest Australian flora. Ann. Rev. Ecol. Syst. 10:399–422.
Karr JR, Roth RR (1971) Vegetation structure and avian diversity in several New World areas. Amer. Nat. 105:423–435.
Lamont BB, Downes S, Fox JED (1977) Importance-value curves and diversity indices applied to a species-rich heathland in western Australia. Nature 265:438–441.
Levyns MR (1964) Migrations and origin of the Cape flora. Trans. R. Soc. S. Afr. 37:85–107.
Loucks OL (1970) Evolution of diversity, efficiency, and community stability. Amer. Zool. 10:17–25.
Lubchenco J (1978) Plant species diversity in a marine intertidal community: importance of herbivore food preference and algal competitive abilities. Amer. Nat. 112:23–39.
Maarel E van der (1975) Man-made natural ecosystems in environmental management and planning. In: van Dobben WH, Lowe-McConnell RH (eds) Unifying Concepts in Ecology: Rept. Plen. Sess., 1st Int. Congr. Ecol., The Hague, 1974, Junk, The Hague, Pudoc, Wageningen.
MacArthur RH (1964) Environmental factors affecting bird species diversity. Amer. Nat. 98:387–397.
MacArthur RH (1965) Patterns of species diversity. Biol. Rev. 40:510–533.
Margalef R (1968) Perspectives in Ecology Theory. Univ. Chicago Press, Chicago, 111 pp.

May RM (1975) Patterns of species abundance and diversity. In: Cody ML, Diamond JM (eds) Ecology and Evolution of Communities, Harvard Univ. Press, Cambridge, pp. 81–120.

McNaughton SJ (1968) Structure and function in California grasslands Ecology 49:962–972.

McPherson JK, Muller CH (1969) Allelopathic effects of Adenostoma fasciculatum, 'chamise', in the California chaparral. Ecol. Monogr. 39:177–198.

Miller PC, Bradbury DE, Hajek E, LaMarche V, Thrower NJW (1977) Past and present environment. In: Mooney HA (ed.) Convergent Evolution in Chile and California, US/IBP Synth. Ser. 5:27–72.

Mooney HA (ed) (1977) Convergent Evolution in Chile and California: Mediterranean Climate Ecosystems. US/IBP Synthesis Series 5:1–224. Dowden, Hutchinson & Ross, Stroudsburg, Pa.

Mooney HA, [a]. (1977) The producers – their resources and adaptive responses. In: Mooney HA, Convergent Evolution in Chile and California. (ed.) US/IBP Synth. Ser. 5:85–143.

Moral R del (1972) Diversity patterns in forest vegetation of the Wenatchee Mountains, Washington. Bull. Torrey Bot. Club 99:57–64.

Morris MG (1973) The effects of seasonal grazing on the Heteroptera and Auchenorhyncha (Hemiptera) of chalk grasslands. J. Appl. Ecol. 10:761–780.

Muller CH, Hanawalt RB, McPherson JK (1968) Allelopathic control of herb growth in the fire cycle of California chaparral. Bull. Torrey Bot. Club 95:225–231.

Murdoch WW, Evans FC, Peterson CH (1972) Diversity and pattern in plants and insects. Ecology 53:819–829.

Naveh Z (1955) Some aspects of range improvement in a mediterranean environment. J. Range Mangt. 8:265–270.

Naveh Z (1967) Mediterranean ecosystems and vegetation types in California and Israel. Ecology 48:445–459.

Naveh Z (1971) The conservation of ecological diversity of mediterranean ecosystems through ecological management. In: Duffey E, Watt AS (eds) The Scientific Management of Animal and Plant Communities for Conservation, Symp. Brit. Ecol. Soc. 11:605–622.

Naveh Z (1975) The evolutionary significance of tire in the Mediterranean Region. Vegetatio 29:199–208.

Naveh Z (1977) The role of fire in the mediterranean landscape of Israel. In: Proc. Symp. Environmental Consequences of Fire and Fuel Management in Mediterranean Ecosystems. Palo Alto, Calif., Aug. 1977, pp. 299–306.

Naveh Z, Dan J (1973) Human degradation of mediterranean landscapes in Israel. In: de Castri F, Mooney HA (eds). Mediterranean Type Ecosystems, Ecol. Stud. 7:373–390.

Naveh Z, Kinsky J (1975) The effect of climate and management on species diversity of Tabor oak savanna pastures. Proc. 6th Sci. Conf. Isr. Ecol. Soc., pp. 284–296.

Naveh Z, Whittaker RH (1980) Measurements and relationships of plant species diversity in Mediterranean shrublands and woodlands. In: Grassle F, Patil GP, Smith W, Taillee C (eds) Ecological Diversity in Theory and Practice, Statistical Ecology, Volume S6. Internat. Coop. Publ. House, Fairland, Md.:219–239.

Odum EP (1969) The strategy of ecosystem development. Science 164:262–270.

Parsons DJ, Moldenke AR (1975) Convergence in vegetation structure along analogous climatic gradients in California and Chile. Ecology 56:950–957.

Peet RK (1974) The measurement of species diversity. Ann. Rev. Ecol. Syst. 5:285–307.

Peet RK (1978) Forest vegetation of the Colorado Front Range: patterns of species diversity. Vegetatio 37:65–78.

Pickett STA (1976) Succession: an evolutionary interpretation. Amer. Nat. 110:107–119.

Raunkiaer C (1934) The Life Forms of Plants and Statistical Plant Geography. Clarendon, Oxford, 632 pp.

Raven PH (1973) The evolution of mediterranean floras. In: di Castri F, Mooney HA (eds) Mediterranean Type Ecosystems. Ecol. Stud. 7:212–224.

Solbrig OT, Cody ML, Fuentes ER, Glanz W, Hunt JH, Moldenke AR (1977) The origin of the biota. In: Mooney HA (ed.) Convergent Evolution in Chile and California, US/IBP Synth. Ser. 5:13–26.

Taylor HC (1972) Fynbos. Veld & Flora 2:68–75.

Taylor HC (1977) The Cape floral kingdom: an ecological view. In: Proc. 2n Natn. Weeds Conf. S. Afr., pp. 19–33.
Taylor HC (1978) Capensis. In: Werger MJA (ed.) Biogeography and Ecology of Southern Africa, Junk, The Hague, pp. 171–229.
Warburg M (1977) Plant and animal species diversity along environmental gradients in a Mediterranean landscape of Israel. (B.S.F. No. 450). Animal species diversity. Technion, Isr. Inst. Technol., Haifa, Israel (Mimeographed).
Warburg MR, Ben-Horin A, Rankevich D (1978) Rodent species diversity in mesic and xeric habitats in the Mediterranean region of northern Israel. J. Arid Environ. 1:63–69.
Werger MJA (1972) Species-area relationship and plot size; with some examples from South African vegetation. Bothalia 10:583–594.
Werger MJA, Kruger FJ, Taylor HC (1972) Pflanzensoziologische Studie der Fynbosvegetation am Kap der Guten Hoffnung (Engl. summ.). Vegetatio 24:71–89.
Westman WE (1975) Pattern and diversity in swamp and dune vegetation, North Stradbroke Island. Aust. J. Bot. 23:339–354.
Whittaker RH (1972) Evolution and measurement of species diversity. Taxon 21:213–251.
Whittaker RH (1975) Communities and Ecosystems, 2nd edn. Macmillan, New York, 385 pp.
Whittaker RH (1977a) Evolution of species diversity in land communities. Evol. Biol. 10:1–67.
Whittaker RH (1977b) Animal effects on plant species diversity. In: Vegetation und Fauna, (ed.) R. Tüxen. Ber. Symp. Int. Ver. Vegetskunde, Rinteln, 1976. Cramer, Vaduz. pp. 409–425.
Whittaker RH, Naveh Z (1980) Analysis of two-phase patterns. In: Patil GP, Rosenzweig M (eds) Contemporary Quantitative Ecology and Related Ecometrics, Statistical Ecology Volume S12. Internat. Coop. Publ. House, Fairland, Md.:157–165.
Whittaker RH, Niering WA (1965) Vegetation of the Santa Catalina Mountains, Arizona. (II). A gradient analysis of the south slope. Ecology 46:429–452.
Whittaker RH, Niering WA (1975) Vegetation of the Santa Catalina Mountains, Arizona. V. Biomass, production, and diversity along the elevation gradient. Ecology 56:771–790.
Whittaker RH, Niering WA, Crisp MD (1979) Structure, pattern, and diversity of a mallee community in New South Wales. Vegetatio 39:65–76.
Willson MF (1974) Avian community organization and habitat structure. Ecology 55:1017–1029.
Zeevalking HJ, Fresco LFM (1977) Rabbit grazing and species diversity in a dune area. Vegetatio 35:193–196.
Zinderen Bakker Sr., EM van (1976) The evolution of Late-Quaternary palaeoclimates of southern Africa. Palaeoecology of Africa 9:160–202.
Zohary M (1962) Plant Life of Palestine: Israel and Jordan. Ronald Press, New York, 262 pp.
Zohary M, Feinbrun N., late Eig A. (1948) Analytical Flora of Palestine (Hebrew) Hamagdir, Jerusalem.

CHAPTER 3

NAVEH Z. (1990)

FIRE IN THE MEDITERRANEAN – A LANDSCAPE ECOLOGICAL PERSPECTIVE.

IN: GOLDAMMER, J.F., JENKINS M.J. (Eds.) FIRE IN ECOSYSTEMS DYNAMICS. PROCEEDINGS OF THE THIRD INTERNATIONAL SYMPOSIUM IN FREIBURG, FRG, MAY 1989. SPB ACADEMIC PUBLISHING BY, THE HAGUE, THE NETHERLANDS, PP. 1–20.

Fire and ash stimulated seedling of *Salvia triloba* seedling compared with control seedling on the right.

Abstract Due to the destructive combination with uncontrolled grazing and other land abuses, fire has been regarded in the Mediterranean as a wholly condemnable element which has to be prevented at all costs. But at the same time wildfires are becoming more devastating in spite of the great efforts and heavy expenditures invested in fire prevention and control. Its important role as a major driving force in the coevolution of the paleolithic man and Mediterranean landscapes has been overlooked and its potential use as a beneficial tool in vegetation management has not been recognized. Because of the irrational ban on fire, even for research purposes, ecologists and land managers are lagging far behind in the systematic study of its effects on the landscape, its biotic and abiotic components and functions.

It is timely and mandatory to study the effects of fire at all levels of the ecological and perceptional hierarchy and to learn how to employ it in a judicious way as a cheap and efficient tool for the prevention of destructive wildfires and the conservation of the biological diversity and attractiveness of our open landscapes in fire-induced plant communities, ecotones and corridors

Keywords Mediterranean landscapes, fire, grazing, fire adaptation, Pleistocene fires, landscape holarchy

INTRODUCTION

The destructive combination of fire, uncontrolled grazing, and other abuses has affected Mediterranean uplands for many centuries. Fire, in general, has been treated in an irrational and non-scientific manner as a curse to be prevented at all costs. Judged, like goat grazing, only by its ill effects fire has been regarded by most Mediterranean ecologists as a negative factor degrading the 'sclerophyll forest and maquis climax' and being detrimental to conservation (Tomaselli 1977). Fire's important role as a selective force in the evolution of Mediterranean vegetation and landscapes and in the maintenance of their productivity and diversity has been completely overlooked.

In spite of the efforts and expenditures on fire suppression, wildfires in the Mediterranean are becoming more frequent and devastating. The situation in southern France, described by Robertson (1979), is typical of most Mediterranean countries. It is the result of the cessation of traditional defoliation practises of grazing, cutting, coppicing, and buring and subsequent fuel accumulation and high fire intensity. This has been coupled with the increased fire hazard from recreation and urban population pressures on these uplands. In many countries it has been aggravated by the planting of highly flammable pine and eucalyptus forests. Thus, in spite of one of the most advanced and best-organized fire prevention and suppression systems, in the Riviera's so-called 'red belt' a pine tree's probability of living 20 years before being hit by fire is only 60%. Local authorities who manage thousands of hectares of forest, can no longer afford to use expensive labor for clearing the dry understory, and prescribed burning is not yet practiced.

In addition to the traditional reasons for arson, forest fires open the protected 'green belt' to construction and development. In Spain it has become good business for land developers and paper manufacturers, to acquire the charred stems cheaply and replace them with fast growing eucalyptus trees. In 1979 at the Costa Brava near the tourist village of Llorat de Mar the pine forests of the nature park burned, killing 22 Spanish summer vacationers. This and other fires were apparently set in order to obtain the land cheaply for construction of hotels and bungalows.

At the same time, because of the total ban on fire, the use of controlled burning even for research purposes has been prohibited. In most Mediterranean countries, anyone considering the possibility that fire could be beneficial was in danger of being regarded as a pyromaniac and jeopardizing his professional career. It is therefore no wonder that research on the role of fire in Mediterranean ecosystems and landscapes has been sporadic because only wildfires could be used for this purpose. We are, therefore, lagging behind other Mediterranean climate regions and biomes.

The far-sighted ecologist and forester H. Biswell (1989) overcame much prejudice and fierce opposition and proved that fire can be turned from a cruel master into a good servant through prescribed burning for vegetation management.

Very comprehensive reviews on the effects of fire and humans on the vegetation in the Mediterranean have been provided by Le Houerou (1973) and Trabaud (1981). In my paper I will stress those aspects which are most relevant for the study and application of fire as a tool in integrated and dynamic landscape management and the conservation of Mediterranean uplands. We must approach it in an objective and rational way and study it without bias or prejudice using an integrative and holistic systems approach.

THE ROLE OF FIRE IN THE EVOLUTION OF MEDITERRANEAN VEGETATION AND LANDSCAPES

Full comprehension of the role of fire in Mediterranean uplands can be reached only by recognizing its significance in the evolution of natural, semi-natural and agro-pastoral landscapes and vegetation. The following is a summary of what has been discussed in detail elsewhere (Naveh 1974, 1984; Naveh and Kutiel 1989).

In addition to other, more or less catastrophic natural perturbations fire, caused by volcanic activity and lightning, acted as a strong selective force during the Pleistocene. This was probably most true during the drier interpluvials when the Mediterranean climate and vegetation patterns became established.

Those woody and herbaceous genotypes developing efficient evolutionary strategies or adaptive traits to avoid or survive fire had the best chances for selection. This is manifested in the Mediterranean chiefly by the positive feedback responses of increased physiological activity by fire-stimulated resprouting and germination. Most sclerophyllous shrubs and trees, as well as climbers, are *obligatory root resprouters*. Most chaemeaephytes, as well as perennial herbaceous plants are *facultative root respouters*, regenerating also by fire-simulated germination and growth. Mediterranean conifers, like *Pinus halepensis* and *P. brutia* do not resprout, but rely solely on proliferous seed germination from cones that burst open from the heat of the fire. They are *obligatory seed regenerators* and fire apparently provides the only opportunity for their natural regeneration under a dense maquis understory (Walter 1968).

Fire regeneration behaviour is closely linked with response to season and water availability. In contrast to the facultative root resprouters, which are *drought evaders*, the obligatory root resprouters are *drought-enduring* and summer active. They rely on deep and well-branched root systems, and begin to resprout by the middle of the summer immediately after the fire. Postfire browsing pressure of young, soft and lush suckers, shoots and leaves has probably also acted as a powerful selective agent. It favors those species and biotypes developing hard, thorny or distasteful leaves and twigs, and those with the highest vegetative regeneration capacities to overcome defoliation stresses of both fire and grazing. This may have also pre-adapted such species to further defoliation 'catastrophes' such as cutting and coppicing. Among the most outstanding examples of such evolutionary maximization of over-all survival potential are *Quercus coccifera* and its east Mediterranean vicariad *O. calliprinos*, dominating maquis and garrigues; and *Pistacia lentiscus* – the Eumediterranean 'mastic tree'. The latter has developed probably the most drought enduring ecotypes of any Mediterranean sclerophyll phanerophyte and responds in the first days after fire by intensive cambial activity from root tips and buds (Shavit 1978).

In most facultative root resprouters, positive feedback responses to fire by vigorous resprouting from fire-avoiding or resistant underground stem bases, bulbs, and corms, and by propagation from seeds, is coupled with morphological and physiological plasticity and aggressiveness in colonization of newly opened habitats.

In certain perennial grasses, an additional regeneration mechanism is provided by reactivation of intercalary meristem and axillary buds along the charred culms. Among these, *Piptatherum miliaceum* is the most prominent pyrophyte, increasing germination from seeds, heated to 90 °C. Its west Mediterranean ecological vicariant is *Brachypodium ramosum*, for which Trabaud (1973) found a 'regeneration index' of 100%.

In therophytes, adaptive responses to fire drought and grazing, involve reproductive and growth behaviour, such as early and prolific seed production, efficient dispersal mechanisms, seed dormancy, poloymorphism and trypanocarpy. The latter is a good example of the coupling of survival mechanisms against fire, grazing and drought. The development of hygroscopic awns, callous tips, and other torsion mechanisms, enable it to drill and bury the disseminules several centimeters deep in the soil or between rock cracks. The seeds thereby escape both fire and grazing and benefit from more favorable moisture conditions for germination. Among these grasses are *Hordeum spontaneum, Triticum dicoccoides* and *Avena sterilis*, the progenitors of domesticated barley, wheat and oats. In some Mediterranean grasses which have been studied in California and in Israel, it is combined with heat tolerance.

The role of fire in the Pleistocene was not restricted to the selection of fire avoiding or resistant genotypes and species, but also to the evolution of fire-induced and diversified Mediterranean landscapes. By creating favorable regeneration niches for germination of light demanding woody and herbaceous plants, including *Pinus halepensis*, fire has favoured not only the evolution of tenaceous maquis dominants, but also the spreading of the opportunistic and fast-growing subordinate dwarf-shrubs, hemicryptophytes, geophytes and therophytes. Many of these, especially the therophytes, evolved in the semi-arid woodland savannas and steppe grasslands, where more frequent, less intense fires played an important evolutionary role in the Mediterranean and elsewhere (Whittaker and Woodwell 1972).

From the middle Pleistocene on, the final stages of geological and biological evolution of Mediterranean landscapes coincided with the major stages of biological and cultural evolution of Mediterranean man. This included the acheulian *Homo erectus* the middle paleolithic Neanderthaloid, and the upper paleolithic mesolithic and neolithic *Homo sapiens*. We can consider these mutual physical, biological, and cultural processes of human development from hunting and food gathering, to intensive food collecting and food producing, and the conversion of natural landscapes and vegetation into semi-natural and agricultural landscapes as a co-evolutionary process. In these closely coupled feedback relationships the use of fire became a major driving force in a very early stage of intentional and unintentional human activities. The creation of gaps in the closed tree cover for food, fuel and habitation, the trampling of paths and digging for bulbs and earth animals and disposing of human and kitchen waste, created favorable conditions for herbaceous colonizers and ruderals which take advantage of the improved light, moisture and fertility regimes.

The results of our studies on postfire vegetation and soil dynamics suggest that such positive feedback couplings between wildfires and the ruderalic fire followers could have caused them to spread in space and time far beyond the natural and man-induced forest gaps and ecotones. The paleolithic gatherer-hunter in efforts to improve his diet of grasses, legumes, bulbs and tuberous plants, and to facilitate hunting and attract game, began to use fire not only as the first extrasomatal energy source for heating, cooking, tool making and socialization, but also as vegetation and landscape management tool. This was suggested by Sauer (1965), the eminent geographer, who regarded speech, tools and fire as the 'tripods of cultural evolution'. He also pointed out that volcanic activity in Italy, the eastern Mediterranean and the Rift Valley was the first source of fires of cultural importance for food collection and hunting.

The earliest archeological evidence of human use of fire in Israel is from such volcanic land in the Upper Jordan Rift Valley. Stekelis (1969) found burned fibia fractures together with stone tools made of basalt. Palynological samples of lacustrine sediments from this period and region show the predominance of sclerophyll phanerophytes, such as *Quercus* and *Pistacia* (Horowitz 1979).

Most Mediterranean paleoecologists underestimate the influence of fire and the paleolithic food gatherer during the Pleistocene. Tectonics have obliterated most archeological evidence of human activities from this period. No ash deposits of forest and brush fires and sparse floral remains have been detected in open in situ habitations. Even recent sophisticated flooding methods have not provided large samples of vegetal relics in shallow and eroded Mediterranean upland soils, especially in terra rossa in which preservation is very poor. In these specific climatic and edaphic conditions, most of the ash of forest and brush fires were washed away by the first heavy rains, and their remnants become intimately mixed with the thin upper layer of humus-rich terra rossa and rendzina soils (Fig. 1). Some of these slopes underwent severe geological erosion and morphotectonic upheavals, and even in caves most traces of hearths and

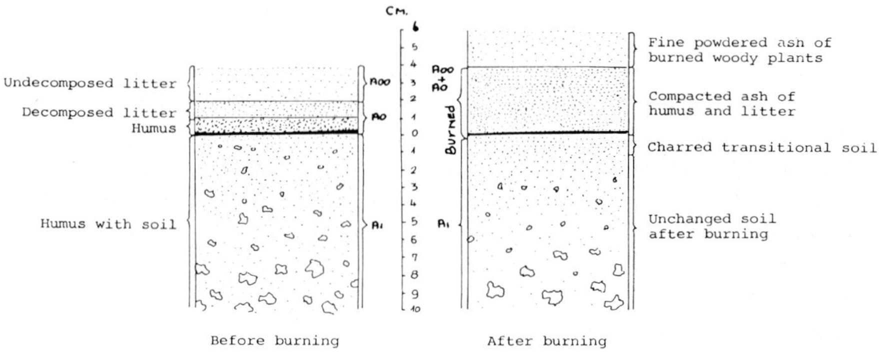

Fig. 1. Schematic representation of the maquis floor and soil before and after burning. Mazuba, Western Galilee. 1952/53.

fire have probably been erased by erosion and by changes in sea level, followed by sedimentation.

Nevertheless, there is sound archeological evidence from the Petralona limestone caves in northern Greece (Ikeya and Poulianos 1979) and the Escalle caves in southern France (Pfeiffer 1969) tracing the human use of fire in the Mediterranean back at least half a million years. These caves and the Tabun cave on Mt. Carmel in Israel are still surrounded by typical sclerophyll maquis. In the upper 70 000 year old Mousterian layers of the Tabun cave, not only reddened earth and mixed ashes from hearths were found, but also accumulations of fine ash deposits (Jelinek 1981). This ash could have been blown inside the cave from burning of the woody vegetation, surrounding it. According to Perles (1977) such Mousterian gatherer-hunters and Neanderthaloids, inhabiting the Tabun cave and adjacent ones of Mt. Carmel, mastered the use of fire and produced torches to carry and ignite fire.

During the last Pluvial, 10–15 000 years ago, the fire-induced proto-agricultural landscape modification was brought to its peak by prospering epipaleolithic cultures, such as the Natufian of Mt. Carmel. Bar Joseph (1984), described the carefully prepared, tiny sharp microliths used to hunt game, especially gazelles, been attracted to the freshly burned open forest and brush fire ‘pastures’. They used flint sickels to cut wild grasses, and mortars and pestles as pounding tools in the preparation of staple food from roasted cereals and acorns and developed a complex and rich communal, cultural and spiritual life. Their intensive, broad spectrum utilization of coastal and upland plant and animal resources had many common features with pre-European Indians of central and southern coastal California. These were the most advanced food-gathering and hunting Indians, living in comparable ecological conditions and using prescribed burning as a major management tool which maintained a dynamic flow equilibrium between the woody and herbaceous vegetation layers (Lewis 1973). Probably the Natufians also set cooler fires at shorter intervals than the natural ones and on smaller patches, creating mosaics of different regeneration stages, frequently burned fire corridors and grassy fields.

Fire also acted as an ecological and cultural trigger in the long process of the conversion of natural and semi-natural landscapes into agro-pastoral ones in the early Holocene. These final stages of co-evolution were initiated by agricultural domestication in the favorable ash seedbeds of forest and maquis burns and in the fire-swept drier woodlands and grasslands. The Jarmukian neolithic cultures in the Jordan Valley bear early evidence of such activity. The use of fire for clearing and cultivation of *Quercus pupescens* forests through slash-burn rotations is indicated by 9000 year old paleobotanical findings, reported by Pons and Quezel (1985), and was probably preceeded several thousand years earlier in the Levant.

The Bible, Talmud, and the classical Greco-Roman sources discuss the important role of natural and human-caused fire in Mediterranean lands in early historical times. Many Biblical citations suggest that damaging wildfires caused by lightning and other natural disturbances were abundant. Fire was also used intentionally for burning stubble fields, fertilizing, weed and pest control and for pasture improvement. This pastoral use of fire has continued throughout history until the present and together with grazing, cutting, terracing and patch cultivating, became an integral part of the multifactorial anthropogenic functions shaping the Mediterranean hill and mountain landscapes and vegetation patterns (Naveh and Dan 1973; Naveh and Lieberman 1984; Naveh 1982a).

FIRE AS PART OF MULTIFACTORIAL LANDSCAPE FUNCTIONS

The so-called 'primitive' cultures – like the Aborigines of Australia have different names for the fires used for vegetation management. We should distinguish between wildfires and prescribed fires, by the timing, frequency, intensity, size, vegetation, fuel, site conditions, metereological conditions during and after the fire, and the history of the burned site before and after the fire. All these factors and each set of specific ecological, pyric and biotic conditions must be considered in evaluating fire effects.

The following are a few examples of fire effects on soil and vegetation, showing how misleading generalizations about fire and fire effects can be. They all point to the need to consider fire as part of the multifactorial landscape functioning.

The Combined Effects of Fire and Grazing on Vegetation and Soil Stability

Fire in the Mediterranean region has been always judged in combination with grazing, because until very recently there were few burned sclerophyll forests, maquis, shrublands, woodlands and grasslands not grazed by cattle and/or goats before and after the fire. We were fortunate in our fire ecology studies in western Galilee to study fire effects for longer periods over larger and more diverse soil, topographic and vegetation types with different grazing histories before and after the fire (Naveh 1960, 1973, 1974).

Our comparisons of protected and grazed brush burns revealed that the greatest damage was inflicted to vegetation and soil by goat and cattle grazing in the first winter and spring after fire. They retarded the regeneration of obligatory sclerophyll root resprouters and especially the most palatable trees, such as *Pistacia palaestina, Ceratonia siliqua* and *Quercus calliprinus*. In these species most of the lower and external twigs were browsed down together with the inner young apical shoots, thus stunting growth and stimulating a shrublike growth habit. The young seedlings and regenerating perennial grasses, especially *Piptatherum miliaceum* and *Stipa bromoides* which were most conspicuous by their rapid spreading in protected sites, were also nibbled off. On the other hand, most chamephytes especially aromatic *Labiatae* and *Cistus* species were rejected. Because of their positive feedback response to fire these species can undergo population explosions with vigorous vegetative and reproductive development. As aggressive post-fire colonizers they become heavy competitors not only for herbaceous plants, but also for all browsed root resprouters.

Thus, selective grazing pressure, is not only deleterious to the temporary post-fire herbaceous dominance, but deflects the 'auto-succession' of the woody species in the direction of less palatable, aggressive shrublike pyrophytes. This happened over large areas in many Mediterranean countries which have been turning many sites into severely degraded, worthless scrubland.

In the case of *Pinus halepensis*, it is not the fire itself that prevents regeneration as claimed by Zohary (1962). It is the postfire grazing of the numerous small seedlings, emerging in the ashes, and the unhibited chamephyte competition. The grazing history before and after the fire can have also detrimental effects on soil stability. However, this is not always the case and we found great differences in this regard between the different upland soils we studied.

The poorer, highly calcareous pale rendzina soils over soft limestone and chalk, which were covered by a low shrub canopy of *Calycotome villosa* and *Sarcopoterium spinosa* and lacked a well developed profile and granular structure, showed indications of run-off and rill erosion after fire if they were grazed previously and had been compacted by trampling of cattle. However, the following winter when the slopes were covered by regenerating shrubs and grasses, no sign of erosion could be found, even along gullies, created in previous years.

On much more fertile, well developed, humus rich terra rosa and brown rendzina soils, covered by dense maquis and garrigue before the fire (Fig. 1), we did not observe any traces of water run-off, soil splashing, creation of rills, soil movement in the first winter following the hot fall fire even on slopes of 30–40%. Even at the time of rainfall intensities of 25 mm per hour, no appreciable soil movement was observed.

This can be explained by the fact that these soils contain more than 50% clay particles and about 12% organic matter in the upper 20 cm layer. They have an excellent granular structure and a high infiltration capacity and stability which are not impaired after burning. We found that even after losing about a fifth of the organic matter, 13–16% are still retained in the upper 4 cm A_1 profile. This has been incinerated by the hot fire together with the dry semi-decomposed and organic matter of the A_{oo} and A_o soil profiles and was converted into a compact ash layer, underlain by a dark, charred soil layer (Fig. 1).

No evidence of run-off or soil erosion after wildfires in dense maquis and mixed oak and pine forests could be observed in our more recent studies on Mt. Carmel. This does not mean, of course, that the absence of grazing before the fire will prevent the occurrence of erosion under different conditions in the Mediterranean or elsewhere.

Effects of Fire on Soil Fertility and on Herbaceous Plants

Another misleading generalization is that hot fires will deplete the soil of nutrients, especially N and P. It may be the case after very frequent brush fires. But our findings (Naveh 1960; Kutiel and Naveh 1987), and those of others in southern France (Trabaud 1983), and in California chaparral (Rundel 1983), demonstrate that the opposite is true. Hot fires mobilize nutrients tied up in the highly lignified wood and slowly decomposing litter and duff of the A_{oo} and A_o profiles. Thus in spite of the loss of total nitrogen, a striking increase in water soluble nutrients in the upper centimeters of the soil occurs in the first winter and spring after the fire. This rather short-termed postfire nutrient flush is utilized by the herbaceous invaders, for forage and seed production. The fire serves thereby as an important link in the recycling of nutrients to the soil for the regenerating, deeper rooted woody plants to use in the following years.

This was demonstrated by growing grasses and legumes in pots in the upper 2 cm layer of a brown rendzina, collected two months after a hot wildfire and from an adjacent unburned site of an old *Pinus halepensis* forest on Mt. Carmel (Kutiel and Naveh 1987). Wheat had 6 times higher phytomass and 12 times higher seed production in the burned soil (Table 1). Of significance for the enhanced nutrient cycling is the 4.5 fold increase in root production, facilitating the manifold increase in nutrient accumulation in the plants.

Following the promising results obtained in California (Love and Jones 1952), we utilized these mineral enriched, ash seedbeds for reseeding of perennial pasture grasses in

Table 1. Production and nutrient accumulation of wheat plants from burned and unburned pine forest of Mt. Carmel (Kutiel and Naveh 1987)

	Burned	Unburned	Ratio
Dry weight (g/m^2) shoots	1142	197	5.3
Dry weight (g/m^2) roots	709	158	4.5
Shoot:root ratio	2.2	1.2	1.8
Dry weight spikes (g/m^2)	512	39	13.0
No. of seeds m^{-2}	13740	1102	12.5
Seed weight (g/m^2)	128	11	11.6
Nutrient acumulation N	0.6	0.2	3.0
Nutrient acumulation P	0.2	0.02	10.2
Nutrient acumulation M_g	0.6	0.08	7.5
Nutrient acumulation H	0.2	0.04	5.0
Nutrient acumulation Ca	0.9	0.1	9.0
Nutrient acumulation Zn	0.02	0.005	4.0
Nutrient acumulation Fe	0.3	0.06	5

a semi-commercial experiment in the western Galilee (Naveh 1960). *Phalaris tuberosa* and *Piptatherum miliaceum* developed very rapidly and covered already 30–40% of the rocky slopes and shallow soils in the second year after the burn. After one year of protection, the slopes were grazed by beef cattle and provided an average of 317 grazing days and 1580 Scandinavian Feed Units per ha in the following three years (one Feed Unit has the caloric value of 1 kg barley or of approximately 1.6 mg metabolizable energy). This compared favorably with the highest pasture production achieved from fertilized dry land or open woodland pastures (Naveh 1982b).

Of great importance for the conservation of biological diversity and attractiveness is the striking increase of flowering geophytes, like *Bellevalia, Ornithogallum, Narcisus, Cyclamen* (Loeb 1960). We also observed it in the western Galilee, and on Mt. Carmel. Among these are many heliophytic orchid species, such as *Serapias vomeracea, Orchis papilionacens* and *Ophris sintenissi* found before the fire only in open and well-lighted niches, if at all.

But neither the post-fire mineral flush nor the grass proliferation should be taken for granted. Nutrient levels are dependent not only on initial soil properties and the prefire vegetation, but also on fire temperatures. The rate and extent of postfire colonization by herbaceous plants is also highly stochastic. It is determined by the occurrence of the few perennial grasses and herbs that survived in the dense tree and shrub cover as shade tolerant relicts, the availability of seeds from these and other invading herbaceous plants and the climatic conditions prevailing in the first and second rainy seasons after the fire.

Important seed sources are small, grassy patch openings and edge habitats, as well as waste heaps near human habitations. The most prolific post-fire invader from such a waste heap on Mt. Carmel was *Hordeum spontaneum*.

An example of annual species, recorded after a hot Maquis burn in the western Galilee is provided in Table 2. These include so-called opportunistic 'anthracopohytes' and 'nitratphytes' (Le Houerou 1973), ruderalic plants, and other species, like *Lotus peregrinus* and *Anagallis coerulea* which seem to be especially responsive to such fires. These herbaceous fire followers are gradually crowded out by the regenerating woody plants and after a few years the latter regain their prefire dominance and complete fire-induced auto-succession. Our studies in Israel and France showed, that after 20–30 years such undisturbed, dense and monotonous stands of tall shrubs and pine tree thickets contain only 20–30 species with 5% or less herbaceous species per 1000 m^2. But at the same time, their small grassy patches and gaps can contain up to 100, chiefly annual species (Naveh and Whittaker 1979 and unpublished data).

This reduction is related to the amount of light reaching the lower plant strata under such dense pine or oak cover as was shown by Woodcock (1986) on Mt. Carmel.

Comparisons of Burned and Clear-cut Maquis Regeneration

The paucity of grasses and herbs, even under lower brush cover of chameaphytes cannot be explained in all cases by lack of light, space and poor soil fertility. It is probably connected with the presence of heat non-stable germination inhibitors, such

Table 2. Abundance of annual plants in the first and second spring after maquis burn (Mazuba, western Galilee, Israel) (Naveh 1974)

	Relative abundance	
	1953	1954
Grasses		
Aegilops peregrina	1	1
Alopecurus myosuroides	1	1
Avena barbata		1
Avena sterilis	2	4
Bromus madritensis	1	1
Bromus scoparius	1	1
Bryza maxima	–	2
Cynosurus coloratus	–	2
Lamarckia aurea	1	1
Lolium rigidum	–	1
Phalaris minor	–	1
Scleropoa rigida	–	1
Stipa tortilis	1	1
Legumes		
Hippocrepis unisiliquosa	–	1
Lathyrus aphaca	–	1
Lotus peregrinus*	1	2
Medicago polymorpha	–	1
Medicago orbicularis	–	1
Medicago tuberculata	1	1
Onobrychis crista-galli	1	2
Ononis natrix	–	1
Scorpiurus muricata	–	1
Trifolium campestre	1	3
Trifolium clypeatum	–	1
Trifolium cherleri	–	1
Trifolium eriosphaerum	1	1
Trifolium purpureum	–	2
Trifolium resupinatum	–	2
Trifolium stellatum	–	1
Trifolium argutum	–	2
Vicia angustifolia	1	1
Compositeae		
Anthemis spp.	1	2
Chrysanthemum coronarium	1	1
Carthamus tenuis	1	2
Cichorium pumilum	1	2
Crepis palaestina	1	3
Centaurea cyanoides	1	1
Rhagadiolus stellatus	1	2
Geropogon globrum	–	1
Silybum marianum	–	1
Diverse Families		
Anagallis coerulea*	1	1
Caucalis tenella	–	1
Cephalaria joppica*	2	3
Chaetosciadium trichospermum	1	1
Chenopodium ambrosioides	1	1
Crucianella macrostachys	1	1
Convulvulus spp.	1	1
Erodium malacoides	1	1
Eryngium creticum*	1	1
Geranium purpureum	1	1
Galium articulatum	1	2
Lagoecia cuminoides	1	3
Lavatera punclata	1	1
Plantago lanceolata	1	1
Plantago psyllium	1	1
Scabiosa prolifera	–	1
Scandicium stellatum	1	1
Scandix pecten-veneris	1	1
Sinapis alba	–	1
Specularia falcata	1	1

1 = rare; 2 = occasional; 3 = abundant; 4 = very abundant; * = sometimes very abundant after burn.

as tannins and etheric oils, in the undecomposed leaf litter and duff, destroyed by the fire. A comparison of the establishment of seeded perennial grasses after cutting or burning the brush cover revealed that in the ash seedbed, 2–5 plants per square meter – mostly *Piptatherum milaceum* – emerged. None of this species and only a

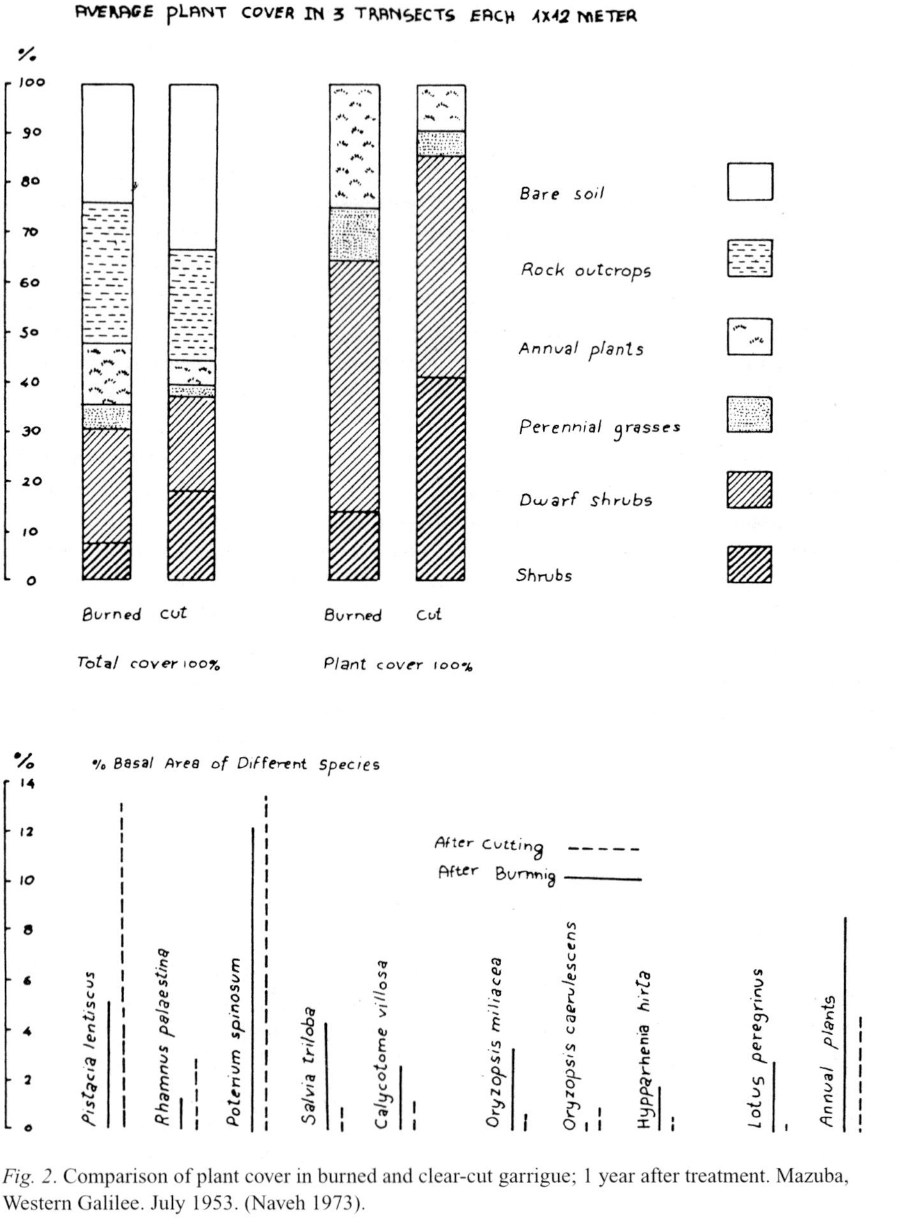

Fig. 2. Comparison of plant cover in burned and clear-cut garrigue; 1 year after treatment. Mazuba, Western Galilee. July 1953. (Naveh 1973).

few other grasses in the clear-cut plots emerged. In burned, but unreseeded plots there were more volunteering herbaceous plants as well as facultative resprouters with fire stimulated germination, such as *Salvia triloba* and *Calycotome villosa*. But the regeneration of *Pistacia lentiscus* was much slower (Fig. 2).

However, much more extensive, long-term field studies, accompanied by laboratory tests are necessary to reach conclusive answers to these important questions.

A HIERARCHICAL VIEW OF FIRE IN THE MEDITERRANEAN

One of the most complex problems in studying natural phenomena is the fact that our field of vision is conditioned by the perceptional window through which we view these phenomena; or in the terms of systems theory, by the spatio-temporal, intensity and complexity scales of the hierarchical systems levels to which these phenomena belong. At the lower physico-chemical levels of this hierarchy we deal with simple systems of very large numbers, undergoing very rapid changes. But at the biological and ecological levels we deal with a wide range of systems, differing greatly in all these properties. Their complexity increases from the single organism and its physical environment to the most complex ecosystem, namely that of the biosphere. Their spatial scales increase from the microsite habitat to the global landscape, and the temporal scales from seconds and minutes to thousands and millions of years of evolution. Together with these, change the rates and intensities of the process driving these systems. Therefore we cannot blindly infer from one level of the ecological hierarchy to the other in our evaluation of fire effects and have to judge them according to the hierarchical window through which they are viewed. Thus, if we look only at a single plant system immediately after the fire in a pine forest or maquis, we will percieve it as a catastrophic event, which destroyed all above-ground parts of the perennial root regenerating plants and killed all annual plants and obligatory seed regenerating perennials, such as *Pinus halepensis*. But if we open the window further and further we will realize that for this species and many others in these plant communities fire may be the only way to ensure their natural rejuvenation. For their forest and maquis ecosystems it may have been the only or the most efficient way to ensure their long-term biological productivity, diversity and nutrient cycling, and for these landscape as a whole, to retain their spatio-temporal heterogeneity, scenic attractiveness and dynamic flow equilibrium in space and time.

But these effects may change also from one time scale to the other. If the interval between two fires is longer than the reproduction and/or regeneration capacity of the burned plants, then the fire can lead to its extinction. In the case of *P. halepensis* this may be a hundred years, but in the case of geophytes in the forest under-story only a couple of years. The same is true also for too short fire frequencies which will not allow the plant sufficient time for reproduction and/or regeneration.

The spatial scale is interrelated with the temporal scale. The more narrow the ratio between the area occupied by this species in the landscape and the burned area, the poorer will be its prospects for survival. But if the species are distributed over a wide range of habitats in a heterogeneous landscape, burning part of this system will not endanger these, even if their regeneration in the burned part is not ensured or may be very slow. Such a burn, controlled in space and time, may, be necessary to ensure the survival of those light requiring species, such as the above-mentioned geophytes, which are smothered out by the dense forest or shrub thicket. It is therefore obvious, that in order to ensure highest attainable plant species diversity, as well as animal species diversity with the help of fire, its management must be highly flexible in space and time, and adopted to the requirements and fire responses of the 'target' species at which management is aimed, and the micro- and macro site heterogeneity of the landscape as a whole.

It should be realized also that there are great differences in our perception of the same ecological reality according to our personal conceptions, beliefs, attitudes and interests. Different people will look through different coloured glasses of the same window: looking at a burned forest or shrubland, the pastoralist's window glasses will let through only the green colours of the increased forage he expects to obtain for his livestock. But the nature lover will see only the black colours of the charred tree stumps and ashed soil and the forester, the tons of timber he lost – or the additional trees he will able to plant in the land cleared by the fire and for the arsonists, these glasses will have a perverse distortion of hostility or greed and easy profits.

However, the landscape ecologist will remove all these coloured glasses and view fire like any other natural or human caused disturbance factor along the continuum of closely interwoven physico-chemical, biological, ecological, historical and socio-economical processes shaping our landscapes. Instead of fragmentary hierarchical and perceptional views, the landscape ecologist will attempt to apply an integrative and multidimensional approach. By this he will view each hierarchical systems level as a 'holon' sensu Koestler (1969): As a *whole* towards its lower levels, but as a *part* of the higher level. Thereby he will be able to overcome the dichotomy between exclusively reductionistic or holistic approaches, merring his view. At the same time he will judge fire within the broader context of integrated land use planning and management and evaluate fire, its hazards and benefits on base of its impacts on all tangible and intangible, or so-called 'hard' and 'soft' landscape functions and values.

As explained elsewhere in detail (Naveh 1979; Naveh and Lieberman 1984), we have attempted this in sensitivity models of multi-purpose upland management strategies. These were aimed at maximizing *bio-ecological landscape functions* as last refuges for organic variety and as 'life-supporting systems' and buffering zones for protection of watersheds and environmental pollution control and reduction of fire hazards; their *socio-ecological, aesthetic and psychohygienic landscape functions* for recreation, and their *socio-economic functions* for plant, livestock, and water production and tourism.

DISCUSSION AND CONCLUSIONS

Contrary to earlier assumptions, there seems now little doubt that natural fires, operating as a major force in the biological evolution of Mediterranean biota, became also a vehicle of cultural evolution of Mediterranean man and his first cultural semi-natural landscapes, which in turn changed these into agro-pastoral landscapes. In these fire also played an important role, which became, however through the combination of fire with grazing and land abuses in the last centuries, mostly destructive. It is now up to the fire specialist, the landscape ecologist and the enlightened land-manager and user to show how its judicious use can turn it again from a curse into a blessing.

From the short review of the results of some of our studies on fire effects on the landscape it became apparent that in their evaluation we must be careful not to jump to too far-reaching conclusions. This is true not only for above-mentioned generalizations based on superficial observations, but also for quantified and unbiased experiments which are carried out in a restricted area and for the short time period of a Master's or Ph.D. thesis. Their systematic, long-term study must take into consideration not only the above-mentioned fire factors but also the great landscape heterogeneity of Mediterranean uplands and the physical, biological, ecological and cultural forces interacting with them. For this purpose fire effects must be treated as multivariate function. By using the functional-factorial approach of Jenny (1961), and distinguishing between controlling landscape state factors and their depending variables, we can apply these fire functions on all levels of the landscape holon hierarchy or 'holarchy':

$$L_{s,v,a,ha\ldots} = f\,F_{t\ldots}\,(H, P, R, C, O, \ldots T) \qquad (1)$$

In this equation $L_{s,v,a,ha\ldots}$ are the dependent post-fire variables of the landscape holon, such as soil, vegetation, animals, human-made artifacts, and other unspecified variables – as a function of the fire (F) and its parameters, such as temperature (t), season etc., and the controlling state factors of the higher holon level, such as human land use (H), soil parent material (P), relief (R) and the flux potentials of climate (C) and organisms (O), other unspecified state factors (. . .) and the time (T) since the fire, during which this function is measured. It will not be easy to gain total control not only of the fire, but also of all other disturbance factors, including grazing of wild and domesticated animal and human uses. But this multivariate equation will help us to realize those factors which play a major role in the postfire landscape functions.

Not all of these parameters can be quantified. For some we are still lacking sufficient quantitative data and others are not quantifiable in the conventional scientific way. However, recent developments in artificial intelligence and knowledge engineering have opened innovative methods to deal with such parameters.

In knowledge engineering human knowledge is integrated in computer systems as state-driven knowledge, telling how the data of a problem can be manipulated to solve it. Most relevant for fire and land-use related knowledge are *expert systems*, and *fuzzy set systems*. As explained by Negotia (1985) in a very lucid way, both can be combined. In expert systems we can pose and answer questions relating to information borrowed from human experts and stored in the systems knowledge base. However this knowledge is usually derived in linguistic and not quantitative form from these experts. The theory of fuzzy sets offers a way to incorporate subjective evaluations in such knowledge bases. As linguistic variables are not numbers but words, fuzzy set theory can be used as a tool of precisely manipulable natural language expressions. This is possible because this method assumes that an element need not be simply true or false as in an ordinary set with a binary choice, but may be *partly true* to any degree. It has therefore a degree of membership in a fuzzy set function as a real number in the interval from 0 to 1. Thus, f.i., if we do not know the exact age of the maquis since the last fire, we can describe it in imprecise terms, such as 'old': The degree of membership of an age of – let us say – 50 years or greater is 1.0 and that of a 'partially' old one of 30 is 0.7. In this way the vagueness of the term 'old' can be captured mathematically and dealt with in algorithmic fashion. The same can be done also with 'hot' or 'cold' fires and any other parameter.

But of special relevance for decision making in the choice of fire and fuel management options is the use of fuzzy linear programming of verbal models for the automatic transfer of knowledge from such linguistic values for *optimization* of the above-mentioned landscape functions and values, as derived from the fire and fuel management.

As we have seen above, the long-lasting prehistoric and historic human and fire impacts and land uses have modified the original mountain and hill landscapes into very heterogeneous and attractive mosaics of semi-natural and agro-pastoral ecosystems. Their vegetation patterns are maintained in a metastable and dynamic flow equilibrium or '*homeorhesis*' – *sensu* Waddington (1975) of different regeneration and degradation stages.

This striking floristic and structural diversity was achieved by a combination of the long evolutionary coexistence of many herbaceous drought, fire and grazing tolerant plants in heterogeneous microsites and open, rocky grass patches and forest and maquis gaps, and the long history of chronic perturbations by periodic defoliation stresses and climatic fluctuations. In such *perturbation-dependent* landscapes it would be futile to attempt the restoration of the self-stabilizing processes by simply stopping all human interventions and fire. On the contrary, we have to conserve and to reestablish their homeorhetic flow equilibrium by continuing or simulating all ecological processes and defoliation pressures, to which they have been adapted throughout their long cultural history (Ricklefs *et al.* 1984).

In the context of vegetation dynamics this means the prevention of competitive reduction and exclusion, the establishment of a new competitive equilibrium (Huston 1979; Pickett 1980), and the insurance of regeneration niches (Grubb 1977) through controlled grazing, cutting and burning.

In the context of ecosystem and landscape ecology this has even more far-reaching theoretical and practical implications, and as discussed elsewhere in detail (Naveh 1987; Naveh and Lieberman 1984), can be explained with the help of information theory and non-equilibrium, irreversible thermodynamics.

In natural ecosystems, high-quality potential and chemical energy (and therefore low entropy producing energy) is derived from solar energy by photosynthetic and other biological production functions. Part of this energy is dissipated into low quality metabolic heat and respiration, and therefore neg-entropy – as a measure of organizational order and information – is built up in the landscape by structural and spatial heterogeneity, high species diversity and complexity in food chains and webs. Simultaneously entropy production – as a measure of homogeneity, and disorder – is also minimized by the protection and stabilization function of the 'living sponge' of vegetation cover. These reduce the rate of kinetic energy and heat flows and their destructive and destabilizing impacts on the landscape. Thereby, in such undisturbed and mature ecosystems a steady state is presumably reached in which high negentropic order and information is maintained through homeostatic self-stabilization. But the thermodynamic behavior and homeorhetic regulation of our Mediterranean sclerophyll forests, maquis, shrublands and woodlands is very different. As perturbation-dependent, metastable and therefore non-equilibrium systems they are behaving like '*dissipative structures*'. These, according to Prigogine (1976), create 'order through fluctuation' and 'order out of chaos' (Prigogine and Stenger 1984). They are maintained and stabilized only by permanent energy/matter and entropy exchange with the environment. Thereby the system maintains its inner nonequilibrium and this, in turn, maintains the exchange process. Driven by positive feedbacks of environmental and internal fluctuations, dissipative structures move to new regimes which generate the conditions of renewal of higher entropy production.

In our case, such new regimes are apparently created by the periodical perturbations of fire, grazing cutting. But this will happen only, if sufficient time has been allowed for the regeneration phase and thereby also for the import of neg-entropy through intensified photosynthetic growth processes. At the same time the system can actually use free energy to reorganize itself with increasing structural complexity, biological diversity and productivity. But if these perturbation cycles are too frequent and severe, then the external entropy exchange may become more and more positive and disorder will remain at a high level. The same is true also, if these perturbations are stopped altogether – as in total nature protection or abandonment. In this case neg-entropy and information rise in the early regeneration phase, but with the lack of further perturbations the rates of entropy production again increase and disorder becomes more and more positive.

This is expressed by the monotony and low structural, floristic and faunistic diversity of undisturbed sclerophyll forest and maquis thickets and by their high inflammability. As opposed to above-described mature ecosystems, these are *aging* and become more and more stagnant and senescent.

This can be illustrated with the help of Prigogine's dissipative function, in which entropy (s) and therefore also disorder (D) grow at the rate ds/dt:

$$D = ds/dt \tag{2}$$

D may be positive, negative or zero. If it is zero then the system is in a stationary state – as in the homeostatic 'climax' state of natural systems. If it is positive ($D > 0$), then it is in a state of progressive disorganization and conversely the rate of neg-entropy and information (info) decreases and it looses its capacity for self stabilization and organization, as in the state of too frequent or no perturbations:

$$D > 0 = d\ info/dt < 0 \tag{3}$$

But if D is negative (D 0), then the system is in a state of progressive organization and increases its neg-entropy and information, as in the case of optimum perturbations, when

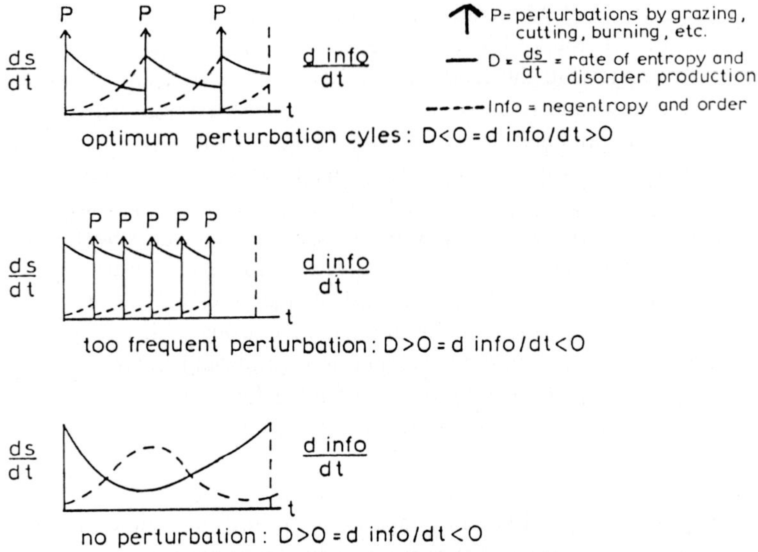

Fig. 3. Mediterranean landscapes as perturbation dependent dissipative structures.

the homeorhetic metastability and therefore the capacity of constant self-organization and stabilization can be maintained:

$$D < 0 = d\ \text{info}/dt > 0 \tag{4}$$

Figure 3 illustrates the results of these three different perturbation regimes in Mediterranean ecosystems and landscapes. The determination of such 'optimum' perturbation regimes for the conservation and enhancement of biological diversity in nature reserves and protected areas requires systematic long-term studies on landscape scales. In the open landscapes this must be combined with other land use goals, such as increase of economic production, recreation amenities and scenic values. As will be shown below, this can be achieved by integrating dynamic conservation management with multiple benefit land use. These will have to provide the answer if and when controlled burning, both for the reduction of fuel and prevention of destructive wildfires and for dynamic conservation management can be replaced by chemical or mechanical means or by grazing.

The results of our studies indicate that fire may be vital for the direct stimulation of germination, growth and regeneration of many species, and especially geophytes and hemicryptophytes, for the removal of heat unstable phytotoxic agents accumulating in the unburned litter and duff and for efficient nutrient cycling.

These contentions for the need of a change from fire suppression to dynamic fire ecology as part of wildlife conservation were supported by many of the studies, reported in the important symposium on fire and fuel management in Mediterranean ecosystems (Mooney and Conrad 1977) and especially by those of Walter (1977) in the Mediterranean, on the indirect beneficial effects of controlled burning on bird populations. These stressed the potentials of prescribed burning rotations in time and space to encourage a mosaic of different stages of postfire regeneration, stimulating the natural ones in 'fire generated' (pyrogeneous) habitats in California and the Mediterranean. But highest diversity and productivity seem to be ensured only by certain optimal frequency and intensity of burning, as indicated by the thermodynamic behaviour of such dissipative structures, described above.

On base of these principles and with the help of fire analysis models and matrices for regional land use planning (Derman and Naveh 1978), a masterplan was prepared by Derman (1985) of fire and fuel management for the conversion and restoration of low valuable and degraded maquis and batha (dwarfshrub) and dense pine forests of the Lower Carmel in Israel into productive, diverse and attractive park-forests with multi-beneficial values. However, its realization will require further research and development of management methods, in addition to the preparation of professional land managers, capable not only to suppress destructive wildfires, but also to apply beneficial fire and fuel management methods as part of integrated landscape management and conservation (Derman and Naveh 1987).

The main object of this and other projects, hopefully undertaken in the Mediterranean Region, should be to use fire and fuel management as one of the tools for a reconciliation between the needs for conservation and reconstitution of our open landscapes and the needs of their populations.

REFERENCES

Bar Joseph O (1984) Near East. In: Neue Forschungen zur Altsteinzeit, pp 232–298.

Biswell H (1989) Prescribed burning in California wildland vegetation management. University of California Press, Berkeley, California. (in press)

Derman A (1985) Ecological aspects of wildfires in regional planning. M.Sc. thesis, Technion, Haifa. (Hebrew with English summary).

Derman A, Naveh Z (1978) Fire analysis in land-use planning in the Mediterranean region. J. Environm. Mgt. 7:1–8.

Derman A, Naveh Z (1987) Fire and fuel management for multipurpose utilization and conservation of Mediterranean uplands. In: Dubinski Z, Steinberger Y (eds), Environmental Quality and Ecosystem Stability, Vol. III A/B. Bar-Ilan University Press, Ramat Gan, Israel, pp 421–427.

Grubb PJ (1977) The maintenance of species richness in plant communities and the importance of the regeneration niche. Biol. Rev. 52:107–145.

Horowitz A (1979) The Quaternary of Israel. Academic Press, New York.

Huston M (1979) A general hypothesis of species diversity. Amer. Nat. 113:81–101.

Ikeya M, Poulianos AN (1979) ESR Age of the trace fire at Petralona. Anthropos 6:44–47.

Jelinek AJ (1981) Middle Paleolithic of the Tabun Cave. In: Préhistoire du Levant CNRS, Paris, pp 265–283.

Jenny H (1961) Derivation of state factor equations of soils and ecosystems. Soil Sci. Proc., pp 385–388.

Koestler A (1969) Beyond atomism and holism – the concept of the holon. In: Koestler A, Smithies JR (eds), Beyond Reductionism: New Perspectives in the Life Sciences, Hutchinson of London, pp 192–216.

Kutiel P, Naveh Z (1987) The effect of fire on nutrients in a pine forest soil. Plant and Soil 104: 269–274.

Lewis HT (1973) Pattern of Indian burning in California: ecology and ethnohistory. Ballena Press Anthropological Papers 1. Ramona, California.

LeHouerou HN (1973) Fire and vegetation in the Mediterranean Basin. In: Proc. 13th Ann. Tall Timbers Fire Ecol. Conf., Tallahassee, Florida, pp 237–277.

Loeb H (1960) Regeneration of maquis shrubland after burning. In: The Western Galilee, Tel Aviv (Hebrew), pp 271–277.
Love RM, Jones BJ (1952) Improving California brush ranges, Univ. Calif. Circ., 371.
Mooney HA, Conrad CE (eds.) (1977) Proc. Symp. Environmental Consequences of Fire and Fuel Management in Mediterranean Ecosystems, August 1977. Palo Alto, California, USDA FS Gen, Tech. Rep. WO-3, Washington, D.C.
Naveh Z (1960) Agro-ecological aspects of brush range improvement in the maquis belt of Israel. Ph.D. Thesis, Hebrew University Jerusalem. (Hebrew with English summary).
Naveh Z (1973) The ecology of fire in Israel. In: Proc. 13th Ann. Tall Timbers Fire Ecol. Conf., Tallahassee, Florida, pp 133–170.
Naveh Z (1974) Effects of fire in the Mediterranean region. In: Kozlowski TT, Ahlgren CE (eds), Fire and Ecosystems, Academic Press, New York, pp 401–434.
Naveh Z (1975) The evolutionary significance of fire in the Mediterranean region. Vegetatio 9:199–206.
Naveh Z (1979) A model of multiple-purpose ecosystem management of degraded Mediterranean uplands. Environm. Mgt. 2:31–37.
Naveh Z (1982a) Mediterranean landscape evolution and degradation as multivariate bio-functions – theoretical and practical implications. Landscape Planning 9:125–146.
Naveh Z (1982b) The dependence of the productivity of a semi-arid Mediterranean hill pasture ecosystem on climatical fluctuations. Agriculture and Environment 7:47–61.
Naveh Z (1984) The vegetation of the Carmel and Nahal Sefunim and the evolution of the cultural landscape. In: The Sefunim Prehistoric Sites Mount Carmel Israel. A. Ronen BAR Intern, Ser. 2310, Oxford, pp 23–63.
Naveh Z (1987) Biocybernetic and thermodynamic perspectives of landscape functions and land use patterns. Landscape Ecol. 1:75–83.
Naveh Z, Dan J (1973) The human degradation of Mediterranean landscapes in Israel. In: di Castri F, Mooney HA (eds), Mediterranean-type Ecosystems, Origin and Structure, Springer Verlag, Heidelberg, Berlin, New York, pp 370–390.
Naveh Z, Whittaker RH (1979) Structural and floristic diversity of shrublands and woodlands in Northern Israel and other Mediterranean areas. Vegetatio 51:171–190.
Naveh Z, Lieberman AS (1984) Landscape ecology – Theory and applications. Springer Verlag, New York, Heidelberg, Berlin, Tokyo.
Naveh Z, Kutiel P (1989) Changes in the Mediterranean vegetation of Israel in response to human habitation and land uses. In: Woodwell GM (ed.), The Earth in Transition. Patterns and Processes in Biotic Impoverishment. Cambridge Press, New York. (in press)
Negotia CV (1985) Expert systems and fuzzy systems. The Benjamin/Cummings Publishing Company, Menlo Park California.
Perles C (1977) Préhistoire du feu. Masson, Paris.
Pfeiffer JE (1969) The emergence of man. Harper & Row Publisher, New York.
Pickett STA (1980) Non-equilibrium coexistence of plants. Bull. Torrey Bot. Club 107:23–248.
Pons A, Quezel M (1985) The history of the flora and vegetation and past and present human disturbance in the Mediterranean. In: Gommez-Campo C (ed.), Conservation of Mediterranean Plants, Dr. W. Junk Publishers, The Hague, pp 25–43.
Prigogine I (1976) Order through fluctuation: Self-organization and social systems. In: Jantsch E, Waddington CW (eds), Evolution and Consciousness: Human Systems in Transition, Addison-Wesley, Reading, Massachusetts, pp 93–130.
Prigogine I, Stengers I (1984) Order and chaos. Bantam Books, New York.
Ricklefs RR, Naveh Z, Turner RE (1984) Conservation of ecological processes. Commission on Ecology Papers No. 8. IUCN Gland, Switzerland.
Robertson IMS (1977) Land use planning of the French Mediterranean region. In: Proc. Symp. on Environmental Consequences on Fire and Fuel Management in Mediterranean Ecosystems, August 1977, Palo Alto, California, USDA FS Gen. Tech. Rep. WO-3, Washington, D.C., pp 293–295.
Rundel PW (1983) Impact of fire on nutrient cycles in Mediterranean-type ecosystems with reference to chaparral. In: Kruger FJ, Mitchell DT, Jarvis JUM (eds), Mediterranean-type Ecosystems: The Role of Nutrients, Springer Verlag, Berlin, pp 192–207.

Sauer CO (1961) Sedentary and mobile bents in early societies. In: Washburn SL (ed.), Social Life of Early Man. Viking Fund. Publ. in Anthropology 31:256–266.
Shavit I (1978) Aut-ecological studies of *Pistacia lentiscus* in Israel. Ph Sc. thesis, Faculty of Agricultural Engineering, Technion, Israel Institute of Technology. Haifa, Israel. (Hebrew with English summary).
Stekelis M (1960) The paleolithic deposits of Jisr Banat Yaqub. Bull. Res. Conc. Israel, Sect. G. Geo-sciences 9:346–367.
Tomaselli R (1977) Degradation of the Mediterranean maquis. In: Mediterranean Forests and Maquis: Ecology, Conservation and Management. MAB Tech. Note 2, UNESCO, Paris, pp 33–72.
Trabaud L (1973) Experimental study on the effects of prescribed burning on a *Quercus coccifera L.* garrigue: early results. Proc. 13th Tall Timbers Fire Ecology Conf. Tallahassee, Florida.
Trabaud L (1981) Man and fire impacts on Mediterranean vegetation. In: di Castri F, Goodall FW, Specht RL (eds), Ecosystems of the World 11-Mediterranean-Type Shrublands. Elsevier Scientific Publishing Company, Amsterdam, Oxford, New York.
Waddington CH (1975) A catastrophe theory of evolution. In: The Evolution of an Evolutionist, Cornell University Press, Ithaca, New York, pp 253–266.
Walter H (1968) Die Vegetation der Erde, Bd. 2: Die Gemaessigten and Arktischen Zonen. G. Fischer, Jena.
Walter H (1977) Effects of fire on wildlife communities. In: Proc. Symp. Environmental Consequences of Fire and Fuel Management in Mediterranean Ecosystems, August 1977, Palo Alto, California, USDA FS Gen. Tech. Rep. WO-3, Washington, D.C., pp 183–192.
Whittaker RH, Woodwell GM (1972) Evolution of natural communities. In: Wiens JA (ed.), Ecosystem Structure and Function. Ann. Biol. Collog, Oregon State Univ. 31:137–159.
Woodcock J (1987) The light regime in the understory of pine and oak forests on Mt. Carmel. MSc. Thesis, Faculty of Agricultural Engineering, Technion, Israel Institute of Technology, Haifa, Israel.
Zohary D (1962) Plant life in Palestine. Ronald Press, New York.

CHAPTER 4

NAVEH. Z (1994)

FROM BIODIVERSITY TO ECODIVERSITY: A LANDSCAPE-ECOLOGY APPROACH TO CONSERVATION AND RESTORATION. RESTORATION ECOLOGY 2: 180–189.

Reclamation of abandoned Nesher limestone quarry with shrubs, planted in micro terraces of soil pockets, hold by wire fence. On top of slope, a mixture of plastic emulsion and grass seedling is sprayed for early soil protection. (Photo, courtesy Judy & Kenny Tel Aviv).

Abstract An environmental revolution is urgently needed that will lead to a post-industrial symbiosis between man and nature. This can be realized only if the present unrestrained biological impoverishment and neotechnological landscape degradation are replaced by the creation of healthy and attractive landscapes. Restorationists can fulfill a vital role in this process. They must broaden their scales from biodiversity restoration in small, protected nature islands to the large-scale restoration of natural and cultural landscapes. To achieve this they must restore not only the patterns of vegetation but also the processes that create these patterns, including human land uses. Their goal should be to restore the total biological, ecological, and cultural landscape diversity, or "ecodiversity," and its intrinsic and instrumental values of highly valuable, endangered seminatural, agricultural and rural landscapes. For this purpose it is essential to maintain and restore the dynamic flow equilibrium between biodiversity, ecological, and cultural landscape heterogeneity, as influenced by human land uses, which occur at different spatial and temporal scales and intensities. Recent advances in landscape ecology should be utilized for broader assessment of ecodiversity, including proposed indices of ecodiversity, new techniques such as Intelligent Geographical Information Systems (IGIS), and Green Books for the holistic conservation and restoration of valuable endangered landscapes. Restoration ecology can make an important contribution to an urgently needed environmental revolution. This revolution should lead to a new symbiosis between man and nature by broadening the goal of vegetation restoration to ecological and cultural landscape restoration, and thereby to total landscape ecodiversity

ECODIVERSITY CONSERVATION AND RESTORATION AS PART OF THE POST-INDUSTRIAL ENVIRONMENTAL REVOLUTION

The most alarming immediate global threat to life on earth is that of biological impoverishment caused by an accelerating downward spiral of biological diversity. This spiral, in turn, is driven by accelerating rates of loss of species and their habitats (Norton 1987). Myers (1979) claimed that these exponential species extinction rates have increased dramatically in the last 50 000 years, from one extinction per 1000 years to about 1000 extinctions per year and may reach 40 000 per year until the end of this century, so that one species will be lost every hour.

The threats to nature and its life-support systems are even graver because the human population is already destroying a great part of all terrestrial photosynthetic production. This destruction includes crop wastes, forest burning and clearing, desertification, and agricultural and urban-industrial land conversion—both of which are adversely affecting much more of our life-support systems through air, soil, and water pollution. We can assume that more species will become more susceptible to extinction and their prospects for survival will be diminished further because of the degradation and fragmentation of their habitats in natural and human-modified landscapes. These are caused by the great variety of intentional and unintentional effects of modern agricultural, urban-industrial, recreational, and other land uses. In the long range, global warming and other adverse atmospheric changes will further aggravate these threats and thereby accelerate these exponential extinction rates.

This process can be called neotechnological landscape degradation (Naveh & Lieberman 1994). At its root is the accelerating expansion of the technosphere that is replacing the biosphere and leading to the creation of a human-controlled, impoverished, and unstable ecosphere. This process is driven by mutually reinforcing, destabilizing feedback loops among unrestrained growth of human populations, energy and material consumption, and technological power. These run-away feedbacks have replaced most of the restraining and regulating natural and cultural negative feedbacks. As a result, larger areas of diverse natural and cultural landscape are converted within shorter time periods into vulnerable monotonous agricultural steppes and manmade technological deserts. Thereby, our open landscapes are losing not only their biological richness and ecological stability, but also their cultural wealth and scenic beauty.

As the forces driving these run-away feedbacks are culturally ingrained, our global environmental crisis is mainly a cultural crisis. Therefore, its remedies should be sought not only in the scientific, technical, socioeconomic, and political spheres, but also in the sphere of spiritual and ethical values and living norms. The appreciation of nature and its intrinsic value should not merely be viewed as a source for our material satisfaction, but as a source of enlightenment, inspiration, and enjoyment.

In their recent challenging book "Beyond the Limits," Meadows et al. (1992) claimed rightly that we need a far-reaching environmental and cultural revolution to lead us to shift from consumption to conservation and from unrestrained quantitative growth to lasting qualitative improvement and development.

One of the most encouraging developments is the recognition of the urgent need for such a global environmental revolution and the feasibility of its realization. This can no longer be ridiculed as a utopian dream of some "radical environmentalists" or so-called "deep ecologists." It is now shared by some of the most prominent, down-to-earth economists, managers, and decision makers, including the present Vice-President of the United States (King & Schneider 1991; Gore 1992; Tolba 1992).

We have discussed this environmental revolution from the point of view of sustainable land use and management strategies (Naveh & Lieberman 1994). We consider it to be the third major wave of global change in human civilization after the agricultural and industrial revolutions. It should lead to a new post-industrial symbiosis between man and nature. Such a symbiosis will be realized if the unrestrained urban-industrial and wasteland expansion of the technosphere landscapes is replaced by their total integration with the natural and seminatural biosphere landscapes, thereby creating a viable and livable global ecosphere. This will require the conservation and creation of healthy and attractive landscapes with maximum attainable ecodiversity. They should provide ecological and socioeconomic services and combine "soft," intangible values such as health and aesthetics of intrinsic non-economic richness with "hard," marketable values.

Restorationists can fulfill a vital role in this process if they are able to broaden their spatial and conceptual scales from the restoration of small, degraded nature islands to the restoration of large, open landscapes with their natural and cultural patterns and processes. For this purpose, biodiversity will have to be reconsidered within the context of ecological and cultural diversity and heterogeneity of total landscape ecodiversity.

BIODIVERSITY AND ECOLOGICAL HETEROGENEITY

Biodiversity is determined at the species level by the number of species of "species richness," and by the relative abundance or cover of each species in a specific site. Whittaker (1965) suggested that diversity be divided into

(1) "alpha (species) diversity," the diversity within a specific habitat;
(2) "beta (habitat) diversity," the diversity between different habitats or cross-habitat diversity, measured as the gradient of change in diversity along different sites or communities; and
(3) "gamma (landscape) diversity," the total diversity of a whole geographically or ecologically defined region or landscape.

The last is the most important index for the determination of biodiversity because it incorporates both other measures and is affected by the ecological heterogeneity within and between different habitats.

In general, the chances for greater gamma diversity are better in more heterogeneous environments; this, in turn, can further increase ecological heterogeneity. In this way total diversity becomes self-augmenting in ecological time. Conversely, reduction of ecological heterogeneity reduces options for species diversity, and this in turn reduces ecological heterogeneity. Therefore, in both directions biological diversity and ecological heterogeneity are coupled by positive, mutually reinforcing feedback loops (Norton 1987).

Our studies in the Mediterranean (Naveh & Dan 1973; Naveh & Whittaker 1979; Naveh 1984; Naveh & Kutiel 1990) have shown that ecological heterogeneity is closely related to long-term land-use practices, which in turn increase macro- and microsite heterogeneity and thereby also biodiversity. This principle of the relationship of land use and heterogeneity also applies to most other human-modified landscapes that have not yet suffered severely from misuse or that are presently suffering from neotechnological landscape degradation.

Because of the long history of co-evolutionary processes and the close interactions between natural and cultural forces of human land use, the relationship between land use and landscape heterogeneity plus diversity is most pronounced in the Mediterranean region. These interactions resulted in the creation of attractive and diverse seminatural and agro-silvo-pastoral landscapes in the mountaineous regions. The flatlands, in contrast, have long been cultivated and have lost both their natural vegetation and their heterogeneity.

The southern Judean Hills of Israel can serve as a good example of this close interaction between natural and cultural processes in the formation of such heterogeneous landscape patterns. Here, mosaics of terraces and patch-cultivated fields and plantations are closely interwoven with the remnants of natural vegetation and their dynamic regeneration and degradation stages on nontillable slopes and rock-outcrop patches. We used the ancient, abandoned, and disintegrating terraces and their stone-wall remnants as our smallest landscape units or ecotopes in our first landscape ecological studies (Naveh and Dan 1973). These cultural artifacts can serve as an important focus for emergent landscape qualities resulting from increasing landscape complexity. They can therefore provide information on ecological heterogeneity. With their help, we could identify spatiotemporal soil and vegetation aggradation and degradation patterns on the southern slope. These patterns are shown in Fig. 1 as an ordination in vegetation cover and composition, from low-demand grasses and dwarf shrubs along the erosion pavement to sclerophyll shrubs occupying the richerniches along the stone-wall remnants, and the colluvial soil pockets formed by eroded terraces of the highly calcareous, pale rendzina soil.

By broadening the spatiotemporal scale as well as the geological, archeological, and historical dimensions, we realized that we dealt in fact with two entirely different landscape patterns (Figs. 3 and 4). The southern slope, presented in Fig. 2, has been terraced for cultivation in the distant past by cutting the soft limestone into a series of flat and leveled terraces protected by stone walls. The north-facing slope, on the other hand, shown in the front of Fig. 3, was untillable and has never been cultivated. Here, the sclerophyll maquis shrubland, dominated by *Quercus calliprinos* (stone oak) and *Pistacia lentiscus* (lentisc of mastic tree) overlies a shallow and rocky brown rendzina, which was formed on a hard, ancient calcrete crust called Nari. This prevented terracing and cultivation. Soil depth and structure indicate that this nontillable, brush-covered slope has not undergone any severe erosive changes and, presumably, will remain so under continued periodic burning, cutting,

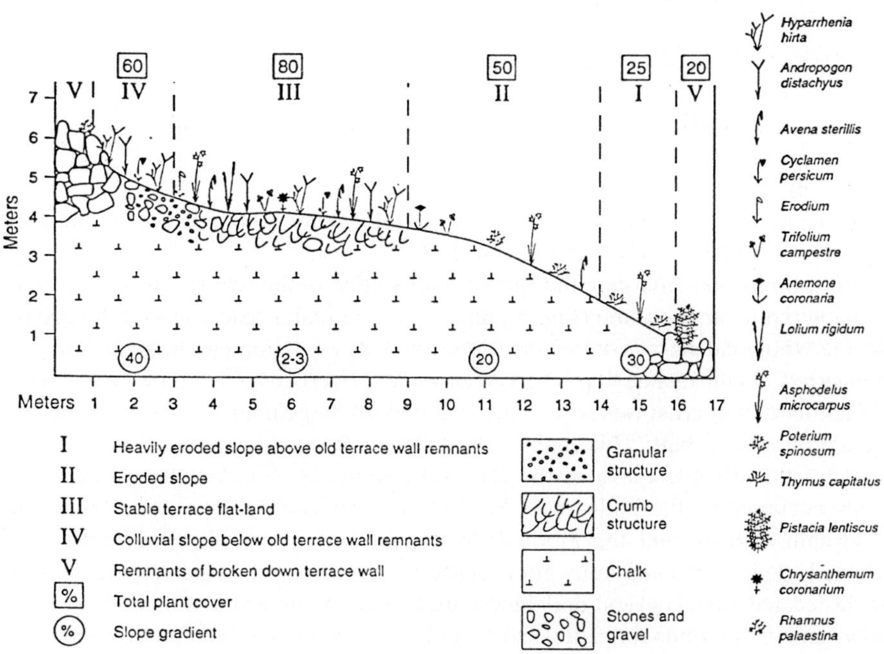

Fig. 1. Soil and vegetation degradation pattern on ecotope of eroded terrace slope of Pale Rendzina soil on chalk in Southern Judean Hills (Naveh & Dan 1973).

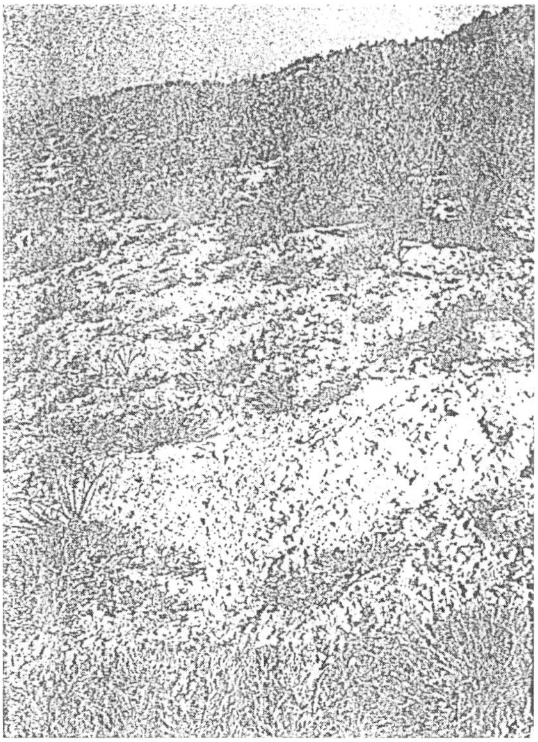

Fig. 2. Ecotope of eroded terrace slope on Pale Rendzina soil in southern Judean Hills.

and moderate browsing and grazing as long as the woody canopy is not uprooted by hand tools or bulldozers (Naveh and Dan 1973).

Finally, by further enlarging the spatial scale to that of the regional landscape system, captured by an air photo (Fig. 5), and with time scales reaching into the distant past, the whole catena of contrasting slopes and its evolution can be revealed. The more narrow, south-facing slopes served as ancient riverbeds in the Late Pleistocene, and their hard Nari crust was apparently stripped off by geological erosion, exposing the underlying soft chalk. This enabled the Judean tribes to terrace these slopes in the Iron Age and to settle down in the forested mountains, because the valleys were already occupied by the Philistines. After the downfall of the Byzantine Empire in the seventh century and the gradual decline of agriculture, these terraces were neglected; their continued cultivation caused the well-known catastrophic erosion that desiccated these agricultural landscapes. After their abandonment they were used as grazing grounds for cattle, sheep, and goats, and in the last hundred centuries typical Mediterranean metastable landscape patterns evolved, which we have studied in the last 20 years.

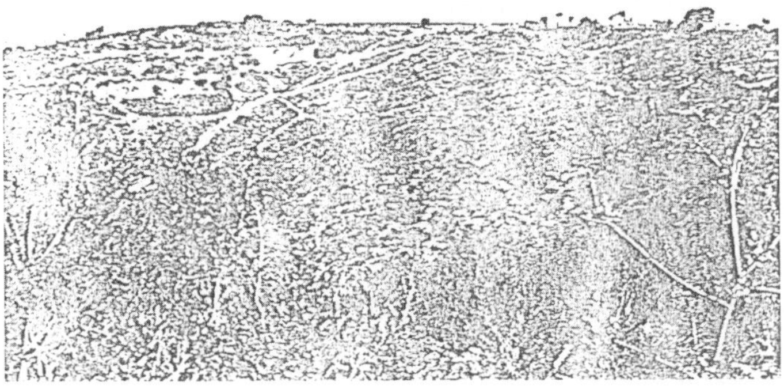

Fig. 3. Two distinct landscape units on contrasting slopes and soils: the southern slope with eroded terrace ecotopes on Pale Rendzina (see also Figs. 1 and 2) and in foreground the northern slope with sclerophyll maquis shrub ecotopes on Brown Rendzina soil overlayed by hard Nari calcerete rock outcrops.

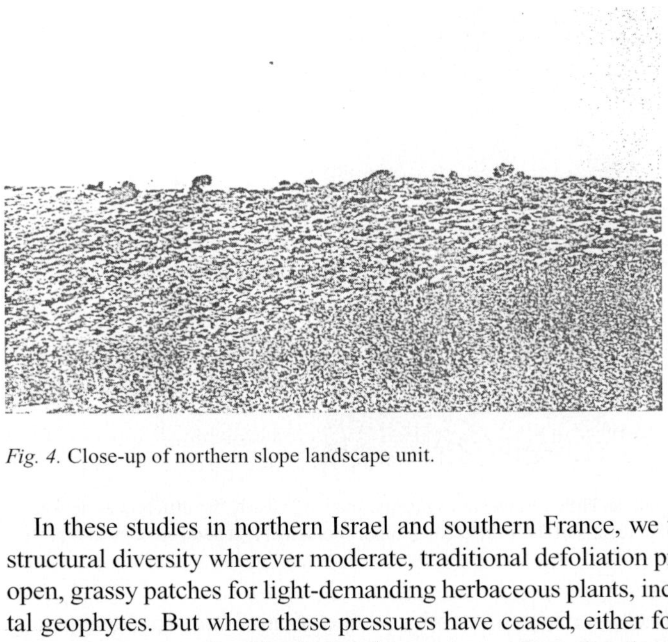

Fig. 4. Close-up of northern slope landscape unit.

In these studies in northern Israel and southern France, we found high floristic and structural diversity wherever moderate, traditional defoliation pressures had maintained open, grassy patches for light-demanding herbaceous plants, including highly ornamental geophytes. But where these pressures have ceased, either for the sake of protecting nature or as a result of land abandonment, or where they have been intensified, both floristic and structural diversity have been reduced considerably. In most cases species richness dropped by 75%—from more than 120 species to less than 30 species per 1/10 hectare—equitability was much lower, and the dominance concentration of the most aggressive and shade-tolerant trees and shrubs increased significantly. At the same time, structural vegetation diversity was also reduced considerably. For instance, on

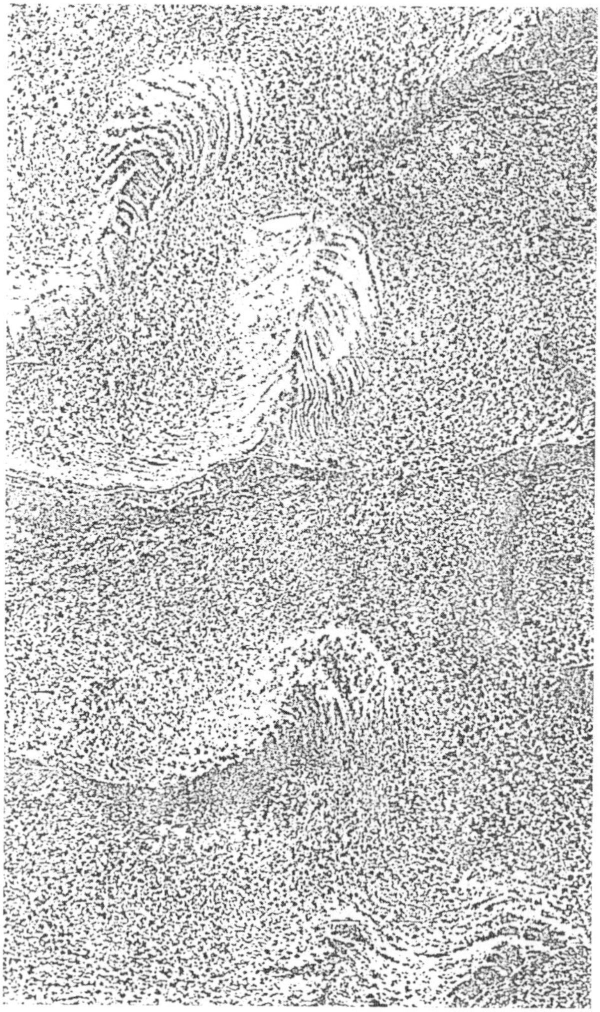

Fig. 5. Air photo of southern Judean Hills catena land systems, showing clearly the differences between the two major land units and the results of their different geomorphological and anthropological history.

Mt. Carmel, the rich herbaceous understory and gap vegetation of semi-open and patchy maquis shrubland contained 93 species, as opposed to only a few shade-tolerant perennial grasses and herbs that could survive under the dense, undisturbed tall-shrub canopy. In simultaneously conducted zoological studies, animal species richness and relative abundance of birds, reptiles, rodents, and isopodes showed similar trends (Warburg 1977; Warburg et al. 1978).

In open oak woodlands, the highest species richness and diversity were reached under moderate grazing pressure and lowest under both heavy and light grazing pressure or complete protection (Figs. 6 and 7). This is also true for California oak savannas (Naveh & Whittaker 1979). In all these complex microsites, a dynamic equilibrium is maintained between grazing intensity and the small-scale heterogeneity created by variable soil depth and microrelief, rock outcrops, tree cover, litter cover, shade, and abundance of different plant species and even ecotypes (Naveh 1991a).

Such rock outcrops and their shallow soil pockets and crevices can serve as refuges for many species and genotypes adapted to these microsite scales (Nevo et al. 1986). Many times these rock pockets are archeological remnants of ancient olive or wine presses or other cultural artifacts, increasing ecological heterogeneity and biological diversity.

Encouraged by the results of our studies and other field observations, the Nature Protection Authorities changed their conservation policy from noninterference to dynamic conservation management. A survey of all nature reserves of northern Israel, in an area of about 30 000 ha of maquis, woodlands, shrublands, and grasslands, confirmed our findings on a much larger scale (Noy-Meir & Kaplan 1991; Kaplan 1992). In these nature reserves the highest attainable landscape diversity was maintained mainly by controlled grazing management of domestic livestock and by

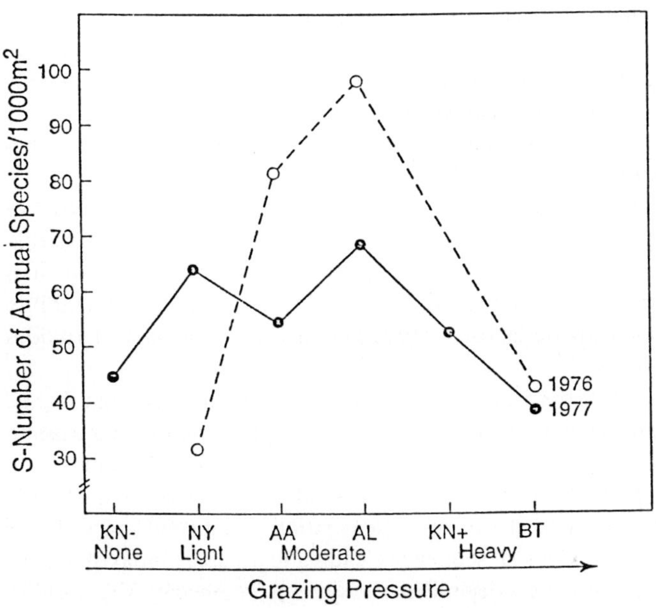

Fig. 6. Effect of grazing intensity on annual species richness of open woodland and shrubland in northern Israel. KN, Kfar Hanoar Hadati, 3 years protection; NY, Neve Yaar, very lightly grazed in 1976 and lightly grazed in 1977; AA, Allone Abba, moderately grazed; AL, Allonim, rotationally grazed; KN, heavily grazed; BT, Bosmat Tivon, very heavily grazed. Spring sampling shortly after grazing reduced S in AA and AL in 1977; more intensive grazing increased S in NY in 1977 (Naveh & Whittaker 1979).

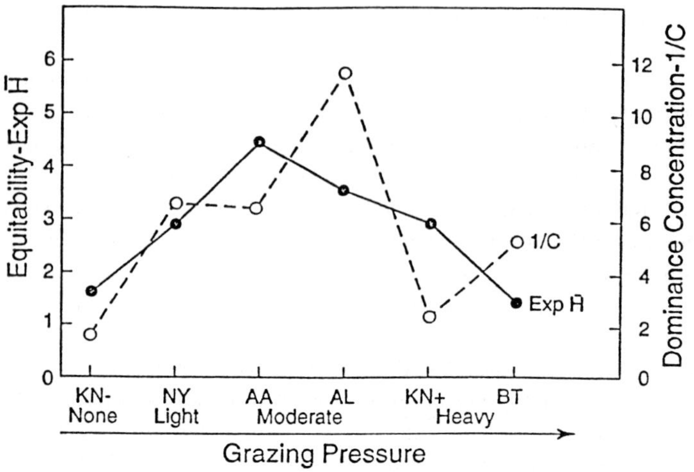

Fig. 7. Effect of grazing intensity on reciprocal Simpson dominance concentration (1/C) index and on Shannon-Wiener equitability (expH') index of vascular plant species of open woodland and shrubland in northern Israel. Plots and grazing pressures as in Fig. 6.

free grazing of wild animals, chiefly mountain gazelles and boars. Light-to-moderate grazing almost always increased plant species richness and diversity. In wooded vegetation types the contribution of controlled goat grazing was most important. In protected and too-lightly grazed sites, species richness and diversity were not only lower, but the frequency of wildfires also was higher than in moderately and heavily grazed ones. Protected herbaceous plant communities, dominated chiefly by tall grasses, also suffered more from a periodic outburst of rodents.

In all other Mediterranean countries, the cessation of human agropastoral activities and defoliation processes is leading to lower structural and floristic diversity of herbaceous plants—including many endemics and ornamental geophytes—and to the loss of landscape heterogeneity. This is, in general, combined with increased vulnerability to fire (Horvat et al. 1974; Ruiz de la Torre 1985; Farina 1989; Gonzalez Bernaldes 1991; Vos & Stortelder 1992).

Throughout the closely interwoven environmental and cultural processes that shaped these landscapes and their vegetation for many thousands of years, the seminatural and agricultural vegetation of open forests, shrublands, woodlands, grasslands, and terraces has all become part of closely interwoven landscape mosaics with highly dynamic vegetation patterns of degradation and regeneration in time and space. These do not fit any deterministic successional sequences leading to a stable climax state. For such a situation the term "perturbation-dependent ecosystems," coined by Vogl (1980), seems to be most appropriate.

It can be assumed that in the traditional Mediterranean pastoral systems the great climatic seasonal and annual fluctuations induced fluctuations in productivity that acted as effective negative regulative feedbacks in preventing overgrazing, similar to the natural regulating of wildlife populations. At the same time, over-utting and over-burning were prevented by coppicing and burning rotations necessary to ensure sustained productivity and sufficient recovery. These regular grazing, burning, and coppicing regimes, lasting over centuries, were incorporated in the landscape on different spatiotemporal scales, according to the rates of regeneration of the different plants and vegetation types and their micro-and macrosites. They led to the establishment of a man-maintained balance between the tree, shrub, herb, and grass layers within and between the different sites in those forests, woodlands, and shrublands that were not overgrazed or overcoppiced or were completely rested for prolonged periods. Combined with the great macro- and microsite heterogeneity of the rocky and rough terrain, they produced the unique ecological and cultural landscape diversity of Mediterranean uplands.

For such a dynamic flow equilibrium, the eminent geneticist Waddington (1975) coined the term "homeorhesis" (from the Greek, meaning "preserving the flow"). In this, the system is not returned to a stationary state of homeostasis, as in traditional climax systems, but is going on to change in the same way as it has in the past, as long as similar perturbations are driving the changes. As has been shown above, in semi-natural Mediterranean landscapes these changes are induced by human disturbances.

Mediterranean landscapes should therefore be considered as nonequilibrium systems dependent on human perturbation that have acquired long-term adaptive resilience and metastability. In these, the maintenance of a homeorhethic flow process of well-controlled but constant interference of varying intensity and timing is crucial not only for biodiversity but also for ecological stability and cultural diversity and attractiveness.

The same is most probably true for all other seminatural and agro-and silvopastoral landscapes depending on human perturbation. Maximum total landscape ecodiversity with fine-grained spatial heterogeneity patterns and vital biological, ecological, cultural and socioeconomic functions and values will therefore be ensured by the maintenance of a dynamic homeorhetic flow equilibrium. This requires the perpetuation, simulation, and/or restoration of all natural and cultural patterns and processes, including well-controlled defoliation pressures by cutting, grazing, and burning rotations in time and space (Naveh 1991b; Naveh & Lieberman 1994).

In conclusion, in all these seminatural landscapes ecodiversity is the result of the interplay of three major landshaping forces, namely biology, ecology, and culture (Fig. 8). These factors are determined respectively by biodiversity, landscape heterogeneity, human land use, and their resulting perturbations.

It is therefore obvious that restoration of ecodiversity cannot be achieved simply re-establishing some native plants in the hope of restoring an illusionary homeostatic, stable climax stage, maintained without human disturbance. On the contrary, it is often a complex, ongoing process of optimizing multibeneficial values through dynamic landscape management. Regionally throughout the world it will require not only much research and field experimentation, but also great legislative, administrative, and educational efforts, and above all new integrative strategies, methods, and tools. These should be based on transdisciplinary

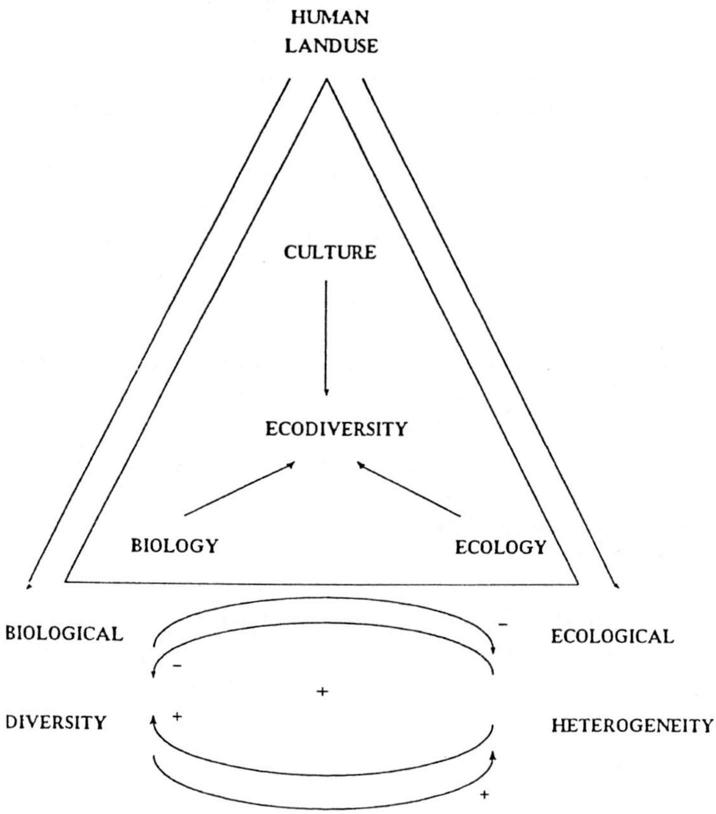

Fig. 8. The landscape-shaping forces of the biology-ecology-culture triangle determining ecodiversity. Biological diversity and ecological heterogeneity are coupled by mutually amplifying feedback loops. Both are effected by human land use.

concepts of landscape ecology and restoration ecology. For this purpose restorationists will have to combine ecological knowledge with ecological wisdom and ethics to learn from the past, comprehend the present, and envisage the future. Society must be persuaded about the need for restoration of intrinsic natural and cultural landscapes.

LANDSCAPE ECOLOGY AND ECODIVERSITY

In recent years landscape ecology has made great strides in the development of integrative methods. These enable us to identify, analyze, synthesize, and—at least partly—quantify in more holistic ways complex natural and cultural patterns and processes on different time-space scales. Of greatest importance among these methods is the utilization of recent advances in computer hardware and software, remote sensing and satellite images, Intelligent Geographic Information Systems—combining artificial intelligence with geographic information systems—and the application of innovative mathematical and system theories, such as information theory, fractal geometry, fuzzy logic, and hierarchy theory, in landscape ecological studies (Naveh & Lieberman 1954).

Landscape ecologists are now also paying much attention to spatial heterogeneity along different scales and its relation to biodiversity. For this purpose they have broadened the spatial scales of the diversity indices from the single-species level to the three-dimensional landscape level. Thus, for instance, Haber (1990) has adopted the Shanon-Wiener eveness diversity index to determine the "ecotope (type) diversity" in order to compare the variability between different ecotypes as affected by land use in Germany within one landscape unit. In Yellowstone Park, Romme (1982) described the "biotic landscape diversity" of different vegetation types as affected by landscape heterogeneity and fire history by transforming the Simpson index into a "landscape eveness" index and into a "landscape patchiness" index. This index was further developed into a "landscape contrast" index by Hoover and Parker (1991) for the determination of the spatial components of biotic landscape diversity in Georgia, as reflected by ecological heterogeneity.

In all these studies, however, vegetation was used as the sole parameter of diversity. For systematic measurement of the multidimensional landscape ecodiversity, these strictly biological parameters must be expanded to incorporate other relevant ecological and cultural parameters at different spatial and perceptional scales, such as micro- and macro-site heterogeneity in soil and topography, as well as cultural artifacts and land-use variables.

For this purpose, species diversity indices should be transformed into spatial, holistic ecodiversity indices:

(1) "alpha ecotope ecodiversity" of biotic, ecological, and cultural components within each ecotype;
(2) "beta ecotope ecodiversity" between ecotopes of each landscape unit;
(3) "gamma land-unit ecodiversity" within each regional land system; and eventually also
(4) "delta land-system ecodiversity" within each regional landscape.

Gamma ecodiversity incorporates an entire watershed, for instance, while delta ecodiversity can refer to larger-scale land systems such as a bioregion or a biome.

Taking further advantage of recent relevant methodological developments, these assessments should serve as guidelines for conservation and restoration strategies in each specific landscape.

As discussed elsewhere in more detail (Naveh 1994), landscape ecologists in western Europe already apply an ecodiversity approach to rural landscapes. In Denmark, for instance, a comprehensive landscape-ecological classification is used for the management of small, uncultivated biotopes (Brandt et al. 1992). This demonstrates the great diversity in biological, ecological, and cultural features of these "nature islands" and their potential for multibeneficial ecodiversity restoration (Fig. 9). For this purpose the appraisal of small biotopes could be

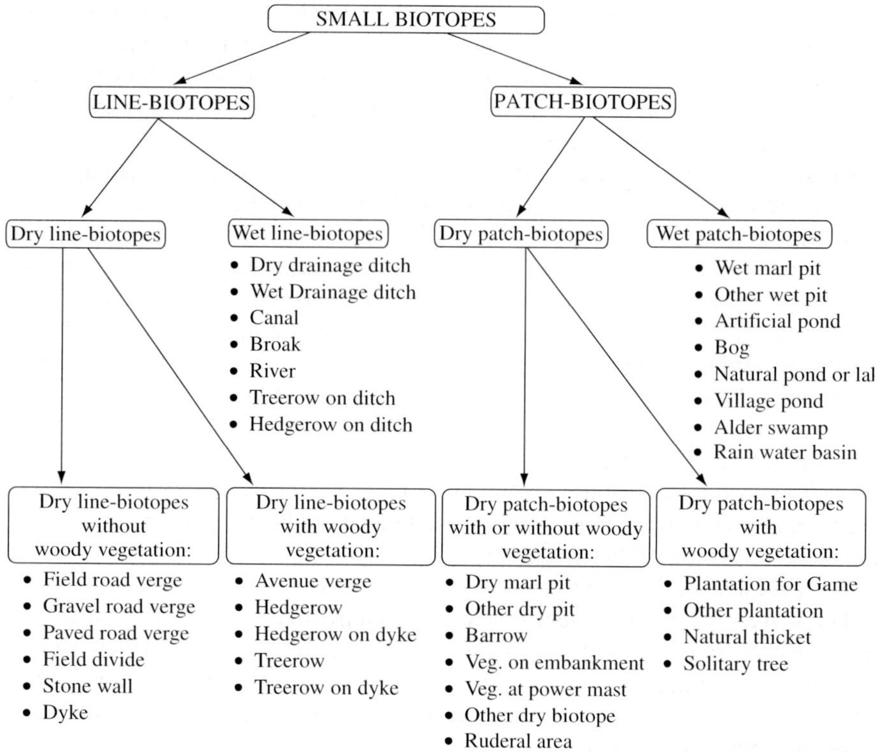

Fig. 9. A landscape-ecology-oriented classification of small biotopes used in the Danish small-biotope management system (Brandt et al. 1992).

complemented by an evaluation of the alpha ecodiversity of each biotope on the basis of its biodiversity and its overall amenity values. Beta ecodiversity could compare these values among different line and patch biotopes in dry and wet habitats, and gamma ecodiversity could take into consideration the number of the most valuable biotope types within each landscape unit to which highest conservation and restoration priority should be given.

TOWARD HOLISTIC LANDSCAPE CONSERVATION AND RESTORATION

The need for broadening biodiversity into ecodiversity has become even more urgent in view of present efforts to broaden conservation of habitat. Currently the emphasis is still on nature reserves, but it is shifting to the most valuable open, seminatural and cultural landscapes. This is clearly reflected in the resolutions of the seventeenth General Assembly of the International Union for the Conservation of Nature and Natural Resources (IUCN)—The World Conservation Union—in Costa Rica in

1988, and in those of the IVth World Congress for Protected Areas "Parks for Life" at Caracas in 1992, urging the establishment and effective management of protected landscapes, such as Biosphere Reserves and World Heritage Sites. The latter have been endorsed by the World Heritage Convention and are implemented by the World Heritage Centre of the United Nations Educational, Scientific, and Conservation Organization (UNESCO) on a world-wide basis. Among other important initiatives in this direction is the proposal for a European convention for the protection of rural landscapes.

For culturally important landscapes, a major role has been designated in the strategies for sustainable living in "Caring for the Earth" (International Union for Conservation of Nature and Natural Resources—The World Conservation Movement/World Wildlife Fund/United Nations Environmental Program 1991). According to Lucas (1992), who provided a useful guide for the planning, implementation, and management of protected landscapes, these should fulfill the following functions:

(1) conserving nature and biological diversity,
(2) buffering more strictly protected areas,
(3) conserving human history in structures and land-use practices,
(4) maintaining traditional ways of life,
(5) offering recreation and inspiration,
(6) demonstrating durable systems of use in harmony with nature.

The benefits that will accrue for those who live and work in and visit these landscapes and to the wider world beyond their boundaries will be determined by their overall ecodiversity. Ultimately, however, the success of such a holistic landscape management policy will depend on the creation of awareness of the value of these landscapes and their ecodiversity, concern about their future, and motivation for active involvement in their sustainable utilization, conservation, and restoration. Such a shift from prevailing one-sided instrumental and exploitative attitudes will be an important step toward the environmental revolution. This educational process should reach all those who care for these landscapes, those who live from them, and those who deal with them at all levels of the decision-making process. We also need much more efficient communication tools to bridge the gap between the academicians and the professionals, between the conservation- and restoration-minded ecologist and the production-minded foresters, agronomists, and economists, and between all these specialists and the public at large.

These communication tools should transform "semantic" scientific information (expressed in words and figures), into "pragmatic" information as actual conservation action. This information, combining confirmation with novelty, becomes meaningful through its feedback on the receiver when it leads to action. In other words, these tools should provide factual updated information on the present status of the landscapes and their ecodiversity, collected by integrated field surveys and remote sensing, dynamic Intelligent Geographic Information Systems, and other advanced landscape ecological methods (and thereby confirmation), and at the same time they should serve as tools for predicting and prescribing the fate of these landscapes—thereby introducing new information as "novelty," leading to their conservation and restoration (Naveh & Lieberman 1993).

For this purpose the Working Group of Landscape Conservation of the IUCN—The World Conservation Union—Commission on Environmental Strategies and Planning has proposed to this organization the preparation of world-wide Red Lists of Endangered Valuable Landscapes and of Green Books for their conservation and restoration. The latter should present in clear, nontechnical terms with ample maps and illustrations not only recent adverse changes endangering both natural and cultural assets and scenic and economic values but should also suggest alternative, sustainable land-use strategies based on holistic landscape planning and dynamic conservation management. By being not only descriptive but also anticipatory, they could help to change the attitudes of politicians and decision makers and provide practical guidance for holistic, sustainable, and multibeneficial land-use planning and management (Naveh 1993). The first of such Green Book case studies has been completed already in western Crete. This study revealed that plant species diversity is closely coupled with cultural diversity. For instance, rare and highly ornamental species that have no other habitats exist on ancient buildings and terrace walls. Detailed recommendations for sustainable land-use strategies show how the demands to safeguard "soft" biological and cultural values could be reconciled with controlled utilization of "hard" values that are vital for socioeconomic advancement of the local population (Grove et al. 1993).

In conclusion, I hope that these efforts for the conservation of total landscape ecodiversity will bear fruit and that they will be joined by restorationists as their major contribution to a new, post-industrial symbiosis between man and nature.

ACKNOWLEDGEMENTS

The author is grateful to Edith Allen and an anonymous reviewer for their comments, which helped to improve the manuscript.

REFERENCES

Brandt J, Holmes E, Larsen D (1992) Conceptual problems in connection with "applied monitoring" of a dynamic agricultural mosaic using detailed spatial landscape database. In Ecosystem classification for environmental policy and conservation. Proceedings of a Workshop, Leiden, The Netherlands, December 17–18, pp 1–17.

Farina A (1989) Recent changes of the mosaic patterns in a montane landscape (North Italy) and consequences on vertebrate fauna. In: Baudry J, Bunce RGH (eds). Land abandonment and its role in conservation. Proceedings of the Zaragossa/Spain Seminar December 10–12. Options Méditerraneénnes Series A, 15, pp 121–134.

Gonzales Bernaldes F (1991) Ecological consequences of the abandonement of traditional land use systems in central Spain. In: Baudry J, Bunce RGH (eds) Land abandonement and its role in conservation. Proceedings of the Zaragossa/Spain Seminar December 10–12. Options Méditerraneénnes Series A, 15, pp 23–29.

Gore A (1992) Earth in balance: forging a new common purpose. Earthwatch Publisher, London, England.

Grove AT, Ispikoudis J, Kazaklis A, Moody JA, Papanastasis V, Rackham O (1993) Threatened Mediterranean landscapes: West Crete. Final Report to European Community. Department of Geography, University of Cambridge, Cambridge, England.

Haber W (1990) Using landscape ecology in planning and management. In: Zonneveld IS, Forman RTT (eds) Changing landscapes: an ecological perspective. Springer-Verlag, New York.

Hoover SR, Parker AJ (1991) Spatial components of biotic diversity in landscapes of Georgia, USA. Landscape Ecology 5:125–136.

Horvat I, Glavac V, Ellenberg H (1974) Vegetation Suedosteuropas. Gustav Fischer Verlag, Suttgart, Germany.

International Union for Conservation of Nature and Natural Resources—The World Conservation Movement/World Wild-life Fund/United Nations Environmental Program 1991. Caring for the earth: a strategy for sustainable living. Gland, Switzerland.

Kaplan D (1992) Responses of Mediterranean grassland plants to gazelle grazing. In: Thanos CA, (ed.) Proceedings of the 6th International Conference on Mediterranean Climate Ecosystems, September 23–27, 1991. University of Athens, Athens, Greece, pp 75–79.

King A, Schneider B (Club of Rome) (1991) The first global revolution. From the problematique to the resolutique, a "blueprint for the twenty-first century" by the Council of the Club of Rome. Pantheon Books, New York.

Lucas PHC (1992) Protected landscapes: a guide for policy makers and planners. Chapman & Hall, London, England.

Meadows DH, Meadows DL, Randers J (1992) Beyond the limits: global collapse or a sustainable future? Earthscan Publications Limited, London, England.

Myers N (1979) The sinking ark. Pergamon Press, Oxford, England.

Naveh Z (1984) The vegetation of the Carmel and Nahal Sefunim and the evolution of the cultural landscape. In: Ronen A. The Sefunim prehistoric sites, Mount Carmel Israel. BAR International Series 2310, Oxford, England. pp 23–63.

Naveh Z (1991*a*) Biodiversity and ecological heterogeneity of Mediterranean uplands. Economia Montana—Linea Ecologica 24:47–60.

Naveh Z (1991*b*) Mediterranean uplands as anthropogenic perturbation dependent systems and their dynamic conservation management. In: Ravera OA (ed.) Terrestrial and aquatic ecosystems, perturbation and recovery. Ellis Horwood, New York, pp 544–556.

Naveh Z (1993) Introduction to the symposium on Redbooks for threatened Mediterranean landscapes. Landscape and Urban Planning 24:241–247.

Naveh Z (1994) Biodiversity and landscape management. In: Kim KC, Weaver RD (eds) Biodiversity and landscapes: a paradox of humanity. Cambridge University Press, New York.

Naveh Z, Dan J (1973) The human degradation of Mediterranean landscapes in Israel. In: di Castri F, Mooney HA (eds) Mediterranean-type ecosystems: origin and structure. Springer-Verlag, Heidelberg, Germany, pp 370–390.

Naveh Z, Kutiel P (1990) Changes in vegetation in the Mediterranean basin in response to human habitation. In: Woodwell G (ed.) The impoverishment of the biosphere. Cambridge University Press, New York, pp 259–300.

Naveh Z, Lieberman AS (1994) Landscape ecology theory and application. 2nd edition. Springer-Verlag, New York.

Naveh Z, Whittaker RH (1979) Structural and floristic diversity of shrublands and woodlands in northern Israel and other Mediterranean areas. Vegetatio 41:171–190.

Nevo E, Geiles A, Kaplan D, Golenberg EM, Olsvig-Whittaker L, Naveh Z (1986) Natural selection of allozyme polymorphism: a microsite test revealing ecological genetic differentiation in wild barley. Evolution 40:13–22.

Norton BG (1987) Why preserve natural variety. Princeton University Press, Princeton, New Jersey.

Noy-Meir E, Kaplan D (1991) The effect of grazing on the herbaceous Mediterranean vegetation and its implications on the management of nature reserves. Interim Report to the Nature Conservation Authorities, Jerusalem, Israel (In Hebrew).

Romme WH (1982) Fire and landscape diversity in subalpine forests of Yellowstone National Park. Ecological Monographs 52:199–221.

Ruiz de la Torre JR (1985) Conservation of plants within their native ecosystems: In: Gomez-Campo C (ed.) Plant conservation in the Mediterranean. Junk, The Hague, The Netherlands, pp 197–219.

Tolba M (1992) Saving our planet: challenges and hope. Chapman & Hall, London, England.
Vogl RJ (1980) The ecological factors that produce perturbation-dependent ecosystems. In: Cairns J Jr., (ed.) The recovery process in damaged ecosystems. Ann Arbor Science Publisher, Ann Arbor, Michigan, pp 63–94.
Vos S, Stortelder AHF (1992) Vanishing Tuscan landscapes: landscape ecology of a submediterranean-Montane area (Solano Basis), Tuscany, Italy. Pudoc Scientific Publishers, Wageningen, The Netherlands.
Waddington CH (1975) A catastrophe theory of evolution: the evolution of an evolutionist. Cornell University Press, Ithaca, New York.
Warburg M (1977) Plant and animal species diversity along environmental gradients in a Mediterranean landscape of Israel. Animal species diversity. Research Report 450. Binational Israel—American Science Foundation. Technion, Israel Institute of Technology, Haifa, Israel.
Warburg M, Ben-Horin RA, Rankevich D (1978) Rodent species diversity in mesic and xeric habitats in the Mediterranean region of northern Israel. Journal of Arid Environment 1:63–69.
Whittaker RH (1965) Dominance and diversity in land plant communities. Taxon 21(2/3):213–251.

CHAPTER 5

NAVEH Z. (1995)

CONSERVATION, RESTORATION, AND RESEARCH PRIORITIES FOR MEDITERRANEAN UPLANDS THREATENED BY GLOBAL CLIMATE CHANGE.

IN: MORENO J, OECHEL WE (Eds.) GLOBAL CHANGE AND MEDITERRANEAN–TYPE ECOSYSTEMS ECOLOGICAL STUDIES VOL. 117. SPRINGER, NEW YORK, PP. 482–508.

North-facing "European" slope with sub-humid flora and South-facing "African" slope with semi-arid flora in "Evolution canyon" Nahal Oren, Mt. Carmel. (Photo, courtesy I.S.C.A).

There seems to be an almost general consensus among climatologists that global warming will occur as a result of rising levels of CO_2 and other anthropogenic polyatomic gases. Atmospheric concentrations of these greenhouse gases expressed in CO_2 equivalence have grown from 280 ppm in preindustrial times to 365 ppm now. At the same time, the mean global surface layer temperature has increased by about 0.5 °C, and in this decade the six warmest years have been recorded. In addition, ice-core studies from Antarctica and Greenland showed that large natural fluctuations in CO_2 and methane were highly correlated with temperature changes over the last 160 000 years. These trends of climate warming have been identified in a number of comprehensive reports by international groups of prominent scientists as one of the most serious threats to the future of life on earth, e.g., Bolin et al. (1986), WCED (1987), Schneider (1989). IPCC (1990), IGBP (1990), and Jaeger and Ferguson (1991). Most of these studies assumed that a doubling of CO_2 could occur until 2050 if no countermeasures are taken. They attempted to predict its impact on the global climatic pattern with the help of General Circulation Models (GCMs) and estimated that consequently the average global surface air temperature will increase by about 2–4 °C. In the Northern Hemisphere the warming in higher latidutes is predicted to the about double that in low latitudes. These models also suggest a global increase in sea surface temperatures and an increase in the average global rate of evaporation and precipitation.

These findings paved the way for the International Convention on Climate Change at the UN Conference on Environment and Development (UNCED) at Rio de Janeiro in 1992. It was generally agreed that the threats of global climate change can only be prevented by the same modern industrialized society that caused them. However, at this Rio Earth Summit conference-as well as at many other occasions-it became apparent that one of the major excuses for not taking any immediate effective action to reduce the emissions of heat-trapping and polluting gases is the fact that there are still large uncertainties with regard to the magnitude and rate of climate change.

FACING THE UNCERTAINTIES OF CLIMATE CHANGES

The main difficulties in determining how changes in concentrations of these greenhouse gases will further affect the climate have been discussed recently by Graber and Cohen (1993). Most of these are related to the fact that the atmospheric warming is followed by other processes that may act as positive feedbacks to further increase temperatures, while others may act as negative feedbacks and inhibit this increase and stabilize it.

One of the main criticisms raised against GCMs are the coarse spatial and temporal resolutions on which their calculations are based. Their grids are too large for the simulation and evaluation of regional climatic changes in specific habitats, and their small-scale topographic and orographic landscape heterogencity and their time scales are unsuitable for predictions of changes in climatic patterns, including short-term variability and extreme meteorological events, not shown by monthly and annual averages. Outlining these and other limitations of regional scenarios in the European Conference on Landscape-Ecological Impact of Climatic Change (LICC), Hulme et al. (1990) concluded that presently it is impossible to predict its future climate. Models at their current state of development are incapable of producing reliable simulations at the regional scale, and empirical studies based on past instrumental data as analogues for the future cannot serve as predictions. They therefore regarded their scenarios as possible climate outcomes for Europe in a warmer world rather than as probable outcomes. For this reason, the "what if" approach was adopted in this conference, focusing on the climate sensitivity of key biotic and abiotic processes operating in terrestrial landscapes (Boer and De Groot, 1990).

However, even such an approach faces the limitations of uncertainty because of our present lack of knowledge of the interactions between these biotic and abiotic processes. According to Woodwell (1992), biospheric positive feedbacks may act to affect the rate of warming. Thus a 1 °C increase in temperature is widely recognized as increasing the rate of respiration by 10% to 30% while having little effect on photosynthesis and thereby stimulating the release of both CO_2 and methane, adding to the greenhouse effect. Biotic influences can be large enought to change the amount of CO_2 in the atmosphere by several percent within a few weeks. The destruction of the biota globally is not only contributing directly to biotic impoverishment but is also speeding the atmospheric warming of the earth. Therefore Woodwell suggests that the safest assumption is that warming will proceed rapidly over the next decades, especially in the higher and middle latitudes. Along with heating will also come increase in the frequency of destructive events such as fires, hurricanes, and drought. These increasing frequencies of destructive events, affecting biotic processes and patterns may be more important than temperature change itself in their threats to life.

The same may be true also for synergistic effects, increasing the risk factor introduced by the uncertainty in the responses of natural and cultivated vegetation to the synergistic interactions with other human-caused disturbances, such as acid rain, photochemical oxidants and other pervasive pollutants, stratospheric ozone, and habitat destruction. Another example of synergism, mentioned by Myers (1992) in reviewing this problem, is the increase in susceptibility to disease of plants under high temperature or reduced water stress. Conversely, diseased or otherwise damaged or weakened plants are less able to cope with the onset of unduly elevated temperatures or reduction in moisture. They could also become subject to pandemic diseases and pathogen-carrying insects and invasion of aggressive competitors, such as might occur through ecological disruptions of a greenhouse-affected world. He concluded that in light of such synergistic interactions, of which we know almost nothing, we could well be facing a greater and more rapid mass extinction than is generally expected to date. There is therefore an urgent need for well-coordinated interdisciplinary research that could be regarded as a creative challenge for conservation biologists and ecologists. Harte et al. (1992), discussing the uncertainties created by the existence of a myriad of indirect linkages between climate change and biodiversity, concluded that "we hedge our bets and implement a wide variety of protective policies, for no single approach is likely to be successful." They maintain that although parks will and should continue to be the core of our efforts, they will not be a panacea for protecting biodiversity in a world of changing climate. If parks become habitat islands, extinctions are more likely than invasions, but the consequences of the species reshuffling cannot be predicted. Given such an uncertainty of future distributions of species, we should minimize species extinctions from existing threats and look for conservation tools in addition to parks. These include the conservation of biodiversity in unprotected seminatural rangeland, forests, and deserts by a multitude of management policies, such as sustainable timbering practices, regulating recreational uses, constraining locations of roads and human dwellings, and confining grazing activities to protect streams and riparian zones. They stressed that the challenges for formulating a viable species conservation policy are not less important than the challenge for forecasting effects. But first of all we will have to overcome counter-productive disciplinary barriers to such research.

Recent unexpected results of deep ice cores in Greenland from the previous interglacial period, lasting around 20 thousand years and ending 115 000 years ago, require a reassessment of climatic control mechanisms (GRIP, 1993). These findings show that at that time Greenland experienced a succession of short-term oscillations of 10–30 years, switching between phases about 2 °C warmer that now and about 5 °C cooler, and indicating a great variability in the duration of the different climatic modes. Although the longest uninterrupted warm period was around 2000 years, there were several cool (and warm) events lasting for only a few hundred years or less. Apparently there was an inherent instability in regional (and global) ocean circulation and ocean transport processes. This introduces a further risk factor, because enhanced greenhouse effects, due to human activities, could trigger abrupt and highly disruptive climatic changes, rather than gradual global warming (Williamson and Oeschger, 1993).

There are some clear indications that in the eastern Mediterranean we have already entered such a period of disruptive changes and of increase in climatological fluctuations and unpredictability. Recent findings of scientists from Greece and Israel identifed a trend of decreasing sea surface temperatures in the eastern Mediterrancan (Lascaratos et al., 1991; Kutiel and Bar-Tuv, 1992), decreasing daily temperatures (Katsoulis, 1987), decreasing daily temperature ranges (Jaffe, 1992), and decreasing potential evapotranspiration rates (Mandel, 1986). There also seems to be a clear indication of the destabilization of the atmospheric circulation patterns and pressure systems affecting seasonal rainfall patterns, which may have far-reaching ecological implications. Sea level pressure fields show a retardation of about 6 weeks in the building of the Siberian High, resulting in a similar retardation in the rainfall regime in Turkey (Wigley and Farmer, 1982) and in Israel. The delay of this anticyclone could be attributed to the warming in the higher latitudes, due to an increase of greenhouse gases in the atmosphere (Kuticl, 1992). This subtropical high pressure belt, in general cause the prolonged summer drought and its decrease will enable some of the tropical late spring and summer monsoon rains to reach the southern parts of Israel. This pattern apparently reinforces the model of a multimodal course of rainfall (Kutiel, 1985), which further increases rainfall variability and decreases its monthly predictability.

These disruptive trends of the climate patterns have become even more apparent in recent years in Israel. They are characterized chiefly by an increase in the frequency of extreme meteorological events and an even clearer shift in spatial rainfall distribution, with relatively larger amounts in the southern part of Israel and smaller amounts in the northern parts, resulting in a further increase in the coefficient of rainfall variability (M. Kutiel and H. Steinberger, personal communication).

Thus, as in the two preceding rain seasons, also in 1990–91, the deficit in total rainfall in northern Israel was larger than in the central part. On the other hand, in southern Israel the rainfall was even above the long-term mean, and here some unusually high amounts of precipitation were recorded during several days in January. Contrary to the normal pattern, rainfall was much higher in March–May than in the early season, which was extremely dry and had some unusual warm days with 3–4.5 °C above the normal measures and up to 45% higher evaporation rates in certain locations. A similar very dry and even warmer early rain season is occurring presently between November 1993 and January 1993–94. However, winter 1991–92 was one of the wettest and coldest winters ever recorded in Israel. It was declared a natural disaster season for agriculture because of the combination of unusually extreme meteorological events and severe conditions of heavy and prolonged rainfall and lack of sunshine, floods, cold waves, frost, snow, and strong winds, causing heavy damages. In that year even the absolute rainfall amounts in central Israel were higher than in the northern parts. They reached 1,500–1,700 mm, 200% to 220% of the long-term mean, as compared to 140% in northern parts (Gat and Rubin, 1993).* In conclusion, judging fom all this sometimes even contradictary evidence for the expected warming and drying tendencies, presently only a clear tendency for the decrease in the predictability of eastern Mediterranean climate patterns seems to prevail, and consequently an increase in climate uncertainly*.

Further warming at the higher latitudes would probably cause an even greater distortion of Mediterrancan rainfall patterns*. At the middle latitudes of the Mediterranean, this could lead to the alarming scenario of increasing aridity with potentially far-reaching and wide-ranging environmental effects and mostly disastrous socioeconomic impacts. At the Mediterranean Working Group of the above-mentioned LICC meeting, it was assumed that an increase of 3–4 °C in average temperatures in the Mediterranean Basin could lead to an increase in potential evapotranspiration of 400 mm/year. Together with a changing frequency and magnitude of rainfall events and soil degradation processes decreasing its water-holding capacity, this could lead to an increase in aridity, even if the annual amount of precipitation were unchanged (Boer and De Groot, 1990).

According to the findings of a special UNEP Mediterranean task force, an assumed temperature increase of 1.5 °C by the year 2025 will cause an increase of potential evapotransporation throughout the Mediterranean, coupled with a possible decrease in

---

* At the time of the final editing of this paper, in March 1995, the same trends were observed of increased climatic unpredictability and of higher rainfall amounts in central Israel, reaching, in Tel Aviv 792 mm (the annual long term mean is 560 mm), as compared with the Galilee in northern Israel with only 676 mm (the long-term meaning is 925 mm). However the distribution was again a very abnormal one, and this year it was concentrated chiefly in the early rainfall season, whereas the usually highest rainfall and coldest months, January and February were much drier and also warmer.

precipitation in the South and an increase in the North. Hot, dry summers and exceptional events of drought or rainfall and floods, marine storms, tidal surges, and water stagnation and eutrophication could become more frequent. However, future adverse impacts of exponential population growth and urban-industrial development may far exceed the direct impacts of climate change. They will cause continuous increase in society's vulnerability to climatic stress, particularly in the South (Jeftic, 1993). In the semiarid Mediterranean region, the resulting desert encroachment is presently estimated at more than 2% per annum, and in combination with climatic changes it will have truly catastrophic consequences (Le Houcrou, 1988; Sestini et al., 1989).

In view of this chaotic and nonlinear behavior of atmosphere–biosphere interactions and of our almost complete lack of knowledge from the Mediterranean on the response of organisms, populations, communities, ecosystems, and landscapes at different time and space scales to these uncertainties, it would be futile, at this stage, to attempt any robust models and scenarios.

On the other hand, we can anticipate with much greater certainly that any increase in climatic stresses would further aggravate the already existing, alarming processes of overall landscape degradation. Therefore, in order to face these challenges of climatic uncertainties and to prevent, as far as possible, their adverse repercussions on the open Mediterranean landscape, we have to act without delay on a broad front of research and conservation issues.

For this purpose, I will first review briefly some of the most relevant ecological and evolutionary features and processes on which these conservation and research priorities should be based, and then outline the holistic ecological strategies and tools by which these priorities could be realized. For most of these issues, I will use Israel as an example, but the same principles apply to the other much more extensive regions, especially in southern Europe and North Africa, located in the drier fringes of the Mediterranean region.

PAST AND PRESENT LANDSCAPE DYNAMICS, STABILITY, AND DIVERSITY

Most of the Mediterranean uplands, too steep and rocky or with soils too poor and shallow for profitable intensive agriculture, make up great proportions of all these countries. Although of low economic value, their vegetation canopy has great ecological importance for environmental and watershed protection of the densely populated valleys and coastal regions and as a genetic source for future utilization. Together with the rapidly vanishing coastal dunes and undrained wetlands and marshes, they are the last refuges of the natural, spontaneously occurring and reproducing Mediterranean plants and animals.

As described elsewhere in more detail (Naveh and Whittaker, 1979; Naveh and Lieberman, 1994; Naveh and Kutiel, 1990), these landscapes are distinguished by their striking floristic, faunistic, and structural diversity and scenic attractiveness. These are the result of the unique combination of both natural and cultural factors, inducing variability in time and space: 1) the great ecological micro- and macrosite heterogeneity.

2) the mild winter rainfall climates with great long- and short-term fluctuations, and 3) millenia-lasting human perturbations and defoliation pressures, first by periodic, prescribed burning and later also by grazing, cutting, and coppieing.

Their final shaping took place from the Middle Pleistocene onward in a coevolutionary process with Mediterranean paleolithic food gatherer-hunters. There were gradually replaced in the early Holocene by agro-pastoralists in closely interwoven natural and cultural processes of agro-pastoral, multifactorial land use functions, which lasted until modern times. These processes created complex and dynamic vegetation regeneration and degradation patterns in which the different woody and herbaceous strata were maintained in a dynamic flow equilibrium. In such a so-called homeorhesis (Greek: maintaining the flow), the system is going on to change along the same perturbation trajectory as in the past, as long as similar perturbations are driving these changes. It is thereby returning to its natural and cultural closely interwoven dynamics but not to an artificial, undisturbed, and stationary state of homeostasis, as assumed in the classical succession-to-climax theory. These perturbation-dependent systems and their landscapes have thereby acquired long-term adaptive resilience and metastability, and their homeorhetic flow equilibrium can be maintained and restored only by the continuation of optimum defoliation pressures. From a thermodynamic point of view, they can be regarded as "non-equilibrium, dissipative systems." According to Prigogine (1976), they create "order through fluctuation." Such dissipative structures are maintained and stabilized only by permanent energy/matter and entropy exchange processes. Driven by positive feedbacks of environmental and internal fluctuations, they move to new regimes that generate the conditions of renewal of higher entropy production. In the case of Mediterranean upland ecosystems, such new regimes are apparently created by the periodic perturbations of fire, grazing, and cutting (Naveh, 1987, 1991).

The outcome of these interactions between nature and culture is still present in a great number of not yet irreversibly despoiled seminatural and agro-pastoral landscapes. These form a matrix of fine-grained, heterogeneous mosaics of forests, woodlands, shrublands and grasslands, interspersed with cultivated field, vineyards, olive groves, and orchards. However, recent dramatic changes in land use have severely distorted the human-maintained flow equilibrium, with mostly adverse and far-reaching undesirable consequences for overall environmental quality and stability. This results from the unfortunate combination of exponential growth of populations and their growing need for land, coupled with deviation-amplifying (positive) "run-away" feedback loops with land-use pressures of even faster-growing mass tourism, together with the combined and synergistic processes of intensification of traditional and modern agricultural land uses and uncontrolled urban-industrial expansion.

As discussed recently in detail in a conference on the future of Mediterranean landscape (Farina and Naveh, 1993), in this process of "neo-technological landscape despoilation," nontillable and marginal uplands are either left to their fate of abandonment and neglect, or are exposed to agro-pastoral over-exploitation and uncontrolled urban industrial encroachment. Its major impact is the rapid loss and fragmentation of unspoiled open landscapes, their soil, plant, and animal resources, and their "life-supporting" production, protection, carrier, and regulatory functions. This process is aggravated by the gradual functional, structural, and visual degradation of the remaining open landscape with concomittant biological impoverishment and ecological disruption by accelerated crosion and soil, water, and air pollution from urban-industrial and agricultural sources.

Other, mostly adverse, impacts on the mountainous regions are the creation of monotonous and highly flammable conifer plantations and the spread of scrub thickets from abandoned terraces and pastures. Because of the accumulation of dry and dead branches and undecomposed litter and the increase in uncontrolled human and recreational activities, there is an alarming increase in more and more devastating and hotter wildfires. In the absence of efficient fire and fuel management policies, including prescribed burning, and in spite of the costly but futile efforts for total fire suppression, wildfire damage and sometimes even human losses are becoming greater from year to year.

In addition to the loss of open landscapes and their fragmentation by urban-industrial developments, the floristic and structural diversity of the natural vegetation, as well as the faunistic species richness and abundance, is constantly being reduced. This impoverishment is on the one hand the result of heavy human and livestock pressures in densely populated hill and mountain areas, chiefly in the Levant and in North Africa, and on the other, that of complete cessation of human interference in depopulated areas, as well as in protected nature reserves and parks. In many of these, moreover, heavy pressures of mass recreation are threatening floristic and structural diversity.

In general, after the release of human and livestock pressures, there is an initially vigorous vegetative regeneration of the woody selerophyll vegetation from stunted shoots and encroachment of the shrub canopy. But lack of light and space leads gradually to the almost total suppression of the herbaceous understory in undisturbed and/or protected maquis and forests. These dense and monotonous thickets are dominated by a few species, mostly oak and/or pine, which become stagnant and senescent and more and more prone to hot and devastating wildfires. Thus, contrary to preconceived climax theories, noninterference is turning from a blessing to a curse.

The possibilities of the above-mentioned global and regional climatic changes in rainfall and temperature regimes may further aggravate these threats. In the more humid western Mediterranean biomes, these could further enhance the encroachment of brush in abandoned and neglected forests and maquis and their floristic impoverishment. Their increased fuel loads will cause even hotter and more destructive wildfires. However, in the subhumid biomes, such as those of northern Israel, increasing climate stresses, and especially the anticipated rise in summer temperatures and in evaporation rates, will have severe repercussions on the most mesic and richest plant communities, their diversity and stability. Most prone would be the more demanding Eumediterranean herbaceous and woody plants in sheltered sites and more mesic western slopes on altitudes above 500–600 m, including evergreen and deciduous forests and maquis such as *Quercus calliprinos-Pistacia palestina galilea,* rich in arboreal species and distinguished by a wealth of rare and endemic species. In Israel, these plants enjoyed the most favorable climatic conditions with the most predictable and highest total rainfall of 800 mm and more. However, as mentioned above, there are already alarming indications of a destabilization of the favorable rainfall pattern and a rise in its variability.

In the drier semiarid ecotones of the Mediterranean, this process of dessication, in combination with heavy human and livestock pressures, may enhance even more the process of land denudation and desertification and may thereby cause the advancement of semiarid and semidesertic formations far into the heart of the Mediterranean biomes.

As documented by Naveh and Dan (1973), the harsher and more fragile the environment, the more far-reaching and irreversible will be the impacts of climatic and man-induced changes and the slower and more difficult will be the recovery processes. Thus, the complete destruction of the vegetation canopy in the semiarid zone at a very early date by grazing and cultivation exposed the soil to the direct impact of the harsh climate and to impeded rainfall and created conditions similar to those of the more arid regions with a scarce plant cover, climatic extremes, and flash floods. However, already at that time, they predicted that if these threatening trends of neotechnological landscape despoilation proceed unhampered, then even without climatic changes a similar fate will befall the more mesic Mediterranean mountains and hill regions, and the few remaining nature islands will be turned into overcrowded recreational slums, like most of the shores and inland waters of the Mediterranean Sea.

EFFECT OF CLIMATIC VARIABILITY ON HERBACEOUS PLANTS

Let us now consider some of the ecological consequences of such climatic fluctuations on plant production and diversity, which seems to be presently the most crucial feature of global changes in the Mediterranean. In spite of the great progress achieved in our knowledge of the physiological basis of climatic stress responses in some important woody Mediterranean plants, we have still very little reliable information on these responses of woody plants and communities in the field.

For herbaceous plants we can utilize the information gathered in our earlier, long-term study in the deciduous open woodlands of Tabor oak (*Quercus ithaburense*) in the foothills of the lower Galilee. With 550–600 mm average annual rainfall, the annual rainfall variability was 20–30%, as measured by the mean deviation from the long-term average. But the monthly variability increases from 46% in the highest rainfall month of January to 77% in November, 88% in April, and 113% in October at the beginning of the rainly season. Like all other Mediterranean oak savannas, these serve as important cattle and sheep pastures. They are distinguished by their very species-rich and diverse herbaceous understory and are dominated chiefly by annual plants (Naveh and Whittaker, 1979).

As reported elsewhere in detail (Naveh, 1982, 1988), the great dependence of their productivity-and consequently also of livestock production-on annual and seasonal climatic fluctuations was demonstrated convincingly during a 6-year

Table 1. Rainfall and temperatures in a dry year (1954–55), a wet year (1955–56), and an "average" rainfall year (1956–57), but with delayed early rains and a cold January

a. Mean temperatures in December-January (°C)

	December			January		
	Minimum	Maximum	Daily average	Minimum	Maximum	Daily average
1954–55	8.5	18.9	13.7	7.9	19.7	13.8
1955–56	9.0	17.7	13.4	7.1	18.4	12.8
1956–57	8.3	18.9	13.4	4.6	16.6	10.4

b. Monthly precipitation (mm)

	1954–55	1955–56	1956–57	Average 1931–60
September				1.7
October		7.0		16.8
November	91.8	125.5	17.2	80.9
December	176.8	247.2	153.9	128.9
January	2.2	164.3	132.5	157.3
February	38.1	44.0	109.0	108.2
March	51.7	108.7	124.6	60.7
April	11.7	12.1	22.7	17.9
May	7.8		7.8	5.8
Total	379.6	708.8	567.7	578.0

study of pasture and livestock production. In this period, the rainfall amplitude ranged from 380 to 708 mm between the driest and wettest year with a coefficient of annual rainfall variation of 23.3%. However, the amplitude of phytomass production was four times greater, ranging from 110 to 448 g/m^2, and its coefficient of variation was 64.6%. The correlation coefficient between total annual rainfall and phytomass production was 0.47. But a much greater climatic dependence of herbaceous phytomass production (as well as livestock production) could be detected by taking into account not only the total annual rainfall but also bimonthly rainfall distribution and minimum winter temperatures. These climatic variables could explain 95% of the variance in pasture and livestock production. Intensive improvements by fertilizing and selective chemical weed control almost doubled average annual meat production from 105 to 194 kg/ha, but they exacerbated these interseasonal fluctuations even more and thereby demonstrated climatic dependence.

Of special interest were the effects of extreme climatic events, observed in this study during three consecutive, contrasting rainy seasons. As shown in Table 1, 1954–55 had the lowest annual rainfall of 380 mm and only 281 mm after December; 1955–56 had the highest and most well-distributed rainfall of 709 mm; and 1956–57 had about "average" rainfall of 568 mm, but with delayed early rains and January

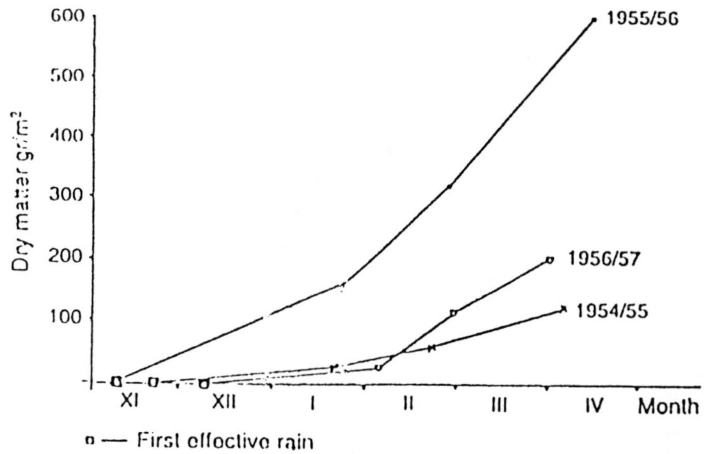

Fig. 1. Herbage production in a dry year (1954–55), a wet year (1955–56), and an "average" year, but with delayed early rains and a cold January.

temperatures 2–3 °C lower. In Fig. 1 the actual above-ground phytomass production during these three seasons is presented, as determined before each grazing cycle.

These not only show the striking differences between a very favorable and a very dry year, but also demonstrate that the total annual rainfall is not the only—and in the case of 1956–57—even not the most important independent climatic variable.

The striking rise in species richness from the driest to the following wettest year and again a drastic drop in species numbers, especially of legumes and other broad-leaved forbs is shown in Fig. 2. These results show that a great number of the less abundant species, which contribute the most to species diversity, are the most sensitive to climatic fluctuations.

Favorable growth seasons with well-distributed rainfall from November through March, such as 1955–56, have apparently ensured the maintenance of seedbanks in the soil, providing rich reservoirs for floristic and genetic diversity and for natural selection to operate on. If such seed pools are available, they will greatly reduce the fitness uncertainty. This was apparently the case during millennia of traditional agro-pastoral land uses. Our studies showed that under moderate, rotational grazing management, such a species-rich, diverse, and productive herbaceous pasture vegetation can be maintained on a sustained basis in the woodland understory, in open, grassy patches in forests and maquis-shrublands, and in open grasslands. However, after short periods of too light grazing or complete protection, aggressive tall grasses and hemicryptophytic thistles will take over. Under heavy, uncontrolled grazing pressures, the least palatable and the most undesirable weedy species will dominate. Thereby both biological diversity and highly valuable economic productivity will be reduced greatly (Naveh, 1957; Naveh and Whittaker, 1979).

Fig. 2. Changes in the number of species in eight permanent quadrats on April 1, in dry, wet, and "average" rain seasons.

It is also obvious that the chances for such rich seed pools will be reduced greatly by an increase in climatic unpredictability. Therefore, climatic instability and stress, in combination with too light or too heavy grazing pressures, could lead to synergistic interactions, severely endangering the biological diversity, productivity, and stability of all Mediterranean subhumid and semiarid woodlands and grasslands in Israel and elsewhere and accelerating the rates of desertification processes in space and time. In view of the above-described climatic trends, it is very possible that such an acceleration is already happening.

ADAPTIVE GENETIC DIVERSITY AND RESILIENCE

Of great relevance for our discussion are the important genetic-ecological studies carried out by Professor E. Nevo and his collaborators from the Institute of Evolution, Haifa University, shedding new light on the evolutionary basis of the Mediterranean biota and its adaptive responses to climatic stress. Their genetic-ecological studies of 21 species of plants and animals of the Mediterranean and steppe region range from the molecular to the landscape scales and include the two most important progenitors of our domesticated cereals, the wild emmer tetraploid wheat *Triticum dicoccoides* and the wild barley *Hordeum spontaneum.* These studies revealed allozyme diversity and DNA polymorphisms which are clearly correlated with ecological factors. All these-factors share a common denominator of increasing aridity stress as a major selective force for short- and long-term survival, and on the transition to the xeric region, increasing climatic fluctuations, both eastward and southward. Nevo (1988) and Nevo and

Beiles (1989) concluded that this was the result of long-term evolution in the widely diverse climatic and edaphic macro- and microsite conditions of Israel.

On landscape scales, the striking spatial topoclimatic and ecological heterogeneity of these upland and mountain landscapes also seems to have induced a great biogeographic, taxonomic, and genetic diversity, even within short distances. Thus, for instance, even in the heart of the Mediterranean territory in Israel, on Mt. Carmel, Nevo et al. (1982), found in opposing north and south slopes, separated by only a few hundred meters, dramatically different populations of European and African floral and faunal elements. With distinctly higher genetic and species diversity on the drier and much more exposed and open southern slope, which is exposed to more than 100% higher radiation (Nevo, personal communication). These differences involve unrelated taxa such as plants, insects, snails, reptiles, birds, and small mammals. These ecotones are presently being studied closely in all their genetic, biological, ecological, climatological, and geographical aspects by an interdisciplinary team led by Prof. Nevo. Long-term monitoring will undoubtely reveal important information on species; population, community, and ecosystem dynamics under increasing climatic stress.

In a recent extensive study of the photosynthetic performance and ecological and genetic predictability in 107 genotypes from 27 wild emmer wheat populations in the Mediterranean territory of Israel, Nevo et al. (1991) found that the highest photosynthetic efficiency was displayed by populations of xeric marginal populations from the Samarian and Judean hills and from Mt. Gilboa. Genotypes and populations with high photosynthetic capacity can be identified by climatic factors and isozyme markers. Climatic variables alone significantly explained a substantial amount of all the variance of all photosynthetic variables, and especially water use efficiency. The main climatic variables were water availability factors, radiation, and temperature variables.

The results of these studies have far-reaching implications for our decisions on research and management priorities. They can serve as important guidelines for in-depth analysis of genetic diversity from the DNA and protein level to the landscape level. Such an analysis is crucial for the understanding of the evolutionary process of adaptation, speciation, and resilience under increasing climatic stress. They will provide the scientific basis for the prevention of the extinction of major herbaceous and woody keystone species, which are essential for the maintenance of ecosystem processes and landscape stability and integrity.

Nevo's studies also point out the great importance of the drier Mediterranean biomes, and especially their semiarid fringes along the 400 mm isohyet. It can be envisaged that under increasing climatic stresses, these most xeric and resilient vegetation types of the drier Mediterranean biomes may become the last refuges of Mediterrancan plant and animal diversity. Here the growth season is not only shorter, but inter- and intraseasonal rainfall predictabilities are much lower, and moisture stress for summer-growing woody plants is more severe. These harsh conditions favored the development of an inherently more drought-tolerant, xeric, and thermophyllous flora which will most probably be better adapted to increasing climatic stresses and therefore could also be more resilient and elastic than those from more mesie habitats. If not abused by overgrazing and cutting or planted with dense monospecies pine plantations, they can be very rich in drought-evading shrubs and dwarf shrubs, grasses, and herbs, including the progenitors of cultivated cereals and other plants with great genetic potential for pharmaceutical, agricultural, and other future economic uses. They also have many sclerophyll woody and other taxa in common with the above-mentioned, more mesic biomes.

In such a semiopen, mosaiclike, low shrub-grass community on Mt. Gilboa, with widely scattered *Ceratonia siliqua* trees and denser stands of *Pistacia lentiscus* shrubs, we encountered the highest vascular plant species richness and diversity of all our study sites in northern Israel: 180 different vascular plant species in 1/10 hectare, including 147 herbs with 102 annuals and many geophytes, including the highly ornamental "Iris of Mt. Gilboa" (*Iris haynei*) and smaller shrubs and dwarf shrubs. It also ranked highest in animal species richness, with 14 reptile, 7 rodent, 31 bird, and 13 isopod species (Naveh and Whittaker, 1979).

With only 400 mm annual rainfall, 60 very dry and hot "sharav" days, and mean temperatures of 33 °C at noon in June, this was the driest site of all our study plots. We explained its great biodiversity by its unique nature as a biological, ecological, and geographical ecotone, serving as a meeting point between the Mediterranean territory of the subhumid Samaria mountains (as their isolated outpost) and the semiarid Mt. Gilboa proper, to which even more xeric species have invaded from the lower-altitude semidesert Jordan plains. Having been exposed to traditional human and grazing pressure for many centuries, it is now grazed chiefly by gazelles. However, it is not included in the Gilboa Nature Reserve, and its fate is not assured at all.

From Mt. Gilboa, where Nevo and his coworkers found genotypes with the highest adaptive survival potentials and photosynthetic capacities, we also obtained the first proof that sclerophyllous woody plants may have undergone similar climatically induced intraspecific ecotypic differentiation. This is indicated by the results of a study on ecophysiological intraspecific variation in the selerophyll evergreen shrub *Pistacia lentiscus* in northern Israel by Shaviv (1978). This shrub is one of the most important Mediterranean "keystone" species, because of its wide abundance, resilience, and effective soil improving and protecting features. In comparison with other populations of this shrub from much more mesic sites, the Gilboa population is distinguished by the most efficient active and passive water saving economy and the capability not only to survive during the hot summer months, but even to continue active photosynthesis and carbon fixation. It reached maximum photosynthesis at 35 °C, as compared to 25 °C for the Mt. Carmel population, and five times higher absolute photosynthetic values. This was the only ecotype to continue photosynthesis at high temperatures of 45 °C, and at the same time it exhibited the lowest increase of respiration rate with increasing temperature, and had still positive values for photosynthesis/respiration at 40 °C. It had the largest number of stomata per unit leaf area, the most rapid decrease in respiration rate when air humidity was lowered, and the lowest transpiration rates even when water was available, with low water potentials and a high albedo in the 500–700 mm region. In *Pistacia lentiscus* we also found striking ecotypic variation in root growth strategies, and those growing in poor calcarcous soils were found most promising for soil erosion control and improvement and for the restoration of denuded slopes (Yogev and Naveh, 1986).

In conclusion, we should regard these xeric Mediterranean ecotones as a major genetic and ecological bank and climatic stress refuge, deserving urgent protection of their genetic and biological diversity.

CONCLUSIONS

The effective prevention of further disastrous increases in greenhouse gases level in the atmosphere, as well as of other detrimental and irreversible human-caused impacts on the ecosphere, will be possible, in my opinion, only by a substantial reduction of wasteful production and consumption in the industrial societies and by drastically curbing the population explosion in developing countries. The outcome of the recent United Nations Conference on Environment and Development in Rio de Janeiro in June 1992 has not given much hope for what di Castri and Hansen (1992) called "a worldwide geopolitical willingness and a kind of new international economic order." This could have become a major driving force for such a cultural and socioeconomic environmental revolution. At the same time, however, we as ecologists and land managers may at least be able to help in reducing and mitigating the damaging effects on our few remaining less despoiled natural and seminatural systems. This can be achieved in the Mediterrancan–as well as elsewhere–by a significant improvement in land use management, aimed at the protection of biodiversity and of other unique natural and cultural landscape assets. Any measure taken now to reduce and counter land despoliation and denudation and to minimize the loss of plant and animal species, especially in the most vulnerable sites in the mesic and drier ecotones, will increase the resilience of these landscapes and have far-reaching beneficial implications.

However, if the heavy and uncontrolled human and livestock pressures prevailing in most of the drier critical Mediterrancah regions in southern Europe, the Levant, and North Africa continue unhampered, their destabilizing, mutual causal feedback loops with increasing climatic stress will accelerate the process of desertification. This could lead to an even more rapid shift of the subdeserts and desert biomes far into the Mediterranean territory in Israel and elsewhere. This will cause catastrophic destabilization of the drier mountain and hill landscapes, similar to those which have already occurred since the neolithic revolution during prehistoric and historic times, but within much shorter time spans.

We may conclude, therefore, that the threat of climatic change is a compelling additional reason to conserve and restore the health, integrity, and diversity of all natural and seminatural landscapes that are not yet irreversibly despoiled. This would be the best insurance policy against climatic uncertainties and environmental peils to rescue their vital ecological, socioeconomic, and evolutionary functions and their intrinsic natural and cultural values:

1. It would ensure sufficient landscape connectivity to allow free plant and animal migrations and nutrient movements across the landscape and to counter further fragmentization, urbanization, agroindustrial pollution, and scenic despoilation of the open landscape.
2. It would provide the greatest options for the sustainable functions of live-supporting ecosystems and for the redundaney of their functional groups and keystone species ensuring all vital ecological processes, sensu Chapin et al. (1992).
3. It would afford the best chances for the survival of rich biotic communities and their free and spontaneously developing and reproducing populations, and thereby ensure also further evolution of plant and animal species.

For this purpose we have to distinguish between two major ecosystem and landscape types:

1. Natural and close-to-natural ecotopes, in which flora, fauna, and eco-system structure and function have undergone very little human modification. These are very rare in the Mediterranean and have remained only in inaccessible sites, such as rock cliffs and steep mountain slopes. Their major conservation object should be the complete protection from any interference and disturbance in order to safeguard not only their biota but also their natural patterns and cycles of self-regulating mechanisms and ecological processes within strict nature reserves.
2. Seminatural ecotopes, in which the largely spontaneous flora and fauna and their self-regulatory mechanisms have undergone the above-described far-reaching modifications through intensive and lasting anthropogenic perturbations. These include the few remaining wetlands, riverside ecotones, and other coupling ecotones, along with aquatic ecosystems in the lowlands and the selerophyllous, coniferous, and mixed forests and the much more extensive woodlands, shrublands, dwarf shrublands, and derived grasslands. These contain most of the natural Mediterranean plant and animal communities in different stages of degradation and regeneration.

Here our primary conservation object is twofold:

1. The prevention of undesirable, destructive processes, including too heavy mass recreation and livestock pressures and hot wildfires, caused by the accumulation of highly inflammable fuel, as well as water, air, and soil pollution.
2. The reestablishment and perpetuation of those multifactorial, homeorhetic flow processes that ensured the fine-grained spatial heterogencity, the biological and cultural diversity and metastability, and the attractiveness of the Mediterrancan landscapes. This can be achieved by developing detailed long-term but flexible management plans in space and time according to local site conditions and requirements. It will include different defoliation pressures of grazing, browsing, cutting, coppicing, pruning, and prescribed burning, rotated in space and time, as well a completely protected "control" treatment. A clear distinction in conservation strategies also has to be made here:
   - In nature reserves the main aim should be the highest attainable biological productivity, diversity, and stability by restoring all ecological processes and

their protection, regulation, and carrier functions through scientific conservation management.

- In nature recreation parks, this aim should be optimized with recreation amenities to allow maximum enjoyment with minimum damage to natural and cultural resources.

This will require, above all, much more efficient protection from external pressure and damage to nature reserves and parks by uncontrolled human interventions and recreation pressure. But it would be a grave mistake to assume that by complete and prolonged protection from any human interference the entirely hypothetical and mostly illusory homeostatic "climax" state could be attained. As has been explained above, most of these protected areas have evolved as human perturbation-dependent ecosystems. Therefore, we can ensure the continuation of all vital ecological processes and their homeorhetic flow equilibrium and evolutionary metastability only by dynamic conservation and restoration management for biodiversity, fuel management, and the maintenance of grassy patches and parklike mosaics. In view of the threat of climatic change, the prevention of dense, monotonous, species-poor, and highly flammable tree and tall shrub thickets becomes one of the most urgent management goals, also in nature reserves. It will open maximum options for woody and herbaccous plants and their structural community diversity, reduce their competition for light, space, and water, and provide suitable habitats for wildlife, even under increasing climatic stress.

Nature reserves and parks, and especially their ecotones, should also serve as the major permanent study and monitoring sites for biological and physical changes through time.

We should, however, also be aware of the possibility that the present locations and sizes of these protected areas may not be adequate to cope with impending climate changes. This implies the establishment of new systems of reserves and especially biosphere reserves to provide more diversified conditions, ranging from networks of small refugia with adequate corridors to larger preservation areas (di Castri and Hansen, 1992).

As emphasized above, highest priority should be given to the protection of the last, richest plant refuges of the semiarid ecotones in Israel and elsewhere. The crucial question is whether there will be a sufficient stock of resilient species and subspecies left from their great genetic and floristic wealth which could serve as such genetic banks for restocking more mesic sites which may undergo a creeping process of desiccation.

In addition, sufficently large areas with highly natural, cultural, and scenic values, and especially those serving as corridors and/or buffering zones with nature reserves and parks, should be declared as managed landscape areas and/or biosphere reserves. In these, nondestructive and nonpolluting economic activities for the utilization of "hard" economic benefits should be continued, and, as far as possible, traditional agro-pastoral land uses should be restored. In highly disturbed sites, multipurpose restoration and afforestation should be carried out, using chielly indigenous plants.

Recent helpful guidelines on protected landscapes for policy makers and planners have been presented by Lucas (1992).

In the remaining open uplands, used primarly for socioeconomic benefits, two major ecotechniques should be applied:

1. Controlled manipulation of the soil-plant-animal complex by controlled grazing, thinning, coppicing, and pruning of trees and controlled fire and pastoral agrotechnical improvements, according to site potentials and socioeconomic and other requirements and conditions.
2. Multipurpose afforestation, restoration, and reclamation wherever land denudation has reached an advanced stage, including roadsides, camping grounds, etc. The aim is the creation of seminatural, multilayered, and stable plant communities and ecosystems with multiple ecological, economic, and social benefits.

In recent years it has become more and more evident that the establishment of fast-growing pine trees in dense monospecies plantations on shallow, rocky, dry, and sleep slopes has neither ecological nor economic justification. The main reasons are their high flammability, their great susceptibility to air pollution and especially to photochemical oxidants, and their synergistic effects with pests. These undesirable features could be further exacerbated with increasing climatic stress, air pollution, and fire hazards.

There is, therefore, urgent need for the diversion of afforestation on nontillable Mediterranean uplands from commercial coniferous and eucalyptus plantations to the establishment of open and parklike multiple layer forests. These should be derived chiefly from local stock and based on the selection and multiplication of woody, multibeneficial keystone species from the drier Mediterranean biomes and ecotones. This should prepare the ground for the planting of special drought-resistant forest reserves in which the biological diversity and productivity of the Mediterranean arboreal vegetation and its rich herbaccous understory and the scenic attractiveness of these landscapes could be safeguarded.

For this purpose, special "climate stress refuges" should be established in which promising drought-resistant ecotypes could be multiplied and stored in long-term seedbanks for restoration projects.

RESEARCH PRIORITIES

It is obvious that all these conservation and restoration challenges cannot be carried out without full support by a comprehensive research program. This should be inspired by a holistic systems approach, bridging the gaps between the manager and researcher, the academicians and the professionals, and between the different subdisciplines and subcultures into which contemporary ecology has been fragmented. This will require a new way of thinking about ecological problems with a transdiciplinary hierarchical view, as presented recently in a very lucid way by Allen and Hockstra (1992). It should be based on long-term, well-integrated, problem-solving-oriented studies embracing all relevant biological and ecological complexity levels and scales from the DNA and subspecies, to ecosystem and landscape-terrain levels. If such a research program is of any practical value, it must be, of course, closely related to the above-described pressing problems of conservative and sustainable land use planning and management, facing uncertainties of climatic change. Its field work should therefore be carried out in representative sites of critical areas, in combination with experimental and larger-scale management and restoration treatments and monitoring.

Highest priority should be given to genetic and ecophysiological research for the study of the genetic-ecological basis and the inter- and intraspecific adaptive plant behavior under multiple stresses in the field, determining the reproduction and competition success and survival potentials of genotypes for their selection as promising woody and herbaceous keystone species. As a model for this research, the above-described studies by Nevo and his coworkers should be used. These should be expanded also to the community and ecosystem level. Greatest attention should be paid also to the edaphic and biotic factors, the root systems, and their rhizosphere and microflora.

There is also urgent need for the systematic study of the potential effects of the rise in greenhouse gases concentration on the Mediterranean vegetation. Not only CO_2 should be considered, but also photochemical oxidants and especially ozone, which is already causing widespread damage by chlorotic decline to *Pinus halepensis, Pinus pinea,* and broad-leaved maquis trees (Naveh et al., 1980, Busotti et al., 1991).

Combined field and fumigation studies in comparable conditions in coastal southern California indicated an aggravation of ozone formation under higher summer temperatures, with far-reaching effects on the natural vegetation (Malanson and Westman, 1991). Such studies could serve as a model for similar research in the Mediterranean proper.

It should be realized, however, that in reality it will be very difficult to separate the effects of climatic stress, as defined by the adaptive tolerance of the different species, and chronic and acute disturbances by grazing, cutting, fire, pests, and air pollution. These are coupled by deviation amplifying (positive) feedback loops, and their interaction determines not only the fate of the individual plant and its community, but that of the landscape as a whole. Their understanding and study requires innovative, interdisciplinary approaches and methods. The same is also true for landscape ecological studies on the spatial and functional organization and the dynamics of landscape patterns and processes and on production, regulation, and carrier functions, as affected by different land use regimes. Such studies have to take into consideration different scenarios of climatic changes.

Such interdisciplinary, long-term multilevel and multidimensional research, combining field and laboratory investigation with integrative ecosystem and landscape studies, are presently completely lacking in the Mediterranean. The creation of special research funds, earmarked for these purposes, from international, European, and national sources will be necessary, in addition to the already existing funds, such as those of the European Union programs.

But even if sufficient funds were available today, there are almost no scientists in the Mediterranean who are acquainted with these methods and there is no proper interdisciplinary organization available for their training. These problems could be solved best by the creation of a special, independent organization or infrasturcture that could initiate and coordinate such studies and train young, promising scientists in these fields. Earlier efforts in this direction have unfortunately failed. However, in view of the combined threats of climatic change and accelerated deterioration of the open Mediterranean landscape, the need for such an International Mediterranean Ecosystem and Landscape Research Institute becomes even more urgent.

THE NEED FOR NEW APPROACHES AND TOOLS

From our discussion emerges the need for an entirely new, holistic land use policy for the conservation and restoration of the Mediterranean landscapes as a whole. It should be based on well-coordinated, simultaneous efforts for better public and professional conservation education, more comprehensive planning and management, and more interdisciplinary and integrated research. For this purpose, neither one-sided agriculture, livestock, or forestry production nor nature conservation or recreation amenity-oriented approaches alone can provide a satisfactory solution. As described in detail elsewhere (Naveh and Lieberman, 1994), this requires a holistic systems approach of landscape-ecological determinism to be applied with innovative methods. These should aim at reconciling the conflicting demands for sustaining and improving the biological productivity, diversity, and scenic attractiveness of the open Mediterranean landscape with the needs of the people and the national economy. It should enable the optimization of all hard and soft landscape values and their long-term multiple benefit functions, by creating productive and attractive multiple use landscape mosaics in comprehensive master plans. In such master plans, all hard and soft values and their marketable and noneconomic richnesses could be expressed as functional parameters of production, regulation, protection, carrier, and information functions. To the latter, instead of monetary values, only relative values should be attached in a matrix of decision making, as suggested by de Groot (1992). For this purpose, innovative expert systems using fuzzy set theory are very suitable. Expert systems are computer programs based on "artificial intelligence" that can pose and answer questions relating to knowledge and information derived from human experts also in linguistic form. Fuzzy set theory enables the combination of such disparate landscape variables and subjective evaluations, including intangible values and functions, which cannot be quantified in a conventional way in numbers, but only in words. As shown by Negotia (1985), such verbal models can be used in fuzzy linear programming for the automatic transfer of knowledge from linguistic values in optimization functions.

GREEN BOOKS FOR THREATENED LANDSCAPES

The realization of these plans will require a change in attitudes and concepts which cannot be initiated merely by interdisciplinary scientific research and by appropriate technological means, but only by a coordinated effort for public education at all levels. It requires the cooperation of all those who care for these uplands and their natural and cultural assets, those who study, plan, and manage them and those who live from their resources. There is also urgent need to provide better tools for the decision making process in land use and conservation, taking into consideration the above-described threats.

As a first practical attempt to fulfill these functions, the preparation of Green Books for the conservation of threatened landscapes has been initiated by a special working group of the International Union for the Conservation of Nature and Natural Resources (IUCN); World Conservation Union Commission on Environmental Strategies, in cooperation with the International Association of Landscape Ecology (IALE). The first Green Book case study has been carried out it western Crete by multidisciplinary and multinational teams from the University of Thessaloniki and the University of Cambridge (Grove et al., 1993). Further case studies are planned in which a uniform methodology of such Green Books will be further developed. At the same time, the proposal for the preparation of worldwide red lists of valuable endangered landscapes has been adopted by the IUCN.

These Green books should serve not only as a monitoring and prediction tool regarding the fate of open landscapes facing climatic change and increasing land pressure. They should also provide guidelines on their conservation, restoration, and sustainable development for decision making. The collection, analysis, synthesis, and interpretation of this information will be carried out with the help of advanced landscape-ecological methods, including integrated field surveys, appraisals, and landscape unit classifications, combined with remote sensing and dynamic Geographical Information Systems (GIS), and with multivariate statistical analysis and clustering methods. Their results, however, will be presented in these Green Books in nontechnical terms with ample illustrations and maps in an appealing and convincing way. Thereby they could also become an important tool for public education and be made available to the public, schools, universities, industries, business, politicians, and officials, and even to foreign visitors and tourists. Their object is not to produce another scientific report to be published and filed away sooner or later, but to become a usable document for professionals and the citizens. In contrast to Environmental Impact Statements, these Green Books should be active and positive iniatives to change already prevailing trends with the help of innovative, holistic, long-term, and multibeneficial planning and management strategies on regional landscape scales. Thereby they would become also a major tool to rescue endangered valuable landscapes included in the Red Lists.

Our hope is that in this way their semantic information will be transformed into pragmatic information, which becomes meaningful by its effects on the receivers and is expressed in their actions.

REFERENCES

Allen TFH, Hockstra TW (1992) Towards a Unified Ecology. Columbia University Press, New York.

Boer MM, De Groot RS (eds) (1990) Landscape-Ecological Impact of Climatic Change. Proe Europ Conf Lunteren. The Netherlands. December 1989, pp. 1–7.

Bolin B, Doos BR, Jaeger J, (1986) The Greenhouse Effect, Climate Change and Ecosystems. Report No 29 SCOPE/ICSU. John Wiley & Sons, New York.

Bussotti F, Ferreti M, Gellini S, Barbolani E, (1991) New type forest damage to Mediterranean vegetation in Southern Sardinia forests. Eur J For Path 21:290–300.

di Castri F, Hansen AJ (1992) The environment and development crises as determinants of landscape dynamics. In: Landscape Boundaries, di Castri F, AJ Hansen (eds) Springer-Verlag, New York, pp 3–18.

Chapin FS, Schulze ED, Mooney HA (1992) Biodiversity and ecosystem processes. Trends Ecol Evol 7:107–108.

De Groot R (1992) Functions of Nature in Environmental Planning, Management and Decision Making. Wolters-Nordhoff. Amsterdam.

Farina A, Naveh Z (eds) (1993) Landscape Approach to Regional Planning: The Future of the Mediterrancan Landscapes. Special Issue July 1993. Landscape and Urban Planning 24:1–4.

Gat Z, Rubin S (1993) Analysis of Weather Conditions During the Winter Season of 1919/92. Israel Meteorological Service. Climatic Service. Bet Dagaan, Israel (In Hebrew).

Graber M, Cohen A (1993) Climate change and air pollution: a review. In: Regional Implications of Future Climate Change. Graber M, Cohen A, Magaritz M (eds) Israel Academy of Sciences and Humanities, Jerusalem, pp 6–25.

GRIP (Greenland Ice-Core Project Members) (1993) Climate instability during the last interglacial period recored in the GRIP ice core. Nature 364:203–207.

Grove AT, Ispikoudis J, Kazaklis A. Moody JA, Papanastasis V, Rackham O (1993) Threatened Mediterranean Landscapes: West Crete. Final Report to EEC. Department of Geography, University of Cambridge, Cambridge, UK.

Harte J, Torn M, Jensen D (1992) The nature and consequences of indirect linkages between climate change and biological diversity. In: Global Warming and Biological Diversity. Peters RL, and Lovejoy TE (eds) Yale University Press, New Haven, CT, pp 325–343.

Hulme M, Wigley TML, Jones PD (1990) Limitations of regional climate scenarios for impact analysis. In: Landscape-Ecological Impact of Climate Change. Proc Europ Conf Lunteren. The Netherlands 3–7 December 1989. MM Boer and RS de Groot (eds). IOS Press, Amsterdam, pp 11–132.

IGBP (1990) The Initial Core Project. Report No 12. The International Geosphere-Biosphere Program–A Study of Global Change. (IGBP) International Council of Scientific Unions (ICSU) Stockholm.

IPCC (1990) Scientific Assessment of Climate Change. Intergovernmental Panel on Climate Change. Cambridge University Press, Cambridge, CA.

Jaffe S (1992) Was there warming in Israel? Meterology Israel 1:92.

Jeftic L (1993) Implications of expected climate change in the Mediterrancan region. In: Regional Implications of Future Climate Change. Graber M, Cohen A, Magaritz M (eds) Israel Academy of Sciences and Humanities, Jerusalem. pp 278–302.

Katsoulis BD (1987) Indications of change of climate from analysis of air temperature time series in Athens. Greece. Climate Change 10:67–79.

Kutiel H (1985) The multimodality of the rainfall in Israel as reflected by the distribution of dry spells. Arch Meteorologic Geophysik Bioklimatologic 36B:15–27.

Kutiel H (1992) Recent climatic variations in the Mediterranean Region. Proc Pan-Hallenic Symp Meteorology. Climatology and Atmospheric Physics. Thessaloniki, 21 May, 1992.

Kutiel H, Bar-Tuv A (1992) Recent trends of decreasing autumnal sea surface temperatures (SST) in the Eastern Mediterranean Basin. Israel J Earth Sci 41:51–53.

Lascaratos A, Gacic M, Oguz T (1991) Recent atmospheric and occanic climatic oscillations in the Northeastern Mediterranean and Black Sea. Proc Intern Workshop Regional Implication of Future Climate Change. April 28, 1991. Rehovot. Israel.

Le Houérou HN (1988) Vegetation and land-use in the Mediterranean basin by the year 2050: a prospective Study. UNEP (OCA)/WG. 2/15.

Lucas PHC (1992) Protected Landscapes. A Guide for Policy Makers and Planners. Chapmann & Hall, London.

Malanson GP, Westman WE (1991) Modeling interactive effects of climate change, air pollution, and fire on a California shrubland. Climatic Change 18:363–376.

Mandel M (1986) Water consumption of Israel according to ratio between the real and the potential evaporation. In: Quantitative and Qualitative Problems in the Present Water Budget in Israel. Proe Annual Meeting Israel H ?? rol Assoc. Jerusalem. 29 October 1986 (in Hebrew).

Myers N (1992) Synergism: joint effects of climate change and other forms of habital destruction. In: Global Warming and Biological Diversity. Peters RL, Lovejoy TE (eds) Yale University Press. New Haven. ??. pp 344–354.

Naveh Z. (1957) Report on Range Research at the Neve Yaar Experimental Station. FAO Working Party on Mediterranean Pasture and Fodder Development. 5th Meeting. Tel Aviv.

Naveh Z (1982) The dependence of the productivity of a semi-a ?? Mediterranean hill pasture ecosystem on climatic Iluctuations. Agric Envion 7: 47–61.

Naveh Z (1987) Biocybernetic and thermodynamic perspectives of landscape funtions and land use patterns. Landscape Ecol 1:75–83.

Naveh Z (1988) The response of the herbaceous vegetation of a semiarid Mediterranean oak savanna to climatic Iluctuations. In: Time Scales and Water Stress. 5th Int Conf on Mediterranean Ecosystems. di Castri F, Floret F (eds) UBS. Paris. pp 361–371.

Naveh Z (1991) Mediterranean uplands as anthropogenic perturbation dependent systems and their dynamic conservation management. In: Terrestrial and Aquatic Ecosystems–Perturbation and Recovery. Ravera O (ed.) Ellis Horwood. New York. pp 545–556.

Naveh Z, Dan J (1973) The human degradation of Mediterranean landscapes in Israel. In: Mediterranean-Type Ecosystems. Origin and Structure. di Castri F, Mooney HA (eds) Springer-Verlag. Heidelberg. pp 370–390.

Naveh Z, Whittaker R (1979) Structural and floristic diversity of shrublands and woodlands in Northern Israel and other Mediterranean areas. Vegetatio 41:171–190.

Naveh Z, Steinberger EH, Chaim S, Rotman A (1980) Photochemical airpollutants–a threat to Mediterrancan coniferous forests and upland ecosystems. Environ Cons 7:301–308.

Naveh Z, Kutiel P (1990) Changes in the Mediterranean vegetation in Israel in response to human habitation and land uses. In: The Earth in Transition. Patterns and Processes of Biotic Impoverishment. Woodwell GM (ed.) Cambridge University Press, New York. pp 259–299.

Naveh Z, Lieberman AS (1994) Landscape Ecology. Theory and Applications. Springer-Verlag, New York. 2nd edition.

Negotia F. (1985) Expert Systems and Fuzzy Systems. Benjamin/Cummings Publishing Company, Menlo Park, California.

Nevo E (1988) Genetic diversity in nature. Patterns and theory. In: Evolutionary Biology. Vol. 23, Hecht MK, Wallace B (eds) Plenum Publishing Corporation, New York. pp 217–246.

Nevo E, Beiles A (1989) Genetic diversity of wild emmer wheat in Israel and Turkey. Theor Appl Genet. 77:421–455.

Nevo E, Carver BF, Beiles A (1991) Photosynthetic performance in wild emmer wheat. *Triticum dicoccoides:* ecological and genetic predictability. Theor Appl Genet 8:334–460.

Nevo E, Bar-El C, Beiles A, Yom-Tov Y (1982) Adaptive microgeographic differentiations in allozyme polymorphism in land snalis. Genetica 59:61–67.

Prigogine L (1976) Order through fluctuations: self-organization and social systems. In: Evolution and Consciousness: Human Systems in Transition, Jantsch E, Wadington CW (eds) Addison-Wesley, Reading, MA, pp 93–130.

Schneider SH (1989) The greehouse effect–science and policy. Science 243: 771–781.
Sestini G, Jeftic L, Milliman JD (1989) Implications of expected climate changes in the Mediterranean region: An overview. MAP Technical Reports Series No 27. UNEP. Athens.
Shaviv L (1978) The autecology of *Pistacia lentiscus* L. in Israel. PhD Thesis, Fac Agr Eng Technion. Haifa. Israel (Hebrew, English summary).
WCED (1987) Our Common Future. World Commission on Environment and Development. Oxford University Press. New York.
Wigley TML, Farmer G (1982) Climate of the Eastern Mediterranean and the Near East. In: Palacelimates Palcoenvironments and Human Communities in the Eastern Mediterranean Region in later Prehistory. BAR International series 133. Bintliff JC, Van Zeist W (eds) British Archacological Reports. Oxford. pp 3–37.
Willamson P, Oeschger H (1993) Climate instability in a warmer world. Climate Change 16:6–7.
Woodwell GM (1992) How does the world work? Great issues of life and government hinge on the answer. In: Global Warming and Biological Diversity. Peters RL, Lovejoy TE (eds) Yale University Press, New Haven, CT, pp 31–37.
Yogev I, Naveh Z (1986) Effects of fertilizing and mulching on the establishment of *Pistacia lentiscus* in limestone debris for landscape reclamation. Proc 3rd Intern Conf. Israel Ecol Soc. pp 403–412.

CHAPTER 6

NAVEH Z. (1998)

FROM BIODIVERSITY TO ECODIVERSITY – HOLISTIC CONSERVATION OF THE BIOLOGICAL AND CULTURAL DIVERSITY OF MEDITERRANEAN LANDSCAPES.

IN: RUNDEL P., MONTENEGRO G., JAKSIC F.M. (Eds.)
LANDSCAPE DISTURBANCE AND BIODIVERSITY IN MEDITERRANEAN – TYPE ECOSYSTEMS. ECOLOGICAL STUDIES VOL. 136 SPRINGER, BERLIN PP. 23–50.

One of the last unspoiled spots along the shores of the Mediterranean Coast in Israel, with flowering *Pancreatum maritinum* in Poleg Nature Reserve.

6.1 INTRODUCTION

Human disturbance in mediterranean-type ecosystems continues to be a subject of serious concern, and nowhere is such disturbance and degradation of natural ecological processes and biodiversity more evident than in the Mediterranean Basin itself. In this chapter I will point out currently prevailing trends in Mediterranean landscapes in the context of our global environmental crisis. This will lead to a discussion of the close interrelations between biological diversity, ecological heterogeneity, and cultural diversity, or in one word, total landscape *ecodiversity* in the Mediterranean Basin, and to recent advances in landscape ecology in the evaluation of this issue. I will conclude by offering some new approaches and tools for holistic ecodiversity conservation as an integral part of the urgently needed environmental revolution.

6.2 THE NEO-TECHNOLOGICAL IMPOVERISHMENT OF THE OPEN MEDITERRANEAN LANDSCAPE

6.2.1 Current Trends

Current trends in Mediterranean landscapes can be summarized in one sentence as the "bad and good news". I will start with the bad news: the quality of the environment and of the open landscape around the Mediterranean Basin is further deteriorating at an exponential rate in time and space. In a recent volume on landscape ecology (Naveh and Lieberman 1994) we have described these threatening trends of *neo-technological* landscape degradation as a major part of the global environmental crisis. It is operating in the Mediterranean Basin as a vicious circle of mutually amplifying and destabilizing feedbacks of exponentional growth of populations and their needs for open land, coupled with the land use pressures of rapidly growing tourism. These are combined with synergistic processes of greater intensity of traditional and modern agricultural land uses and urban-industrial expansion on one hand, and depopulation and land abandonment on the other.

If the scenarios of The Blue Plan of the United Nations Environmental Program (UNEP 1988) come true, then the total population around the Mediterranean Basin will grow from about 360 million to more than 550 million in the next 30 years, nearly two-thirds of this in countries south and east of the basin. The main threats to environmental quality are in the coastal regions with the largest urban industrial concentrations and the fastest growing tourism developments. Here, population will increase from about 140 million at present to 195–217 million in 2025, an increase of 45–62%. Its urban populations alone will rise from 82 to 145–170 million. In addition to around 200 existing energy and industrial installations of oil ports, refineries and thermal power plants, many more will be required by 2025, in addition to cement plants, steel works, fertilizer plants, etc. At the same time, the tourist flow to the Mediterranean coast will increase from 95 million today to 220 million tourists in 2025 and may even treble. About 40% of these holiday makers are merely seeking relaxation, sun, sand and the sea, without caring much about natural and cultural values. Therefore, their mass-accommodation close to the beaches is the main socio-economic and cultural force driving further uncontrolled development of the rapidly diminishing open shorelines and their intrinsic biological, scenic and other *soft* landscape values over much larger areas. Thus, for instance, Marcuzzi (1991) has used the disappearance of typical sand dune tenebrionid beetles from the Adriatic coast of Italy as a sensitive bioindicator of the detrimental impacts of mass tourism on coastal dune areas.

The resulting rapid loss of almost all natural coastal sand dune habitats and wetlands and the advanced degrees of air and water pollution in the lowlands are combined with a rapid process of homogenization, fragmentation, denudation and destabilization in the uplands. The loss of the vital protection and carrier functions of the vegetated mountain watersheds have far-reaching and many times even catastrophic implications for the intensively cultivated and/or densely populated valleys and coastal areas. The final result is accelerated and widespread decline of the unique biological, ecological, and cultural ecodiversity, productivity and stability, leading to more frequent and more destructive wildfires, and to accelerated soil erosion and flooding.

Unfortunately, the ecological, social, and economic damage inflicted by such land abuse and unrestrained developments are not included in the narrow and short-sighted market-based cost/benefit calculations which are still guiding the decisions on land uses in most Mediterranean countries.

One of the major causes for this ecological debacle is the rapid and careless shift from diversified and stable traditional agriculture to intensive large-scale, agro-industrial farming. Chiefly in the western Mediterranean, such farming has caused widespread land abandonment. But in all Mediterranean countries it is combined with indiscriminate planting of pine and eucalyptus monocultures creating large stretches of monotonous and highly flammable vegetation. These developments have destabilized and despoiled rich and unique landscapes. In spite of their detrimental ecological impacts and their highly dubious economic justification, such practices have been promoted by local governments and were heavily subsidized by the European Unit (EU). These high technology and high-input agricultural practices are further increasing rates of air, water and soil pollution, soil erosion, salinization, and siltation.

A typical example is Extremadura, one of the poorest regions in southern Spain (CEPA 1992). Here, the EU has allocated great sums from its European Structural Funds, in order "to develop and to bring this region into line with the rest of Europe". In reality, however, these developments have undermined the local economy by destroying thousands of hectares of Mediterranean woodlands, thereby altering traditional agricultural practices and degrading the natural, scenic and cultural resources of this region. In addition to the great investments in the lowlands, huge sums have been spent on large-scale afforestation projects, and on so-called conservation and erosion control projects. For these, heavy machinery and bulldozers are used to clear existing maquis vegetation and build roads and terraces. Instead of controlling water run-off and soil erosion by vegetative stabilization and appropriate engineering, these projects have accelerated processes of landscape degradation by leveling large areas and removing their protective natural vegetation cover. In this way, the upper and richest soil profiles which had taken hundreds of years to build have been destroyed. Severe sedimentation occurs as a result, with more than 30 tons of soil per hectare washed away every year.

Among those threats whose danger has not yet been recognized sufficiently is the alarming rise of photochemical oxidant pollutants and especially of tropospheric ozone oxidant pollutants and acid deposition. Such atmospheric pollutants may be playing a major role in the progressive decline of coniferous forests in the Mediterranean Basin (Ferretti et al. 1995; Milan et al. 1996). As in California, these threats from photochemical air pollutants are not restricted to the coastal urbanized zones close to the emission sources, but spread also farther inland (Naveh et al. 1981). In 1987, 82% of *Pinus pinea* in Sardinia, and 78% of these trees in Tuscany showed evidence of pollution damage. Other woody species, especially evergreen oaks and other broad-leaved maquis trees, were also declining (Bussotti et al. 1991, 1992).

In their thorough study of the transformation of agricultural landscapes in Europe, Meeus et al. (1990) predicted that in the future, the *montados* of the Iberian Peninsula and the Italian *coltura promiscua* landscapes will no longer exists, except in history books or national parks and outdoor museums, because of marginalization, abandonment, deterioration and the futile attempts of reforestation of the mountain slopes.

Such a doomsday scenario seems to approach closer every year, and landscape disturbance is now affecting remote mountainous uplands. The speed and extent of human encroachment can be demonstrated in the western Galilee of Israel (Fig. 1). Here, within 14 years, most of the open agro-pastoral (so-called rugged land) has been lost or fragmented by monospecific pine plantations and intensive agriculture, as well as by urban-industrial developments.

As a result of large-scale emigration and agricultural intensification (encouraged by EU policies and subsidized wheat prices) over large areas of these uplands in the western Mediterranean, the rich mosaics of coppiced and managed oak and chestnut forests, mountain grasslands, and the terraced mixed vine/olive and crop cultures are being replaced by intensive monocultures of wheat and grapes, or by olive orchards on non-terraced slopes, and by pine plantations and secondary scrub on abandoned pastures and fields. Since the soils of non-terraced slopes are cultivated now by heavy machinery they are suffering from erosion and losing their fertility.

This breaking down of fine-grained patterns into coarser landscape patterns with impoverished floras and faunas has been documented in the Solano Basin of Tuscany (Vos and Stortelder 1992). If this process of biological, cultural and scenic impoverishment continues unimpeded, then, the four richest out of ten major land units will vanish completely by the year 2035. The remaining land units will be chiefly intensively cultivated, heavily sprayed and fertilized fields, with planted coniferous forests spreading over the whole uncultivated area, and secondary communities on abandoned pastures, thereby creating the same species-poor oak and pine forest types over extensive areas (Fig. 2). Many more examples of this closely interwoven biological, ecological and

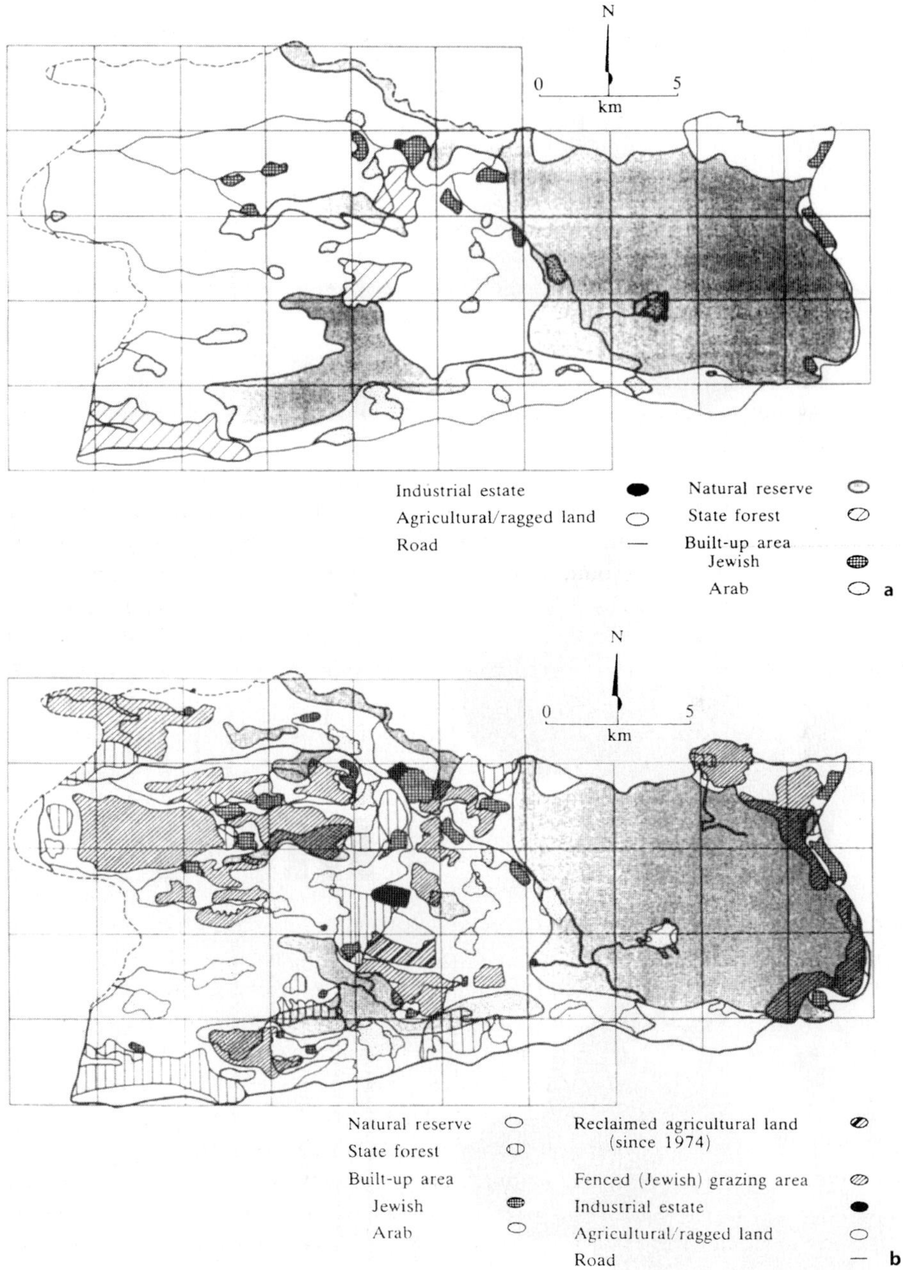

Fig. 1. Major land uses in the Yehiam region western Galilee, Israel: *a* 1974; *b* 1988 (Yiftach-El and Law-Yone 1995).

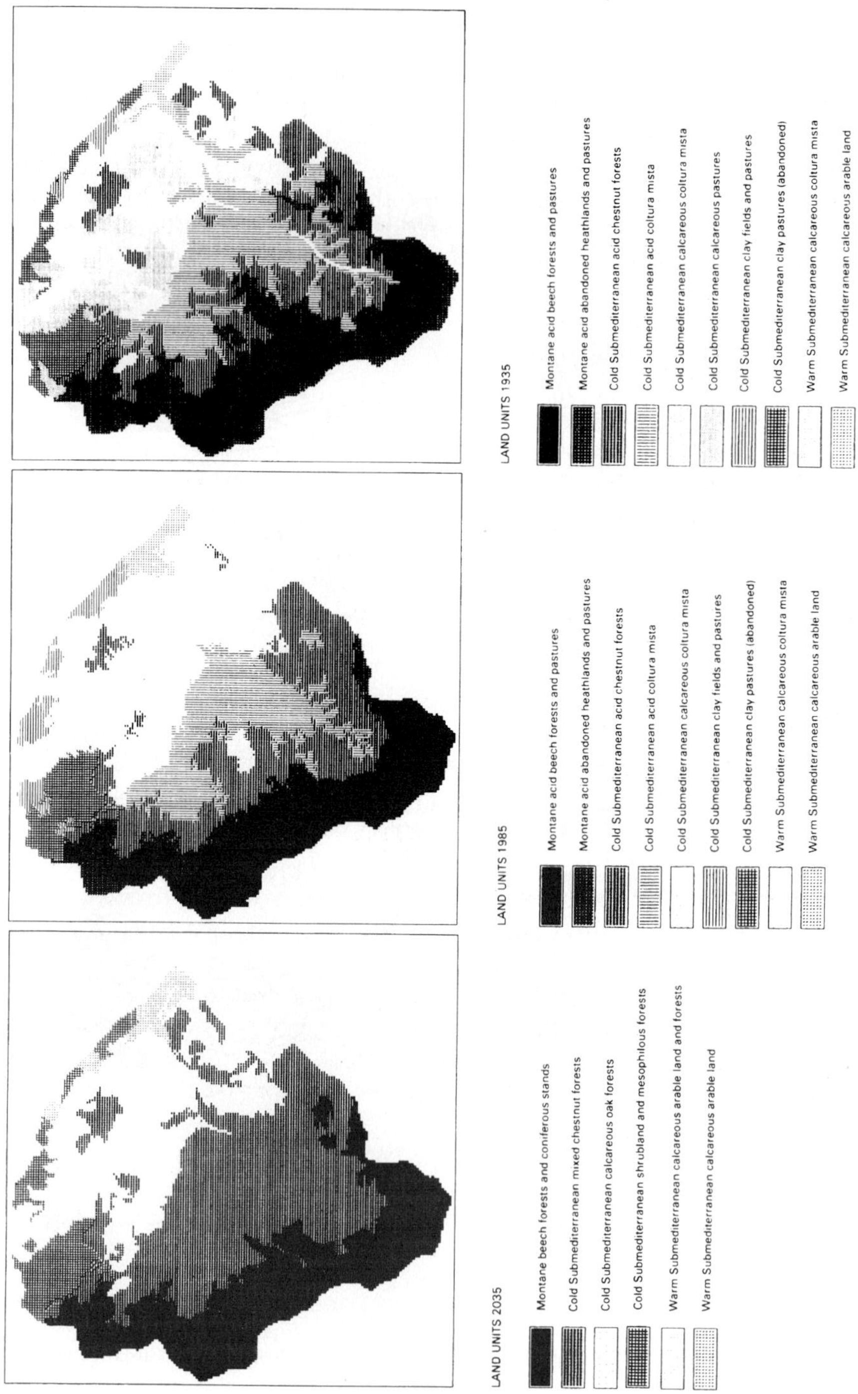

Fig. 2. The vanishing occurrence of Tuscan landscape units between 1935 and 2015 (Vos And Stortelder 1992).

cultural impoverishment have been provided recently in a symposium dealing with land abandonment (Baudry and Bunce 1989), and in conference proceedings dealing with the future of Mediterranean landscapes (Farina and Naveh 1993).

One of the best examples of planning efforts toward the resource management of Mediterranean landscapes comes from an interdisciplinary study of landscape planning and ecology in western Crete. This study, described in detail in Chapter 7 of this book, was initiated by the Working Group for the Conservation of Landscapes of the IUCN Commission on Environmental Strategies and Planning (WGCL-CESP) as the first case study of *Green Books for Landscape Conservation*, and was funded by the EU EPOCH program (European Program on Climatology and Natural Hazards). The study was carried out by a multinational and interdisciplinary team of landscape ecologists and planners, botanists, physical and cultural geographers, forest and range, and remote sensing specialists.

However, the most severe problems of environmental degradation come from the fastest growing semi-arid countries of the eastern and southern Mediterranean Basin. Here, the major causes for land degradation and desertification are overgrazing and excessive removal of fuelwood. According to FAO studies, in the year 2000 about 140 million people in these countries will suffer a deficit of close to 30 million m^3 of wood to supply their annual needs for cooking and heating. Taking into consideration the low wood production of these forests and woodlands, extreme deforestation of twindling woodlands will likely affect very large areas.

6.2.2 Impacts of Global Change

Predicted regional warming and the destabilization of rainfall and temperature regimes resulting from global change may have severe repercussions on the biological diversity and stability of inland and upland landscapes in the Mediterranean Basin. In addition to the destructive impact of rising sea levels on coastal shores, and the far-reaching effects on hydrological regimes and on agricultural crops, global change could further aggravate the above-mentioned processes of landscape degradation (Jeftic 1993).

In the more humid western Mediterranean ecosystems these predicted climatic changes could further enhance brush encroachment in abandoned and neglected forests and maquis, and thus encourage their floristic impoverishment and increase wildfire hazards. In the subhumid ecosystems such as those of northern Israel, the expected rise in summer temperatures and in evaporation rates will have severe repercussions on the diversity and stability of the most mesic and diverse plant communities (Naveh 1995a).

In the driest semi-arid ecotones of the Mediterranean these climatic stresses would even further aggravate the existing processes of landscape desertification induced by heavy human and livestock pressures. This may cause the advancement of semi-arid and semi-desert formations far into the heart of the Mediterranean biomes. Considering that the desert encroachment into the semi-arid Mediterranean region is presently estimated at more than 2% per annum, this would have truly catastrophic consequences (Le Houerou 1992).

Because of the chaotic and non-linear behavior of atmosphere-biosphere interactions, and their complex and mostly synergistic interactions with other environmental stresses, our present knowledge is not sufficient for any detailed, robust prediction models. Therefore, at this stage the only certain prediction which can be made is that we will face a period of increasing uncertainty in spatial and temporal climatic trends and their ecological effects. We can also anticipate with great certainty that any increase in climatic stress will further aggravate the process of overall landscape degradation (Naveh 1995a).

6.3 NEW HOPES FOR MEDITERRANEAN LANDSCAPES

The good news is that there is now also a rapidly growing public awareness of these perils. There are presently many more research activities and better conservation management policies leading in the right direction; and important international, national and non-governmental initiatives are instilling some hope that this situation can be changed for the better before it is too late in at least in some of these countries.

Among the most outstanding international examples is the Mediterranean Action Plan (MAP) of the United Nations Environmental Program (UNEP), which was approved in 1975 by 16 Mediterranean countries. It called for a series of legally-binding treaties to be drawn up and signed by the Mediterranean governments, the creation of a pollution monitoring and research network, and a socio-economic program to reconcile development priorities with a healthy Mediterranean environment. Up to now, the most important product of this group has been the Blue Plan (UNEP 1988), as the first prospective study on the relationship between the environment and development in the Mediterranean. It has reviewed present trends and also recommended scenarios for sustainable integrated social and economic development of the Mediterranean Basin.

Another promising development is related to the ongoing peace talks in the Middle East which began in 1992, and brought together ministers and top governmental delegates from all Arab countries and Israel. A constructive dialog on environmental issues and their solution at the level of specialists and decision-makers has resulted from these talks. If this and other important international and regional activities are taken seriously by law-makers, planners, politicians and decision-makers it may lead the way for a better future for these landscapes in all participating countries. This will be especially the case if the new EU agricultural policy recognizing the natural, cultural and socio-economic values of marginal lands is put into practice. The same is also true for other important initiatives such as a convention for the protection of Europe's rural landscapes.

The importance of holistic studies on landscape planning was in studies on vanishing Tuscan landscapes (Vos and Stortelder 1992). Based on this report, the regional Tuscan government requested that there be a comprehensive landscape assessment of the ecological impacts of the planned Farva River dam and impoundment. The conclusions of this report led to cancellation of ambitious plans for intensive agricultural development and irrigation of the biologically significant and scenic Grosseto Plains, which would have had disastrous environmental consequences (Pedroli et al. 1988). The regional Tuscan government has supported further ecological landscape studies, as well as conferences, workshops and training courses.

6.4 BIODIVERSITY AND ECODIVERSITY IN THE MEDITERRANEAN BASIN

It is now generally recognized among all those who care for the fate of our planet that the most alarming immediate global threat to life on earth is that of biological impoverishment. The rapid reduction of the biotic resources threatens the integrity of the global biosphere landscapes and their life-support systems (Norton 1987; Kim and Weaver 1994a). The severity of this threat can be measured by the predicted per capita shrinkage of all human-modified and used landscapes in relation to population growth. According to the latest report of the Worldwatch Institute (1994), a predicted increase in the world population by 33% by the year 2010 will result in the drop of per capita forest cover by 30%, grazing and pasture land by 22%, cropland by 21%, and irrigated land by 12%. Although representing only a very small portion of all the terrestrial landscapes on earth, the Mediterranean Basin is of special significance because its unique evolutionary and cultural history makes it one of the richest biological regions outside the tropics.

As described elsewhere in detail (Naveh 1984, 1990b; Naveh and Lieberman 1994), we can assume that the final geomorphologic, climatological and biological evolution of Mediterranean landscapes in the Pleistocene coincided with the major phases of Mediterranean human evolution from the Acheulian hunter-gatherer *Homo erectus* to the food-producing neolithic *Homo sapiens*. In this closely physical, biological and cultural environment, both natural and anthropogenic fires became a major driving force in human culture. Human use of fire can be traced back several hundred thousand years in Israel, thanks to findings of stone tools, artifacts, and the charred and fractured bones of 80 different animal species in the fire-swept volcanic soils of the ecotones between semiarid tropical and Mediterranean forests and savannas of the Jordan Valley (Haas 1966; Stekelis 1966). This region was also one of the first cradles of cereal and stock farming about 10 000 years ago, and its drought, fire and grazing-resistant annual grasses and legumes were among the first domesticated plants. This process of ecotypic variation and speciation was continued and gradually intensified during the long period of human occupation. Through this process, those Mediterranean uplands which are steep and rocky or have soils which are too poor and/or shallow and for intensive, mechanized agriculture, have become the largest refuge of spontaneously occurring and reproducing plants and animals in Europe, the Near East and North Africa. Because of their longer history of intensive human modification, these habitats have high degrees of structural, floral and faunal diversities compared to mediterranean-type landscapes in Chile and California (Naveh and Whittaker 1979).

The Mediterranean Basin flora is distinguished by its great number of herbaceous plants, including many annuals and exceptionally rich and colorful flowering

Asteraceae and geophytes with ornamental value. Among these are many rare and endemic plants. For example, there are more than 1000 different species of flowering plants within a 10 km radius from Jerusalem, while Greece contains more than 6000 different species of flowering plants. Much attention is presently devoted by geneticists and plant breeders to important cereals pasture grasses and legumes from the Mediterraean Basin such as Emmer wheat (*Triticum diccocoides*), the progenitor of cultivated wheat. Among woody plants, and especially in the Labiatae, there are many species with great value for pharmaceutical, cosmetic, spice, balsam, and other uses. These are now collected, grown and utilized with increasing intensity for commercial production. A number of deeply rooted, evergreen and hardy shrubs are now widely used for erosion control on steep and denuded slopes.

The genetic diversity on the subspecies level is indicated by the results of a study of 21 plant and animal species from the Mediterranean and steppe region of Israel which revealed high levels of allozyme diversity and DNA polymorphism. This genetic diversity has been hypothesized to correlate with the broad ecological plasticity shown by many of these species (Nevo 1988).

Mediterranean landscapes provide an ecological bridge between Europe, Africa and Asia, and their wetlands serve as vital breeding and feeding stations for hundreds of resident and migrating bird species crossing the Mediterranean Sea between Europe and Africa.

Contrary to a widely held misconception, Mediterranean peoples did not always degrade their habitats and deplete their flora and fauna. They reclaimed these uplands with great skill, utilized their biological productivity and preserved at the same time their organic diversity. This happened when ecological, socio-economic and cultural peaks were reached during biblical and classical times, and much later again during the Renaissance chiefly in the western Mediterranean.

It is not by chance that in these periods landscapes were perceived as the total natural and human living space, combining beauty and utility. In the Bible, the Hebrew word *noff* (etymologically related to *jafeh* meaning beautiful) symbolizes the overall view of King Solomon's Jerusalem and its surrounding green hills. The lofty mountains in which the Greeks erected their temples, symbolized the closeness of God to man in his natural, most beautiful and harmonious surroundings, and the graciously terraced slopes of olive groves and vineyards and interspersed with majestic cypresses and the dark green oak forests of the Tuscan landscape have served as a source of inspiration for many generations of artists and poets.

Di Castri (1981) has clearly described the closely interwoven natural and cultural multidimensional and multiscale heterogeneity of Mediterranean ecosystems and landscapes. This diversity was the result of the closely coupled natural and anthropogenic feedback processes which induced the co-evolution of Mediterranean peoples and their cultural landscapes.

There is a close interdependence between biological and cultural diversity and the threats both by neo-technological landscape degradation and by despoliation. Special efforts were made by a research team studying resource management issues in western Crete to distinguish genuine threats from changes that are part of the inherent cycle of changes, and to link ecological conservation to other branches such as archeological and historical conservation (Grove et al. 1993, 1994). This enabled the recognition of the threats to total landscape ecodiversity by the declining biological, historical, and scenic values. As a result, from the long interplay of a rugged topography with human activities and wild plants and animals, the White Mountain region of Crete continues to be a fine-grained and harmonious Mediterranean landscape, rich in cultural features.

6.5 MEDITERRANEAN LANDSCAPES AS PERTURBATION-DEPENDENT NON-EQUILIBRIUM SYSTEMS

Several prominent American ecologists have proposed a new paradigm in ecology on which conservation and restoration ecology should be based. This paradigm replaces the metaphor of a balance of nature maintained without human interference with the flow of nature through dynamic change induced by disturbances including humans and their effects (Pickett 1994) or by the recognition of *discordant harmonies* (Botkin 1990).

It was shown more than 20 years ago that the mechanistic and deterministic concepts of a homeostatic stable state equilibrium, presented in classical ecology by the succession-to-climax paradigm, is not suitable for conservation management in the Mediterranean Basin (Naveh 1974; Naveh and Dan 1973). We introduced new insights of systems theory and system dynamics into landscape ecology, especially the new thermodynamic principles of self-organization in non-equilibrium systems (Prigogine 1976; Prigogine and Stengers 1984) and the concepts of homeorhetic flow equilibrium developed by Waddington (1975) as a catastrophe theory of evolution (Naveh 1982; Naveh and Lieberman 1994). Any new paradigm in ecology with relevance for conservation management cannot ignore these and other new insights into systems dynamics.

As has been discussed previously in more detail (Naveh 1987, 1991a), Mediterranean landscapes, as well as most other cultural semi-natural and agro-silvo-pastoral landscapes, behave in thermodynamic terms like dissipative structures. They are maintained and stabilized only by permanently interchanging energy and entropy with their environment. Driven by positive feedbacks of environmental and internal fluctuations, they move to new intermediate states that generate conditions of renewal of higher internal entropy production, while undergoing short- and long-term cyclic fluctuations. Thus they are far from a homeostatic equilibrium stage.

Such cyclical perturbations have been introduced for centuries by rotational grazing, browsing, burning, cutting and coppicing regimes, together with cultivation and other human land uses. These human perturbations were superimposed on the seasonal and annual climatic fluctuations, and their resulting defoliation pressures were incorporated into the landscape, together with these and other natural perturbations at different spatio-temporal scales. They resulted in the establishment of a human-maintained and dynamic long- and short-term flow equilibrium or *homeorhesis* (from the Greek, meaning preserving the flow) between the tree, shrub, herb, and grass layers and the cultural artifacts of forests, woodlands, shrublands and grasslands.

In homeorhesis, the system is not returned to a stable state as in homeostatic climax systems, but moves along the same trajectory of change, as long as these cyclic perturbations are continued with similar intensities and frequencies. In this manner, these perturbation-dependent systems have acquired long-term adaptive resilience and evolutionary metastability. The long-term maintenance of such a homeorhetic flow equilibrium, operating within the great macro- and microsite heterogeneity of such rocky and rough terrain, producing the fine-grained agro-silvo-pastoral land use patterns, apparently played a major role in inducing the unique ecological and cultural ecodiversity of these landscapes.

A typical example for such a human-perturbation dependent landscape and the present neo-technological distortion of homeorhetic flow equilibrium, is the *montado* in southern Portugal and the *dehesa* in southwestern Spain. As described in more detail by Pinto-Correia (1993), this agro-silvo-pastoral system was well adapted to the low edaphic and climatic potentials of these uplands in the complementary utilization of the open evergreen woodlands of oaks (chiefly *Quercus suber*), olives and chestnut trees for cork, fruit, wood and livestock production, and cereal cultivation. Managed extensively for many decades with minimum human inputs, this land use created a fine- grained heterogeneous and attractive landscape with relatively high biodiversity. At the same time it furnished also the necessary *hard* economic outputs for the dispersed rural population. However, today there are highly intensive and extensive land uses that are distorting this human-maintained homeorhetic flow equilibrium and endangering not only the biological and economic productivity, but also the ecodiversity and stability of these unique cultural landscapes (Figs. 3–4).

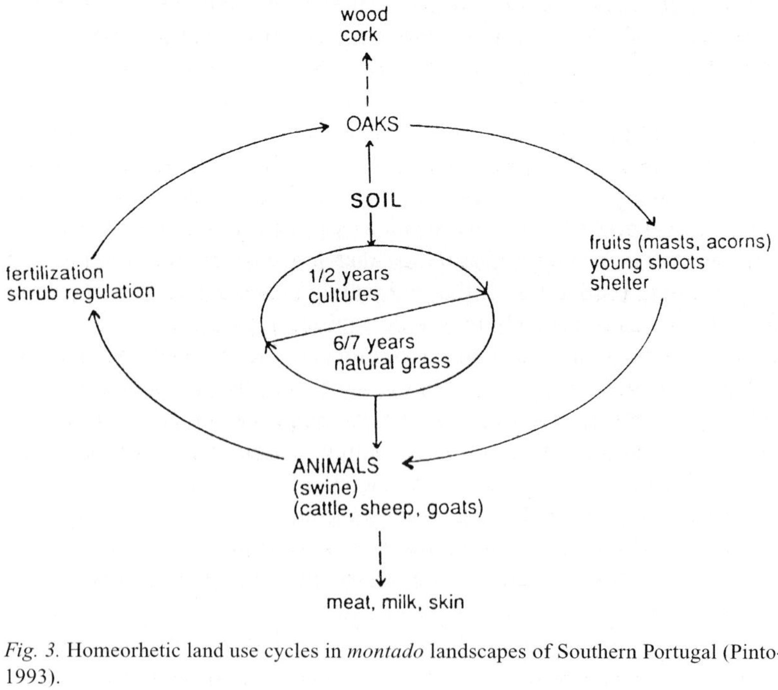

Fig. 3. Homeorhetic land use cycles in *montado* landscapes of Southern Portugal (Pinto-Correia 1993).

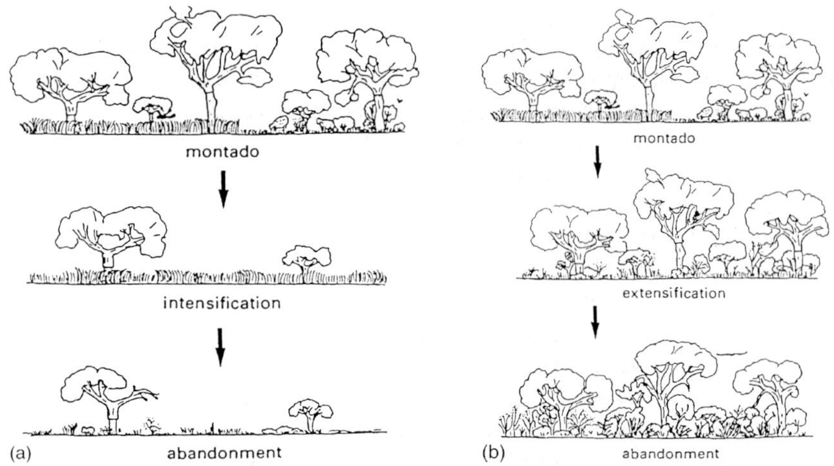

Fig. 4. (*a*) Degradation of the *montado* by intensified land use (Pinto-Correia 1993). In order to intensify crop production, oak density was reduced, the soil work mechanized, and grazing animals excluded. Trees were severely damaged, soil erosion increased, and crop yields remained too low. The fragile relationships of the *montado* system were broken and the resources deteriorated. The final result was abandonment, since the land no longer had production potential and there was no capacity for regeneration. In a further phase this area will probably be planted with eucalyptus trees for production of wood fiber. (*b*) Degradation of the *montado* by extensification (Pinto-Correia 1993). Due to the end of pig raising as a result of African swine fever and in general the effect of changing socio-economic factors, the soil in the *montado* has been progressively used in a more extensive way. Cultivation and sometimes grazing disappear. The density of the undercover vegetation increases, until it forms a new assemblage of maquis where tree production must also be abandoned. Such shrub formations correspond to a simplification of the landscape, increased risk of fire, and little or no possibility of exploitation.

The results of our studies of maquis forests and shrublands in northern Israel and southern France (Naveh and Whittaker 1979; Naveh 1982, Schreiber and Naveh, unpubl. data) can serve as a further indication of the disruption of this homeorhetic flow equilibrium. Wherever moderate, traditional defoliation pressures had maintained open, grassy patches for light demanding herbaceous plants, including also highly ornamental geophytes, we found a very high floristic and structural diversity. But where these pressures had ceased, either for the sake of nature protection or as a result of land abandonment, or where they had been intensified, both floristic and structural diversity had been reduced considerably. In most cases, species richness dropped by 75%, namely from more than 100–120 species to less than 30 species per 0.1 ha. Equitability was much lower and the dominance concentration of the most aggressive and shade tolerant trees and shrubs increased significantly. At the same time, structural vegetation diversity was also reduced considerably. Thus, for instance, on Mt. Carmel, the rich herbaceous understory and gap vegetation of semi-open and patchy sclerophyll forests and shrublands contained 93 species, as opposed to only a few shade tolerant perennial grasses and herbs which could survive under the dense, undisturbed tall shrub canopy, dominated by *Quercus calliprinos*. This is shown in Fig. 6, comparing the dominance-diversity curves of the Muhraqa site which at that time had not been disturbed for more than 30 years, and the tall Forty Oaks Grove, which had been grazed and coppiced in the past. Here the woody species are well partitioned between the different strata and species, leading to a much lower dominance concentration and higher structural and floristic diversity contributed to by many small and densely packed annuals (therophytes) and a wealth of flowering geophytes, as well as hemicryptophytes. In simultaneously conducted zoological studies, animal species richness and relative abundance of birds, reptiles, rodents and isopods showed similar trends (Warburg 1977; Warburg et al. 1978) (Fig. 5).

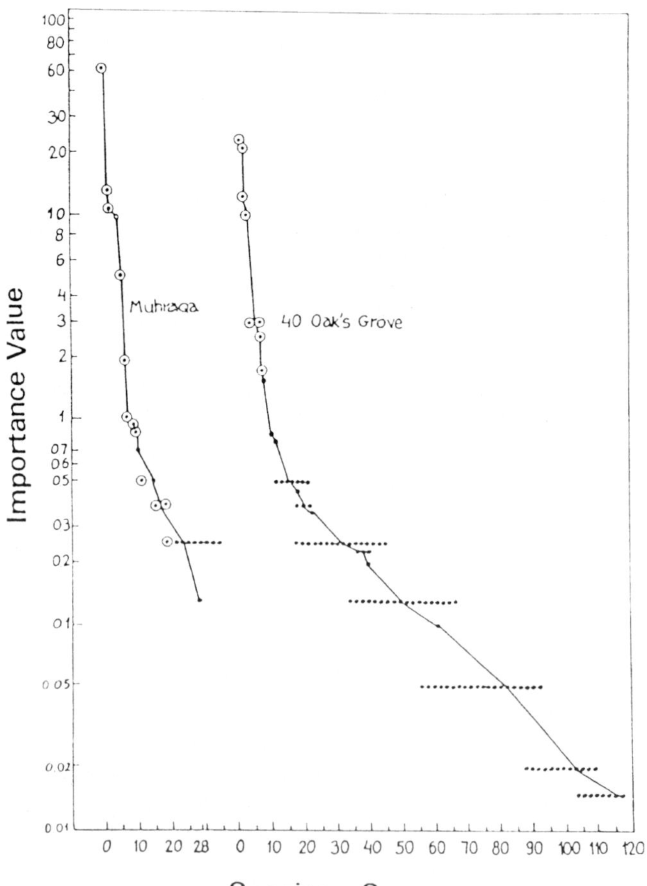

Fig. 5. Dominance-diversity curves for all vascular plant species in 1000 m^2 communities of closed and semi-open maquis on Mt. Carmel, Israel (Naveh and Whittaker 1979). Species percentages of mean cover are used as the importance values in a logarithmic scale on the ordinate, and species numbers in the sequence from the most important to the least important on the absicca. *Dots* are herb species and *circled dots* are woody species.

In open oak woodlands of Tabor oak (*Quercus ithaburense*), which can be considered as the eastern Mediterranean counterpart to the *montados* and *dehesas*, highest species richness and diversity were reached under moderate grazing pressures and lowest under both heavy and light grazing pressure or complete protection. This is true also for comparable open *Q. dumosa* woodlands in California (Naveh and Whittaker 1979).

The effect of such different grazing pressures on herbaceous plant species diversity is shown in Fig. 6a–c. In such complex microsites, a dynamic flow equilibrium is maintained between the small-scale heterogeneity created by variable soil depth and micro-relief, availability of moisture and nutrients, rock outcrops, tree cover, litter cover, shade and the abundance of different plant species, as affected by grazing and trampling (Mann 1986; Naveh 1991a; L. Olswig-Whittaker, M. Blumler, and Z. Naveh, unpubl. data).

Terrace walls, rock outcrops and their shallow soil pockets and crevices can serve as refuges for many species and genotypes which are well adapted to these microsite scales. The latter is indicated by the distinct allozyme polymorphism of adjacent populations of *Hordeum spontaneum*, growing either in the open, deeper soil patches or nearby in very shallow rock pockets. For such a rock refuge population, higher drought tolerance is essential to ensure survival and viable seed production (Nevo et al. 1986). In many cases such rock niches are archeological remnants of ancient olive or wine presses, or other cultural artifacts, increasing ecological heterogeneity and biological diversity.

A recent survey of all nature reserves in northern Israel, over an area of about 3000 ha of mixed vegetation (maquis, woodlands, shrublands, and grasslands), corroborated some of our findings on much larger scales. According to Noy-Meir and Kaplan (1991), and Kaplan (1992), highest vascular plant richness and diversity was maintained by light to moderate grazing of domestic livestock, and by free grazing of wild animals, chiefly mountain gazelles and boars. In protected and too lightly grazed sites, species richness and diversity was not only lower, but the frequency of wildfires was also higher. Protected herbaceous plant communities dominated chiefly by tall grasses, also suffered more from periodic outbreaks of rodent populations.

This pattern of diversity is true also for all other Mediterranean countries. As has been described previously, the cessation of all human interference by land abandonment or by complete protection can lead to the impoverishment of structural and biological diversity. Most affected are the light demanding herbaceous species, including many rare endemics and ornamental geophytes (Ruiz de la Torre 1985). Also influenced are many vertebrates, especially bird species, whose loss is closely coupled with the loss of overall landscape heterogeneity and diversity (Farina 1989).

6.6 ECODIVERSITY AND LANDSCAPE ECOLOGY

At present, there are strong tendencies to shift some of the major conservation efforts from the protection of rare and endangered species and their natural habitats and ecosystems in restricted nature reserves and parks to the protection of most valuable open semi-natural and rural landscapes. Therefore, the need for developing a sound conceptual and methodological foundation for the conservation of total landscape eco-diversity has also become acute on a global scale. On the international level, these efforts are carried out by IUCN and UNESCO, chiefly through the establishment of

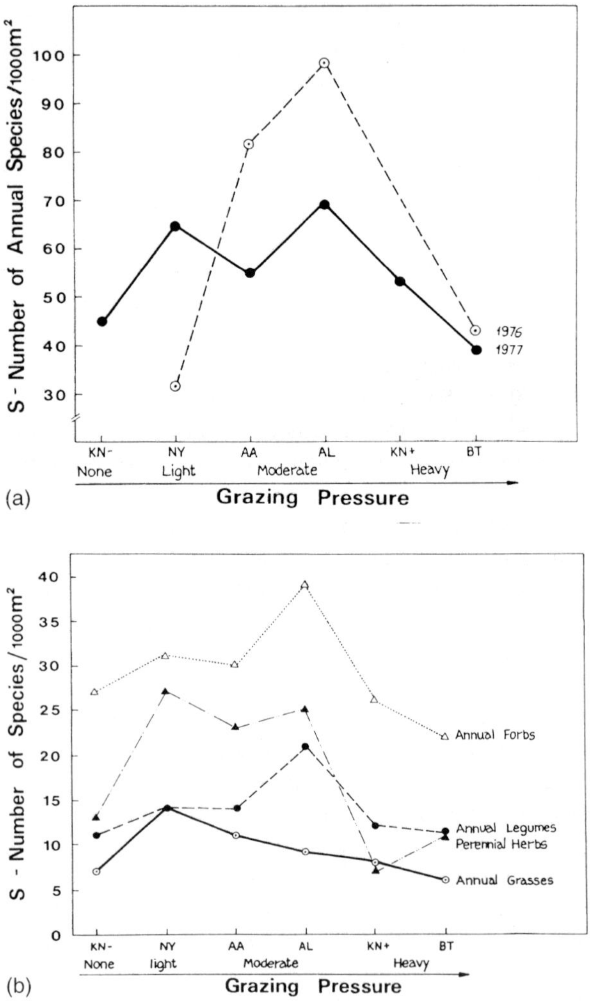

Fig. 6a–c. Species richness and diversity of open woodlands and shrublands in the lower Galilee, Israel, as related to grazing pressure (Naveh and Whittaker 1979): (*a*) Annual species richness in 1976 and 1977. *KN-3* years protection; *NY* very lightly grazed in 1976 and lightly grazed in 1977; *AA* moderately grazed; *AL* rotationally grazed; *KN+* heavily grazed; *BT* very heavily grazed (spring sampling shortly after grazing reduced *s* in *AA* and *AL*: in 1977, and more intensive grazing increased *s* in *NY* in 1977). (*b*) Species richness of major herb groups, 1977. Pasture plots and grazing intensities as in a; (*c*) Dominance concentration reciprocal (1/c) and equitability (EXP H) of vascular plant species, 1977. Pasture plots and grazing intensities as in *a*.

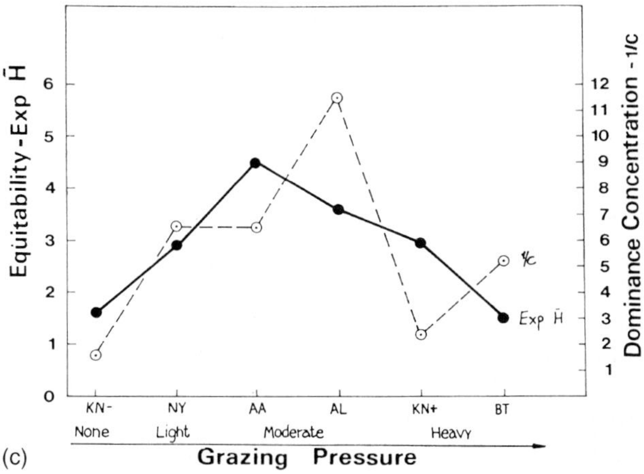

(c)

Fig. 6. Continued.

Biosphere Reserves and World Heritage Sites. But as has been described recently in detail by Lucas (1992), there are also many other regional and local success stories of protected natural and managed cultural landscapes. Lucas emphasized their important role in the strategies for sustainable living as outlined in the important document *Caring For the Earth* (IUCN/WWF/UNEP 1991), and provided a very useful guide for the selection, implementation, legalization and management of such protected landscapes. In this context, a clear distinction has to be made between:

1. Solar-powered biosphere landscapes or *bio-landscapes*, which are driven mainly by solar energy and its biological conversion and are regulated to greater or lesser extents by natural, bio-physico-chemical information. In addition to the natural and near natural landscapes, these also include the two most important cultural biosphere landscape classes for ecodiversity conservation, namely: (a) semi-natural and agro-silvo-pastoral (or semi-agricultural) landscapes in which their natural (spontaneously occurring and reproducing) biota has been modified by human land uses; and (b) low-input traditional agricultural landscapes and their agro-ecosystems. The latter are subsidized only to a low degree by fossil energy and chemicals, but are still controlled to a large degree by natural information and processes. To the latter belong the traditional agro-ecosystems as well as planted forests and sown pastures in rural landscapes, and we may also include here the rapidly expanding organic farming systems.
2. Intensively cultivated, fertilized, and chemically treated *agro-industrial* landscapes which depend on solar energy for their production. They are, however, heavily subsidized by large inputs of low-grade energy of entropy and pollution-producing fuels and chemicals, and are controlled almost entirely by human agro-technology. As such they have no specific conservation values, and may even have detrimental environmental impacts on the remaining open landscapes and their biological and ecological diversity and stability (Naveh 1987).

3. Fossil-fuel powered technosphere landscapes, with highways, bridges and power lines that are entirely artificial, are human-made and controlled. They are driven by the technological conversion of high-grade solar energy into low-grade energy, which produces great amounts of entropy, pollution, and waste products. The overall environmental impact of these landscapes on our total human ecosystem is indicated by the huge amount of energy that flows annually through great urban-industrialized centers. According to Odum (1993), this flow can be measured in millions of kcal per square meter, as compared to only thousands of kcal in solar powered landscapes.

The exponential and generally uncontrolled expansion, and chiefly adverse role, of the partly and wholly fuel-powered landscape classes is one of the major causes of our global environmental crisis. However, as has been shown in successful examples of rural conservation management in Great Britain (Green 1985) and of other rural landscapes elsewhere (Lucas 1992), there is still hope that the antagonistic relations between these two major cultural landscape classes can be transformed into mutually beneficial, and even symbiotic relations.

The broadening of conservation perspectives from species and natural ecosystems to cultural bio-landscapes and their intrinsic and instrumental values requires the broadening of spatial scales and conceptual and methodological dimensions from the realm of natural sciences to the realm of humanities, dealing with human land uses. The uniqueness of these cultural bio-landscapes as the tangible meeting point between nature and mind cannot be conceived, studied and managed by conventional, mechanistic and mostly reductionist bio-ecological approaches and methods. They require a multidisciplinary landscape-ecology approach to land use, helping to bridge the gaps between bio-ecology and human ecology. These issues have been discussed elsewhere in detail (Naveh 1990a, 1991b, 1994, 1995b). Here I can point only to some of the methodological and practical aspects of ecodiversity conservation.

In recent years landscape ecology has made great strides in the development of integrative and cross-disciplinary methods, enabling workers to identify, analyze and synthesize and, at least partly, to quantify in more holistic ways complex natural and cultural patterns and processes at different temporal and spatial scales. This has been achieved with the help of recent advances in computer hardware and software, remote sensing and satellite images, and Intelligent Geographic Information Systems (IGIS) combining artificial intelligence with geographic information systems. As has been reviewed recently in more detail (Naveh and Lieberman 1994), these have been utilized in landscape ecological studies together with the application of innovative mathematical and system theories, such as information theory, fractal geometry, fuzzy logic, and hierarchy theory.

A first important step for ecodiversity conservation has been made in California by Davis et al. (1990). They proposed the use of comprehensive GIS models to organize existing biodiversity data and to improve the spatial aspects of their assessment for the US National Biodiversity Center along multilevel spatial scales from the local site to the regional and the biogeographical scale. In each spatial level the specific cultural features of land use and their effect on environmental quality and landscape structure were also taken into consideration. The model shows clearly the complexity of biodiversity assessments for management and conservation purposes. It is meant to facilitate the needed cooperation and coordination across administrative and political borders.

A practical ecodiversity approach has already been applied in Denmark for conservation management of small and locally distributed communities (e.g. natural or artificial lakes and pools, rivers, canals, drainage ditches, hedges, road verges, tree rows, brush thickets, bogs, and marl pits) as natural and cultural biosphere islands in the monotonous agro-industrial steppe (Agger and Brandt 1988).

Some landscape ecologists are also paying attention to ecological heterogeneity along different spatial scales and its relation to biodiversity. Among others they have broadened the spatial scales of the diversity indices from the single species level to the three-dimensional landscape level by transforming the Simpson dominance-diversity index into a *landscape eveness* index, and a *landscape patchiness* index (Romme 1982), and into a *landscape contrast* index (Hoover and Parker 1991).

In Germany, Haber (1990) has broadened the Shannon-Weiner species diversity index into an *ecotope diversity index* to compare the variability between different ecotopes or communities as affected by land use within one landscape unit. Such a *gamma ecotope diversity index* should be used as a guideline for nature conservation and biodiversity conservation in agricultural landscapes through a differentiated land use strategy. Haber proposed to preserve at least 10–15% of the area within each regional natural unit for natural, semi-natural and semi-agricultural ecotopes (communities) of extensively managed pastures and forests.

However, in all these studies the vegetation served as the sole diversity parameter. For systematic measurement of the multidimensional landscape ecodiversity, such strictly biological parameters have to be expanded. They should also incorporate other relevant ecological and cultural parameters at different spatial and perceptional scales, such as micro- and macrosite heterogeneity in soil and topography, as well as cultural artifacts and land use variables.

For this purpose these species diversity indices could be transformed into even more holistic ecodiversity indices following Whittaker's (1965) classification: (1) *alpha ecotope ecodiversity* of biotic, ecological and cultural components within each ecotope; (2) *beta ecotope ecodiversity* between ecotopes of each landscape unit; (3) *gamma land unit ecodiversity* within each regional land system; and eventually; (4) *delta land system ecodiversity* within each regional landscape.

Gamma ecodiversity incorporates for instance an entire watershed, while delta ecodiversity can refer to larger scale land systems such as a bioregion or a biome.

Taking further advantage of recent methodological developments, these assessments could be used as quantitative guidelines for holistic ecodiversity conservation and restoration strategies in each specific landscape. Highest priority should be given to those endangered cultural landscapes in which there are the greatest potential for maximum biological and cultural preservation and restoration.

Ultimately, however, the success of such a holistic landscape planning and management policy will depend on the creation of awareness and understanding of the value of these landscapes and their ecodiversity, and concern about their future and motivation for active involvement in conservation activities. In many Mediterranean countries this is probably the greatest difficulty to overcome. In some of these nations, the clashing demands of traditional users of Mediterranean uplands, the modern private and public developers, and the classic conservationists/protectionists have led to a special version of the Mediterranean *tragedy of the commons* (Hardin 1968). In this situation, each side seeks to maximize his own profit or institutional interest, regardless of the fate of the *common* of the open landscapes as a whole. Unfortunately, in many of these countries the lack of public awareness, understanding, and motivation leads to options which are short term (from election to election) and economically driven (for those with the most powerful pressure group), and therefore also the most harmful and irreversible.

The application of trans-disciplinary concepts and of innovative, holistic methods of landscape-ecological planning and management could help to bridge the communication gap between academicians and professionals, conservation-minded ecologists and production-minded foresters and agronomists, economists and engineers. Landscape ecology can also serve as the scientific and practical basis for the education of a new breed of interdisciplinary landscape managers with a broad ecological, economical, sociological and technological basis who can serve as *integrators* (Naveh and Lieberman 1994).

To achieve these goals, there is urgent need for more effective communication tools. These should be used in order to transform *semantic information*, expressed in words and figures in scientific publications, reports, and lectures, into *pragmatic information* which leads to action through its feedback on the receiver. These will be effective only if their contents can reach all those who care for, those who live from and those who deal with these landscapes at all levels of decision making. Such a tool could be the Green Books for Landscape Conservation (GBLC), developed by the above-mentioned IUCN-CESP Working Group for Landscape Conservation. For this purpose they should fulfill the following functions:

1. In contrast to the IUCN Red Data Books for endangered plants and animals, which deal with the abstract taxonomic species level, these Green Books should deal with the concrete space/time defined landscape level. Its scales range from the smallest mappable landscape ecotope to regional landscape scales of 200–400 km^2. It is on this level of terrain that decisions on land uses and conservation measures are made, which determine the fate of plant and animal populations, and of all other landscape values and functions. It should also be realized that the threats to such tangible and familiar landscape units have much more meaning and public appeal than threats to species or to spatially vaguely defined and even intangible ecosystems. This will be even more the case if these Green Books present relevant information not only on endangered natural assets but also on all other crucial issues and perils to cultural, socio-economic, historical and scenic landscape assets, comprising the total landscape ecodiversity.
2. In contrast to the species Red Data Books, and to the common type of land studies and surveys, GBLCs should recommend alternative and more sustainable land

use practices and conservation strategies, including zoning and protecting, for each specific landscape unit within the regional landscape systems.

3. In order to avoid the "top-down" syndrome of conservation plans prepared by experts and imposed by administrators, efforts should be made for maximum involvement of the local populations from the early planning stages. With the help of such open dialogue and by approaching conservation "from the roots", their fullest comprehension and cooperation should be ensured. They should demonstrate above all, how demands to safeguard intrinsic biological and cultural "soft" landscape values can be reconciled with controlled utilization of "hard" values, which are vital for sustainable socio-economic advancement.
4. To arouse awareness of the dangers to these landscapes, and apprehension about their future, comprehension of the complex problems and their best solutions, and motivation for active involvement in their realization, these Green Books should not be only descriptive but anticipatory. This could be achieved with the help of different optional scenarios on the fate of these landscape units under alternative land use options. However, although based on advanced scientific methods, this information should be presented to the planner, land manager and user, the political decision maker and the public at large in clear nontechnical language with ample illustration by maps, figures and photos.
5. In contrast to the fate of most scientific reports, which are filed away, GBLCs should be used as a practical background document for planning, policy and management guidelines, usable by professionals, politicians and the public. They should serve not only as a first database and dynamic landscape model, but should lay out the foundations for a long-term impact evaluation and forecasting tool.
6. Although containing information on present and anticipated environmental impacts, such Green Books for threatened landscapes differ from the regular environmental impact statements. The latter are merely defensive and passive reactions to planned, urban-industrial or other economic development projects, limited to specific sites and their potential threats. But GBLCs should be active and positive initiatives to change already prevailing undesirable trends and to prevent their continuation with the help of innovative, holistic, long-term and multi-beneficial planning and management strategies on regional landscape scales.
7. Published under the auspices of IUCN in English, as well as by the local authorities in the language(s) of the country and/or the ethnic region, they should be as widely distributed as possible, to reach all those directly or indirectly concerned with these threatened landscapes and to all those who can utilize their contents. With appropriate modifications, they could have relevance for public and school education in conservation, as well as for tourists visiting the region.
8. GBLCs should be used as the major tool for "rescuing" endangered landscapes from Red Lists of Endangered Valuable Landscapes, to be prepared according to the guidelines of IUCN on a worldwide basis in each country. But they should be regarded also as a emergency tool wherever necessary, even before such Red Lists are ready.

The IUCN General Assembly at Buenos Aires in January 1994 adopted a resolution, drafted by the WGCL-CESP recommending actions to be taken for the conservation of threatened landscapes and for the promotion of interdisciplinary approaches between natural and social sciences in the planning and management of threatened landscapes. It urged the undertaking of a pilot study to identify regionally and nationally significant landscapes under threat, to establish means and processes to promote their conservation, and to bring together researchers and practitioners to prepare case studies as a vehicle for developing strategies and methodologies for holistic landscape conservation.

Now, the IUCN is working toward Red Lists for Endangered Landscapes, and the experience gained in the western Crete study will be utilized for the development of a uniform Green Book methodology and for further case studies in Italy, Portugal, Israel, Costa Rica, and other countries, depending on the financial support available for each project.

6.7 DISCUSSION AND CONCLUSIONS

One of the major conclusions to be drawn from this discussion is that holistic conservation strategies for cultural biosphere landscapes should be guided by the recognition that ecodiversity is the tangible expression of the interplay of biological, ecological and cultural processes and patterns. These are determined by biodiversity, ecological micro- and macrosite heterogeneity, human land uses and cultural artifacts. The last two are either enriching or impoverishing both biodiversity and overall landscape heterogeneity. As demonstrated in Fig. 7, these are coupled by positive, mutually amplifying and self-augmenting feedbacks: the greater the ecological landscape heterogeneity, the greater the chance for biological diversity and at the same time floristic, faunal and structural vegetation diversity enhances ecological heterogeneity. But the opposite is also true. Landscape homogeneization fragmentation, and despoliation diminishes biodiversity which in turn further reduces landscape heterogeneity.

I have attempted to show that in perturbation-dependent Mediterranean cultural bio-landscapes, total landscape ecodiversity is determined by the maintenance of a dynamic homeorhetic flow equilibrium between these three major forces. In all these Mediterranean uplands, the disruption of this dynamic homeorhetic flow equilibrium is the result of neotechnological landscape degradation. It is characterized by the unfortunate combination of the cessation of human agro-silvo-pastoral activities and disturbance processes, following depopulation on one hand, and on the other hand by the rapid and careless shift from diversified and stable traditional agriculture to large-scale agro-industrial farming, coupled with overgrazing and indiscriminate planting of monocultures of pine and eucalyptus. High landscape ecodiversity and vital biological, ecological, cultural and socio-economic functions and values can only be ensured by the maintenance of a dynamic homeorhetic flow equilibrium on both micro-and macroscales.

Further neo-technological landscape degradation and despoliation cannot be prevented simply by protective measures. Maximum attainable biodiversity, within a fine-grained matrix of ecological heterogeneity patterns, fulfilling all above-mentioned functions, requires the perpetuation, simulation, and/or restoration of all natural and cultural patterns and processes. These include strictly controlled disturbance pressures

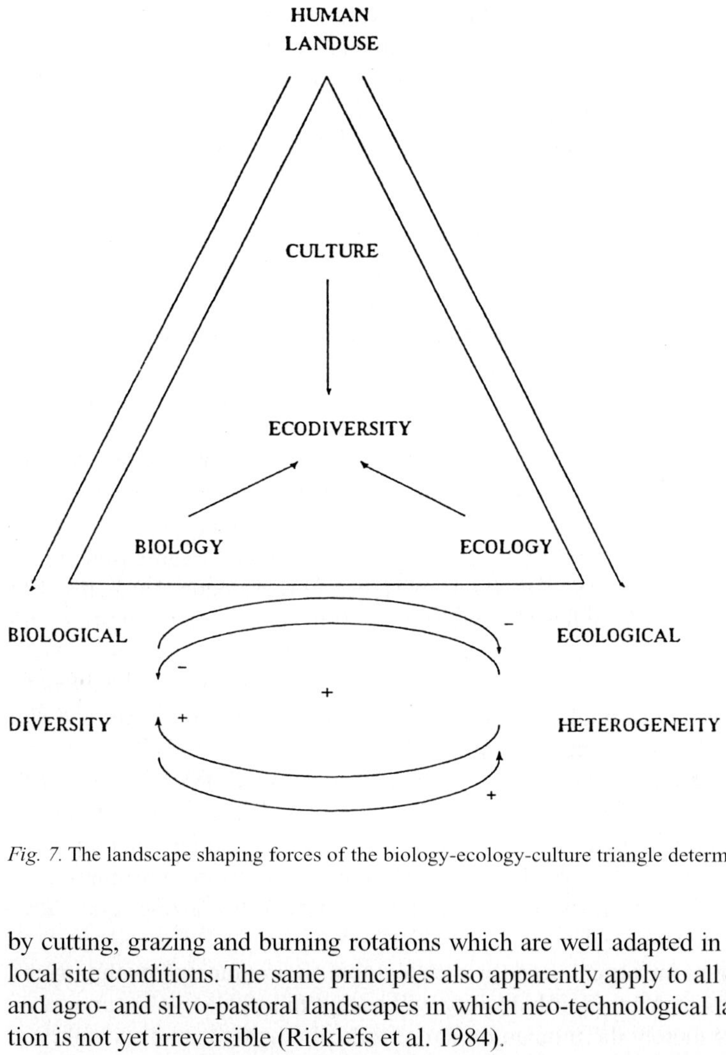

Fig. 7. The landscape shaping forces of the biology-ecology-culture triangle determining ecodiversity.

by cutting, grazing and burning rotations which are well adapted in time and space to local site conditions. The same principles also apparently apply to all other semi-natural and agro- and silvo-pastoral landscapes in which neo-technological landscape degradation is not yet irreversible (Ricklefs et al. 1984).

In view of the above-described exponential speed and extension of this process, the unique co-evolutionary biological and cultural richness of Mediterranean uplands and their intrinsic and instrumental values are gravely endangered. The climatic uncertainties and their synergistic couplings with land degradation and desertification are an additional compelling reason to double our efforts and to conserve and restore the health, integrity and ecodiversity of as many as possible of these most valuable landscapes. This would be the best insurance policy. It would ensure sufficient landscape connectivity, and counter further fragmentation, homogenization, pollution and scenic despoliation. It would provide options for sustainable life-supporting and other vital ecosystem and landscape functions and keystone species, and afford best chances for the survival of rich biotic communities and their further evolution (Naveh 1995a).

However, it has to be realized that the conservation of ecodiversity has to face many strong cultural, demographic, political and socioeconomic forces, all driving in the direction of the cumulative loss of biodiversity, ecodiversity and stability, and thereby enhancing neo-technological landscape degradation. Unfortunately, the dominant cultural values and recognized models of nature and landscapes cannot constrain or may even encourage the vicious circle of exponential population growth and unrestrained growth of consumption and acquisition.

It is therefore obvious that the remedies for our environmental crisis should be sought not only in the scientific, technological, political and socio-economic spheres, but also in the spheres of cultural, spiritual and ethical values, and living norms. Biosphere landscapes should not merely be viewed as a source for our materialistic satisfaction, but also as a source of enlightenment and enjoyment, and as has been shown recently in a convincing manner by Roszak (1992) thereby also as a source of mental health.

In a recent publication (Kim and Weaver 1994a), biodiversity and landscapes were approached from a broad interdisciplinary viewpoint that has much relevance to these issues. Much of this publication dealt with the paradoxical conflict that human life depends for survival on biodiversity and landscapes, yet at the same time, modern human activity threatens the survival or existence of these natural systems. The editors defined its roots in human values and the limits of human cognition and pleaded for an "imperative for change" from myopic and hedonistic values toward a new *Green* equilibrium in the relationship between human society and biodiversity. These could be reached through education and a new consensus on local, national and international scales (Weaver and Kim 1994).

In their recent challenging *Beyond the Limits* (Meadows et al. 1992), three world-renowned systems analysts and modelers went even farther. They claimed, rightly, that in order to avoid the collapse of our earth system, we need a far-reaching environmental revolution which should lead to a radical shift from consumption to conservation and from unrestrained quantitative growth to lasting qualitative improvement and development.

This recognition of the urgent need for such a global environmental revolution and the feasibility of its realization, if humanity will accept this challenge in time, is no longer regarded as merely the utopian dream of radical environmentalists and *deep* ecologists. This view is now shared by some of the most prominent economists, managers and political decision makers (Gore 1991; King et al. 1991; Tolba 1992).

As outlined elsewhere (Naveh and Lieberman 1994), this environmental revolution should lead to a new post-industrial symbiosis between human society and nature which will ensure a viable and healthy global ecosphere. This can be realized only, if the unrestrained expansion of technosphere and agro-industrial landscapes will be replaced by their total integration with the natural and cultural biosphere landscapes.

In addition to many other encouraging initiatives, the conservation of the total ecodiversity of our highly valuable and most cherished cultural landscapes can be regarded as one important step toward this cultural revolution. In order to ensure its success, we will have to integrate scientific ecological knowledge with ecological wisdom and ecological ethics. This should help us learning from the past, comprehending the present and envisaging the future of our total human ecosystem landscapes, and translating these insights into practical, large-scale conservation activities in the field.

REFERENCES

Agger P, Brandt J (1988) Dynamics of small biotopes in Danish agricultural landscapes. Landscape Ecol 3: 227–240.

Baudry J, Bunce RGH (eds) (1989) Land abandonment and its role in conservation. Options Mediterraneenes Ser A.15.

Botkin DB (1990) Discordant harmonies: a new ecology for the twenty-first century. Oxford University Press, New York.

Bussotti F, Ferretti M, Cenni E, Gellinni R, Clauser F, Grossoni P, Barbolani E (1991) New type forest damage to mediterranean vegetation in Southern Sardinian forests (Italy). Eur J For Pathol 21: 290–300.

Bussotti F, Gellini R, Ferretti M, Cenni E, Pietrini R, Sbrilli G (1992) Monitoring in 1989 of Mediterranean tree condition and nutritional status in southern Tuscany, Italy. For Ecol Manage 51: 81–93.

CEPA (Coordinadora Extremena de Proteccion Ambiental) (1992) Dealing with disparity. European structural funds in southwest Spain. Ecologist 22: 91–96.

Davis FW, Stoms DM, Estes JE, Scepan J (1990) An information systems approach to the preservation of biodiversity. Int J Geogr Syst 4: 55–78.

Di Castri F (1981) Mediterranean-type shrublands of the world. In: Di Castri F, Goodall DW, Specht RL (eds) In Mediterranean-type shrublands. Elsevier, Amsterdam, pp 1–52.

Farina A (1989) Recent changes of the mosaic patterns in a montane landscape (North Italy) and consequences on vertebrate fauna. In: Baudry J, Bunce RGH (eds) Land abandonment and its role in conservation. Proc Zaragossa/Spain Seminar 10–12 December 1989. Options Mediterraneennes Ser A 15: 121–134.

Farina A, Naveh Z (eds) (1993) Landscape approach to regional panning: The future of Mediterranean landscapes. Landscape Urb Plann 24. Special Issue, July 1993.

Ferretti M, Bussotti F, Cozzi A, Cenni E (1995) Forest decline and environmental pollution in Italy. Agr Medit.

Gore A (1992) Earth in balance: forging a new common purpose. Earthwatch, London.

Green B (1985) Countryside conservation. The protection and management of amenity ecosystems, 2nd ed. Unwin Hyman, London

Grove AT, Ispikoudies J, Kazaklis A, Moody JA, Papanastasis V, Rackham O (1993) Threatened Mediterranean landscapes: west Crete. Research Report to EU. Cambridge University, Department of Geography, Cambridge

Grove AT, Ispikoudies J, Kazaklis A, Moody JA, Papanastasis V, Rackham O (1994) Green Book: Threatened Mediterranean landscapes of western Crete. Adapted and translated into Greek by V. Papanastasis. Integrated Mediterranean Project of Crete, Aristotele University of Thessaloniki, and Mediterranean Agronomic Institute of Chania, Crete

Haas G (1966) On the vertebrate fauna of the Lower Pleistocene site Ubeidiya. Israel Academy of Science and Humanities, Jerusalem, pp 33–68

Haber W (1990) Using landscape ecology in planning and management. In: Zonneveld IS, Forman RTT (eds) Changing landscapes: an ecological perspective. Springer, Berlin, Heidelberg New York, pp 217–232

Hardin G (1968) The tragedy of the commons. Science 162:1243–1248

Hoover SR, Parker AJ (1991) Spatial components of biotic diversity in landscapes of Georgia, USA. Landscape Ecol 5:125–136

IUCN/UNEP/WWF (1991) Caring for the earth. A strategy for sustainable living. IUCN, Gland, Switzerland

Jeftic L (1993) Implications of expected climate change in the Mediterranean region. In: Garber M, Cohen A, Magaritz M (eds) Regional implications of future climate change. Israel Academy of Sciences and Humantities, Jerusalem, pp 278–302

Kaplan D (1992) Responses of Mediterranean grassland plants to gazelle grazing. In: Thanos CA (ed) Proc 6th Int Conf Mediterranean Climate Ecosystems, September 1991. University of Athens, Athens, pp 75–79
Kim KC, Weaver RD (eds) (1994a) Biodiversity and landscapes: a paradox of humanity. Cambridge University Press, New York
Kim KC, Weaver RD (1994b) Biodiversity and humanity: paradox and challenge. In: Kim KC, Weaver RD (eds) Biodiversity and landscapes: a paradox of humanity. Cambridge University Press, New York, pp 3–27
King A, Richter G, Schneider B (1991) The first global revolution: from the problematique to the resolutique, a "blueprint for the twenty-first century" by the Council of the Club of Rome. Pantheon Books, New York
Le Houerou HN (1992) Change in vegetation and land-use by the year 2050: a prospective study. In: Jeftic L, Miliman JD, Sestini G (eds) Climate change and the Mediterranan. Edward Arnold, London, pp 175–232
Lucas PHC (1992) Protected landscapes. A guide for policymakers and planners. Chapman and Hall, London
Marcuzzi G (1991) Man-made heterogeneity in some Italian habitats – beaches of the Puglia region and Matera province. Econ Montana Lin Ecol 4:21–24
Meadows DH, Meadows DL, Randers J (1992) Beyond the limits: global collapse or a sustainable future. Earthscan, London
Meeus J, Wijermans MP, Vroom MJ (1990) Agricultural landscapes in Europe and their transformation. Landscape Urban Plann 18:289–352
Milan M, Salvador R, Mantilla E, Artnano B (1996) Meteorological and photochemical air pollution in southern Europe: experimental results from EC research projects. Atmos Environ 30:1909–1924
Naveh Z (1974) The ecological management of non-arable Mediterranean uplands. J Environ Manage 2:351–471
Naveh Z (1982) Mediterranean landscape evolution and degradation as multivariate biofunctions: theoretical and practical implications. Landscape Plann 9:125–146
Naveh Z (1984) The vegetation of the Carmel and Nahal Sefunim and the evolution of the cultural landscape. In: Ronen A (ed) Sefunim prehistoric sites in Mount Carmel, Israel. BAR Intern Ser 230, Oxford, pp 23–63
Naveh Z (1987) Biocybernetic and thermodynamic perspectives of landscape functions and land use patterns. Landscape Ecol 1:75–85
Naveh Z (1990a) Landscape ecology as a bridge between bio-ecology and human ecology. In: Svobodova H (ed) Cultural aspects of landscape. PUDOC, Wageningen, pp 45–58
Naveh Z (1990b) Ancient man's impact on the Mediterranean landscape in Israel: ecological and evolutionary perspectives. In: Bottema S, Entjes-Nieborg G, Van Zeist W (eds) Man's role in the shaping of the eastern Mediterranean landscape. AA Baalkema, Rotterdam, pp 43–50
Naveh Z (1991a) Mediterranean uplands as anthropogenic perturbation dependent systems and their dynamic conservation management. In: Ravera OA (ed) Terrestrial and aquatic ecosystems, perturbation and recovery. Ellis Horwood, New York, pp 544–556
Naveh Z (1991b) Biodiversity and ecological heterogeneity of Mediterranean uplands. Econ Montana – Lin Ecol 24:47–60
Naveh Z (1995a) Conservation, restoration and research priorities for Mediterranean uplands threatened by global change. In: Moreno JM, Oechel WC (eds) Anticipated effects of a changing global environment on mediterranean-type ecosystems Springer, Berlin Heidelberg New York, pp 482–508
Naveh Z (1995b) Interactions of landscapes and cultures. Landscape Urb Plann 32:43–54
Naveh Z, Dan J (1973) The human degradation of Mediterranean landscapes in Israel. In: di Castri F, Mooney HA (eds) Mediterranean-type ecosystems: origin and structure. Springer, Berlin Heidelberg New York, pp 370–390
Naveh Z, Whittaker RH (1979) Structural and floristic diversity of shrublands and wood-lands in northern Israel and other mediterranean areas. Vegetatio 41:171–190 (see also chapter 2)
Naveh Z, Steinberger E, Chaim S (1980) Photochemical air pollutants: a new threat to Mediterranean conifer forests and upland ecosystems. Environ Conserv 7:301–309
Naveh Z, Lieberman AS (1994) Landscape ecology – theory and applications. 2nd edn. Springer, Berlin Heidelberg New York

Nevo E (1988) Genetic diversity in nature; patterns and theory. Evol Biol 23:217–246

Nevo E, Geiles A, Kaplan D, Golenberg EM, Olsvig-Whittaker L, Naveh Z (1986) Natural selection of allozyme polymorphism: a microsite test revealing ecological genetic differentiation in wild barley. Evolution 40:13–22

Norton BG (1987) Why preserve natural variety? Princeton University Press, New Jersey

Noy-Meir E, Kaplan D (1991) The effect of grazing on the herbaceous Mediterranean vegetation and its implications on the management of nature reserves. Interim Report to the Nature Conservation Authorities, Jerusalem. (in Hebrew)

Odum EP (1993) Ecology and our endangered life-support systems. 2nd edn. Sinauer, Sunderland, Massachusetts

Papanastasis V (1993) Legal status of land tenure and its implication for open landscapes of western Crete. Landscape Urb Plann 24:273–278

Pedroli GMB, Vos W, Dijkstra H, Rossi R (1988) The Farma River barrage effect study. Giunta Regionale Toscana. Marsilio, Firenze, Italy

Pickett STA, Parker VT (1994) Avoiding the old pitfalls: opportunities in a new discipline. Restor Ecol 2:75–79

Pinto-Correia T (1993) Threatened landscapes in Alentejo, Portugal: the *montado* and other agro-silvo-pastoral systems. Landscape Urb Plann 24:43–48

Prigogine I (1976) Order through fluctuation: Self-organization and social systems. In: Jantsch E, Waddington CH (eds) Evolution and consciousness: human systems in transition. Addison Wesley, Reading, pp 93–130

Prigogine I, Stengers I (1984) Order out of chaos. Bantam Books, New York

Ricklefs RR, Naveh Z, Turner RE (1984) Conservation of ecological processes. Commission on Ecology Paper No. 8. IUCN, Gland, Switzerland.

Romme WH (1982) Fire and landscape diversity in subalpine forests of Yellowstone National Park. Ecol Monogr 52: 199–221.

Roszak T (1992) The voice of the earth: an exploration of ecopsychology. Simon and Schuster, New York

Ruiz de la Torre JR (1985) Conservation of plants within their native ecosystems. In: Gomez-Campo C (ed) Plant conservation in the Mediterranean. De Junk, The Hague: pp 197–219.

Stekelis M (1966) Archeological excavations at Ubaidiya. Israel Academy of Sciences, Jerusalem, pp 1–32.

Tolba M (1992) Saving our planet: challenges and hope. Chapman and Hall, London.

UNEP (1988) Mediterranean blue plan regional activity centre: futures of the Mediterranean basin. Executive summary and suggestions for action. United Nations Environmental Programme, Mediterranean Action Plan, Mediterranean Blue Plan Regional Activity Centre, Sophia Antipolis, France.

Vos S, Stortelder AHF (1992) Vanishing Tuscan landscapes: landscape ecology of a submediterranean-montane area (Solano Basin), Tuscany, Italy. Pudoc, Wageningen.

Waddington CH (1975) A catastrophe theory of evolution: the evolution of an evolutionist. Cornell University Press, Ithaca New York.

Warburg M (1977) Plant and animal species diversity along environmental gradients in the Mediterranean landscape of Israel. Animal species diversity. Research Report 450. Binational Israel-American Science Foundation. Technion, Israel Institute of Technology, Haifa.

Warburg M, Ben-Horin RA, Rankevich D (1978) Rodent species diversity in mesic and xeric habitats in the Mediterranean region of northern Israel. J Arid Environ 1: 63–69.

Weaver RD, Kim KC (1994) Biodiversity and humanity: toward a new paradigm. In: Kim KC, Weaver RD (eds) Biodiversity and landscapes: a paradox of humanity. Cambridge University Press, New York, pp 393–423.

Whittaker RH (1965) Dominance and diversity in land plant communities. Taxon 21: 213–251.

Worldwatch Institute (1994) State of the World 1994. Norton, New York and Earthscan, London.

Yiftach-El O, Law-Yone H (1995) Regional policy and minority attitudes in Israel. Environ Plann 27: 1261–1301.

CHAPTER 7

NAVEH Z. (2005)

THE ROLE OF FIRE IN THE EVOLUTION OF THE MEDITERRANEAN CULTURAL LANDSCAPE IN THE PLEISTOCENE AND EARLY HOLOCENE WITH SPECIAL REFERENCE TO MT. CARMEL. (BASED ON: THE EVOLUTION OF THE CULTURAL LANDSCAPE IN ISRAEL AS AFFECTED BY FIRE, GRAZING, AND HUMAN ACTIVITIES. Z. NAVEH AND Y. CARMEL. IN: WASSER S. P. (Ed.) PAPERS IN HONOUR OF EVIATAR NEVO. EVOLUTIONARY THEORY AND PROCESSES: MODERN HORIZONS. KLUWER ACADEMIC PUBLISHERS, DORDRECHT PP. 337–409).

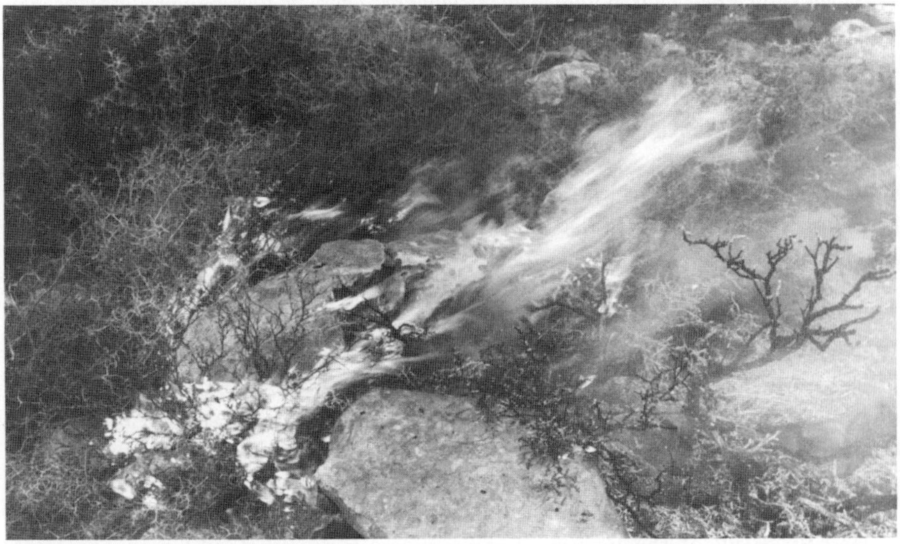

Wild fire.

INTRODUCTION

To support our contentions on the important role of fire in the evolution of the cultural Mediterranean landscapes, selected parts of the above study on the evolution of these landscapes in Israel are presented with special relevance to archeological findings on Mt. Carmel and their interpretations in the light of our fire ecology studies and the ethno-ecological equivalent use of fire for vegetation management by the pre-agricultural Coastal Californian Indians in comparable conditions to those of the pre-agricultural Epipaleolithic Natufians of Mt. Carmel.

SOME ECOLOGICAL AND EVOLUTIONARY CHARACTERISTICS OF THE MEDITERRANEAN LANDSCAPES AND THEIR VEGETATON

All five Mediterranean climate regions of the world are typical for their evergreen shrublands, dominated by evergreen species with tough and leathery sclerophyllous leaves. In a comprehensive ecological comparison, Di Castri (1981) showed that these are replaced in general along environmental gradients of moisture and nutrient availability by other vegetation types. However, these gradients are also greatly influenced by human impacts on these landscapes and their vegetation. A recent overview of both natural and anthropogenic disturbance regimes in these so-called "Mediterranean-Type Ecosystems" has been provided by Rundel (1998).

Because of the unique location of the Mediterranean Basin between Europe, Asia and Africa, it has served as a meeting point and melting pot for species of varying origins. As the southernmost outpost of the Eastern Mediterranean, Israel is the best example of such a bridge and corridor of the different biogeographic elements for these regions. Axelrod (1958) has shown that sclerophylly is an ancient characteristic that long predates the evolution of summer-dry climates. He traced scleropyllous vegetation in North America-Eurasia back to Madro-Tertiary and Mediterranean – Tertiary geoflora (the fossil flora with a common geological history), respectively, which had their origin in the southwestern parts of the continents in the late Cretaceous age. The Mediterranean geoflora has been derived from Indo-Malesian, Paleo-African, and xerothermic Mesogen stock (Zohary 1974). Seasonal aridity appeared sporadically already in Middle Eocene times, but the true Mediterranean climate pattern was established only in the Pleistocene. The sclerophyll woody species were apparently best pre-adapted to climatic patterns of increasing drought and lower winter temperatures. that developed during the Pleniglacial in the Levant. The shift toward aridity in the Quarternary became a major factor in the extinction of many Afrotropical and Palearctic elements and for the increased separation of tropical Africa and Eurasia. The impact of glacial episodes and the proximity of large desert tracts also played a major role in the distribution of Levantine plants, animals, and humans (Horowitz, 1992). According to Pignatti (1978), large numbers of chiefly herbaceous and annual plant taxa which evolved during the Pleistocene in the Mediterranean Basin and were accompanied by further speciation of woody plants and endemism.

Presently in all of these regions, but especially in the Mediterranean Basin, with the exception of the rapidly vanishing coastal dunes, wetlands, and marshes, the noncultivable uplands have become the last refuges of natural (that means) spontaneously occurring and reproducing plants and animals. Wherever these uplands have not yet been converted into dense pine or eucalyptus forests or depleted into scrub and rock deserts, they are distinguished by their great ecological heterogeneity and biological diversity. Although covering only a tiny 1.2% of the earth surface, their contribution to species diversity of vascular plants, far exceeds their relatively small area of coverage. The largest number of these species can be found in the Mediterranean Basin (25 000), making 20% of the world's total (Cowling et al. 1996). Therefore, it has been recognized as one of the 18 most important biodiversity "hot spots". On the basis of species/area ratio, the Mediterranean territory of Israel is, after Cyprus, by far the richest, containing more than 1500 species with a ratio of 0. 15 – as compared to 0.01 in Turkey, 0.04 in Greece and even fewer in all other Mediterranean landscapes (Naveh and Kutiel 1990). This striking richness is also most apparent from the results of our biodiversity study in northern Israel, described in Chapter 2 in this anthology. The Mediterranean flora is especially rich in herbaceous plants, including many annuals and exceptional colorful flowering compositae and geophytes with ornamental value. Among these are many rare and endemic plants. Many grasses and legumes are outstanding pasture plants and are cultivated widely in improved pastures, especially in Australia. Among the woody plants, and especially in the Labiatae family, there are many species with great value in pharmaceuticals, cosmetics, spices, balsams, and other uses with considerable economic potentials. Theses are now collected, grown and utilized with increasing intensity for commercial production. The richness of the Carmel vegetation in usable plants, both for foraging humans and herbivores is clearly indicated in Table 1, showing all vascular species growing near one of the important archeological sites, that is representative for most of the Carmel vegetation.

The Mediterranean zone of northern Israel is the center of distribution in the Near East "Fertile Crescent" of the wild tetraploid Emmer wheat *Triticum dicoccoides*, the progenitor of most tetraploid and hexaploid cultivated wheat, and it will play an important role in further wheat improvement. Its rich genetic resources have been studied by Nevo and his co-workers at the Institute of Evolution of the University of Haifa, since 1971. At the same time it also served these scientists as a major model organism for evolution (Nevo, 2001; Nevo et al., 2002).

CONTROVERSIES ABOUT THE TIMING AND THE EFFECT OF HUMAN IMPACTS ON THE MEDITERRANEAN LANDSCAPE

The Mediterranean Region has endured a very long and intensive period of human-induced perturbations, and has suffered more than most other regions from a combination of a fragile environment and a long history of land abuses with adverse effects on the land and its people. However, the denuded Mediterranean uplands have shown the resilience and the soil-building capacities of the native vegetation in a striking way. It should be recognized that the human role in the Mediterranean has

Table 1. Exploitable plant species of the surroundings of Nahal Sefunim. Determination for human consumption according to Dafni (1984). (Modified after Naveh, 1984 by Weinstein-Evron, 1998)

Woody plants

Tr bvcees		***Shrubs***		***Dwarfshrubs***	
Arbutus andrachne	F W	*Calicotome villosa*	Fl Br	*Cistus salviifolius*	L
Ceratonia siliqua	F! P! Br	*Genista fasselata*	W	*Cistus creticus*	L
Cercis siliquastrum	W	*Pistacia lentiscus*	W	*Coridothymus capitatus*	L
Crataegus aronia	Fl F Br W	*Rhamnus alaternus*	Br W	*Majorana syriaca*	L
Laurus nobilis	F Br W	*Rhamnus palaestina*	Br W	*Melissa officinalis*	L
Olea europaea	L Br W	*Ruscus aculeatus*	F Br W	*Micromeria fruticosa*	L
Phyllirea latifolia	F! Br W!	*Asparagus aphyllus*	L	*Salvia fruticosa*	L
Pinus halepensis	Br W!	*Smilax aspera*	Sh	*Sarcopoterium spinosum*	Br W
Pistacia palaestina	W	*Tamus communis*	Sh	*Satureja thymbra*	L
Styrax officinalis	F Br!W		Sh	*Teucrium capitatum*	L
Quercus calliprinos	W				
	F Br! W				

Herbaceous plants

Geophytes		***Legumes***		***Miscellaneaous herbs***	
Arisarum vulgare	L	*Anthyllis tetraphylla*	P	*Alcea acaulis*	L S
Arum dioscioridis	L	*Coronilla cretica*	P	*Alcea setosa*	L S
Asphodelus ramosus	B	*Hippocrepis unisiliquosa*	P	*Anagallis arvensis*	L
Crocus hyemalis	B	*Hymenocarpos*	S P	*Caspella bursa-pastoris*	L
Cyclamen persicum	L Fl	*circinnatus*	S P	*Convolvulus*	P
Ophrys umbilicata	B	*Lathyrus blepharicarpus*	P	*caelesyriacus*	R
Ophrys bornmuelleri	B	*Lotus peregrinus*	P!	*Daucus carota*	P
Ophrys israelitica	B	*Medicago orbicularis*	P!	*Erodium gruinum*	P
Ophrys galilaea	B	*Medicago scutellata*	P!	*Erodium moschatum*	L
Ophrys transhyrcana	B	*Medicago polymorpha*	P	*Foeniculum vulgare*	P
Orchis caspia	B	*Onobrychis squarrosa*	S L P	*Geranium molle*	P
Orchis galilaea	B	*Pisum elatius*	P!	*Geranium purpureum*	P
Orchis tridentata	B	*Scorpiurus muricatus*	P!	*Geranium rotundifolium*	L
Serapias levantina	B	*Tetragonolobus*	P!	*Isatis lusitanica*	P
Tulipa agenensis	B	*paleastinus*	P	*Kicksia spuria*	F
		Trifolium campestre	P!	*Mandragora autumnalis*	S
		Trifolium clusii	P	*Nigella arvensis*	S
Grasses	S P	*Trifolium clypeatum*	S P!	*Papaver carmeli*	L
Aegilops ovata	P	*Trifolium stellatum*		*Plantago cretica*	L
Brachypodium	P!	*Vicia hybrida*		*Plantago afra*	L
distachyon	P!		L P	*Salvia hierosolymitana*	L
Andropogon distachyus	P	***Asteraceae***	L	*Salvia pinnata*	L P
Avena sterilis	P	*Calendula arvensis*	L	*Sanguisorba minor*	L
Bromus alopecurus	P	*Carduus argentatus*	P	*Sinapsis arvensis*	
Bromus syriacus	P!	*Carlina involucrata*	L S P		
Catapodium rigidum	S B P!	*Catananche lutea*	C		
Dactylis glomerata	S P!	*Cichorium pumilum*	P		
Hordeum bulbosum	P	*Gundelia tournefortii*	L C		

(*Continued*)

Table 1. Continued

Hordeum spontaneum	P	*Hedypnois cretica*	P
Hyparrhenia hirta	P	*Inula viscosa*	P
Lopochloa phleoides	P!	*Notobasis syriaca*	B P
Phleum subulatum	P!	*Rhagadiolus stellatus*	P
Piptatherum miliaceaum	P	*Scorzonera papposa*	P
Piptatherum blancheanum		*Senecio vernalis*	P
Stipa bromoides		*Tolpis virgata*	
		Thrincia tuberosa	

F – fruits, S – seeds, B – bulbs, corms, etc., Fl – flowers, L – leaves, Sh – shoots, R – roots, C – capitulum. P–pasture for livestock and browsers (Br). W – wood. (!– high value).

not always been destructive and that the inhabitants of the Mediterranean uplands have utilized the regeneration capacities judiciously in their traditional farming practices (Naveh and Dan 1973; Naveh and Kutiel 1990). Di Castri (1981) stated that human beings have coevolved with Mediterranean ecosystems and coevolutionary features are present in a number of ecological and cultural characteristics in this region.

This view is contrary to the widely shared idea of an entirely negative and destructive human impact (McNeil 1992), expressed strongly also by Zohary (1983) who claimed that humanities effects far exceeded the destruction from Pleistocene climatic changes. On the other hand, Horvat et al. (1994) called humanity the "codesigner" of the Mediterranean vegetation, and Pignatti (1983) emphasized the human role in the evolution of new habitats, and in stimulating evolution of flora. More recently, Grove and Rackham (2001) in their comprehensive account of the ecological history of European Mediterranean landscapes stated rightly that the pervasive "ruined landscape theory" in the Mediterranean is far too simplistic and should not be taken literally. They also refuted the deterministic, preconceived theory of a Mediterranean forest "climax", which is automatically degraded by human interference into lower woody successional stages of maquis, garrigue and batha (or phrygana) and finally into steppe grassland. This corroborates our findings on the dynamics of Mediterranean landscapes in Israel and in the Mediterranean in general (Naveh 1971; Naveh and Dan 1973; Naveh and Kutiel 1990). We offered an alternative theory in which this deterministic succession and homeostatic climax paradigm is replaced by the recognition of the non-cultivable uplands as perturbation-dependent semi-natural landscapes, maintained until very recently, by short- and long-term fluctuations in a metastable state of a homeorhetic flow equilibrium (Naveh 1991; Naveh and Lieberman, 1994, and also discussed in several chapters of this anthology).

Until recently there was also a general tendency to deny any significant human influence on the pristine vegetation, and to underestimate the impact of human habitation and food gathering on the Pleistocene environment and its vegetation. Thus, for instance, Pons and Quezel (1985:35), in reviewing human impact in the Mediterranean, claimed: "Early man was a hunter and gatherer and had relatively little influence on natural vegetation." McNeill (1992) has this view uncritically accepted, maintaining that until the Neolithic revolution, hunters and gatherers played almost no role because of their slender numbers. Although he recognized their early use of fire, he claimed that their activity scarcely affected vegetation complexes. But at the same time, he maintained that prehistoric fires started the "long history of anthropogenic erosion". However, Grove and Rackham (2001) in their review of prehistoric and historic causes of erosion could not find any well-founded evidence that human activities were its major cause, but erosion revealed a much better correlation with climatic long- and short-term events.

Blondel and Aronson (1999) reduced the evolutionary impact of livestock grazing and burning by pastoralists and agriculturist only until the last 10 millennia. Contrary to this view, we claim that humans started to modify the pristine landscape much earlier. They even co-evolved together during their biological and cultural evolution in the Pleistocene and the geological, climatic, and other natural forces and stresses of this long geological period, and especially those caused by fire and foraging of wild herbivores. By these forces, together with humans, the landscapes were transferred gradually from a natural into sub-natural and semi-natural landscapes, and in the Holocene, part of these were finally converted into agro-pastoral, agricultural, and rural cultural landscapes. The ensuing closely interwoven natural and cultural processes and patterns contributed much to the great ecological heterogeneity, biological diversity and cultural richness of the remaining non-arable Mediterranean uplands.

The Mediterranean landscapes of Israel, and especially of Mt. Carmel on which we have chiefly focused, can serve as a suitable example. Here we have ample archeological evidence of human habitations from the Middle Pleistocene onwards. As part of the Southern Levant and the "Fertile Crescent", Israel was one of the first locations to undergo the transition from food collection to food production, marking the beginning of plant and animal domestication during the so-called "Neolithic (agricultural) Revolution". As one of the older and better-known human cultural centers it served also as the cradle of the monotheistic Judean-Christian religions and Western civilization.

THE ROLE OF FIRE IN THE COEVOLUTION BETWEEN THE PALEOLITHIC HUNTERER–GATHERER AND THE MEDITERRANEAN LANDSCAPES OF ISRAEL

Naveh and Dan (1973) and Naveh and Kutiel (1990) distinguished three major periods of human-induced changes in the landscapes and their vegetation of Israel:

1. A very long period during the Pleistocene, which marked the major phases of the coevolution of Mediterranean peoples with their landscapes.
2. A long prehistoric and historic agricultural period in the Holocene, during which the agropastoral cultural landscape was shaped, reached its peak, then gradually declined and seminatural vegetation formations were maintained in a metastable stage of homeorhetic flow equilibrium.

3. A recent, short modern "neotechnological" period in the 20th century of increasing heavy human pressures on these solar energy-powered, semi-natural and agricultural biosphere landscapes, and their conversion into fossil-energy-powered agro-industrial, rural and urban-industrial technosphere landscapes, causing the biological and cultural deterioration of the Mediterranean landscape and the formation of the disorganized "Total Industrial Landscape".

According to Axelrod (1958, 1989), the evolution of the Mediterranean Basin in general has much in common with California. In both cases, fire and drought apparently played important evolutionary roles. Many wide-ranging wildfires caused by volcanic eruptions and by lightning have presumably raged throughout the Pleistocene. Therefore, in addition to other, more or less catastrophic natural perturbations and to increasing drought, fire caused by volcanic event and by lightning, most probably acted as a strong selection force. As explained in Chapter 2 and in more detail by Naveh (1975), and will be discussed further below, we can safely assume that those woody and herbaceous genotypes which developed the most efficient physiological and morphological evolutionary strategies for active and passive vegetative and reproductive regeneration mechanisms had the best chances to overcome any fire stresses.

The frequency of such volcanic outcrops on Mt. Carmel, shown in Fig. 1, can serve as a good indication of the occurrence of many hot wildfires during volcanic outbreaks in the Pleistocene, in addition to those wildfires caused most probably by lightning.

Presently, lightning out of a clear sky may occur in rare cases on several days in April–June, causing wildfires. However in ancient times, such a single lightning fire on a dry day could catch the undisturbed, dense and highly inflammable woody and herbaceous vegetation and spread rapidly over vast areas. Such fires on wildlands and pastures were mentioned in the Bible, with lightning as the "fire of God" and "the heat of summer drought" in the Book of Job 9 (1:6). They gained also special symbolic importance for Mt. Carmel where the Prophet Elijah fought the Baal prophets and "the fire of the Lord fell and consumed the burnt sacrifices, and the wood, and the stones, and the dust and licked up the water that was in the trench" (Kings 18, 38).

Such hot fires most probably destroyed the woody above-ground vegetation from time to time. Therefore, only those plants and animals with efficient adaptive resilience could survive such recurring fire stresses. Anderson (1956) showed that even at present times, in California, fire-induced shrub openings could provide ideal opportunities for further speciation, hybridization and genotype recombination of such woody fire-followers (or "pyrophytes").

The Levant remains apparently the only potential corridor for human migrations out of Africa. According to Tchernov (1988), the main route of biotic and hominid dispersal, from Africa to the rest of the world took place through Israel, as the southern Levantine corridor. These were as a rule associated with biotic dispersal events, mainly large mammals. Tchernov et al. (1994) summarized recent findings and the Quarternary chronosequence and faunal remains of the earliest sites that mark the dispersal routes of *Homo erectus* into Eurasia in the

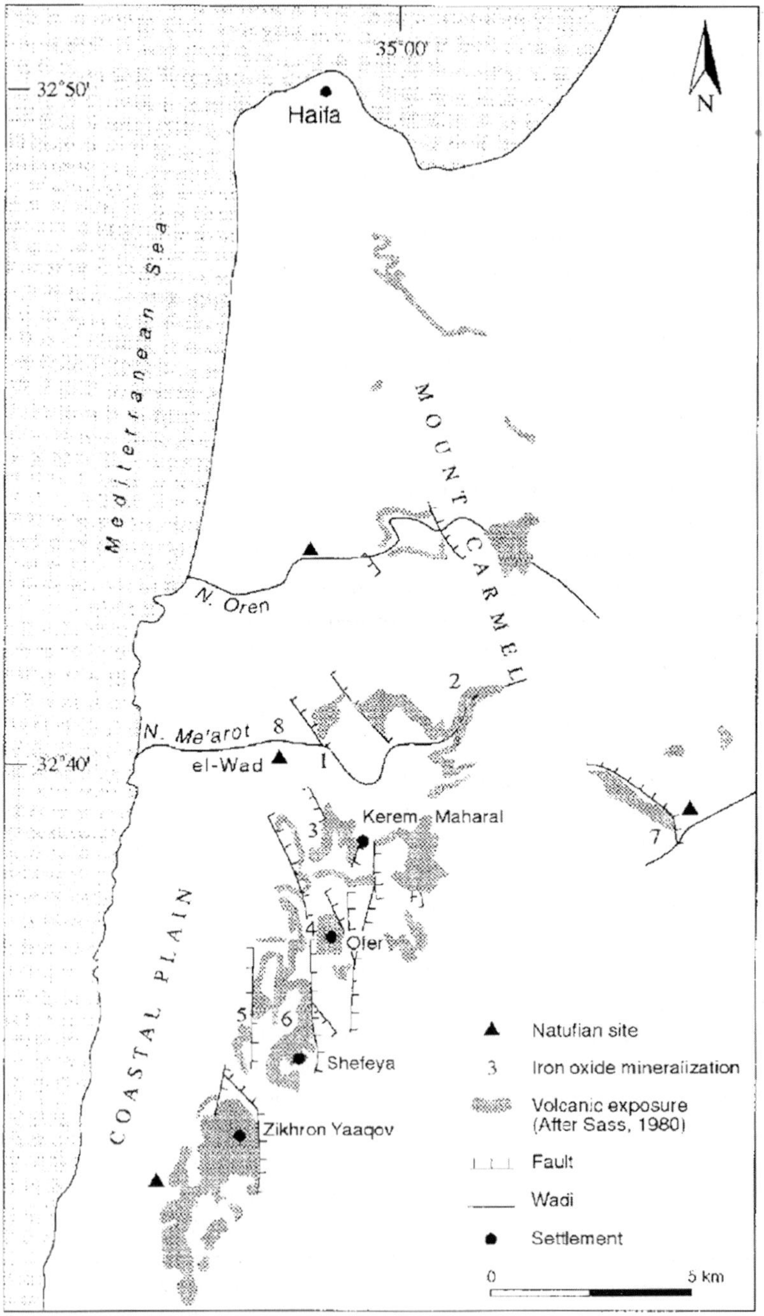

Fig. 1. Location map of the volcanic outcrops and iron mineralization sites with natufian sites of Mt. carmel (Weinstein-Evron, 1998).

Early Acheulian Bone Bearing Beds in northern Israel from Ubeidiya, in the Central Jordan Valley, near Lake Kinnereth, and Gesher Benot Yaakov, South of the Hule Lake. In this Acheulian occupancy Gesher Benot Yaakov, Stekelis (1960) found burned fibia fractures and within the bifacially worked artifacts, cleavers and hand axes made of basalt, dated around 0.800 myr. This is our first archeological evidence of fire not only in Israel but according to Clark (1966) also one of its first use by *H. erectus* in general.

From the Middle Pleistocene onward, the geological and biological landscape evolution coincided with major phases of the biological and cultural evolution of humans. During this long period of more than a million years, the Lower Paleolithic *H. erectus* was replaced by more advanced food gatherers and hunters such as Middle Paleolithic Neanderthaloids and the first *H. sapiens* around 100 000 years ago, and subsequently by the intensive food collecting Late Paleolithic and Epipaleolithic *H. s. sapiens,* including the Natufians, and the food-producing Neolithic *H. s. sapiens* in the early stages of the Holocene, around 12 000 to 10 000 years ago.

As explained in more detail in the above study, in this first very long of the coevolutionary phase of the Pleistocene land use, the natural landscape of Mt. Carmel was transformed very gradually into a subnatural one, changing the floristic composition, but largely retaining the natural vegetation structure and formation. In the second phase, intensive vegetation management transformed presumably by a major bifurcation most of the natural and subnatural landscape into seminatural vegetation, altering also their structure. In this period the coevolution reached its peak, and foundations were laid for the metastable homeorhetic flow equilibrium on which the vegetation dynamics of these semi-natural landscapes are based until present times.

For this hypothesis, I applied the definition of the eminent evolutionarist Ledyard Stebbins (1982), regarding coevolution as the simultaneous evolution of two genetically independent, but ecologically interdependent lines via both biological and cultural templates. This coevolution on Mt. Carmel has been described first by Naveh (1984) in Israel, and by Naveh and Vernet (1991) in the context of the palaeohistory of the Mediterranean biota in general. As explained in Chapter 14 of this anthology, the closely coupled evolutionary processes of the Mediterranean peoples and their landscapes and vegetation have been driven by loops of auto- and crosscatalytic networks and their hypercycles through a series of bifurcations. We assumed that both natural and human-set fires played an important role in this process. Some of its crucial stages have been documented by archeological findings in the caves of Mt. Carmel, especially in those of Nahal Hame'arot ("The River of the Caves").

Bar-Josef (1984) has pointed out already that tectonics and erosion have obliterated most of the direct archeological evidence in the Pleistocene. No ash deposits, and sparse floral remains have been detected in open in situ habitations. Even recent sophisticated flooding methods have not provided large samples of vegetal relics in shallow and eroded Mediterranean upland soils, especially Terra Rosa in which preservation is very poor. In the specific climatic and edaphic conditions of Mediterranean uplands most of the ashes of forest and brush fires are washed away by the first heavy rains, and remnants become intimately mixed with the thin upper layer of humus-rich

Terra Rosa or Rendzina soils (Naveh, 1973). We can therefore also hardly expect to find archeological evidence of such fires, especially since the Carmel slopes underwent severe geological erosion and above-described morphogenetic upheavals.

However, Mt. Carmel is endowed with karstic caves and many of our contentions on this coevolution are supported chiefly by the archeological findings and in the sequences from the Paleolithic layers of Quaternary pollen spectra in the Tabun cave (Horowitz 1979) and those from the el-Wad cave, summarized more recently by Weinstein-Evron (1998). Thanks to the findings in these caves the coevolutionary process and human uses of fire can be traced back here to their Lower Paleolithic layers even if most traces of hearths have been erased by erosion and by changes in sea level followed by sedimentation.

Tatskin et al. (1995) investigated wind-blown sedimentary fill of lithic assemblages in the Tabun and Jamal caves. With the help of micromorphological methods they could recognize also anthropological sediments of charred organic material, incorporated into microaggregates, scattered pieces of charcoal 0.5–0.9 mm across. In the sediment of layer G of the Tabun cave they detected indications of anthropogenic burning by ferruginized chips of bones. These findings indicate human occupation and the use of fire from the Middle Pleistocene onward in Israel for more than a half million years.

We can assume that more than any other feature, fire as their first extrasomatal energy source affected not only their environment and shaped their landscapes, but also their life, behavior, economy, and culture of the inhabitants of these caves. This has been claimed already by Sauer (1956: 54–55). "Speech, tools and fire are tripods of culture and have been so, we think, from the beginning . . . About the fireplace, social life took form, and the exchange of idea was fostered. The availability of fuel has been one of the main factors determining the location of clustered habitation."

Sauer (1961) also pointed to volcanism in the Eastern Mediterranean and the Rift Valley as the first source of such fires. These contentions by Sauer have been verified by Perles (1977) in her comprehensive review of the prehistoric use of fire, claiming that "Mankind could have evolved into *Homo habilis* without fire but it would never have become *Homo sapiens* without fire".

The gradual intensification of human interferences was also accompanied most probably by more efficient use of fire by Mousterian gatherers-hunters. According to Perles (1977), the mastering of fire happened about 100 000 years ago by these Neanderthaloids who produced lamps to light their caves and torches to carry fire. They could open dense forest and bush thickets to facilitate hunting and food collecting and to increase edible food by encouraging the lush regeneration or trees and shrubs, invading grasses, bulbs, and tuberous plants. Evidence for the use of fire by such Mediterranean Mousterian cultures has been provided also at the Kasitstra Caves near Lake Ionina in Greece (Higgs et al., 1967). From this period, Vernet (1973) reported findings of charcoal specimens of sclerophylls and phanerophytes such as *Phyllirea* and *Quercus* in southeastern France.

The Tabun Cave bears clear evidence of human use of fire by reddened earth and mixed ashes from hearths. According to Jelinek (1981), the upper Mousterian layers indicate even repeated burning of the whole cave surface, reminiscent of the practices of Australian aborigines and Bushmen for clearing their caves by fire. Ronen (personal communication) who explored these caves intensively speculated that this could have been caused also by accumulation of wind-blown fine ash deposits, originating from the woody vegetation canopy surrounding the cave. It seems, therefore, reasonable to assume that from the early phases of the Late Acheulian and Middle Paleolithic cultures onward, fire became a major driving force in the coevolutionary symbiotic feedback relations between these Mediterranean ancient people and their environment. Accepting the claims of Sauer (1956) on the cultural importance of human-set fires for food collection and hunting, we assume that after observing the beneficial effect of wildfires they could have realized that it can serve as the major energy source not only for heating and cooking but also for opening of dense forest and bush thickets. Intentionally set spot fires could have been used to create more accessible and richer ecotones for food collecting and hunting and to increase edible food by encouraging the lush regenerating woody plants and the invading herbaceous plants. This hypothesis is also supported by the results of our fire ecology studies in the Western Galilee and in the more recent studies on Mt. Carmel region, which will be reported below in this chapter.

Considering the interrelations of the Paleolithic humans with plants as the first step of a coevolutionary process of "incidental domestication", Rindos (1984: 137) stated that

> The habitual destruction or preservation of species will have major effects on the floristic structure of the region, and eventually on the directions open in plant evolution. Such habitual activities, passed as a cultural trait, are inseparable from human language.

Such interactions included the opening of gaps in the closed vegetation canopy for food and fuel, the clearing of land for habitation, the trampling of paths, the digging for bulbs and burrowing animals, and the disposing of human kitchen waste. All these intentional and unintentional interferences opened new regeneration niches senso Grubb (1977), creating favorable conditions for light demanding herbaceous colonizers.

The interferences of the widely scattered and scarce Pleistocene populations with the vegetation canopy, its litter, humus, and upper soil layer were only very slight and very patchy and, as pointed out above, have therefore been neglected. However, in spite of these slight initial anthropogenic perturbations of such scarce populations could have great impacts on their landscapes, we base our hypothesis on the following considerations: these widely scattered and very small landscape modifications could have been greatly enhanced by positive feedback loops with much greater, accidental forest gaps, opened by wildfires and presumably also by intentional, human-set ones, together with foraging and hunting. Therefore, these cross-catalytic mutual amplifying networks could have carried far-reaching implications through time on plant evolution and vegetation dynamics, and on the transformation of the pristine, natural landscape into a semi-natural and human-modified, cultural one.

In this process, the unique combination of natural raging wildfires with human set more frequent, much cooler and less fierce isolated fires could have become major landscape-shaping factors also later on, and from the Neolithic agricultural revolution onward, many of these semi-natural biosphere landscapes were further converted into agro-pastoral landscapes. This occurred most probably during the drier and warmer interpluvials in which the Mediterranean climate patterns became established. It created favorable conditions for the germination of light demanding woody plants, such as *Pinus halepensis* and most chamaephytes, as well as for above-mentioned herbaceous plants, facilitating their spreading over vast areas.

During the last Pluvial, 10 000–15 000 years ago, occurred the second major bifurcation, in which the fire-induced landscape modification from natural and subnatural into seminatural landscapes of Mt Carmel presumably reached its peak. We assume that this was achieved by the intensive and rational vegetation management of the economically prospering and culturally advanced Epipaleolithic economies of the Natufians, described in the comprehensive review of recent archeological findings by Mina Weinstein-Evron (1998) in the el-Wad cave and the adjacent terrace, connected with the organization of the Natufian living place. In these hamlets they constructed houses and thus developed a complex and rich communal, cultural and spiritual life. (A brief review on the Natufian culture and economy, and their advanced hunting, food collecting and preparation technologies can be found in Chapter 3.)

In the context of advanced technologies for the use of fire, it is worthwhile mentioning that amongst the findings of a rich assemblage of lithics, ground stones and bone implements, including decorated sickle hafts, Weinstein-Evron and Belfer-Cohen (1993), red ochre fragments composed of hematite were also found on pestles in this cave which were used for grinding of both yellow goethite and red hematite rocks. These pigments could have been produced by grinding natural hematite iron oxide extracted from the veins of rock outcrops or alluvium clays or by heating the more common goethite. Such red colors also appear on burned iron oxide, containing limestone and dolomites of volcanic outcrops of Mt. Carmel after hot brush fires with which the Natufians were probably well acquainted.

Already in 1937, Bankroft (1937) reported findings of a few pieces of wood charcoal of *Quercus* sp. and *Olea europea* in Upper Paleolithic layers of the el-Wad cave, indicating the presence of burned Mediterranean sclerophylls in the surroundings. More recently, Lev-Yadun and Weinstein-Evron (1994) identified 32 pieces, 5–10 cm long of Early Natufian Epipaleolithic wood charcoal of *Q. calliprinos, Q. ithaburensis, Myrtus communis, Cupressus sempervirens, Salix* sp., and *Tamarix* sp.

Tchernov (1975) interpreted the appearance of typical desert rock dweller rodents, such as *Acomys russatus, A. cahirinis,* and *Gerbillus dasyuru,* as indicators of a principal climatic trend of desiccation during the Late Pleistocene. However, Bottema and van Zeist (1981) suggested that the desiccation from this period onward may not necessarily have been climatically determined, but could have been caused by increasing human interference with the natural vegetation, and – as we assume – by the increasing exposure of rock outcrops, caused by such fires as a result of the opening of the dense woody cover. The term "desiccations" should be therefore only interpreted as a reduction in the woody cover at the expense of the increase of a more open, herbaceous vegetation cover and rock outcrops. The increasing burning activities, causing such a "desiccation", most probably also encouraged also the invasion and the colonization of such rock outcrops of the more xeric *Q. ithaburensis* woodlands by herbaceous plants by their more drought-tolerant genotypes. This could be a much more plausible explanation for this rodent invasion into the newly created more open and rocky sites than climate changes. Our assumptions are supported by the contentions of the distinguished Italian botanist and ecologist, Pignatti (1983) on the important human role not only in the stimulation of the flora evolution but also in the evolution of new habitats in the first seminatural cultural landscapes created in the Mediterranean Basin.

In his paleo-ecological interpretations of faunal remains from Mt. Carmel, Tchernov (1984) described these landscapes as "a kind of constant balance between open country and woodland". It is possible that the rocky and less fertile Terra Rosa and light Rendzina soil, covered by sclerophyll trees and shrubs with a fire-induced and therefore rapidly spreading light demanding herbaceous understory, was maintained by rotational burning in such "a constant balance". We consider this as the result of the homeorhetic flow equilibrium between the woody and herbaceous vegetation. These grasses could have become even more dominant in the fertile brown Rendzina soils, and their denser and more productive grass stands could have been the most preferable habitats for cutting the grasses and collecting of their seeds. This could have been also accomplished by collecting their scorched seed dispersal units directly from the ground after burning their dry stands. According to Harlan (1967), these primitive glumed cereals needed to be parched anyhow before they could be thrashed and winnowed.

In conclusion, through intensive vegetation management the Natufians created the first proto-agricultural seminatural Total Human Landscape. This was their most efficient way for channeling high-quality chemical energy of the spontaneously occurring wild plants, through their foraging herbivores into human food production on a sustainable basis.

THE ROLE OF FIRE IN THE ETHNO-ECOLOGICAL EQUIVALENCE OF THE NATUFIANS WITH CALIFORNIAN PRE-EUROPEAN INDIANS

Thanks to the extensive studies by archeologists, anthropologists, and ethnologists, a rich amount of information has become available on the life of these pre-European Indians, lending strong support to our claims on the ecological, economic, and cultural resemblance of the coastal Californian Indians to the Natufians of Mt. Carmel and its coast. Much of this has been summarized by Heizer and Elsasser (1980). For our comparison of greatest interest are those tribes living in comparable ecological conditions to the Natufians in the ecotones between the coast, the foothills, and the coastal mountain. Such ecologically equivalent landscapes stretch from the California Carmel to the Santa Barbara regions, with closest resemblance to the mild Mediterranean sub-humid climate of North Israel and the Carmel (see also Chapter 1). These were occupied chiefly by the Esselen, the Salinan, and farther South by the Chumash. As fishermen, gatherers, and hunters, those closest to the coast and the coastal foothills and mountains enjoyed the richest and greatest variety of food. The ethnological equivalence with the Natufians is expressed by their material wealth, as reflected in the elaborate round houses with wooden frames of their well-organized villages with several hundred inhabitants, their spectacular religious cult ceremonies, and the great efforts and skill invested in the ornamentation of their tools and especially their baskets. Their highly versatile and healthy year-round diet helped them to overcome drought periods and famine better than populations of the drier inland valleys and the South, enabling to reach the highest population densities of any North American Indians (Baumhoff, 1963) with 4 persons/km^2. This is most probably comparable with the estimated population density of the Natufians along the Carmel coast of about 500 people (Naveh 1984).

Heizer and Elasser (1980) described their elaborate ways of making fire by various types of fire "drills" and of the many other uses of it. These included the cutting down of trees, the leaching out of the tannin of acorns – their major staple food – with hot water, and the catching of grasshoppers by firing a circular area and then roasting them in a kind of earth oven with hot rocks. It is not too far-fetched to assume that similar methods could have been used also by the Natufians for the preparation of their staple food.

Most relevant for our comparison is the Indian's sophisticated use of intentionally set fires, as described by these authors and in more detail by Lewis (1973). In order to improve the visibility for hunting, grazing conditions for deer, and to increase the yields of wild grasses, fruits and berries harvest, and to uncover acorns and nuts, they set cooler fires in shorter cycles and on smaller patches than the natural wildfires. Thereby they created heterogeneous mosaics of open forests and woodlands, in different regeneration stages, dominated to a great extent, like in the Mediterranean by sclerophyll phanerophytes with a rich herbaceous understory. It resulted in the establishment of what Lewis described as "a dynamic balance of natural forces". This is comparable to the fire-induced homeorhetic flow equilibrium, maintained in the seminatural Epipaleolithic landscapes.

Lewis (1985) carried out extensive anthropological and ecological research on the "pyrotechnics" by native North Americans and Aborigines in North Australian, who still use fire to increase their plant and animal food resources. He concluded that the general aim of such intentional use of fire in some areas, while excluding it from others is to enhance and maintain an overall fire mosaic. This means a complex, more productive, and stable environment than would have derived from natural fires in terms of seasonality, frequency, and intensity.

He emphasized rightly that habitat burning was but one component in the total system of foraging adaptation. According to Bean and Lawton (1973), these semi-agricultural Indians of the Central California Coast reached a very efficient "interlocking of energy extraction processes". They suggested that Lewis' findings could provide new perspectives and ideas for other cultural evolutionarists and ecologists, showing how high levels of cultural integration and adaptation could have been reached by hunting, fishing, and food gathering communities in intensive proto- and semi-agricultural utilization of their natural resources. This may be true also for the advanced Epipaleolithic Natufians.

If these findings on the sophisticated vegetation management with the help of prescribed burning by the Californian Coastal Indians are true also for the Natufians, it would lend strong support to our contentions on their active environmental role, as opposed to the misconception of the passive role of the paleolithic preagricultural foragers. The Coastal Californian Indians could also give us some cultural clues as to how the pre-agricultural Natufians could reach such high levels in their coevolution with their landscapes. There is now ample archeological evidence that the Natufians could develop a complex and rich communal, cultural, and spiritual life, based on an advanced, intensive and broad-spectrum, metastable proto-agricultural food collecting-hunting-fishing economy. In many ways this resembles that of the Californian Coastal Indians at the time of their first closer contacts with Europeans. Presumably, in the gradual conversion of the Carmel slopes and the open coastal plains and hills into a mosaic of semi-natural ecotopes, driven by auto- and cross-catalytic feedback loops, both natural and intentional fire also operated as what Renfrew (1979) has defined as "multiplier effect". It induced mutually beneficial cross-catalytic couplings of the Natufians with their landscapes through the conversion of denser forests and maquis into more open, diversified, and fine-grained vegetation patterns, richer in herbaceous fire-followers. A similar process of landscape conversion has apparently taken place along the Central Coast of California. In both cases, close symbiotic relations with nature ensured sustainable landscapes management during which fire was incorporated as a larger-scale cultural – that means anthropogenic – perturbation factor, closely allied with humans and wild animals foraging. In both cases, thanks to longer-term and shorter-term rotationally shifting defoliation pressures such homeorhetic flow equilibrium between the woody and herbaceous vegetation layers and their postfire regeneration stages could have been maintained.

THE ROLE OF FIRE IN THE FINAL PHASE OF COEVOLUTION DURING THE EARLY NEOLITHIC PERIOD

Diamond (2001: 111) has pointed out rightly that the prerequisites for planting cereals as crops were the technological innovations adopted by the Epipaleolithic Natufians for the exploitation of wild cereals, such as sickles of flint blades for harvesting wild grains, baskets in which to carry the grains home from the hillsides, mortars and pestles, and underground storage pits. I maintain that their harvesting was also conditioned by the greater abundance of these cereals, on those Mt. Carmel hillsides for which the earlier technological innovation of controlled burning, and rational vegetation management was at least partly responsible.

According to Rindos (1984), this was the final coevolutionary stage of "specialized domestication of plants and animals". Broadening the ecological and cultural scales of these symbiotic relationships we can regard this as the culmination of the self-organizing process operating in this human–landscape coevolution. In this process, new evolutionary patterns of greater efficiency in solar energy and in the use of fire were channeled from suitable plants (including pyrophytic grasses) on restricted areas into human food production and livelihood was favored, and therefore succeeded in gradually replacing the former patterns of gathering-hunting.

The third major bifurcation occurred during the Neolithic Revolution together with the gradual conversion of larger and larger pieces of arable seminatural landscapes into agropastoral ones, simultaneously replacing the coevolutionary symbiotic relationship between people and their landscapes and nature by human dominance.

As shown below, fire could have played a further important role in this Neolithic revolution. This has also been claimed by Lewis (1972), who assumed that in these early phases of the "broad spectrum revolution" in the oak-pistachio area, the intensification and seasonal extension of man-made fires, coupled with increasing grazing of either wild or domesticated animals, could have provided the necessary "shock stimulus" for the emergence of agriculture. The sudden appearance of numerous larger-seeded Ceralia-type grasses (such as the above-mentioned species), which appear to be clearly domesticated, in the Proto-Neolithic period at the ecotones between the Mediterranean oak-pistachio woodland belt and the semi-arid Assyrian steppe could have been caused by increasing human activities. It was directly correlated to the fire-induced reduction of trees and to the findings of extensive, multicolored dry and dusty ash beds in the Shanidar cave. As mentioned above, according to Ronen, such ash deposit findings by Jelinek (1981) in the Tabun cave of Mt. Carmel could have also been caused by accumulation of wind-blown fine ash deposits, originating from the woody vegetation canopy surrounding the cave. Lewis (1972: 209) maintained that "in burns of grasslands and brushlands great amount of fine ash become airborne which, in the less disturbed air of a cave, would have settled like ordinary house dust". I have indeed observed in our controlled burning and reseeding studies in the Western Galilee that the fine ash of brush fires was soon carried away by wind or water erosion. From the much hotter fires of densely growing and much taller trees, the amount of such fine ash would have been much greater, and could be blown into these and other caves.

Kislev (1984: 62–63) described this early stage of the emergence of cereal agriculture as "the agrotechnical revolution", initiated by the Natufians with their intensive collection and consumption of seeds, including wild diploid or tetraploid wheat and barley. He considered the invention of the sickle as a major component of this agrotechnical revolution. It was twice as efficient as manual reaping and allowed the grain to ripen fully before gathering without a loss of yield. He assumed that in the next stage of the "domestication revolution", the invention of fields by Neolithic people was inspired by the observation of indigenous natural grasslands of annuals, such as wild emmer and barley occurring on open forest belts of deciduous oaks. He further speculated that they may have burned off unwanted grasses and used the cleared space for sowing grain and then, later on, they cleared also dense woods of evergreen sclerophyll low trees and shrubs. Kislev (1984) assumed that this transformation of the landscape from evergreen vegetation to fields with annual winter grasses started in the Prepottery Neolithic from 8500–7600 BC. As stated above, this transformation could have commenced gradually much earlier by the Natufians, using repeated fires, to induce the postfire grass flush.

Our fire ecology studies (Naveh: 1973) have shown that burning such wild grass stands does not prevent their germination in the following rain season, and may even stimulate it, thereby preventing the germination and development of sown grasses. A very different situation is created by the burning of denser tree and shrubs stands. The striking fertilizing effect of their ash seedbeds on the increase in grain yield could have acted as a major cultural trigger and multiplier effect for initiating the cultivation of these cereals in the early Holocene on Mt. Carmel and elsewhere. As indicated by the demonstrations of Iversen (1971), flint axes could be used for the felling of well-developed oaks and other tall trees in favorable sites, burning them and exploiting their ash seedbeds for the cultivation of these cereals. I have suggested (Naveh 1984) that this could have been also the case on the deeper and more fertile soils of the broader riverbeds, wadis and terraces on Mt. Carmel, implied by the findings of the great assemblages of such flint axes in the Sefunim cave by Ronen (1984). Such slash-burn rotations have been repeated several thousand years later by the Neolithic farmers in Europe (Narr 1956). Archeological evidence for such Neolithic land clearing is indicated also by paleobotanical findings in West Mediterranean *Q. pubescens* forests by Pons and Quezel (1985). They showed that in southern France in the Rhone Valley early slash-and-burn agriculture is suggested by charcoal dated 7.350 BC. There was a simultaneous decrease in the percentage of deciduous oak pollen, and a higher percentage of *Labiatae* and *Leguminosae, Plantago, Compositae* and other species considered to be weeds of cereal cultivation. Cereal pollen appeared at the same time.

THE FINAL STAGES IN THE CONVERSION OF MEDITERRANEAN LANDSCAPES INTO THE PRESENT TOTAL INDUSTRIAL LANDSCAPE

In this third major bifurcation, at the dawn of the Holocene, about 10 000 years ago, in which part of the seminatural landscapes were converted into agropastoral ones, fire has played a major role. As mentioned earlier, the coevolutionary symbiotic relationship between people and nature, based on the productivity of the spontaneous flora and fauna of autopoietic biosphere landscapes was replaced by human dominance. The reliance on a much larger but less reliable agro-pastoral biomass production from field crops and livestock enabled a much greater efficiency of energy extraction from the same size of land, enabling a much greater population density. However, this more intensive land exploitation did not ensure more sustainable natural renewable resources utilization. As described in detail by Naveh and Kutiel (1990), in Israel and in other Mediterranean countries it caused severe environmental upheavals and landscape degradation and accelerated soil erosion from denuded slopes, partly prevented by terracing of the cultivated slopes. It was followed by long prehistoric and historic agricultural land uses. During the Classical Greek, Roman, and Israel period the agro-silvo-pastoral cultural landscape was finally shaped, reached its peak, but then gradually declined and the seminatural vegetation formations were maintained in a metastable stage of homeorhetic flow equilibrium, in which fire still played an important ecological role. However when followed by heavy goat grazing and browsing, it contributed in the Mediterranean uplands to landscape degradation and is therefore presently regarded only as an entirely negative factor, and – like the goats – a curse of the land.

In the fourth major bifurcation of the industrial revolution, starting in Europe at the 18th century, but reaching the Mediterranean and the Levant much later, fire lost its importance as a major extrasomatal energy source. Driven by the most dramatic technological advances, this bifurcation caused a split between the solar-energy-powered, seminatural, and traditional agricultural biosphere landscapes and the fossil-energy powered, high input and throughput urban-industrial technosphere landscapes. The formation of the disorganized "Total Industrial Landscape" was completed in the second part of the 20th century by rapidly expanding agro-industrial and urban-industrial landscapes. As described in Chapter 6, increasingly heavy human pressures and exponential urban spread, followed in the Mediterranean by an unfortunate combination of land-abandonment on one hand, and misuse on the other, distorted the homeorhetic metastability of the fragmentizing and rapidly vanishing autopoietic biosphere landscapes, and their rich biological and cultural assets.

The following figure can serve as a summarizing demonstration of the evolution of the cultural Mediterranean landscape and its major bifurcations. Inspired by the models of cultural evolution and their bifurcation by Laszlo (1987 and 1994), it illustrates the close connection between the different above-described levels of cultural and landscape evolution. In these, the pristine, natural biosphere landscapes of the Middle Pleistocene were transformed through four major bifurcations into the Holocene cultural Total Industrial Landscape. The latter is replaced presently by the emerging fifth bifurcation toward the post-industrial Information Age. Because of the chaotic nature of this bifurcation, we cannot predict the future of our landscapes, but we can take part in shaping their future. We can provide human society with scientific and pragmatic information, through landscape and restoration research and action, helping to divert the present evolutionary trajectory from breakdown into a breakthrough toward the sustainable future for both landscapes and society and their utilization in the emerging information age.

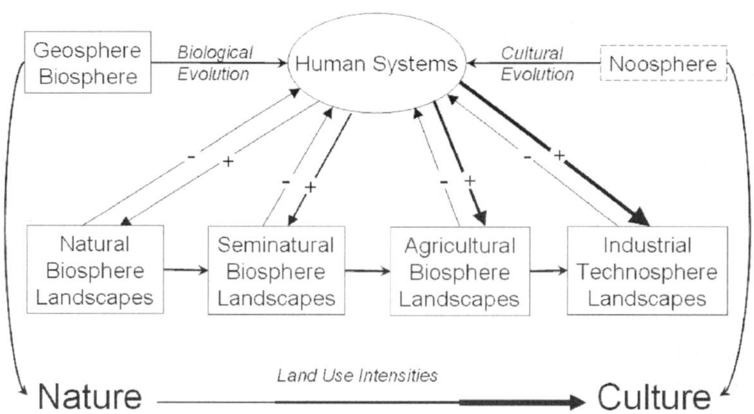

Fig. 2. The evolution of the cultural Mediterranean landscape in Israel, as part of the cultural evolution of Homo erectus into Homo sapiens industrialis and their major bifurcations.

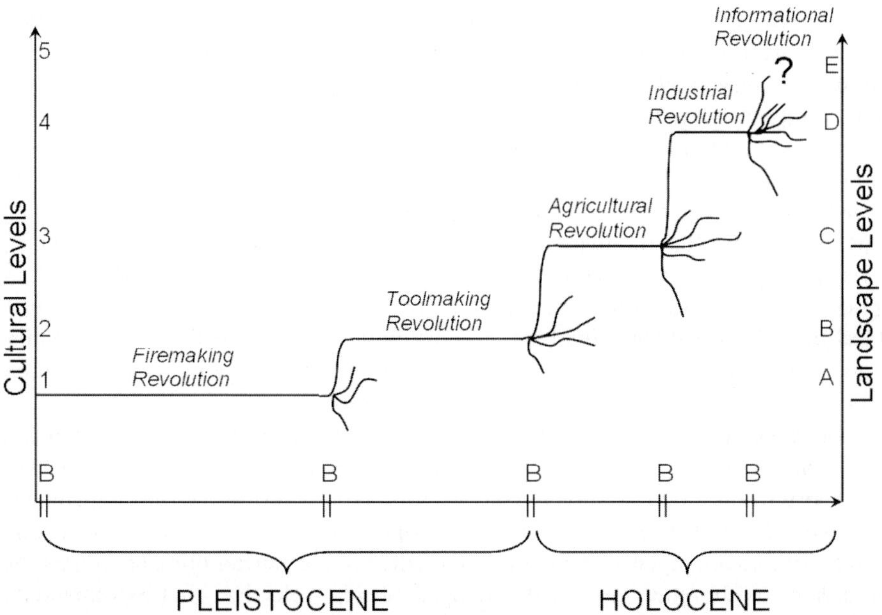

Left: CULTURAL LEVELS: 1) Lower Paleolithic Acheulian – *Homo Erectus* 2) Middle Paleolithic Mousterian *Homo Neanderthalis, Homo sapiens* 3) Epipaleolithic Natufian *Homo sapiens sapiens* 4) Neolithic *Homo sapiens sapiens* 5) Industrial *Homo sapiens industrialis* 6) Post-industrial-Information Age? Right: LANDSCAPE LEVELS: A) Natural (pristine) – Middle Pleistocene Biosphere B) Subnatural – Middle and Upper Pleistocene Biosphere C) Semi-natural – Late Pleistocene "Total Natufian Landscape" Biosphere D) Agro-silvo-pastoral – Holocene Biosphere E) Industrial-agro-urban – "Total Industrial Landscape" Technosphere F) Post-Industrial?
Fig.2. Continued

POSTFIRE REGENERATION STUDIES AND THEIR ECOLOGICAL SIGNIFICANCE

As a final subchapter, I will present some further results of the fire ecology studies in the Western Galilee and on Mt. Carmel. These confirm the great postfire regeneration powers of the sclerophyll woody plants, owing not only to their adaptive resilience to fire, but also to the striking temporary postfire flush of herbaceous plants, and especially of grasses. These lend support to our contentions on the role of fire as a major evolutionary driving force. The importance of the most recently published postfire regeneration study of the vegetation of Mt. Carmel (Naveh 1999) is the fact that it was carried out in near vicinity of the paleolithic archeological sites of Nahal Me'arot (The River of the Caves), which contain most plants mentioned in Table 1. This study included a comparison between an open pine forest with a dense shrub understory, burned by a hot wildfire several years earlier, and an adjacent unburned site in summer 1973.

Almost all "obligatory root resprouters" (Naveh 1975), namely sclerophyllous trees and shrubs (or phanerophytes), and climbers, as well as most "facultative root resprouters" (Naveh 1975), namely dwarfshrubs (or chamaephytes), geophytes, and hemicryptophytes regained close to a third of the original cover determined in the adjacent unburned plot. As explained in Chapter 3, these plants can regenerate both from root crowns and from fire-stimulated seed germination, followed by vigorous regrowth. We observed also an almost dramatic, increase in the herbaceous plant abundance and in floristic diversity, rising from fewer than 7–52 species. The striking rise of the economic value of the burned site, both for humans and for wildlife is implied by the fact that out of these 52 herbaceous species, 15 have high pasture values and 25 are valuable for human consumption; thanks to their edible bulbs, shoots, leaves, fruits, or seeds. In addition, 5 woody species increased highly their browsing values, because of their lush young leaf growth.

On the basis of extensive fossil records in California, Axelrod (1989) concluded that the above-mentioned evolutionary role of lightning and volcanic fires was combined with the increasingly stressful environments during the late Pleistocene and Quaternary in preadapted sclerophyllous taxa, and on fire-favored speciation. He, therefore, maintained that in countries with Mediterranean climates it was the elimination of summer rain that imposed strong selective factors, including increased fire frequency on sclerophyllous vegetation.

This is certainly true for the obligatory resprouting sclerophylls. Their fire-stimulated vegetative regeneration is closely linked with their hydro-ecological behavior as drought-tolerant and summer active plants. Since they rely on a deep and well-branched root systems, they are capable of starting to resprout immediately after the fire even in the middle of the summer by the mobilization of stored carbohydrates and possibly also of metabolized water in the roots. Due to their year-round intensive photosynthetic postfire activity, they can recover their former ground cover most probably even after 10–15 years, depending on site and climate conditions and on prevailing postfire grazing pressures.

We assume also that postfire foraging and especially browsing pressures could have acted, together with gradually increasing drought periods, as additional powerful selective agents, favoring those woody species and biotypes which very soon after the fire develop hard, thorny, or distasteful leaves and twigs, together with the highest vegetative regeneration capacities to overcome these defoliation stresses. Outstanding examples of such successful evolutionary strategies **for maximization of overall survival potentials** on Mt. Carmel are the East Mediterranean Kermes oak – *Q. calliprinos* – and the Eu-Mediterranean Pistachio shrub – *Pistacia lentiscus*. The Kermes oak is distinguished by vigorous resprouting from root crowns and suckers and adventative roots, but *P. lentiscus* branches also off laterally from prostrate leafy twigs which send roots and soon forms a dense and compact shrub canopy with high soil and water conserving features. As one, if not the most tenacious, evergreen shrub it responds already in the first days after fire by intensive cambial activity from root tips and buds (Naveh 1960). In the drier ecotones of the xero-thermo Mediterranean zone in Israel, on the semi-arid slopes of Mt. Gilboa, it has developed an even more drought-enduring ecotype than those on Mt. Carmel (Swarzboim 1978). It can withstand very heavy browsing pressures, and because of these high recuperative powers, it has remained on Mt. Carmel and elsewhere as the last woody survivor in frequently burned and heavily grazed pine forests, maquis, and woodlands.

On the other hand, facultative root resprouters, such as *Calycotome villosa, Cystus salvofolius, C. villosus*, and all other chamaephytes, including *Sarcopoterium spinosum* produce seeds from resprouting plants already in the first year after the fire. This explains their relatively large cover values in the burned site. Perennial herbaceous plants, namely hemicryptophytes and geophytes and perennial grasses, have similar dual vegetative and reproductive postfire regeneration mechanisms. Like these chamaephytes, they are typical drought evaders, adapting to the dry summers by more restricted physiological activity and especially by reducing their transpiration surface. They commence resprouting only after the first winter rains from fire-avoiding underground stem bases, bulbs, tubers, and corms, as well as from fire-stimulated growth from seeds. These adaptive traits are coupled with morphological and physiological plasticity and aggressiveness in colonization of fire-denuded and mineral-rich patches (Naveh, 1960).

Pinus halepensis is the only indigenous coniferous tree occurring in Israel, and Mt. Carmel is one of its most important last refuges. It relies solely on vigorous seed germination from cones that burst open from the heat of the fire, and is therefore an obligatory seed regenerator for which fire provides almost the only opportunity for natural regeneration under a dense maquis understory. This lack of resprouting ability is fully compensated by postfire germination of the great number of seeds benefiting from the temporary removal of the competition in the dense maquis understory. This is followed in general by a process of natural thinning out, similar to that of most chamaephytes seedlings. It leaves a scattered, rejuvenated stand of pine trees under the regenerating shrub canopy, or in the case of burned planted pine forests, more or less dense, even-aged pine stands. In the above-mentioned study we found several tall (natural) *P. halepensis* trees in the unburned site, and even more dead ones in the burned site and also numerous seedlings. It can, however, be expected that only few seedlings, growing in suitable regeneration niches, will survive and develop into taller trees.

In general, the rate and extent of the postfire herbaceous plant colonization is determined chiefly by the few perennial shade-tolerant herbaceous plants that survived in the dense brush cover as shade-tolerant relicts, together with the availability of seeds from annual grass fire-followers and their seed banks, as well as from other invading plants. Important seed sources for these postfire colonizers of Mt. Carmel were small, grassy patch openings and edge habitats, as well as adjacent fields and waste heaps near human habitations. We counted close to a hundred annual plant species, chiefly in shrub openings of relatively small grassy patches on rocky and shallow soil on Mt. Carmel, in our biodiversity study, reported in Chapter 2.

After a hot wildfire on Mt. Carmel in 1986, these ruderalic plants spread from the waste heaps of Kibbutz Beth Oren to several adjacent forests gaps and dominated the herbaceous fire flush and especially by *Hordeum spontaneum* and *Piptatherum (Oryzopsis) miliaceum*. The latter is the most prolific shade-tolerant fire-follower perennial grass, and is one of the last herbaceous survivors in the Maquis thicket regenerating after such fires on Mt. Carmel (Naveh, 1984) and the Western Galilee (Naveh 1974). Its plentiful millet-like seeds can be baked and used as staple food, in a similar way to its American counterpart *O. hymenoides*. The latter is called "Indian Rice" because the Indians of the drier Western Ranges gathered its seeds in quantities by cutting wild stands. The same practice may have been applied also to *P. miliaceum*, growing in dense and stout bunches, which can be cut shortly before ripening, like annual grasses. For *H. spontaneum*, this has been demonstrated by Harlan (1967) with the help of a flint sickle of the type found in the Natufian layers of the el-Wad cave on Mt. Carmel. The same may be true also for *T. dicoccoides*, the most important progenitor of wheat, of which several relict populations can be found on open, grassy sites and near rock outcrops on Mt. Carmel. In its major natural habitat on the fire-swept slopes of the Eastern Galilee hills, facing the Jordan Valley, this grass, together with *Avena sterilis*, recolonize after fire that may occur year after year. The latter is also the most important grass component of the open Tabor oak savannas and other grasslands, whose germination is stimulated by the heat shock of the grass fire (Naveh 1973). In these plants, adaptive responses to fire and its avoidance, as well as to drought and grazing stresses, are centered naturally around reproductive and growth behavior. Thus, early and prolific seed production, early seed shedding and distribution by efficient dispersal mechanisms and seed dormancy and polymorphism – especially in legumes – increase the chances of escaping fire and environmental rigor. A good example for the coupling of such survival mechanisms is "trypanocarpy" – namely the development of hygroscopic awns, callous tips, and other torsion mechanisms, that enable drilling and burying itself several centimeters deep into the soil. Thereby the seeds escape fire, grazing hazards and at the same time benefit from more favorable moisture and temperature regimes for germination. Among these grasses are *H. spontaneum, T. dicoccoides* and *A. sterilis* – the progenitors of domesticated barley, wheat, and oats, all endowed with big seeds. In these woodlands and grasslands, perennial grasses and geophytes also demonstrated most successful strategies for maximizing overall drought, fire, and grazing survival potential and resilience.

This temporary postfire flush of herbaceous plants is very similar to that of the California chaparral, where it has been aptly called *autosuccession* by Hanes (1971). In northern California, controlled burning of dense chaparral is a common practice to raise its carrying capacity for sport hunting of deer, resulting in a manifold increase of their carrying capacity (Biswell, 1989).

As reported in Chapter 3, our fire ecology studies further revealed that these herbaceous fire-followers serve as an efficient sink for the follow-up postfire flush of nutrients released in the first winter rain season. In Brown Rendzina soil, collected two months after a hot wildfire in 1983 on Mt. Carmel of an open pine forest with a dense and well-developed maquis shrub cover, we found a striking increase in water-soluble nutrients, especially nitrogen and phosphate in the first winter and spring after the fire in the upper centimeters of the soil. This postfire nutrient flush was rather short termed, but it could be utilized by the herbaceous fire-followers for proliferous forage and seed production. These plants, and probably also the rapidly regenerating dwarf shrubs, thus, serve as an important link in recycling these nutrients to the soil from which the resprouting, deeper-rooted woody plants can benefit in the following years. Of special significance for the enhancement of nutrient cycling is the 4.5 times increase in root production, facilitating the manifold increase in nutrient accumulation in the plants. The striking postfire increase in seed production lends strong support to our hypothesis that such favorable ash seedbeds served as triggers for domestication of cereals and their incipient cultivation in slash-burn rotations on Mt. Carmel and elsewhere, in the early stages of the agricultural revolution. We found also a striking increase in water-soluble nutrients, especially of nitrogen and phosphate in the first winter and spring after the fire in the upper centimeters of the soil (Kutiel and Naveh 1987; Kutiel et al. 1990).

CONCLUSIONS

In this chapter, I have attempted to show the evolutionary and ecological significance of fire in the shaping of Mediterranean landscapes. In addition to other, more or less catastrophic natural perturbations and to increasing drought, as well as human and wildlife foraging, fire originating from volcanic activities, lightning, and from repeated and long-lasting human interventions probably acted as a strong selection force. Those woody and herbaceous genotypes which developed the most efficient physiological and morphological evolutionary strategies for active and passive vegetative and reproductive regeneration mechanisms had the best chances to overcome the natural and human-induced fire stresses. These occurred during the drier and warmer interpluvials in which the Mediterranean climate patterns became established. It created favorable conditions for the germination of light demanding woody plants, such as *P. halepensis* and most chamaephytes, as well as for above-mentioned herbaceous plants, facilitating their spreading over vast areas.

Following the arguments of Axelrod (1958, 1989), then both the Mediterranean and the Californian sclerophylls are adapted to fire and this may have also preadapted them to grazing and browsing. This is also true for the Mediterranean herbaceous plants whose adaptive defense mechanisms against fire have been also effective against grazing and cutting. As described above, in annual grasses, as well as in legumes, and other highly valuable pasture plants, all these adaptive responses are centered naturally around early and prolific seed production, early seed shedding and distribution by efficient dispersal mechanisms and seed dormancy and polymorphism – especially in legumes – increasing their chances of escaping fire, grazing, and environmental rigor. Of special selective advantage in this respect is the above-described capacity to drill and to bury their dissimulates several centimeters deep in the soil by trypanocarpy and this may be one of the reasons why *Avena* and *Erodium* species became such successful invaders of California grasslands, in spite of their high palatability.

The same is true for their ability to compensate fully or partly for the lost tissue of biomass production both to grazing and to fire, by increased tillering, protected ground level meristem and in perennial plants also with vigorous vegetative reproduction (Stebbins, 1972). Both natural and human-set fires have presumably extended over the whole landscape the initially widely scattered and small gaps, opened by the Paleolithic food gatherers and hunters in the pristine Pleistocene vegetation canopy, together with grazers and browsers. This created many favorable niches for the colonization of the evolving herbaceous plants, including pyrophytic grasses which could have served later on as the progenitors of the first domesticated cereals. Therefore, fire may have played a most important role in the coevolutionary process and its cross-catalytic networks.

REFERENCES

Anderson E (1956) Man as a maker of new plants and new plant communities. In: Thomas WL (ed.) Man's Role in Changing the Face of the Earth. University of Chicago Press, Chicago, pp 763–777.

Axelrod DI (1958) Evolution of the Madro-Tertian geoflora. Botanical Review 24: 433–509.

Axelrod DI (1977) Outline history of California vegetation. In: Barbour MG, Major J (eds) Terrestrial vegetation of California. John Wiley & Sons, New York, pp 139–193.

Axelrod DI (1989) Age and origin of Chaparral In: Keeley SC (ed.) The California Chaparral Paradigms Reexamined. Natural History Museum of Los Angeles County, Los Angeles, pp 7–20.

Bankroft H (1937) Report on charcoal fragments. In: Garrod DAE, Bate DMA (eds) The Stone Age of Mount Carmel I. Excavations at the Wadi Mughara. Clarendon Press, Oxford, pp 129.

Bar-Josef O (1984) Near East. In: Neue Forschungen zur Altsteinzeit. pp 232–208.

Baumhoff MA (1963). Ecological determinants of aboriginal California populations. University of California Publications in American Archeology and Ethnography, 49:144–263.

Bean LJ, Lawton HW (1973) Some explanations into the rise of cultural complexity in Native California with comments on proto-agriculture. Introduction to Lewis HT Pattern of Indian Burning in California; Ecology and Ethnohistory. Anthropological Papers I. Ballena Press Ramona, California.

Biswell H (1989) Prescribed Burning for Wildland Management in California. University of California Press, Berkeley.

Blondel J, Aronson J (1999) Biology and Wildlife of the Mediterranean Region. Oxford University Press, Oxford.

Bottema S, van Zeist W (1981) Palynological history of the Near East. In: Prehistoire du Levant CNRS, Paris, pp 111–123.

Clark JD (1966) Acheulian occupation sites in the Middle East and Africa, a study in cultural variability In: Clark JD, Howell FC (eds) Recent Studies in Paleanthropology. American Anthropologists 68:394–412.

Cowling RM, Rundel PW, Lamont BP, Arroyo MK, Arianoutsou M (1996) Plant diversity in mediterranean-climate regions Trends Ecology and Evolution, 11:352–360.

Dafni A (1984) Edible Wild Plants. The Society for the Protection of Nature, Tel Aviv.

Diamond J (2001) Guns, Germs, and Steel. Norton & Company, New York.

Di Castri F (1981) Mediterranean-type shrublands of the World, In: Di Castri F, Godall DW, Specht RL (eds) Ecosystems of the World, II- Mediterranean- Type Shrublands. Elsevier Scientific Publishing Company, Amsterdam, pp 1–52.

Di Castri F (1991) An ecological overview of the five regions in the world with a mediterranean climate. In: Groves RH, Castri Di F (eds) Biogeography of Mediterranean Invasions. Cambridge University Press, Cambridge, pp 3–17.

Grove AT, Rackham O (2001) The Nature of Mediterranean Europe. An Ecological History. Yale University Press, New Haven and London.

Grubb PJ (1977) The maintenance of species richness in plant communities and the importance of the regeneration niche. Biol. Rev. 52:107–145.
Harlan JR (1967) A wild wheat harvest in Turkey. Archeology 20:197–201.
Heizer RF, Elsasser AB (1980) The Natural World of the California Indians. University of California Press, Berkeley and Los Angeles.
Higgs ESC, Vita-Finci D, Harris R, Fagg AE (1967) The climate, environment and industries of Stone Age Greece Part III. Prehist. Soc. 33:1–29.
Horowitz A (1979) The Quarternary of Israel. Academic Press, New York.
Horowitz A (1992) Palynology of Arid Lands. Elsevier, Amsterdam.
Horvat I, Glavac V, Ellenberg H (1974) Vegetation Suedosteuropas. Gustav Fischer Verlag, Stuttgart.
Iversen J (1971) Forest clearance in the Stone Age. In: Ehrlich PR, Holdren JP, Holm RW (eds) Man and the Ecosphere. Freeman and Company, San Francisco, pp 26–31.
Jelinek AH (1981) Middle Paleolithic in the Tabun Cave. Prehistoire du Levant. CNRS, Paris, pp 265–283
Kislev ME (1984) Emergence of wheat agriculture. Paleorient. 10:61–70.
Kutiel P, Naveh Z (1987) The effect of fire on nutrients in a pine forest soil. Plant & Soil 104:269–274.
Kutiel P, Naveh Z, Kutiel T (1990) The effect of wildfire on soil nutrients and vegetation in an Aleppo pine forest on MT. Carmel in Israel. In:Goldammer JG, Jenkins MJ (eds) Fire in ECosystem DYnamics. Mediterranean and Northern Perspectives. SPB Academic Publishing. The Hague, pp 84–94.
Lev-Yadun S, Weinstein-Evron M (1994) Late Epipaleolithic Wood Remains from El-Wad Cave, Mount Carmel, Israel. New Phytologist 127:391–396.
Lewis HT (1972) The role of fire in the domestication of plants and animals in Southwest Asia: A hypothesis. Man, 7:198–222.
Lewis HT (1973) Pattern of Indian Burning in California; Ecology and Ethnohistory. Anthropological Papers I. Ballena Press Ramona, California.
Lewis HT (1985) Why Indians burned: specific versus general reasons, Symposium and workshop on wilderness fire. Intermountain Forest and Range Experiment Station, U.S.D.A Forest Service, Ogden, Utah, pp 75–80.
McNeill JR (1992) The Mountains of the Mediterranean World. An Environmental History. Cambridge University Press, Cambridge, New York.
Narr KJ (1956) Early food producing populations. In: Thomas WL (ed) Man's Role in Changing the Face of the Earth. University of Chicago Press, Chicago, pp 134–151.
Naveh Z (1960) Agro-ecological aspects of brush range improvement in the maquis belt of Israel. Ph.D. Thesis, Hebrew University, Jerusalem. (Hebrew with English Summary).
Naveh Z (1971) The conservation of ecological diversity of Mediterranean ecosystems through ecological management. In: Duffy E, Watt AS (eds) Symp. Brit. Ecol. Soc., 11. Blackwell Scientific Publications, Oxford, pp 605–622.
Naveh Z (1973) The ecology of fire in Israel. Proceedings 13th Annual Tall Timbers Fire Ecology Conference, Tallahassee, Florida, pp 133–170.
Naveh Z (1974) Effects of fire in the Mediterranean regions. In: Kozlowski TT, Ahlgren CE. Fire and Ecosystems. Academic Press, New York, pp 401–434.
Naveh Z (1975) The evolutionary significance of fire in the Mediterranean region. Vegetatio 9:199–206.
Naveh Z (1984) The vegetation of the Carmel and Nahal Sefunim and the evolution of the cultural landscape. In: Ronen A (ed.) Sefunim Prehistoric Sites in Mount Carmel, Israel. B.A.R. International Series 230. Oxford, pp 23–63.
Naveh Z (1991) Mediterranean uplands as anthropogenic perturbation dependent systems and their dynamic conservation management. In Ravera OA (ed.) Terrestrial and Aquatic Ecosystems, Perturbation and Recovery. Ellis Horwood, New York, pp 544–556.
Naveh Z (1999) The role of fire as an evolutionary and ecological factor on landscape and vegetation of Mt. Carmel, Journal of Mediterranean Ecology, 1:11–26.
Naveh Z, Dan J (1973) The human degradation of Mediterranean landscapes in Israel. In: di Castri F, Mooney HA (eds) Mediterranean- type Ecosystems. Origin and Structure. Springer Verlag, Heidelberg, pp 370–390.
Naveh Z, Whittaker RH (1979) Structural and floristic diversity of shrublands and woodlands in Northern Israel and other Mediterranean areas. Vegetatio, 47:171–190.

Naveh Z, Kutiel P (1990) Changes in the Mediterranean vegetation in Israel in response to human habitation and land uses. In: Woodwell GM (ed.) The Earth in Transition. Patterns and Processes of Biotic Impoverishment. Cambridge University Press, Cambridge, pp 259–299.

Naveh Z, Vernet JL (1991) The palaeohistory of Mediterranean biota. In: Groves RH, Di Castri F (eds) Biogeography of Mediterranean Invasions. Cambridge University Press, Cambridge, pp 19–31.

Naveh Z, Lieberman AS (1994) Landscape Ecology Theory and Applications. (2nd Edition) Springer, New York.

Narr KJ (1956) Early food producing populations. In: Thomas WL (ed.) Man's Role in Changing the Face of the Earth. University of Chicago Press, Chicago, pp 134–151.

Nevo E (2001) Evolution of genome-phenome diversity under environmental stress. Proc. Natl. Academ. Sci. USA. 98:6233–6240.

Nevo E (2001b) Genetic resources of wild Emmer, *Triticum dicoccoides* for wheat improvement in the third millennium. Israel Journal of Plants Sciences 49:77–91.

Nevo E, Korol AB, Beiles A, Fahima T (2002) Evolution of Wild Emmer and Wheat Improvement Springer, Berlin.

Perles C (1977) Prehistoire du feu. Masson, Paris.

Pignatti S (1978) Evolutionary trends in Mediterranean flora and vegetation. Vegetatio 37:175–185.

Pignatti S (1983) Human impact on the vegetation of the Mediterranean. In: Holzer W, Werger MJA, Kosmas I (eds) Man's Impact on Vegetation. Junk, The Hague, pp 151–162.

Pons SA, Quezel M (1985) The history of the flora and vegetation and past and present human disturbances in the Mediterranean. In: Gomez- Campo C (ed.) Conservation of Mediterranean Plants Dr. Junk, The Hague, pp 25–43.

Renfrew C (1979) Problems in Prehistory. Edinburgh University Press, Edinburgh.

Rindos D (1984) The Origin of Agriculture. An Evolutionary Perspective. Academic Press, New York.

Rundel PW (1998) Landscape disturbance in Mediterranean-type ecosystems of the world In: Rundel PW, Montenegro G, Jaksi FM (eds) Landscape Disturbance and Biodiversity in Mediterranean–Type Ecosystems Springer, Berlin, pp 3–22.

Ronen A (1979) Guide to the Prehistoric Caves of Mount Carmel. Regional Council of Hof Carmel, Haifa.

Ronen A (1984) The Sefunim Prehistoric Sites, Mt. Carmel. B.A.R. International Series 230, Oxford.

Sauer C (1956) The agency of man on earth. In: Thomas WL (ed.) Man's Role in Changing the Face of the Earth The University of Chicago Press, Chicago, pp 46–69.

Sauer CO (1961) Sedentary and mobile bents in early societies In: Washburn SL (ed.) Social Life of Early Man Viking Fund. Publ. in Anthropology 31, pp 256–266.

Stebbins GL (1982) Darwin to DNA, Molecules to Humanity. W.H. Freeman and Company, San Francisco.

Stekelis M (1960) The Paleolithic deposits of Jsr Banat Yaqub. Bull.Res. Conc. Israel, Sect G. Geosciences 9:346–367.

Swarzboim I (1978) The Autecology of *Pistacia lentiscus* L. in Israel. Ph D. Thesis, Technion, Israel Institute of Technology, Haifa (Hebrew with English summary).

Tchernov E (1975) Rodent faunas and environmental changes in the Pleistocene of Israel. In: Pratksh G, Ghosh PK (eds) Rodents in Desert Environments. Junk, The Hague.

Tchernov E (1984) The fauna of Sefunim cave, Mt. Carmel. In: Ronen A (ed.) Sefunim Prehistoric Sites in Mount Carmel, Israel. B.A.R. International Series 230. Oxford, pp 401–422.

Tchernov E (1988) The paleobiogeographical history of the Southern Levant. In: Yom-Tov Y, Tchernov E (eds) The Zoogeography of Israel W. Junk, Dodrecht, pp 159–250.

Tchernov E, Horowitz LK, Ronen A, Lister A (1994) The Faunal remains from Evron Quarry in relation to other Lower Paleolithic hominid sites in the southern Levant. Quaternary Research 42:328–339.

Tatskin A, Weinstein-Evron M, Ronen A (1995) Weathering and pedogenesis of wind-blown sediments in the Mount Carmel Caves, Israel. Quarternary Proceedings 4:83–93.

Vernet JL (1975) Les charbons de bois des naiveaux mindelliens de Terra Amata Nice Alpes-mariitimes. Comptes Rendaus de l'Aca'demie Sciences, Paris 208D:1535–1537.

Waddington CH (1975) A catastrophe theory of evolution. The evolution of an Evolutionist. Cornell University Press, Ithaca, pp 253–266.

Weinstein-Evron M (1993a) El-Wad Cave. In: Stern E, Lewinson-Gilboa A, Aviram AJ (eds) The New Encyclopedia of Archeological excavations in the Holy Land The Israel Exploration Society Carta, Vol. 4. Jerusalem, pp 1498–1499.

Weinstein-Evron M (1998) Early Natufian el_Wad Revisited. ERAUL 77.Service de Pre'histoire, Universite' de Lie'ge.

Zohary M (1974) Geobotanical Foundations of the Middle East. Gustav Fischer Verlag, Stuttgart.

Zohary M (1983) Man and vegetation in the Middle East In: Holzer W, Werger MJA, Kosmas I (eds) Man's Impact on Vegetation. Junk, The Hague, pp 287–296.

B. GLOBAL ISSUES

Snow-capped Mt. Kilimanjaro and Masailand in 1963. As a result of global warming, the snow is retreating rapidly and will vanish in a couple of years.

CHAPTER 8

NAVEH Z. (1966)

THE DEVELOPMENT OF TANZANIA MASAILAND. A SOCIOLOGICAL AND ECOLOGICAL CHALLENGE.

AFRICAN SOILS: 499–518.

Masai cattle, herded by young Masai "warrior". (Photo courtesy by Kenya Information Service).

INTRODUCTION

Tanzania Masailand is a vast, undulating semi-arid peneplain of 23 250 square miles with scattered, higher and wetter mountain areas. It extends from the Kenya border about 300 miles to the south and from the Pare mountains about 100 miles to the west (Fig. I).

It is sparsely populated by less than 60 000 pastoral Masai of Southern Nilo Hamitic stock (Huntingford 1953) with about 1 000 000 cattle and more than a

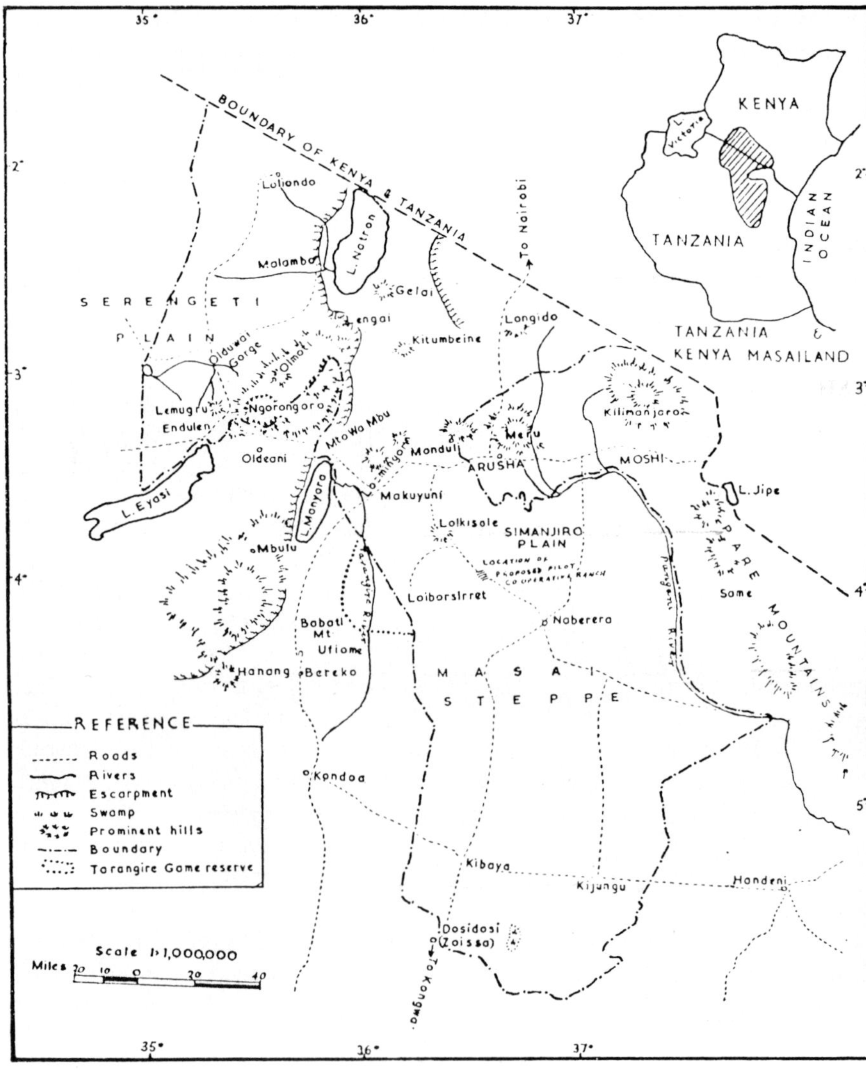

Fig. 1. Map of Tanzania Masailand Carte du Pays Masai de Tanzanie.

million sheep and goats. The Masai ranges are increasingly encroached upon by cultivators from neighbouring Bantu tribes who inhabit the surrounding smaller and wetter mountainous areas, which have dense agricultural populations of 500 people and more per square mile. This situation is typical of the dilemma with which many of the developing African States are faced at present.

Can this large and almost empty land be opened at all for settlement and agricultural development? If not, if the rainfall in this area is too low and erratic, and if water is not available for large-scale irrigation to ensure a stable and profitable cash crop economy, is there any alternative development possible? Should it be left to its fate as a barren sore in the heart of Tanzania and should a process of deterioration be even enhanced by this unplanned encroachment of cultivators, who try to scratch their livelihood from a few acres, but have to rely on famine relief in the drought years—as will most of the Masai themselves, after the loss of their grazing lands to the cultivators?

What should be done in order to prevent a progressive deterioration of Masailand and to utilise its natural resources for the most lasting benefit of its inhabitants and of Tanzania as a whole?

The following is an attempt to answer these questions after careful consideration and study of all factors involved.

PHYSICAL AND BIOTIC ENVIRONMENT OF TANZANIA MASAILAND

Climate

With the exception of the higher, sub-humid regions of Mt. Meru, Kilimanjaro, the Crater Highlands and the isolated mountains rising from the 4 000 feet high Masai Steppe, Masailand has a typical tropical semi-arid climate.

In the drier parts—chiefly in the western and north-western rainshadows of the mountain masses, the reliable rainfall is less than 500 mm. (20 in.) per annum and in the wetter parts not more than 600 mm.

Effective rains occur chiefly between November and May in two more or less pronounced concentrations of an earlier "short rains" and a later "long rains" period and as much as 125–150 mm. is apt to fall within a period of 24 hours. However, there seem to prevail extreme fluctuations in yearly and monthly distribution, as well as topographically conditioned spatial variation and a high incidence of convectional rain showers. These variations render almost meaningless, from the ecological point of view, any broad generalisations, since they can only be based on averages of a few widespread stations.

Mean maximum temperatures of 30 °C (87 °F) to 25 °C (78 °F) and mean minimum temperatures of 18 °C (65 °F) to 12 °C (54 °F) with absolute maximum of 40 °C (104 °F) in January and an absolute minimum of 4 °C (40 °F) in July, measured at the Tarangire Game Reserve (Lamprey 1963) are probably typical for most parts of the Masai Steppe. However, higher temperatures may be expected in the lower parts—chiefly in the Rift Valley—and lower temperatures in the higher elevations.

More accurate hydrological measurements from Kongwa—at the southern, dry corner of Masailand—showed, that under an annual rainfall regime of 200–525 mm. (8–21 in.) and with values of 2 250 mm. (90 in.) for Penman's estimates of open water evaporation, conditions of potential evapo-transpiration could be reached only for short periods (Pereira and McCulloch 1962). Deep rooted perennial grasses, such as *Cenchrus ciliaris*, as well as trees and shrubs are making full use of this annual rainfall, so that no surplus water can be expected for soil storage and perennial rivers. However, if part of this moisture is lost by run-off from bare soil, or by rapid evaporation as a result of reduced water penetration and soil infiltration capacity, this delicate water balance is shifted rapidly from semi-arid to arid growth conditions. This was the case over large areas in the semi-arid Karamoja region in Uganda, resulting from overgrazing, trampling and overburning as well as cultivation (Wilson 1962). Such "man made deserts" may be also the inevitable outcome of land misuse in Masailand.

Water and Soils

The greater part of Masailand has very little permanent water. The catchment areas of the northern mountain masses are already under intensive cultivation with irrigation and the only large river, the Ruvu, flows along its extreme eastern border. The other mountains give rise only to small streams or springs, most of which are only seasonal. An assessment of available water sources (Kametz, 1963) showed that even if all these sources could be fully developed, there would remain about 10 000 square miles with inadequate water for domestic and livestock use. This area covers chiefly the Rift Volcanic Zone, which extends from the Rift Valley and the Crater Highlands and Mt. Meru and Kilimanjaro into the Central Masai Steppe, south of Arusha. Most of these soils are calcareous, black or grey/brown clays and loams, derived from alkaline rich volcanic rock. The only analytical data available are from the Central Masai Steppe. These show a high exchangeable base capacity, fairly high phosphate reserves, but only a low humus and nitrogen status. Their drainage is slow and their clays have a tendency to swell and deflocculate, when wet. The major drawback to mechanical cultivation is the high amount of surface and subsurface stones in many locations.

In Southern Masailand, most soils are senile, basement—complex derived red to dark reddish and brown loams. These soils are less fertile, in general deep and well drained, but susceptible to erosion.

Vegetation

With exception of the more humid montane grass and forest formations, Masailand is part of the great xerophyllous Acacia Savanna belt, extending from Sudan and Northern Kenya and merging in the South—in the Central Region of Tanzania and South Masailand—with the "Miombo" (Brachystegia) woodland belt.

In Tanzania Masailand this vegetation is characterised by an over-storey of several Acacia and Commiphora and other woody species, occurring in varying densities and combinations together with a large number of chiefly perennial grasses as a mosaic of open, treeless grasslands, open parklands, wooded grasslands and denser woodlands, and brushlands. The main grasses are *Themeda triandra, Pennisetum mexianum, Cenchrus ciliaris,* and many other perennial and annual species of *Eragrostis, Digitaria, Bothrichloa, Panicum, Cynodon, Spropulus,* e.o.

One of the most prominent features of Masailand are the regularly repeated, topographically and pedologically conditioned variations in soil and vegetation patterns from well drained, drier slopes, to hardpans of calcareous concretions and to shallow depressions of open grasslands with impeded drainage. The latter, so called "Mbugas", which become more or less waterlogged during the rainy season and their dark grey clay-vertisols become sticky and cracking, while drying off.

These Masailand Rain-Pond Catenas (Gillman, 1949) have a decisive effect on present grazing and burning patterns, because of their greater inherent production potentials. They will be therefore of greatest significance for the future management and improvement of Masailand ranges.

Wildlife

The most important natural asset of Masailand is, without doubt, its wildlife: within its boundaries are three of the most outstanding game refuges and tourist attractions of East Africa: the Serengeti Plains, the Ngorongoro Crater and Lake Manyara. The Masai Steppe itself is still considered one of the finest hunting grounds for big game in the world.

Wildlife and its habitats are the main subject to which systematic and long-term research has been devoted in Masailand. The ecology of the mammal population in the Tarangire Reserve and the adjacent Masai Steppe was studied by Lamprey (1963) and in the Serengeti National Park comprehensive studies on the ecology and population dynamics of the chief plains game and their predators are being carried out by a team of ecologists and zoologists under the auspices of FAO.

Lamprey (1963) concluded that Masailand is serving as a wet season dispersal area for a great proportion of those animals who seek refuge in the dry season in the parks and reserves and where permanent waters from rivers, lakes and springs are available. Thus, for instance, out of a total of 10 400 herbivors of twenty-two species, counted by Lamprey in the Tarangire Reserve in the dry season, and out of 618 carnivors of eight species only 4 500 herbivors and 195 carnivors remained there in the wet season. The chief immigrants to the adjacent Masai Steppe in an area of about 8 000 square miles were elephants (*Loxodonta africana*), buffalos (*Syncerus caffer*), zebra (*Equusquagga boehmi*), wildebeest (*Connochaetus taurinus*), eland (*Taurotragus oryx*) and others, accompanied by most of their predators. Impala (*Aepyceros melampus*), giraffe (*Giraffa camelopardis*), Grant gazelle (*Gazella granti*), hartebeest (*Alcelaphus buselaphus cokii*), lesser kudu (*Strepsiceros imperbis australis*), and others were residents. However, the total density over the whole dispersal area, east of the Rift Valley, is very low (2.5 animals per square mile), as compared with the dispersal density in similar vegetation, altitude and rainfall conditions in the Serengeti Plains (80–140 animals per square mile).

According to Lamprey this present stocking rate is well below the potential carrying capacity and is limited chiefly by available water and fodder in the dry season concentrations. The vulnerability of these animals to predators, poachers and hunters during this critical period is also a limiting factor. There is strong evidence that the drastic reductions in numbers in the last thirty years is due mainly to the loss of dry season water supply from irrigations, drinking water for livestock, cultivation and hunting and other activities by European farmers in the vicinity of Arusha and Moshi (Lamprey, 1963).

It is therefore obvious that any further modification of these wet-season dispersal habitats and of the perennial water and fodder reserves in the dry season concentrations will affect the fate of wildlife in Masailand as well as in its parks and reserves. These considerations should be an integral part of any development plans of Masailand.

THE ECOLOGY OF THE DECLINE OF THE NOMADIC MASAI ECOSYSTEM

The Disruption of the Nomadic Grazing and Burning Pattern

The present, transitional stage of the Masai society from a completely nomadic to a more and more sedentary way of life is not an encouraging symptom of socio-economical progress from a primitive or "arrested" (Darling, 1963) stage of development to a more advanced one. On the contrary, it is a stychic process of insidious decline and degeneration. This decline is the result of the disruption of the ecological well-balanced nomadic system of rotations between wet-season pastures with temporary waters to dry-season holdings with permanent waters and between different site-vegetation types and differently burned and grazed pastures.

In addition the severe environmental resistance of limited permanent water and the control of livestock and human populations by disease and war restricted their biotic potentials. Therefore their pastures could only be stocked below the potential carrying capacity and thus, as in the wild ungulate savanna ecosystem, constant energy flow and sustained productivity were safeguarded. However, since then, some of the most important dry-season holdings and permanent waters on the western slopes of Kilimanjaro were alienated for European farming, and from all the wetter fringes the better grazing lands have been steadily encroached upon by cultivators. As a result these nomadic rotations were more and more restricted, and the frequency of burning and grazing pressure increased. At the same time the medical and veterinary care and the development of water points, boreholes and dams removed some of the biological controls and thus greatly disturbing the ecological equilibrium. The greater reliance of the people on the permanent waters also in the wet season resulted in the stocking of nearby pastures to **their upper limits** of carrying capacity in the wetter years and therefore above the threshold of irreparable destruction in the drier years. Now in these areas the only restriction on livestock numbers—and without famine relief also of people—remains the "drought-starvation control".

The Drought-Starvation Control

This control is effective in the critical drought years after all pasture resources have been exhausted and those pastures with permanent water on which survival depends have been turned into "thornbush deserts". Once the environmental resistance of the modified nomadic ecosystem was removed, there is no self-regulating mechanism left which could ensure stability in production and constant energy flow (Odum, 1959). The vicious circle of decline in such short food-chains as the pastoral subsistence economy from the pastures through the livestock and the people, is being closed rapidly. The constant lowering of the efficiency of energy conversion by overgrazing and burning, as well as the lowering in efficiency of water utilisation (Pereira, 1962) aggravates more and more the impact of climate and the resulting seasonal fluctuation in pasture productivity. In these pastoral subsistence economies the possibilities of negative entropy and energy influx from outside are extremely low and there are no motivations of economical returns on which quality versus quantity concepts could be based. Therefore, with the accelerated down-grading of the habitat, more and more animals are needed to overcome these oscillations and in return, their output is further decreased.

It should be stressed that this acute stage of drought-starvation has been reached at present in Tanganyika Masailand only in those parts which suffered most from the above described process—namely the Longido District. But it is much more widespread in Kenya Masailand and other semi-arid regions with pastoral Nilo-Hamitic tribes in Northern Kenya—chiefly in Turkana, and in Uganda—chiefly in Karamoja.

Nevertheless, with growing population pressure and with the alarming increase of uncontrolled cultivation in Central Masailand, this stage will be reached very soon also by those who up to now were able to maintain a fairly stable and well-balanced grazing and burning pattern.

The failure to control stocking rates in water development and grazing schemes in these areas cannot be explained solely by the insatiable desire to build up livestock populations for bridal wealth and prestige. After the disruption of their nomadic ecosystem, these psychological stimuli are overshadowed more and more by the basic biological need for survival, which can be ensured only by larger and larger herds. The trauma of the often repeated experience of disaster and starvation is passed on from generation to generation and the development of water points and disease control have not improved things, but made them worse.

The Masai society lacks a powerful central authority and the age-set structure leaves all the responsibilities to the elders. Grazing is communal and there are few chances that a more far sighted and independent individual will be able to change the present socio-economical pattern and its inherent quality-quantity concepts. He has not only to convince his family and elders and all those which make joint use of pastures and water; he must also prevent all other Masai, who will come from far away in the search for water and fodder, from using the reserves he had kept for his own reduced herd.

During the last severe drought in 1961, when losses reached 80% and more of the herds in certain regions, it was almost impossible to persuade the Masai even in the worst stricken places to cull their poorest cattle in time and thus save their pastures for the better animals. They believed that with the loss of their cattle all hope for life and survival would vanish. One young educated and outstanding Masai from the Longido Region told me how he was almost stoned by the elders, even after he convinced his father to take part of his half-starved cattle to the market in order to trade them in for food.

It seems therefore futile simply to superimpose principles of better livestock and range management, of controlled land use and stocking rates and even rotational grazing and conservation systems, which were developed in a sophisticated "cash crop" economy, on top of a tribal, pastoral subsistence economy. After all, even many "modern" ranchers do not adhere to these principles. Unfortunately this has not been recognised in the past and this was one of the main reasons for the failure to readjust these disrupted ecosystems to their modified environment in the above-mentioned grazing and development schemes in semi-arid East African range lands. In the case of the Masai this was marred in addition either by uncritical admiration of the "wonderful nomadic way of life" which should not be spoiled by modern civilisation ("Masaitis") or by its wholesale condemnation and despisal ("Anti-masaitis"). In both cases the result was the same. The Masai were left alone with their disrupted, starvation-controlled ecosystem, without hope for change and improvement.

THE PLACE OF THE MASAI SOCIETY IN THE DEVELOPMENT OF MASAILAND

In contrast to these unbalanced attitudes, Branagan (1962) proposed the consolidation of the existing family clusters—the "enkudotos"—as co-operative ranching units around present and newly developed water points. Land tenure and permanent water rights should be legalised for the co-operative as a whole, but at the same time strictly controlled land use, range and livestock management should be enforced by law. This law should be imposed by a "Natural Resources Committee" which would use a system of fines and possible expulsion of the individual from the co-operative and loss of water right in cases of disobedience. These suggestions and the detailed proposals on legalisation and administration by Fallon (1963) are serving now as a basis for the development plans of the Tanzania Government. Two senior officers have been charged to carry out an exploratory reconnaissance survey of people, livestock, grazing and water use pattern and of the vegetation, soils and tsetse densities.

This assessment followed by the legal recognition of successive groups, will complete the first phase of the plan. However, this is also a crucial stage for the subsequent development, which will take place within this frame-work. It is therefore of the utmost importance to avoid past mistakes and to show how it will be possible to overcome the problems facing the implementation of these plans.

The first and most decisive factor, which has to be considered, is the Masai society. The Masai need to be convinced of the advantages of these plans and be induced to co-operate in their planful consolidation and stabilisation. But, even if they do agree and co-operate fully, how can we be sure that such a transition from the present dispersed position to a more sedentary concentration will not further accelerate the drought-starvation-deterioration cyclus, whereby more and more Masai will have to be rescued by famine relief funds?

On basis of the above described situation, there is no doubt, that without careful control of cultivation and land use, together with an overall effort to raise productivity of all trophic levels of the Masai-livestock pasture ecosystem **as a whole,** there is little hope of preventing this vicious circle of decline.

The socio-economic, cultural, educational and medical advancement of the Masai society should be, therefore, an integral part of this effort, so that the improvement of their living conditions could proceed together with the rise in productivity of their livestock and pastures, within these co-operative ranching units.

I do not share the pessimistic view of those, who claim on the basis of their experience with the Masai under the British Administration, that it might take 20 years and more, before such efforts would yield results. Many Masai realise already that in an independent Tanzania, striving for progress and advance, they will not be able to remain masters of their vast land, if they do not make better use of its natural potentials. They are worried about increasing encroachment of their better pastures by cultivators, which endangers their very existence, and they are more willing, than at any previous time, to co-operate with the authorities on any development plans, if they can be convinced of the benefit of these plans.

The Masai are well known as people with a high, natural intelligence and alertness and if they could be approached in the right way and with the right means, appealing to their basic attitude and needs as pastoralists and to their pride and integrity, much could be achieved in a short time. It seems to the writer, therefore, that if they can only be shown that there is an alternative to their present danger of progressing deterioration in an **improved pastoralism,** in which the best of their traditional values can be preserved, they will very soon follow this example.

It is obvious that such an all-embracing development project will require a large amount of money and professional help, which cannot be provided solely by the Tanzania Government. If such support could be obtained from international sources, the development of Masailand could serve as an example for all those other vast semi-arid lands in Africa, populated by pastoral tribes with disrupted nomadic subsistence economies and the resulting decline of their ecosystems.

THE RISE IN PRODUCTIVITY OF MASAILAND

How can such a rise in productivity and economical output be achieved under Masailand conditions? If the Masai could be induced to make better use of their intellectual and physical potentials, what should be their enterprise and how can the natural resources of their lands be utilised for this means?

From all previous discussion it is obvious that the natural limitations of a semi-arid, tropical climate, and the lack of water for irrigation will not enable any large scale, profitable and stable crop cultivation economy. We also stressed the socio-cultural and psychological importance of pastoralism in the Masai society and tradition. Therefore, all efforts should be spent not in its removal and replacement by other ways of life and enterprises, less suitable to the natural and human conditions of Masailand, but to the modification, modernisation and intensification of pastoralism in its widest sense.

Therefore, the rise in productivity will be based chiefly on the rise in livestock production from the Masailand ranges.

This can be achieved in two ways:

(1) Increase in efficiency of conversion of primary productivity (the plant material)—into animal products and their recurring revenue.
(2) Increase in the primary production of the ecosystem itself.

In Masailand conditions both ways are complementary and interdependent and should be, therefore, attempted and achieved at the same time.

This could be started in a combined research, development and demonstration scheme in the first co-operatives to be established in Southern Masailand, assisted by a closely working team of range, livestock and wildlife specialists, a community development and a co-operative leader. Such a proposal was submitted to the Tanganyika Government (Naveh and Branagan, 1964). Its results could be applied after its socio-economic and agro-ecological evaluation and after completion of the ecological survey, to further development schemes in Masailand, and eventually also to other semi-arid range lands in East Africa.

At the same time such a pilot scheme could serve as a focus for social, educational and medical improvement, which should reach all layers of the Masai society and especially the younger generation—the "Ol-Murrani" (unmarried, young "warriors") and the "En-Titi" (unmarried girls) whose potentials are wasted at present on a carefree and unproductive life. For this purpose all modern and scientific tools of communication, persuasion and indoctrination should be used.

In this way, such a scheme could have the greatest impact on the Masai and at the same time yield maximum results in the shortest possible time.

Here, only the basic principles and premises can be elucidated.

THE INCREASE OF LIVESTOCK PRODUCTION IN MASAILAND

The Importance of a Diversified Livestock and Wildlife Economy

Studies conducted in recent years in semi-arid, tropical savannas in East and Central Africa and summarised by Lamprey (1963) provide conclusive proof, that if ecological efficiency is expressed in terms of standing crop and biomass and apparent stability in energy flow and productivity, there is a definite advantage in wild ungulates over domestic livestock. This is true, even if the latter are kept under modern range management conditions by European ranchers. Talbot, et al. (1961), working in Kenya and Tanzania Masailand, found that wild ungulates have the following advantages over domestic—both local and exotic breeds:

(1) Higher efficiency of food utilisation because of the greater spectrum of species and feeding habits.
(2) Higher digestive efficiency and killing-out percentages.

(3) Lower water requirements—especially in certain gazelle species.
(4) Greater growth rates and live-weight gains.
(5) Earlier reproduction.
(6) Higher resistance to disease—especially trypanosomiasis.

On the basis of this and further studies in East Africa, as well as on the basis of the practical experience gained from commercial "game ranching" in Central and Southern Africa, Ledger (1964) concluded that on these marginal lands game can provide more and cheaper meat with less destructive effects on soil and vegetation than domestic livestock. At the same time game ranching schemes may provide a useful stepping from subsistence to a productive economy.

Wildlife as a Stimulus for Tourism

At present, however, the greatest economical asset of wildlife in Masailand is probably as a potential stimulus of tourism. In Kenya, tourism is already reckoned the second largest source of the national income, being worth £8.5 millions in 1962 (Brown, 1963). In the first six months of 1963, tourism in East Africa increased at the rate of 26% as against an average rise of 15.5% in the last fifteen years and it is anticipated that the numbers of visitors will have trebled by 1970, spending about £24 000 000 per annum (Oneko, 1963).

Without doubt, more and more people in the U.S.A. and Europe will travel farther and farther in order to escape—at least for a short vacation—the increasing pressure of urbanisation. A realistic description of the over-crowded motor-roads and vacation centres in summer 1963 in Europe, recently in *Time Magazine* indicates that it will be more and more difficult to find this escape in Europe. At the same time a steadily growing part of the national income of the affluent societies is spent on leisure time and recreation and fares are getting cheaper and air transport easier. This makes areas in East Africa, like Masailand, which still offer a unique, vast and unspoiled wilderness more and more attractive for larger circles of people all around the world.

However, if we wish to make wildlife an integral part of the development plans of Masailand and to reconcile its needs with those of the socio-economical advancement of the Masai people, we should know more about the direct benefit, which could be derived from planned conservation and management of wildlife and its habitats. The establishment of a special Masai tribal game park—the Lol Kisale Controlled Area and Tarangire Game Reserve—and the inclusion of the adjacent Central Masai Steppe as a controlled hunting area, with part of its revenue diverted directly to the Masai co-operatives in this area, could be of greatest significance in this respect. In addition, an anthropological museum of Nilo-Hamitic culture, handicraft, art and history, could be established. In this way, not only could the tourist, game viewing and hunting "industry" be combined and diversified, but this could be also a source of "profit, pride and prestige" (Huxley, 1961) for the Masai people.

However, even if wildlife could be made a desirable part of the Masai economy, the main enterprise will still be based on domestic livestock.

The Economic Feasibility of Beef Cattle Ranching and Range Improvement in Masailand

A frequent assumption, that in semi-arid, extensive areas with scattered populations, the only profitable livestock industry is beef cattle ranching, should be viewed in the special conditions of Masailand. Here—as in other semi-arid regions in East Africa, the intensification of the beef cattle industry is faced with handicaps. There is the late maturing of the indigenous Zebu cattle—especially under extensive range conditions—the absence of any feeder industry to allow earlier marketing and fattening of weaned calves and yearlings, the low prices of meat and low demand for high quality meat, the expenses involved in frequent dipping and disease control for higher quality and quantity production. In these conditions, at least four-fifths of the herd will have to be kept throughout the whole year and flexibility in marketing and profitable higher take-off in critical drought periods, which is so important in Masailand conditions, will be very difficult.

At present the annual gross revenue per acre from semi-arid range land in East Africa is less than Shs. 5/- under tribal subsistence pastoralism like in Masailand and up to Shs. 15/- under most advanced, large-scale ranching operations (Brown, 1963). These revenues could be obtained only in few places, where capital and high profesional skill was available to achieve overall daily gains of 0.75 lb. per head from land developed by paddock fencing, water distribution, etc., to carry 1 000 lb. beast to 12 acres (Ledger, 1964).

These low revenues attainable in Masailand conditions set a definite limit to the economical feasibility of investments and expenditures, such as water development, fencing, bush clearing and tsetse fly eradication. Such expenditures might be justified only on land of high potentials, where improvement of these pastures would result in a considerable long-term rise in carrying capacity, to at least 4 acres per stock unit, as well as in improved livestock production (Hutchinson, 1964). Carrying capacity is therefore the most important economic factor to be considered and before expending large amounts on improvement, not only the most efficient ways of long-term reclamation, but also the agro-ecological potentials of the improved site and its most efficient utilisation should be known.

As mentioned already it might be possible to obtain from low potential lands the same, or even higher revenues from wildlife conservation, without the need of such heavy capitalisation and without the danger of progressive decline of production potentials, which is so often the case with extensive cattle ranching on marginal lands.

The Importance of Increased Milk Production

In order to convince the Masai that animal husbandry would not be only a mode of life and philosphy, but also a source of profit, the consideration of the human factor is most important. Beef cattle ranching is probably the most commercialised form of animal husbandry and lacks the element of personal attachment to the individual animal, which is of so great significance in these pastoral systems.

Milk production, on the other hand, might better facilitate implementation of economical principles and a change in the Masai attitude towards quality. It is their most important staple food and wherever marketing is possible—already to-day a most desirable trading commodity. If processing, storage and marketing of dairy products could be improved, the increase in milk production would have not only an immediate impact on daily nutrition, but also on the cash economy. Furthermore, the immediate response of the animal to better husbandry and nutrition is much easier to observe than with beef cattle and the cause-effect relationship will be more obvious. The fact that the Masai will be able to obtain more milk for longer periods from fewer, but better cows on smaller, but improved pastures might release them from the fear of starvation and make them more receptive to control of livestock numbers and further improvements.

The present lactation of most Masai cows is as low as half gallon and less per day. With the exception of a short flush period after calving, the demands for milk of both the family and the calves can hardly be satisfied, so that first the latter and then both are deprived of it. Therefore, improved milk production will not only help to improve the nutrition of the family, but also produce better calves with higher weaning weights and greater disease resistance. In this way a sound foundation could be laid for a healthy, hardy and more productive herd both with regard to milk and meat. Such a dual purpose cow should be the chief goal of cattle improvement in Masailand.

At the same time the possibilities of improvement to the Masai goats and sheep should not be overlooked. Increased milk production of the goats is especially important in the most degraded and bush-encroached range lands, where the prospects of milk production from cattle are currently very low. In suitable areas also the improvement of meat and wool production from sheep and the possibility of introduction of Angora goats to produce mohair and better quality meat should be considered.

THE INCREASE OF RANGE PRODUCTIVITY IN MASAILAND

Pasture Improvement

The level to be reached in the improvement of the Masai-stock by better husbandry, feeding and selection and eventually also by breeding and grading up with genetically higher potential breeds, will depend on the success achieved in removing the chief bottleneck for higher livestock and especially milk production: the great fluctuation in pasture production and its low nutritive value in the dry-season. Of greatest importance in this respect are the dry-season pastures, which are closest to the permanent water and bomas. For six months—and in drought years for much longer periods—the survival of the herds depends mainly on these pastures and they are therefore the most likely to suffer from the greatest grazing and burning pressure. Naturally, they will be also the first candidates for accelerated deterioration and for the resulting bush encroachment when the nomadic and transhumance systems will be replaced by more and more sedentary systems of land use.

As far as possible, the Masai try to leave these pastures as "standing hay" for the dry season. This is especially true for the "mbugas" and other favourable sites of the "Rain Pond Catenas" which are too wet in the rainy season for efficient utilisation. However, at present much of the high production potentials of these pastures is wasted on low value grasses and sedges, which become coarse and poor in the dry season, when their fodder is needed most.

In general, nitrogen is considered the major limiting nutrient for the growth and nutritive value of tropical pastures (Henzell, 1962). This is true not only for the humid tropics, but also in more arid conditions, where free availability of nitrogen and other nutrients will enable the pasture plants to make more efficient use of the limited water sources.

Although we are lacking any basic information on this subject, we may assume that after the replacement of the ungulates by domestic graziers and with increased grazing and burning frequency, the nitrogen removed from the ecosystem could not be compensated because of the low nitrification status of these grasslands. We must realise also that in Masailand great amounts of nitrogen are lost by night corralling and by the building of dung huts, which are burned after their abandonment and also by the tracking of the cattle to and from distant water points. The effect of nitrogen deficiencies on plant growth and composition in Masailand are apparent even by superficial observation and they were also proved recently in fertiliser studies at Kongwa—at the southmost, dry end of the Masai Steppe (Owen, 1964).

In the dry tropics of Australia, in Queensland and the Northern Territory, the reseeding of suitable areas with grass-legume mixtures, increased pasture output, carrying capacity and live-weight gains many-folds (Norman, 1962). This could be achieved also in conditions similar to the better sites of Masailand and with the use of improved material from grasses and legumes, indigeneous also in Masailand, such as *Cenchrus ciliaris*, and *Glycina javanica.* As in these remote and difficult regions in Australia the whole outlook and scope of the livestock might be changed with the help of these improved leguminous pastures also in Masailand: the improvement of mbugas and other favourable sites with the help of nitrogen-fixing, protein rich legumes, may be the only possibility for a decisive rise in the productivity of the Masai-livestock-pasture-soil complex, as long as the use of supplemental fodder and concentrates, the conservation of high quality wet-season fodder as hay and silage and the application of nitrogen fertiliser remain impracticable.

By raising the carrying capacity of well watered pastures and using their sustained productivity chiefly for milk production, fattening of beef cattle near central abattoirs and as emergency fodder, not only would the expenditures involved herein be justified, but much greater ones for additional water development, bush clearing, etc., could be saved. If these pastures were to be fenced and protected from predators and vermin, so that they could be used also for night grazing, they would solve a ajor sanitary problem for the Masai: they could replace the present habit of night corralling within the living quarters. In addition to the great advantages of grazing throughout the cooler hours, the return of fertility through the manure and urine to these pastures would prevent also decline in productivity on long-term basis.

To these subjects should be given, therefore, the highest priority in any research and pasture development programme. The experience from Australia and other countries on the means and methods of investigation and the establishment of legume/grass pastures, after correction of possible soil nutrient deficiencies together with agro-technical improvement of soil-moisture and drainage, should be utilised.

Range Management

Whyte (1962) in an important paper has stated that limited funds should not be spread thinly for the development of wide areas, but spent on intensive improvement of selected, favourite sites. This is also most relevant for Masailand. With the exception of the above proposed intensive improvement of small, but most vital pastures, further prevention of decline and rise in productivity of most Masai ranges should be achieved only by the cheapest ecological means.

The most important one is the control of livestock in numbers, in space and in time, and adjustment of stocking rates, to carrying capacity of the ranges. Additional, most important tools are control of burning and distribution of water points. As pointed out by Heady (1960) in his excellent summary of range management in East Africa, the key problem there is the careful planning and distribution of these water points, according to present and potential carrying capacity of the pastures they serve. This will enable the walking distances and the sizes of the herds which have to rely on each water point to be reduced.

Such ecological management excludes any large-scale and expensive fencing, bush clearing and water development projects in areas of low potential. If efficient and economic utilisation of pastures for domestic livestock cannot be ensured without these heavy development expenditures, they should be devoted solely to wildlife. In general, even in areas of higher potential, large-scale indiscriminate bush clearing operations should not be carried out, until sufficient knowledge has been gained on the value of trees and shrubs in the savanna ecosystems and management of these cleared bushfields and the prevention of their re-infestation (Naveh, 1964).

Fire is by far the cheapest means of bush control and it is also an efficient way to re-mobilise part of the nutrients, bound in the woody tissues and leaves. It might be also important for the control of termites and ants. Therefore most emphasis should be given to this means and to the possibilities of creating less frequent, but hotter and more efficient fires. This was achieved in Rhodesia, in the "summer rain, dry Bushveld" by application of a simple and practical "four paddocks three herds" system in which one paddock is protected from grazing for the whole year and burned—or in case of severe drought—utilised as emergency pasture (West, 1962).

Unfortunately we are lacking almost any experimental and critical basis for the determination of carrying capacity of Masailand, as well as other semi-arid ranges, in East Africa. In semi-arid environments the danger of over-estimation of carrying capacity is especially grave. Here, domestic livestock may for considerable length of time do nearly as well on overstocked ranges, as on correctly managed ones. In long-term trials in Rhodesia, Kennan (1962) found that on overstocked ranges, beef production and returns per acre were even higher, until the complete breakdown of the productivity in the first drought year.

The proper determination of carrying capacity in different climatical zones and site-vegetation types is therefore the most crucial ecological and economic prequisite for any more intensive development plans. This should be based on a dynamic and quantitative approach of range condition and trend and corrected by systematic observations of changes in vegetation and soils and their relationships (Naveh, 1964).

As a basis for the final evaluation of agro-ecological potentials the estimation of livestock production and revenue should be used.

For the improved pastures, improved livestock with higher potentials and for the lower potential, bush and tsetse infested and water deficient areas, wildlife should be the main tool.

We may conclude therefore that a rise in the productivity of Masai ranges should be attempted in the following way:

(1) The intensive improvement of selected, favourite sites, closest to permanent water and living centres as high quality legume-grass pastures. Soil fertility should be ensured by biological synthesis of nitrogen through the legumes and by return of manure to the pastures.
(2) The ecological management of the open ranges for highest, sustained productivity by better grazing and burning management, better water distribution, and by adjustment of stocking rates to carrying capacity and to changes in range condition and trend—as defined by long-term, systematic observations.
(3) The management of lower potential and water lacking sites, as well as most bush and tsetse infested areas, in which large-scale clearings are uneconomic, chiefly for increased game production and their recurring revenue.

The exact definition and delineation of these site types and management areas, their integrated and harmonised utilisation in the whole system, should be the most urgent objects of the ecological survey and the research in the proposed pilot scheme.

However, in order to ensure, at a later stage, the implementation of these methods of modern livestock and wildlife range management and improvement in Masailand at large, and their continuity in time and space, the training of able and willing Masai as range technicians and game wardens will be most vital. Without such a speedy creation of a competent, professional leadership there is little hope that these development plans will be fulfilled.

SUMMARY

Tanzania Masailand is a vast semi-arid tropical savanna with a sparse population of Nilo-Hamitic, pastoral Masai. Because of its low and erratic rainfall, it is unsuitable for stable and profitable crop cultivation, but it has great potentials for livestock and wildlife production.

Further deterioration of Masailand can be prevented only by an integrated socio-economical, educational and agro-ecological effort for a planned transition of the disrupted and declining nomadic subsistence economy of the Masai into a stable pastoral cash economy of self-contained co-operative ranches. This should be attempted by a rise in the level of productivity of the Masai-livestock-pasture ecosystem as a whole, in a combined development, research and demonstration pilot scheme in one of the first Masai co-operative ranches in Central Masailand. At the same time also an intensive educational programme should be commenced, reaching all layers of the Masai society and especially the young, unmarried men ("warriors") and girls.

Increased livestock production and economical output should be obtained by increase in milk, meat, wool, mohair and hide production and marketing of improved cattle, sheep and goats and by conservation and management of wildlife for sport hunting, game viewing, game ranching and tourism. Increased range productivity should be obtained by the establishment of high quality legume-grass pastures on high potential and well watered sites and by better grazing, and burning management, better water distribution and adjustment of stocking rates to carrying capacities of the open ranges. Low potential, water lacking, bush and tsetse-infested areas should be devoted chiefly to wildlife.

These problems should be explored first in the pilot scheme and their practical solution applied in large scale economical operations in this and in further development schemes.

ACKNOWLEDEGEMENTS

Grateful acknowledgment is made of all the help received from the Ministry of Agriculture, Forests and Wildlife, Tanzania and the Northern Research Centre, Tengeru, the Ministry of Foreign Affairs and the Embassy of Israel, Dar es Salaam. I am especially indebted to Mr. B.C. Akehurst, formerly Director of Research, for his interest, encouragement and most helpful comments, as well as to Dr. H. Lamprey, Mr. D. Branagan, Father E. Hillman, Mr. A. Segal and Mr. A. Elman for their valuable comments on the manuscript.

REFERENCES

Branagan D (1962) A discussion of factors involved in the development of Tanganyika Masailand. Unpublished Report.

Brown LH (1963) Kenya's range resources. *The Kenya Farmer*, **85:**1–3 (Editorial).

Darling FF (1963) The unity of ecology. "The advancement of Science 20."

Fallon LE (1963). Development of the range resources, Republic of Tanganyika. U.S.-A.I.D. Mission, Dares Salaam.

Gillman C (1949) A vegetation-types map of Tanganyika. *Geogr. Rev.*, **39:**7–37.

Heady HF (1960) Range Management in East Africa. Government Printer, Nairobi.

Henzell EF (1962) The use of nitrogen fertilizer on pastures in the sub-tropics and tropics. In "A review of nitrogen in the tropics with particular reference to pastures, symposium". *C.A.B. Bull.*, **46:**161–172, Farnham Royal, England.

Hutchinson HG (1964) Ranching. First FAO Afric. meeting on Animal Production and Health, Addis Ababa, March 1964.

Huxley J (1961) The conservation of wildlife and natural habitats in Central and East Africa. UNESCO, Paris.
Huntingford GWB (1953) The Southern Nilo-Hamithes. Intern. Afric. Inst., London.
Kametz H (1963) Assessment of known water sources and suggestions for further development in Masailand. Dep. Water Development and Irrigation, Ministry of Agric., Arusha.
Kennan TCD (1962) Veld management in the semi-intensive, semi-extensive and extensive farming areas of Northern Rhodesia with special reference to research findings. In: Proc. First Interd. Conf. Pasture Workers, Bulawayo, S. Rhodesia, pp 73–91.
Lamprey HF (1963) A study of ecology of the mammal population of a game reserve in the Acacia Savanna of Tanganyika, with particular reference to animal numbers and biomass. Ph.D. thesis, University of Oxford.
Ledger HP (1964). The place of wildlife in African agriculture. First FAO Afric. meeting on Animal Production and Health, Addis Ababa, March 1964.
Naveh Z (1964) The need for integrated, long-term and quantitative research for the management and improvement of East African range lands. East Afric. Conf. of Range, Pasture and Livestock Specialists, Kitale, June 1964.
Naveh Z Branagan D (1964) A proposal for a combined range, livestock and wild-life research and development scheme in Tanganyika Masailand. Ministry of Agriculture, Forests and Wildlife, Tanganyika, Dar es Salaam.
Norman MJT (1962) The utilization of Townsville Lucerne pasture for beef cattle fattening at Katherine, N.T. In: Proc. North Queensland Agrost. Conf. South. Johnstone, Australia, Session 12/4.
Odum EP (1959) Fundamentals of ecology. 2nd edition Sauders, Philadelphia and London.
Oneko A (1963) Importance of tourism. Cited in *Reporter—East Africa's News Magazine*, 7 December 1963.
Owen MA (1964) Annual Report of the Kongwa Pasture Research Station, Ministry of Agriculture, Forests and Wildlife, Tanganyika.
Pereira HC (1962). Hydrological effects of changes in land use in some East African catchment areas. *E. Afric. J.*, 27:126.
Pereira HC, McCulloch JSG (1962) Water requirements of East African crops. In: *The natural resources of East Africa* Russell EW (ed.) Nairobi, pp 88–91.
Talbot LM, Ledger HP, Payne WJA (1961) The possibility of using wild animal production on East African range lands based on a comparison of ecological requirements and efficiency of range utilization by domestic livestock and wild animals. Eighth Intern. Cong. Animal Prod., Hamburg, pp 205–210.
West O (1962) Factors affecting the carrying capacity of Veld grazing. Proc. First Interdep. Conf. Pasture Workers, Bulawayo, S. Rhodesia, pp 23–43.
Whyte RO (1962) The myth of Tropical grassland. *Trop. Agric.*, 31:1–11.
Wilson JG (1962) The vegetation of Karamoja District, Northern Province of Uganda. Uganda Dep. Agric. Memoirs of Res. Div., Ser. 2:Vegetation, No. 5.

CHAPTER 9

NAVEH Z. (1978)

THE ROLE OF LANDSCAPE ECOLOGY IN DEVELOPMENT*

ENVIRONMENTAL CONSERVATION 5: 57–63.

Goats grazing near Arab village in Lower Galilee in *Sarcopoterium spinosum* thicket. In the past this highly inflammable dwarf shrub was been cut as fuel for lime kilns, but now the only usage is for goat grazing in spring and after burning. Note the neglected terrace walls and the removal of soil.

* Based principally on a paper presented at the Fourth ITCC World Congress of Engineers and Architects: Dialogue in Development—Concepts and Actions, held in Tel Aviv, Israel, in 1976.

INTRODUCTION

In a lecture at the Second World Congress of Engineers and Architects, organized by the International Technological Cooperation Center (ITCC), the present Author (Naveh, 1970b) summarized his views on the major ecological imperatives for development as follows:

1. The realization that, without effective birth-control, all achievements in development will soon be nullified.
2. The recognition that natural environmental resistance factors, which are also constraints in development—such as water deficiency and disease—operate as negative feedbacks which ensure ecological balance in undeveloped systems. Therefore, their removal by technological innovations necessitates their replacement by man-induced socioeconomic and cultural controls which should be part of rational and conservative land-use for the highest possible sustained productivity.
3. The need for environmental conservation and protection to avoid further deterioration and contamination, and to ensure the highest possible quality of life.
4. The implementation of these imperatives in development can be ensured only by the active participation of ecologists and ecological expertise, supported by ecologically oriented and trained technologists, by provision of greatly increased funds for ecological research and education, by better appreciation of Nature and its non-economic richness, and by better legislation and law-enforcement for environmental protection.

In recent years the same views have been expressed by many others, including the Club of Rome, who have given substantial weight to these statements on the need for population control and environmental conservation as the first imperative for development—*see* their Second Report, entitled 'Mankind at the Turning Point' (Mesarovic and Pestel, 1974). The disillusionment from the great hopes for dramatic increases in food production from the 'green revolution' has shown again that no technological miracle can solve these problems. On the contrary, the recent large-scale starvation in the drought-stricken Sahel region in northern Africa provided again sad proof of the soundness of earlier warnings that such one-sided technological and agricultural developments remove natural negative feedbacks and act as positive feedbacks by accelerating population and livestock growth-rates. They may, therefore, even worsen the situation in the case of drought—as had happened in Masailand 10 years earlier.

Another disappointing sign is the desire of many political leaders of the less-developed, countries to strive for industrial expansion at all costs, without any consideration of environmental problems. This was demonstrated in the 1972 Stockholm Conference on the Human Environment, organized by the United Nations. They were not ready to spend even 3% of all investments for industrial developments, on the prevention of undesirable environmental impacts arising thereby. But who could blame the leaders of the poorer countries, if those of the richer industrial countries continue to give such examples of environmental irresponsibility as has been demonstrated for example in the recent accidents in Italy? (Renzoni, 1977).

At the same time, however, there are also many encouraging indications of growing awareness of the urgent need for new approaches, based on ecological determinism rather than on short-sighted economic considerations, in development and land-use planning. This is well reflected in the growing number of internationally-sponsored and -coordinated research activities, such as MAB and those of SCOPE and UNDEP, as well as various multi- and bi-national development schemes with relevant inputs by ecologists and ecologically trained specialists. This is especially true in the field of land appraisal and land-use planning, as summarized recently in the important book by R. O. Whyte (1976).

Without doubt, much more could be achieved in the implementation of ecological determinism in development if only principles and practices of landscape ecology, which are confined so far chiefly to advanced industrial countries of Western Europe and North America, were recognized as a major tool in development.

The object of this paper is to present a short discussion of the conceptual basis of landscape ecology and its place in modern ecology, and to show, with the help of a practical example, how the principles of landscape ecology can be applied in the development of eastern Mediterranean uplands.

LANDSCAPE ECOLOGY AS AN INTERDISCIPLINARY 'HUMAN ECOSYSTEM' SCIENCE

In spite of the fact that many ecologists, in their approach and work, could be regarded as landscape ecologists, the term itself is still almost unknown in the English-speaking world and especially in the United States. Thus in a recent, very comprehensive and critical inquiry concerning the bases of human ecology, covering over a hundred pages and with more than 400 references (but, alas, not a single relevant non-English one!), landscape ecology was not mentioned at all (Young, 1974), despite a long discussion of landscape architecture, land-use planning, and conservation![*]

As one of the youngest branches of modern ecology, landscape ecology is concerned with the study of the interrelationships between Man and both natural and built-up landscapes. For this purpose, a landscape should neither be viewed merely as an aesthetic visual-perceptional connotation of Nature, as is done by artists, landscape architects, and designers, nor considered as a synonym for a set of geomorphological landforms, as is done by many geographers. Its definition should also include all visible biotic and man-made components and therefore, in the broadest sense, landscapes are the spatial and visual integration of the geosphere with the biosphere and man-made artefacts.

---

[*] A referee writes: 'To differentiate clearly between the old-fashioned purely "aesthetic" approach to landscape planning and current holistic ecological landscape or land-use planning, the reader is refered to Whyte (1976) and to cited papers of the present Author.—Ed.

However, if landscapes are approached from a holistic, system-theoretical point of view, their definition can be made much more concise, while at the same time it will become apparent that landscape ecology should be considered as an emerging *human ecosystem science.*

This approach is based, first of all, on the recognition of the hierarchical organization of Nature in general and of the living world as open systems with increasing complexity through newly-emerging qualities, from the sub-atomic and atomic physico-chemical levels through the sub-organismic molecular and cellular and organismic biological levels up to the supra-organismic ecological levels of populations, communities, and ecosystems (Tansley, 1935; Bertalanffy, 1968; Weiss, 1969). As the highest level of ecological integration, the *Total Human Ecosystem* should be recognized—integrating, in the words of Egler (1970), 'man-plus-his total environment into a single whole'.

Following Muller (1975) we should also distinguish abstracted, conceptual systems from concrete systems that are real entities which can be defined by space/time relationships. Thus concrete living systems are maintained by inputs of material/energy and information and regulated by biocybernetic feedback controls. *Ecosystems* can now be redefined as the *basic functional units of Nature,* integrating living systems with their living space or, in the terms of ecologists, the *biocoenoses* with their *biotopes.* The smallest, concrete ecosystem is an *ecotope* and the largest, global ecosystems is the *biosphere* (Ellenberg, 1973).

However, a further distinction should be made between '*bio-ecosystems*' and '*techno-ecosystems*'. The former are natural or man-modified bio-ecological systems, composed of autotrophic producer and heterotrophic consumer and decomposer levels and their natural, physical environment and driven mainly by solar energy. In contrast, techno-ecosystems are man-made and -maintained, rural and urban techno-ecological systems of human civilizations. They are driven by inputs of mainly fossil energy and man-made products, and their regulation depends on cultural, scientific, technological, and political, information—instead of the biophysical information ensuring homeostatic control of natural bio-ecosystems. Farms, villages, towns, etc., are thus the smallest, concrete techno-ecosystems, while the largest, global one is the *technosphere.*

Now we can also propose a new, holistic interpretation of *landscape units as the concrete space/time entities of the total human ecosystem,* with the *ecosphere* as the largest, global landscape. As shown in Fig. 1, these landscape units contain both the spatial-visual biotic and abiotic elements of natural and agricultural bio-ecosystems and the artificial, man-made elements of rural and urban techno-ecosystems. According to the kind and amount of material/energy and information inputs from these total-human ecosystem sub-systems, we can distinguish different types of open, cultural, and built-up, landscapes. With increasing human impact of accelerated, exponentional urban and industrial expansion, increasingly poorer and monotonous cultural landscapes are created, and the replacement of the biosphere by the technosphere is accompanied by a process of neo-technological landscape and environmental degradation of the ecosphere (Naveh, 1973).

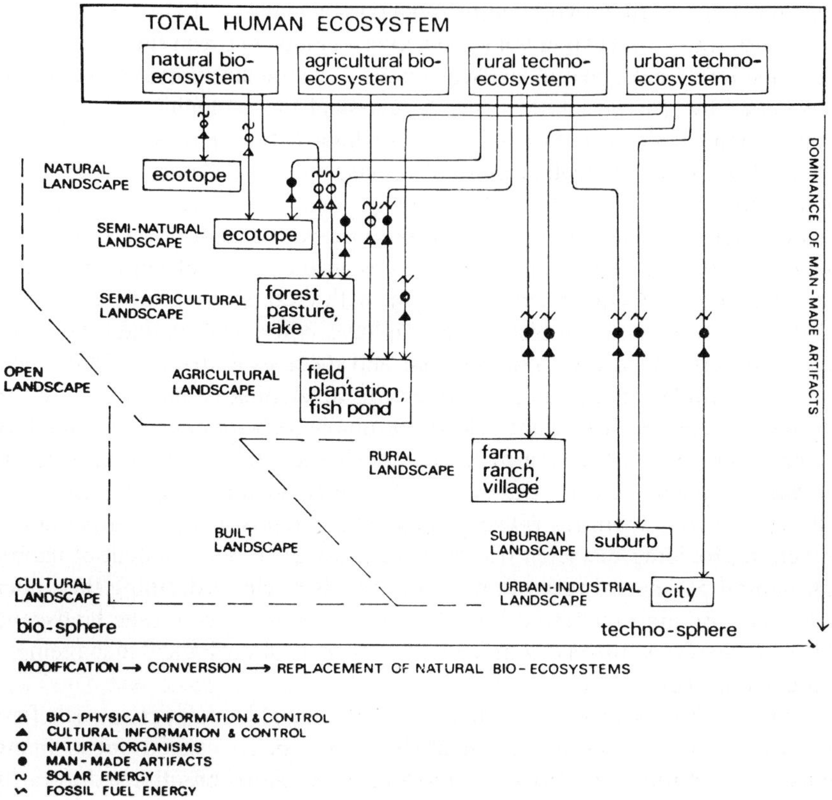

Fig. 1. Diagram of landscapes in terms of spatial units of the Total Human Ecosystem

Bio-ecosystems have proved their viability and stability by already functioning successfully for thousands and even millions of years, due to their homeostatic self-regulation by negative feedback loops and the maximization of negentropic order. The techno-ecosystem, on the other hand, driven on positive feedbacks and increasing entropy and disorder, are already showing alarming symptoms of disintegration, characterized by the urban crisis syndrome (Vester, 1976). Therefore, one of the crucial questions facing mankind is how to ensure the functional and structural integration of these bio- and techno-ecosystems into a self-sustained total human ecosystem or *super-system,* with emergent qualities of ecological stability and resilience, together with socio-economic and cultural values for human societies.

This can be achieved only by coupling the biosphere and the technosphere through scientific, technological, and cultural, negative feedback loops, such as controlled growth of population and economy, recycling of waste, conservation of natural ecosystems, etc.

In this process of cultural evolution or 'noogenesis' (Jantsch, 1975), ecology can play an important role by dealing with these interrelations between the total human ecosystem and its concrete spatial landscape units. As a human ecosystem science it has overstepped the purely natural realm of classical bio-ecological sciences and entered the realm of the man-centred fields of knowledge of those sociological, psychological, historical and other cultural aspects of the noosphere, involved in modern land-use.

Because of their widespread preoccupation with natural and open landscapes, the background of most landscape ecologists is in the fields of applied botany, forestry, range management, or agronomy. But with their growing involvement in regional and urban planning and land-use, emphasis in the outlook and training of landscape ecologists should be shifted more and more to those interdisciplinary aspects. This could be of special importance for developing countries, where the outcome should play an important role by producing environmental parameters for regional planners and decision-makers on desirable land-uses with the help of integrated landscape inventories and evaluation, which are already being used successfully in industrial countries (Olschowy, 1975). Interdisciplinary work should be involved in providing the scientific basis for ecological management of nature reserves, natural parks, and recreation areas, and also help in defining the upper limits of urban and agro-industrial loads which can be imposed on the landscape as a whole. It can also help very widely in other aspects of ecological management as exemplified in Figs. 2 and 3.

As has recently been shown by Whyte (1976), ecological land-appraisal, as a first step towards socio-economic development, is already being carried out in many developing tropical countries. However, most of these efforts are directed towards the most productive agricultural areas, for which the ratio between investment in

Fig. 2. Ecological management of Mediterranean annual grassland for pasture improvement in Lower Galilee: airplane application of fertilizer.

Fig. 3. Ecological management by pasture improvement in Lower Galilee: selective spot-spraying of perennial thistles with 2-4D.

research and resources, and the expected monetary benefits in the shortest time, are most favourable. On the other hand, less fertile and marginal lands are deprived and neglected and often left to their fate. It is, therefore, one of the greatest challenges for the landscape ecologist to include these untillable landscapes, covered mostly by natural vegetation, in the development process. This can be done by finding practical ways and means for a compromise between the needs for the conservation and reconstitution of the open living landscape and its natural ecosystems, and the socio-economical needs of the human inhabitants and of their national economy. One of the first steps towards this goal is the transformation of the above-mentioned 'non-economic richness' contained in these ecosystems, into workable parameters for the land-use planner, the decision maker, and the user, and their integration with other economic land-uses.

In the second part of this paper the development of Mediterranean uplands is used as an example of such an attempt.

A MODEL OF MULTIPLE-USE STRATEGIES FOR THE DEVELOPMENT OF MEDITERRANEAN UPLANDS

Mountainous and hilly uplands—too steep or rocky for profitable cultivation—make up around 50% of the total area in the Mediterranean part of Israel and not less in some other countries of the Mediterranean Basin. These uplands are mostly covered with more or less depleted shrub and dwarf-shrub communities—maquis, garigue,

Fig. 4. Reclamation of abandoned limestone quarry with drought- and limestone-tolerant (garigue) shrubs, planted on steep slope in soil pockets held by chicken wire: eastern Mediterranean region.

or batha (phrygana)—of very low economic and scenic value. Their ecological stability and biotic diversity are threatened by the combined impact of 'traditional'—chiefly pastoral—and 'modern' neotechnological landscape despoliation (as described in detail by Naveh and Dan, 1973). Any reclamation is liable to be very costly (Fig. 4).

In contrast with adjacent arable land, often undergoing a radical transformation into intensive agricultural uses, these often extensive 'wildlands' are badly neglected and up to now the only alternative to their further deterioration has been their protection as nature reserves or their conversion into—chiefly mono-species—planted pine forests. In Israel, as well as in other Mediterranean countries, these untillable uplands remain as last refuges for a wide variety of plant and animal species, serving as sources of genetic stock for future evolution as well as for future economic uses. Their ecosystems also fulfil important ecological functions as natural 'life-supporting' systems (Odum, 1971) and as buffering zones for protection of watersheds, while aiding in the control of floods, erosion, and environmental pollution, in the more densely-populated coastal regions. At the same time they must also fulfil the socio-ecological, aesthetic, and psycho-hygienic, functions of wildland outdoor recreation, introducing a vital component of open landscape in the industrial society through their enrichment of scenic values and recreation amenity. Their economic function lies in utilization of the biological productivity for livestock, forest, and industrial plant products, as well as for tourism. Importance should also be attached to the possibilities of increase of water yields for aquifers through brush conversion, and to the creation of fire-protection buffering zones.

Earlier studies on management and improvement of natural grass or shrub pastures indicated the high production potentials of these uplands for meat, wool, and milk, production under sound scientific management (Naveh, 1970a). Recent work on the effects of fire, grazing, and human disturbances, on maquis-shrub-land, woodland, and grassland, showed the importance of a certain 'optimum' defoliation pressure for maintenance of the highest plant and animal species diversity (Naveh, 1971; Naveh *et al.*, 1976). The promising results of present studies on multipurpose environmental reafforestation, carried out jointly with the Jewish National Fund Forestry Department (Naveh, 1975), point to the great scope of local and exotic trees and shrubs for biological, scenic, and economic, enrichment of these degraded Mediterranean landscapes.

In line with this work, a concept of *dynamic ecosystem* management was proposed (Naveh, 1974), based on flexible multipurpose-use management strategies—according to ecological site potentials, and meeting local and regional requirements.*

A flow-chart of these strategies is presented in Fig. 5. The currently misused, low-value uplands (A) have been divided into 3 major subsystems—according to the density of the woody cover and the degree of misuse and site conditions. With continued protection they can be converted into dense, impenetrable and monotonous maquis reserves (B), or by planting of pines they can develop into dense, highly flammable forests (C).

The two new options proposed are based on two main ecotechniques, represented in the D and E subsystems:

1. *Ecological management* of the soil–plant–animal complex with vegetation manipulation by thinning, coppicing, and pruning, of trees, and controlled fire—also selective brush and weed control, and reseeding with perennial grasses in shrublands or with annual legumes in grasslands. The latter operations can be with or without fertilization, but with controlled (preferably rotationally-deferred) grazing in fenced subplots (D systems).
2. *Multi-purpose reafforestation and revegetation* (E systems) for the creation of semi-natural, multi-layered parklands and woodlands with preferably local trees and shrubs of high cover, fodder, or ornamental, values. This will be necessary wherever land denudation has reached an advanced stage, as well as in Man-created habitats of roadsides, camping grounds, recreation sites, and fire-protection areas. But both techniques can be combined, and D systems, as well as the C forests, can also be converted into such E systems.

Both D and E are divided into three subsystems, according to their major object of utilization—either for recreation (D_1 and E_1), or for fodder production for livestock (D_3 and E_3), or for maximum compatible multiple-use for recreation, wood and

* This is very reminiscent of the earlier work and proposals by Roelof J. Benthem, which he published in the first-ever issue of our former Journal, *Biological Conservation* (Vol. 1, No. 1, pp. 11–12, 1968), and which he called 'Creative Conservation'.—Ed.

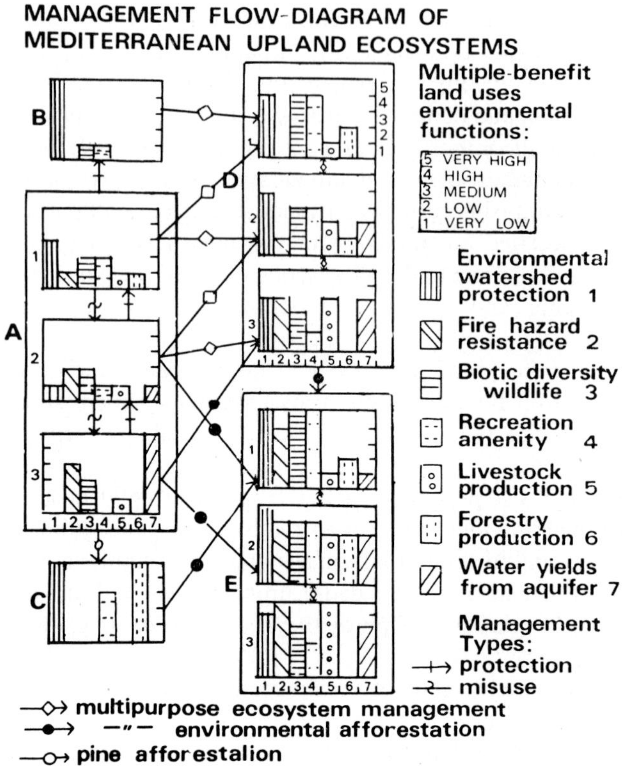

Fig. 5. Management flow-diagram of Mediterranean upland ecosystems.
A = Low-value natural upland ecosystems
A1 = natural forests, maquis, and woodlands
A2 = degraded forests, maquis, and woodlands
A3 = degraded batha, garigue, and derived grasslands
B = Protected forests and maquis reserves
C = Planted pine forests
D = Improved natural upland ecosystems
D1 = improved recreation forests, maquis, and woodlands
D2 = improved multipurpose forests, maquis, and woodlands
D3 = improved fodder-shrub and grass pastures
E = Semi-natural, intensively improved upland ecosystems
E1 = planted recreation forests and parklands
E2 = planted multipurpose forests and parklands
E3 = planted fodder forests and shrublands

fodder production, wildlife, or livestock. At our present stage of knowledge, such D_2 and E_2 systems should be the main goal, leaving options open for future, more one-sided land-uses.

The *multiple-use benefits* have been indicated in each management system by relative ratings from 5 (very high) to nil, and it is immediately obvious that these values rise from A to B and C systems, but are considerably higher still in D, and reach their maximum value in the most intensively improved and managed E systems. The consideration of these evaluations have been described elsewhere (Naveh, 1977).

A major task for future work will be to express these relative values by actual quantitative economic and socio-ecological parameters, enabling cost/benefit analysis and optimization for dynamic planning of land-uses. The inputs necessary for these management options can be provided only on a national level as part of comprehensive landscape development plans. As a first step, combined research and demonstration schemes could be carried out in nature reserves, forests, and recreation parks, where improved scientific management is urgently needed.

In multi-purpose land-use planning, a major problem is the assessment of compatibility of different land-uses on the basis of their mutual influence. This has been attempted by a *cybernetic sensitivity model,* presented in Fig. 6, in which each land-use and functional category is represented as affecting or being affected by all others in the bio-socio-economic ecological network, as expressed in 4 rates in the matrix

A CYBERNETIC SENSITIVITY MODEL OF LANDUSE FACTORS

Effect from ↓ on →		A	B	C	D	E	F	G	AS	Q
Environmental & watershed protection	A	•	0	0	3	1	0	3	7	0.8
Fire hazard resistance	B	1	•	1	1	1	3	3	10	1.1
Biotic diversity & wildlife	C	1	1	•	3	1	1	1	8	0.9
Recreation amenity	D	2	3	1	•	1	2	1	10	0.7
Livestock production	E	2	2	3	3	•	2	2	14	1.7
Forestry production	F	3	3	3	3	3	•	3	18	2.2
Water yields from aquifer	G	0	0	1	1	1	0	•	3	0.2
	PS	9	9	9	14	8	8	13		
	P	63	90	72	140	112	144	39		

AS = Active Sum
PS = Passive Sum
Q = Quotient AS:PS

0 = NO INFLUENCE
1 = SLIGHT INFLUENCE
2 = MEDIUM INFLUENCE
3 = STRONG INFLUENCE

ACTIVE ELEMENT
highest Q: F, (E)

PASSIVE ELEMENT
lowest Q: G, (D)

Fig. 6. A cybernetic sensitivity model of land-use factors.

(Vester, 1976). This enables determination of the most active and critical variables—in our case forestry production and to a lesser degree livestock production and recreation—and the most passive one which changes greatly under the slightest influences of other variables, namely water yields, with other environmental functions remaining intermediate. Therefore, because of their overwhelming impact, management aimed at forestry or livestock production should be weighed carefully against other expected benefits and, on the other hand, increase in water yield can hardly be regarded as a major management goal, but only as a desirable by-product in certain land-use types.

SUMMARY

Landscape ecology is concerned with the interrelations between the Total Human Ecosystem—integrating natural ecosystems and human techno-ecosystems—and its concrete, spatial landscape units. It deals not only with natural and semi-natural landscapes but also with cultural–rural and urban–industrial landscapes and their inputs of fossil energy, artefacts, and cultural information and control. As an emerging branch of human ecosystem science with an interdisciplinary outlook on modern land-uses, it could play an important role in developing countries. Here, one of its greatest challenges is to find a compromise between the needs for conservation or reconstitution of open landscapes and the socio-economic needs of society. This can be achieved by transforming the non-economic richness of the local ecosystem into workable parameters for the land-use planners and decision-makers.

As an example of such an attempt, the development and improvement of non-tillable Mediterranean uplands is presented by a model of multiple-use ecosystem management strategies and their benefits. Highest overall benefit for nature conservation, wildlife and recreation amenities, protection of environmental quality as well as livestock, forestry production, and water yields, can be expected by multiple-purpose reafforestation and revegetation.

For the assessment of compatibility of these different land-use factors and their mutual influences, a cybernetic sensitivity model is proposed and the most active and critical variables—foresty, livestock production, and recreation—and the most passive one—water yields—are determined.

In order to make possible cost-benefit analyses and achieve optimization for dynamic planning, these relative values should be replaced by actual quantitative ecological, economic, and socio-ecological, parameters—to be implemented on a national level as parts of comprehensive landscape master-plans.

REFERENCES

Bertalanffy L. von (1968) General System Theory. George Braziller, New York: xv 1 289 pp.

Egler FE (1970) The Way of Science. A Philosophy of Ecology for the Layman. Hafner, New York: xii 1 145 pp.

Ellenberg H (1973) Ekosystemforschung. Springer, Heidelberg: xiii 1 280 pp.

Jantsch E (1975) Design for Evolution, Selforganization and Planning in the Life of Human Systems. George Braziller, New York: 322 pp.

Mesarovic M, Pestel E (1974) Mankind at the Turning Point. (The Second Report to the Club of Rome.) E. P. Dutton, New York: xiii 1 210 pp., illustr.

Muller JG (1975) The nature of living systems. Behavioral Science, 20 pp 234–365.

Naveh Z (1970a) Effect of integrated ecosystem management on productivity of a degraded Mediterranean hill pasture in Israel. Proc. XI International Grassland Congress, Brisbane, Australia, 1970: pp 59–63.

Naveh Z (1970b) Ecology and development. Pp. 387–95 in Dialogue in Development (Second World Congress of Engineers and Architects). Association of Engineers and Architects in Israel, Tel Aviv, Israel: viii 1 487 pp.

Naveh Z (1971) The conservation of ecological diversity of Mediterranean ecosystems through ecological managegement. Pp. 605–22 in The Scientific Management of Animal and Plant Communities for Conservation (Ed. E. Duffy & A. S. Watt). Blackwell Scientific Publications, Oxford–London–-Edinburgh–Melbourne: xvi 1 652 pp., illustr.

Naveh Z (1973) The neo-technological landscape degradation and its ecological restoration. In: Pollution Engineering and Scientific Solutions E S Barrekette. (ed.) pp 168–81 Plenum Press, New York: xv 1 784 pp., illustr.

Naveh Z (1974) The ecological management of non-arable Mediterranean upland. Journal of Environmental Management, 2, pp 351–71.

Naveh Z (1975) Degradation and rehabilitation of Mediterranean landscapes. Landscape Planning, 2, pp 133–46.

Naveh Z (1977) Reconstitution of Mediterranean uplands by ecosystem management and multiple-purpose environmental afforestation. FAO Technical Consultation of Forest Fires in the Mediterranean Region, May 1977, St Maxim, France. FAO, Rome: 10 pp., mimeogr.

Naveh Z, Dan J (1973) The human degradation of Mediterranean landscapes in Israel. In: Mediterranean Type Ecosystems, Origin and Structure di Castri F, HA Mooney (ed.) (Ecological Studies 7.) pp. 373–390 Springer, Berlin–Heidelberg–New York: xii 1 405 pp., illustr.

Naveh Z, Burmil S, Mann A (1976) Plant species richness in Mediterranean shrubland and woodland of Israel. Proc. 7th Scientific Conference of the Israel Ecological Society, October 1976, Tel Aviv: pp 150–70.

Odum H T (1971) Environment, Power and Society. Wiley, New York: ix 1 331 pp., illustr.

Olschowy G (1975) Ecological landscape inventories and evaluation. Landscape Planning, 2, pp 37–44.

Renzoni Aristeo (1977) The increasing number of environment-degrading accidents in Italy. Environmental Conservation, 4(1), pp 21–6, 5 figs.

Tansley AG (1935) The use and abuse of vegetational concepts and terms. Ecology, 16, pp 284–307.

Vester F (1976) Urban Systems in Crisis—Understanding and Planning of Human Living-spaces: the Biocybernetic Approach. Deutsche Verlags-Anstalt, Stuttgart: 89 pp., illustr.

Weiss P (1969). The living system: determinism stratified. Pp. 3–55 in Beyond Reductionism Koestler A, J R Smithies (ed.). Hutchinson, London: ix 1 435 pp.

Whyte RO (1976) Land and Land Appraisal. Junk, The Hague: xi 1 370 pp., illustr.

Young GL (1974) Human ecology as an interdisciplinary concept: A critical inquiry. Pp. 1–105 in Advances in Ecological Research Macfadyen A (ed.). Academic Press, London & New York: 174 1 418 pp., illustr.

CHAPTER 10

NAVEH Z. (1998)

CULTURE AND LANDSCAPE CONSERVATION: A LANDSCAPE-ECOLOGICAL PERSPECTIVE.

IN: GOPAL B.P., PATHAK P., SAYENA K.G. (Eds.) ECOLOGY TODAY: AN ANTHOLOGY OF CONTEMPORARY ECOLOGICAL RESEARCH INTERNATIONAL SCIENTIFIC PUBLICATIONS, NEW DELHI, PP. 19–48.

Ruins of Jewish fortress from Roman times in the Golan Heights. In the front, ancient olive trees. (Photo, courtesy by N.eches).

Abstract The conservation of culture in landscapes has to be based on a transdiciplinary systems approach. By this approach landscapes are recognized as the tangible Gestalt systems of our Total Human Ecosystem, integrating humans and their total environment at the highest co-evolutionary level of the ecological hierarchy. Cultural landscapes are the tangible meeting points between nature and mind, and the conservation of their cultural assets has to be an integral part of holistic and dynamic landscape management. This has to be aimed at the protection, conservation and restoration of total landscape ecodiversity, namely their biological, ecological and cultural diversity and their intrinsic and instrumental values

Keywords Cultural landscape, Landscape conservation, Landscape ecology, Landscape dynamics, Landscape evolution

INTRODUCTION

In recent years there has been a distinct tendency to broaden conservation efforts from the protection of rare and endangered species and their habitats in restricted nature reserves and parks to the protection of the most valuable seminatural, agricultural and rural land-scapes and their overall biological, ecological and cultural assets.

These trends have been described in more detail by Lucas (1992) who emphasized their important role in the strategies for sustainable living as outlined in "Caring for the Earth" (IUCN/UNEP/WWF 1991). He provides a very useful guide for the selection, implementation, legalization and management of protected landscapes and brings successful examples from all continents. Of greatest significance in this respect is the endorsement by UNESCO of the biosphere concept in its Man and Biosphere Program (MAB) and the recognition of landscapes with outstanding natural and cultural universal values by the UNESCO World Heritage Centre.

A further significant step in this direction has been taken at the recent General Assembly of the International Union for the Conservation of Nature and Natural Resources (IUCN) – The World Conservation Union – at Buenos Aires, in 1994. In its resolutions it recognized even more specifically the importance of holistic conservation management strategies of highly valuable human-modified and used landscapes and acknowledged the Working Group on Landscape Ecology and Conservation of the IUCN Commission for Environmental Strategies and Planning (IUCN-CESP) for furthering these goals.

As a result of these developments, we are witnessing the widening of the scope of conservation from closed islands of nature to the open landscape. Here, as will be further discussed in more detail, the interaction of people and nature over time has produced closely interwoven natural and cultural patterns and processes. It has thus also broadened the biological and ecological landscape elements by cultural — that is human-made and maintained — artifacts and has thereby enriched what could be coined as "**total landscape ecodiversity**" (Naveh 1993b). Its conservation will require a transdisciplinary systems approach and comprehensive and integrative strategies to be based on holistic landscape planning, management, and restoration.

In addition, innovative and more efficient tools are required for the guidance of land managers and users and for persuading the public and the political decision-makers. For this purpose, landscape ecologists have been developing and are already applying such holistic methods and tools. Some of these have been reviewed in our book on landscape ecology (Naveh and Lieberman 1994), and others by Leser (1991), Farina (1997), Bastian and Schreiber (1994), Pignatti (1994), and most recently by Zonneveld (1995).

A vital prerequisite for developing these strategies, methods and tools is a better understanding of the nature of these cultural landscapes, their perception and evolution, and their dynamic interactions with the natural and cultural forces which have shaped them in the past and are driving them presently toward their uncertain future. This is not an easy undertaking at all. By introducing the notion of culture into landscape conservation we have to broaden also its conceptual and methodological scope from the realms of natural sciences to the realms of the humanities, from strictly bio-ecological issues to the even more complex human-ecological issues.

This paper will attempt such broadening of scope with a holistic, landscape-ecological systems approach extended by new insights into order, complexity and trans-disciplinarity, which are most relevant for cultural landscapes. From these conceptual issues, the need for holistic and dynamic conservation strategies will be derived and practical examples will be drawn — chiefly from the Mediterranean region — which have a much broader application for almost all other cultural landscapes.

SOME BASIC PREMISES OF LANDSCAPE ECOLOGY AS A TRANSDISCIPLINARY SCIENCE

A Holistic Systems Approach to Nature and Culture

In his important contribution to the holistic view of nature, the world-renowned, late, theoretical physicist Bohm (1980) lucidly analyzed the deeply ingrained roots of our tendency to fragmentize and take apart what is whole and one in reality. This is especially true of the prevailing views of human-nature relations and the unfortunate rift between exclusively biocentric and anthropocentric schools of thought. It is also most relevant for the disastrous intellectual, professional, academic and institutional fragmentation among those dealing with environmental problems.

We have attempted to overcome these problems at the conceptual level by providing a holistic and transdisciplinary foundation for landscape ecology with the help of General Systems Theory, hierarchy theory, biocybernetics, and additional concepts of what could be called the newly emerging "science of wholeness and complexity" (Naveh 1982, Naveh and Lieberman 1994). This holistic landscape paradigm was further developed in the context of cultural landscapes (Naveh 1990a). It can be presented here only in a very condensed way.

General Systems Theory and its contemporary concepts are considered as a transdisciplinary, overarching metatheory and holistic view of the world which can help to overcome academic and professional barriers between the "two cultures" and link quantitative formal approaches with qualitative descriptive approaches (Grinker 1976). Its basic paradigm is the view of a hierarchical organization of nature as ordered wholes of multileveled stratified open systems. In this natural hierarchy each higher level organizes the level below it with newly emergent qualities. This higher level displays "lower-frequency behavior" and is spatially larger or more constant over time and serves thereby as the context of the lower level.

According to recent developments in holistic systems science, the evolution of the cosmos, matter, life, man and consciousness should be perceived as one of ordered complexities of an interacting "Gestalt" system. In such a Gestalt, as in a symphony, every part is intrinsically related to every other part, and all parts together make up an irreducible whole of what the world-renowned systems-philosopher and prominent member of the Club of Rome, Erwin Laszlo (1987) has named "The Grand Synthesis of Evolution", and what another great systems thinker and planner, the late Erich Jantsch (1980) has described as "The Evolution of the "Self-Organizing Universe".

By this world-view the biological and ecological levels of life on earth are considered as a microhierarchy within the macrohierarchy of the cosmos, which ranges from the space-time field, energy continuum, and quarks, up to the entities of astronomy – planets, stars, star clusters, galaxies, and galaxy clusters (Laszlo 1972).

This ecological hierarchy is composed of living organisms and their environment, functioning together as ecological systems. As such, they are more than the sum of their biological and physical components and their energy and material fluxes, because their interaction creates new, structural and functional system qualities. In contrast to the mixture of gravel and sand in an unorganized heap, they have become thereby ordered wholes with multiple space-time dimensions. These systems are defined, in general, as ecosystems and are regarded by most systems ecologists as the highest level in the ecological hierarchy, above organisms and populations (O'Neill et al. 1987). However, this hierarchical organization is based on the conception of natural ecosystems in which humans are merely external "disturbance" factors. A more realistic conceptualization of the presently prevailing global ecological hierarchy has to take into account that human-modified and managed cultural semi-natural and agricultural landscapes are making out by far the greatest majority of the total open landscape area on global scales. Even the few remaining natural and close to natural landscapes are affected directly and indirectly by human impacts and they are shrinking rapidly from year to year. Their fate – like that of all other land- and seascapes on earth – is depending for good or worse almost solely on the decisions and actions of human society. Consequently, we have to include humans and their sociocultural dimensions above the ecosystem level as the highest bio-geo-anthropo-level of the global ecological hierarchy. At this level, man and his total environment are conceived as the highest co-evolutionary ecological entity on earth. Following one of the first truly holistic and inspiring ecologists

Frank Egler (1974) who died recently, we suggested to call this level the *Total Human Ecosystem* (THE) (Naveh 1982, Naveh and Lieberman 1994).

The Place of Landscapes in the Ecological Hierarchy

The first scientific and holistic definition of landscapes had been provided by the great geographer Alexander von Humboldt 200 years ago as "the total character of a region". Following this definition and that by the German biogeographer Carl Troll (1971), the "father" of landscape ecology, who defined landscapes as "the total natural and human living space", we can regard landscapes as *the concrete, space-time defined, three-dimensional entities of this Total Human Ecosystem* (THE).

For a full comprehension of the uniqueness of landscapes, we have to make a clear distinction between landscapes, as mixed natural and cultural interaction systems and ecosystems as functional systems. This has also been done by Allen and Hoekstra (1992) in their challenging book on ecological hierarchies. They rightly presented *ecosystems as functional systems which are diffuse in space and intangible.* On the other hand, *landscapes are the tangible, spatial matrix and living habitat for all organisms – including humans – and their populations, communities and ecosystems.* They are therefore concrete ecological–geographical systems and ordered wholes – or "Gestalt" systems – of our Total Human Ecosystem, within different scales. These scales and their functional and spatial dimensions have to be studied and managed in their own right. As such, they are more than spatially heterogeneous areas and repeated patterns of ecosystems. They range *from the ecotope as the smallest mappable landscape unit, to the ecosphere, as the largest global THE landscape.* The ecotope is used chiefly by European landscape ecologists as the basic unit for landscape study (Leser 1991, Zonneveld 1995) and is also regarded as the actual "site" of an ecosystem (Haber 1990). It is a much more rigorously defined term than the vague "patch", as generally used by American landscape ecologists.

For the development of suitable strategies of landscape conservation, the usual distinction between natural and cultural landscape is not sufficient, because it does not take into consideration the basic functional differences between cultural biosphere and technosphere landscapes. These include their energy inputs from solar or fossil energy, their living organisms, their control by natural or human information, and their self-organizing and regenerative capacities. Therefore we have to distinguish between the following human-ecosystem landscape classes and their ecotopes:

(1) *The natural and close-to-natural biosphere ecotopes,* or in short bio-ecotopes which are driven by solar energy and its biological conversion and are regulated by natural, biological, physical and chemical information.

(2) *The human-modified semi-natural and traditional agro-silvo-pastoral ecotopes,* which are also "solar-powered", but controlled by a mixture of cultural and natural information. These can be regarded therefore collectively as *cultural biosphere landscapes.* They include also *organic farming systems* operating without chemical

fertilizer, herbicides and pesticides, and those which are further developed into *regenerative systems* by utilizing with the help of human cultural and "neotechnological" inputs the natural biosphere functions of energy/material conversion, distribution, filtration, assimilation, storage and human thought. These should become, together with the other above-mentioned biosphere landscapes, our future life-supporting systems.

Like all naturally evolving systems, natural landscapes, and to a certain extent all other cultural biosphere landscapes, including traditional agricultural and rural landscapes, organize themselves in a coherent way, by maintaining their structural integrity in a process of continuous self-renewal and self-regulation or *autopoeisis* (from the Greek "self-creating").

(3) *The intensive agro-industrial ecotopes:* Although their productivity is dependent on photosynthetic conversion of solar energy, this energy is subsidized to a great extent by fossil energy, and the natural control is replaced almost entirely by heavy chemical inputs and by human agrotechnological information. These landscapes, having thereby lost all self-organizing properties, and cannot anymore be regarded as true cultural biosphere landscapes. They are much closer to the cultural technosphere landscapes and their detrimental environmental impacts.

(4) *The urban-industrial technosphere ecotopes* and their techno-ecosystems, and technological artifacts, such as highways, bridges, mines, quarries and powerlines, which are human-made and driven by fossil and nuclear energy and their technological conversion of solar energy into fuel. These entirely artificial cultural technosphere landscapes completely lack the self-organizing and regenerative capacities of biosphere landscapes. They result in high outputs of entropy, waste and pollution with far-reaching detrimental impacts on the remaining open landscapes and their biological productivity and ecological diversity and stability (Naveh 1987).

This classification is presented in Fig. 1 as a functional ecotope ordination from the biosphere pole to the technosphere pole with increasing modification, conversion and replacement of natural elements, functions and controls, by cultural, human-made technological and artificial ones.

All these ecotopes are interlaced spatially, but they are not structurally and functionally integrated into one sustainable ecosphere system, because of the dominating and largely adverse role of the agro-industrial and urban-industrial technosphere landscapes on the biosphere landscapes. This is one of the major causes of our present global environmental crisis.

It is worthwhile to realize that in rural landscapes, as has been shown by the successful examples of countryside management in Great Britain and elsewhere, these antagonistic relations can be transformed into mutually beneficial, and even symbiotic relations by sound, holistic and dynamic landscape planning and management (Green 1985, Lucas 1992). However, as shown by Mansvelt and Mulder (1993), at the long range future sustainable rural development, can be ensured only, if the intensive agro-industrial farming systems well be replaced by more diversified and balanced local or regional farming systems, based on

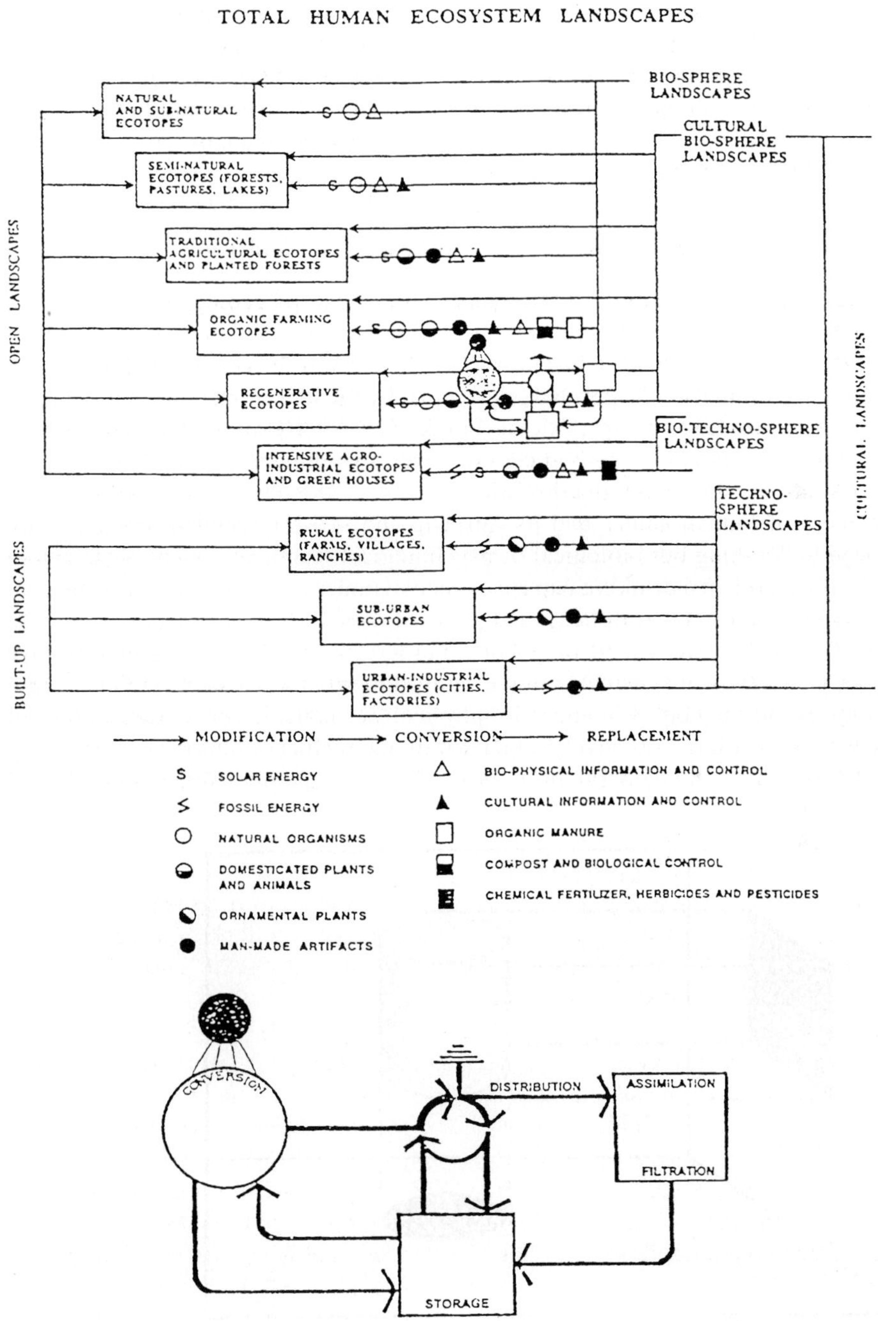

Fig. 1. Functional ordination of landscape ecotopes according to energy, matter and information inputs from the biosphere and technosphere poles. Achievement of a proper balance between these poles should become one of the major goals of landscape ecology.

the principles of the International Federation of Organic Agricultural Movements. Post-industrial "neotechnical" landscapes, functioning as organic farming systems and, even more, as regenerative systems (Lyle 1994), could fulfill thereby an important role in the solution of the ecological crisis. They could present best examples of the new, cybernetic symbiosis between humans and nature, as discussed further below.

The Transdisciplinary Nature of Cultural Biosphere Landscapes

The transdisciplinary notion of landscapes and landscape ecology emerging from this holistic systems view has recently been further elaborated in more detail in the context of environmental education (Naveh 1995a). It can be illustrated by adopting the dimensional approach, developed by the late eminent psychotherapist and founder of Logotherapy, Victor Frankl (1976). He used the metaphor of projecting three-dimensional bodies into lower, two-dimensional ones in order to demonstrate that by such reductionistic projections, the unique multi-dimensional wholeness of human nature and its intrinsic and self-transcendent openness are reduced to "nothing but biological or psychological reactions". Thus, as shown in Fig. 2, if we project a drinking cup as an open cylinder out of its three-dimensional space into the closed two-dimensional plane of the outline of its layout or the side view of its profile, we receive only a circle or a rectangle. The same happens if we project landscapes out of their unique multi-dimensional Gestalt wholeness into their lower "nothing but geological, biophysical, aesthetically, or socio-economical dimensions", dealing exclusively either within the realms of biology or geography and natural sciences in general, or in those of the arts and humanities. In each case

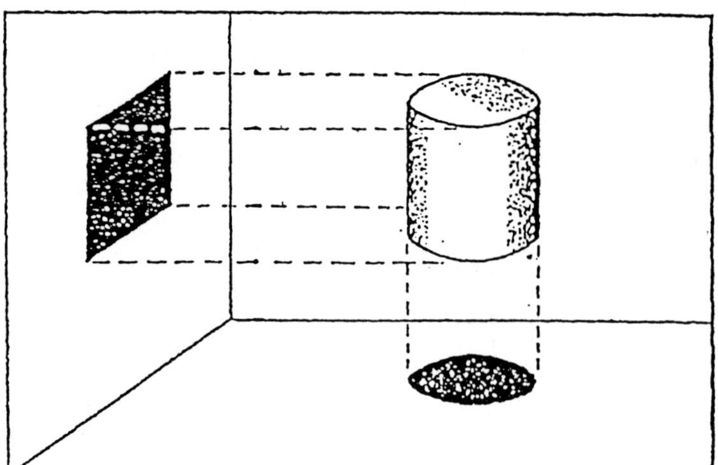

Fig. 2. Projection of three-dimensional space of a drinking glass or an open cylinder into the two-dimensional planes of its profile layout (After Frankl 1969).

we would lose their unique multi-dimensional nature as self-organizing Gestalt systems with intrinsic, self-transcendent openness.

Higher Orders of Landscapes Beyond the Cartesian Mechanistic Orders

Among the more recently emerging, exciting new ideas and theories concerned with the notions of wholeness and order, those of Bohm (1980), and of Bohm and Peat (1987), are of special relevance for our discussion. Bohm (1980), showed that modern physics and especially quantum field theory have opened the way for the realization of further, richer, subtler and higher levels of order, organizing the lower ones beyond the Cartesian mechanistic order. These are *implicate* or *enfolding* orders where the enfoldment process is related to the whole. He presented an entirely different sort of basic connections of elements: In the enfolded order, space and time are no longer the dominant factors determining the relationships of dependence or independence of different elements. From this, our ordinary notions of space and time, along with those of separately existing material particles, are abstracted as forms derived from the deeper order. These ordinary notions appear in what he called *explicate* or *unfolded* order, a special and distinguished form contained within the general totality of all implicate orders. Thus, any element contains, enfolded within itself, the totality of universe matter – life as well as consciousness.

Bohm and Peat (1987) have carried this holistic paradigm further. They recognized *implicate order as a special case of generative order.* This order is not concerned with the outward side of development and evolution in a sequence of successions – like in sequential order – but with a *deeper and more inward order, out of which the manifest form of things can emerge creatively.* They showed that this order is fundamentally relevant in nature, as well as in consciousness and in the creative perception and understanding of nature. They viewed implicate orders as organized by superimplicate order, opening the way for an indefinite extension into even higher levels of implicate orders, as a very rich and subtle generative order. Therefore they reached an entirely new view of consciousness as a generative and implicate order, throwing light on nature, mind and society and opening the door to a kind of dialogue that, in their own words, "may meet the breakdown of order that humanity is experiencing in its relationships to all these fields" (Bohm and Peat 1987).

Within the context of cultural landscapes, this means that further and deeper insight into the holistic nature of landscapes can be gained only if we are ready "to free our minds of rigid commitments to familiar notions of order, so that we may be able to perceive new hidden orders behind the simple regularity and randomness. As context changes, so do categories" (Bohm and Peat 1987: 115).

Such a change in context occurs when we perceive landscapes not as formal, spatial geometric structures and mosaics (Forman and Godron 1986, Forman 1995) which are described by conventional Archimedian geometric configurations, but as *self-transcendent natural and cultural Gestalt systems.* These are imbedded in a hierarchy of subtle, generative, implicate orders, in which human mind, consciousness and creativity play an important role. The recognition of such subtle orders has been initiated already in landscape ecology by the application of *fractal geometry and fractal dimension,* formalized by Mandelbrot (1982). Fractals are the generative order which underlies the geometric regularity of self-similarity. As an innovative method for the study of organized landscape complexity and multi-scale dynamic processes, it allows the quantification of the shape and texture of landscape features and the prediction of multiscale dynamic landscape processes. Fractal dimensions are enhancing our comprehension of the complex interaction between geomorphological, biotic, and anthropogenic factors, operating at different space-time scales, and thereby also the interactions between biodiversity and ecodiversity, which will be discussed future below (see also Burrough 1981, Milne 1991, Allen and Hoekstra 1992, and many others).

Fuzzy Logic and Cultural Landscapes

Of great promise in this context are also recent developments in the application of advanced computer programs of artificial intelligence, based on the mathematical theory of fuzzy sets and logic (Zadeh 1965, Kosko 1993), In contrast to the generally accepted Aristotelian bivalent logic of things being either entirely true or entirely false, beauty or ugly, healthy or sick, fuzzy logic is multivalent and uses "approximate reasoning", also considering things only partly true or false. In mathematical terms this means that instead of the binary choice of two elements of "crisp" sets, in fuzzy sets, weighted membership with more than two elements are used. As a result, a crisp set can be described only by its characteristic function: $f(x) = 1$, if x is an element of M; and $f(x) = 0$, if x is not an element of M. On the other hand, a fuzzy set can be described by a characteristic function with the generalization that $0 < f(x) < 1$, if x is an element of M.

Applying this fuzzy logic as fuzzy sets, we can achieve the goal of having precisely manipulable natural language expressions for these parameters. They enable us to capture qualitative, aesthetic, historic and other "soft" cultural landscape values mathematically as fuzzy sets and to deal with them quantitatively in algorithmic fashion as regular numbers (Negoita 1985). This is of great practical importance because these values by their very nature are fuzzy and their judgment must be based on subjective approximate reasoning along gradients ranging from very high to very low. Fuzzy logic is used already widely for commercial engineering and control systems, ranging from subways to dishwashers. Among the most recent important developments for management of complex ecological and socioeconomic systems is the combination of fuzzy logic with neural networks in expert systems (Kosko 1993). There are already promising beginnings of applying fuzzy assessment in landscape ecology studies (Burrough et al. 1982, Syrbe 1996, Steinhardt (1996, in press)). Li (1993), rightly emphasised the value of fuzzy logic facing the uncertainties of ecology. These are greatest when we deal with human-influenced and modified landscapes, which are influenced and modified by human land uses and culture.

In conclusion, fuzzy logic provides a unifying framework and an innovative tool to integrate numerical data, linguistic statements and expert experience together and to deal with vague and qualitative variables, such as landscape heterogeneity, and cultural diversity -as an integral part of ecodiversity and aesthetics. It thereby greatly enhances the prospects for such a transdisciplinary unification and opens the way for efficient integration of complex ecological, cultural and socio-economicai quantitative as well as qualitative landscape-ecological information. These are essential to an overall holistic assessment of landscape values and priorities for conservation, which planners, researchers and experts have to propose to the decision-makers in land use.

CULTURAL DIMENSIONS OF LANDSCAPE EVOLUTION AND PERCEPTION

Cultural Landscapes – the Tangible Meeting Point Between Nature and Mind

Culture refers to the total way of life of any society; it is therefore such a complex concept that no single definition can give it full justice. In 1949, Californian anthropologists Kroeber and Kluckholm (1949) collected more than 150 definitions from English and American literature, but many more could have been added now. These definitions and approaches to culture, applied by anthropologists, sociologists, human geographers, and philosophers, were well summarized in the 1981 edition of the American Heritage Dictionary as follows:

"Culture means the totality of socially transmitted behavior patterns, arts, religion, beliefs, institutions and all other products of human work and thought, characteristic of a community or population".

Culture, in essence, is what distinguishes us as humans from other animal creatures. It is a result of our unique mental ability for "symbolizing", namely of assigning to things and events certain meanings that cannot be grasped with the senses alone. From a systems-theoretical point of view this ability is a product of the noosphere (from the Greek noos = mind), namely the sphere of the human mind and consciousness. The noosphere can be viewed as an additional natural envelope of life in its totality and in its interaction with the biosphere and geosphere. It is the result of the evolution of the human brain neocortex and the consequent greater mental capacity of *Homo sapiens*. who he has become a mighty geomorphological agent, acting on the landscape in both constructive and destructive ways.

In his last, seminal study on "The Self-Organizing Universe", Jantsch (1980) has lucidly summarized this process of the evolution of our "triune" brain from reptilian brain to the paleo-mammalian brain and finally into the human neo-mammalian brain, endowed with the capacity of abstraction as the creative process of the transformation of the outer world. He called these activities of the neural mind "mentation", with self-reflexive mentation, capable not only of the inner world, but also of the outer world, and with symbolic representation as the highest level of development.

In other words, as cultural human beings we live not only in the three-dimensional physical-geographical space of our land and water systems, but also in this conceptual, noospheric space. This is the realm of our feelings, imagination and understanding, perception and conception.

Another important evolutionary interpretation of culture has been provided by Laszlo (1987), in presenting cultural evolution as an integral part of the biological and social evolution within the above-mentioned "Grand Synthesis" of cosmos and of life on earth. He viewed culture as inclusive of the characteristic features of all human behavior, i.e., not merely the "high culture" of science, art and religion, but as a basic information pool possessed collectively by all members of society. As such, it contains the sum of rules that code its essential operations and determines to a great extent the stability of society. He maintained that:

> The dynamics of social evolution concerns the progressive yet discontinuous development of this cultural, collective information pool. The processes of history act on actual societies but they select cultural information pools, much as the processes of biological evolution act on actual organisms but select genetic information pools. This, then, is a deeper meaning of the term "cultural evolution" (Laszlo 1987: 105).

In his most recent challenging book, Laszlo (1994) describes "The Grand Transition", from the industrial to the post-industrial society – and into the age of global informatization, as a crucial stage in our cultural evolution in which our global survival will be determined by our choice between "extinction or evolution". He views the major steps of the cultural evolution of human societies as progressively higher organizational levels, which are driven by the widespread adoption of basic technological innovations through selection of information pools. This sudden emergence of successive levels of organization by acquiring a fresh range of possibilities in structures of growing complexity is achieved by a destabilizing shift or *Bifurcation*. Today, the adoption of new information and communication technologies drives the process to the global level. This is demonstrated in Fig. 3.

Our cultural landscapes have been shaped also by these bifurcations through interactions between the cultural evolutionary processes and the natural and socioeconomic environment in which they occurred. They evolved in a quasi coevolutionary process between the biosphere-geosphere space of the natural environment and the noospheric space of the human mind. Following Bohm (1980), we may assume that at the same time explicate orders were unfolded from the deeper, more inward, implicate order. In this process both biological and cultural information was selected and utilized for technological advances, such as improving food production, habitation and other living conditions. In evolutionary system terms, all human habitated, used and/or modified landscapes have thereby become the *tangible meeting point between nature and mind.*

Taking the Mediterranean Basin as an example for the evolution of cultural landscapes, such as that of Mt. Carmel in Israel, this co-evolutionary process can be assumed as a result of the close interactions between the natural environment and these noospheric processes of the paleolithic food gatherer and hunter from the Middle Pleistocene onwards, more than half a million years ago. In this mutual-causal

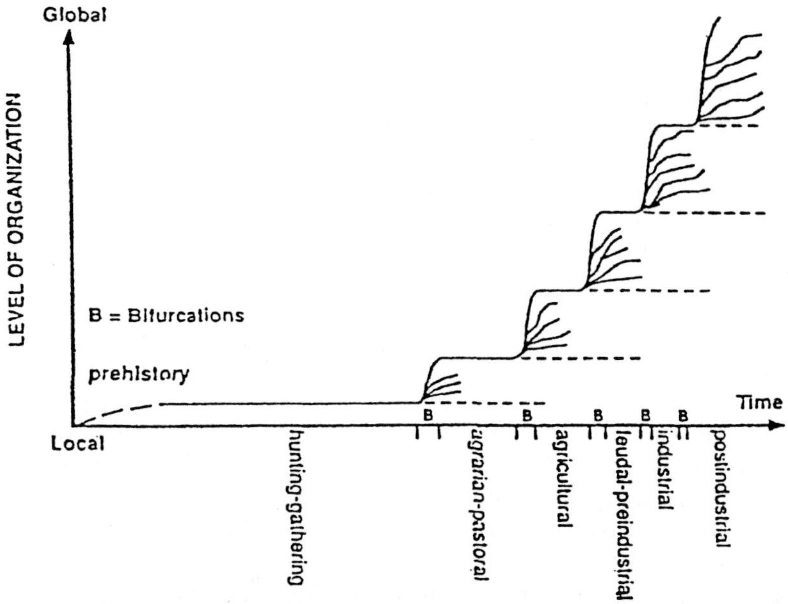

Fig. 3. Steps in cultural evolution as leaps to higher organizational levels (After Laszlo 1994).

feedback relations fire played a major driving force (Naveh 1984, Naveh 1990b also published as Chapter 3 in this anthology, Naveh and Lieberman 1994).

From the Late Pleistocene onwards, the growing human self-reflexive mentation capacities enriched the cultural information pools, and enhanced thereby the social organization and the creation of more sophisticated tools and more efficient use of fire and solar energy for food collecting and hunting, and finally for domesticating plants and animals. This led to the great cultural leap and technological transformation, of the so-called neolithic agricultural revolution. From the early Holocene onwards, about 10, 000 years ago, the Mediterranean agro-pastoralist gradually modified and converted most of the pristine forested landscapes into semi-natural and agro-silvo-pastoral landscapes. During historical times the traditional fine-grained and diverse Mediterranean mountainous rural landscapes evolved. They created closely interwoven mosaics of rocky and untillable forests, maquis shrublands, woodlands, and grasslands in different regeneration and degradation stages, together with terraced fields, vineyards, olive groves and mixed orchards, interspersed by scattered villages and small towns. However, since the industrial revolution, and especially since World War II, these cultural biosphere landscapes are replaced gradually by agro-industrial and urban-industrial technosphere landscapes and are suffering from severe degradation, resulting in their biological and cultural impoverishment.

Whether the trajectory of bifurcations of the traditional shift from the industrial to the post-industrial society will lead our global cultural landscapes to further degradation and instability and therefore into global extinction, or to a higher, sustainable organizatory level of further biological and cultural evolution, will depend entirely on human society and the proper use of these new information technologies.

Cultural Perceptions of Environment and Landscape

Not only have cultural impacts shaped our landscapes but our view of landscapes is also a product of culture, and these reciprocal and mostly cybernetic relations are in turn affecting our relation to these landscapes and determine their fate. These interactions have been discussed elsewhere in more detail (Naveh 1995b), and here only some of the most relevant aspects for our discussion are mentioned.

As tangible space-time-defined entities, landscapes are perceived by our senses. Perception does not depend upon the physiological details of the eyes and ears alone but on a much wider context of the overall disposition of the mind to apprehend objects in particular ways. It plays and important role in helping to select and give form to what is seen. Sense perception is therefore strongly determined by the overall disposition of both mind and body. But, in turn, this disposition is strongly related to the whole general cultural and social status of the individual and his community (Bohm and Peat 1987).

Thus, for instance, in the case of wildfires in the Mediterranean, the fate of the open landscape is determined to a great extent by the different preconceived cultural landscape perceptions by which these fires are viewed, and the resulting reactions to the same ecological reality by farmers, foresters, pastoralists, hunters, nature lovers, land speculators, arsonists, It is upon the landscape ecologist to study these fires and their effects without any preconceptions, like any other ecological factor and within the proper spatio-temporal and conceptual scales (Naveh 1990b, also published as Chapter 3 in this anthology).

The American landscape ecologist and planner, Joan Nassauer (1992) has provided good examples of the cultural ambivalence in these landscape perceptions and their effect on environmental policies in the USA. It seems to be very typical of many American nature lovers to identify "Nature", in which humans have no part, with "Beauty", and "Healthy" ecological systems and with "pristine landscapes" which supposedly have not been despoiled by humans. Such landscapes do not exist anywhere in the USA (or elsewhere) and these identifications are the result of a narrow mental disposition, out of any true scientific context. Instead she suggests the use of local vernacular cultural concepts of landscape beauty as a label to good stewardship and as a standard language for landscape perception. The appearance of landscapes, from wilderness preserves to small environ-mental patches, should be managed to show that they are not left over or simply undeveloped, and that someone stands behind their existence.

The American cultural geographer, Yi-Fu Tuan (1974) distinguishes between perception, as a response of the senses to external stimuli and purposeful activity, attitude, as a cultural stance, implying experience and value, and world view. These represent the conceptualized experience – being partly personal and largely social, and an attitude or belief system. The interaction of all three together with the environmental setting, determine what he called *Topophilia* – the affective bond between people and place or setting, or, in the words of Frank Golley (1990), former editor of *Landscape Ecology,* "loving their landscapes".

Tuan's many illuminating examples were drawn both from illiterate, traditional, and from present-day urban cultures. They show the enormous complexity and diversity and – amongst others – the ambivalence and changing of attitudes towards wilderness, the countryside and the city, in the American cultures. He summarized his original insights with a general remark:

> Human beings have persistently searched for the ideal environment. How it looks varies from one culture to another, but in essence it seems to draw on two antipodal images: the garden of innocence and the cosmos. The fruits of the earth provide security as does the harmony of the stars which offers, in addition, grandeur. So we move from one to the other: from the shade of the baobab to the magic circle under heaven; from home to public square, from suburb to city; from a seaside holiday to the enjoyment of the sophisticated arts, seeking for a point of equilibrium that is not of this world (Tuan 1974: 248).

In this respect, an important perceptive analysis of cultural landscape perception has been provided by the Canadian ecologist Pierre Dansereau (1975). He conceived man-landscape relations as cyclic or even cybernetic "inscape-landscape" interaction in which there is a filtering inward from nature to man, upward from the unconscious to the conscious, and from perception to design and implementation in planning and management.

Of great relevance for our discussion are the writings of the ecological anthropologist Rappaport (1979) on the discrepancy between our cultural images of nature and landscapes, as screened through our filters of beliefs, knowledge and purposes as cognized models, and their actual structure and function as operational models. He claimed, rightly, that the cultural images of cognized models of our "developed" societies are regulated only in accordance with reference values of sustained economic growth. As shown convincingly, with the help of dynamic global systems modeling, by Meadows et al. (1992): Because of these dominating values of sustained quantitative growth coupled with unrestrained population growth, but not with sustainable qualitative growth, our spaceship earth has already reached the limits of growth and is close to the brink of global collapse.

In such a "modern" model, "postulating economic rationality", a forest ecosystem, is composed only of three factors: those that qualify as "resources", those that are neutrally useless, and those that may be regarded as pests, antagonists, or competitors. Presently, by imposing such "economic rationality" models on the few indigenous cultures which have kept their adaptive cognized models, such as the Indians of the Amazonas or the Maasai and other Nilo-hamitic pastoralists of East Africa, we are destroying these resources, together with their cultures.

The common root of most of these approaches lies in the distinction made by Kant on "things as they are perceived and . . . things as they are". This became the philosophical basis for the phenomenological approach after World War I in Germany as conceived by Husserl and further developed by Heidegger, and more recently in France by Merleau-Ponty. These phenomenological philosophers suggest an entirely different way of looking at things, at ourselves in relation to the world and to nature, as compared to the mechanistic viewpoint of Descartes, which had such a great influence on natural sciences. Heidegger, as cited by Steiner (1978), changed Descartes' assertion; "I think, therefore I am" into something like "I care, therefore I am".

As shown by Allesch (1990) in a conference on cultural aspects of landscape, the phenomenological approach, especially Merleau-Ponty's (1962) phenomenology of perception, has also been adopted by environmental psychologists who were critical of the predominant tendencies in behavioral sciences, of fragmentation and objectification of man's inner situation. Allesch refers to the German-American psychiatrist Erwin Straus who distinguished between the abstract space of geography and the sensory space of landscape, which cannot be explained by the structure of stimuli, but only by the specific way of human experience. This experience is lost in our modern leisure culture of air traveling of vast distances in short vacations with on-the-clock planned nature experiences. "The more life is dominated by technology, the greater is our yearning for the landscape, the more forced is the effort to regain it – oddly enough – by means of this very technology" (Straus 1964: 12).

Allesch concludes by stating that behavioral sciences should study the phenomena of landscapes not in the abstracted space of geography but where it occurs: in the primary world of the senses to which Erwin Straus pointed in his phenomonlogy of perception. It is noteworthy that the phenomenological approach has been introduced recently also in a landscape ecological study of East Brabant in the Netherlands by Pedroli (1990). He claims, like us, that the empirical objectivity alone will never touch the holistic nature of the landscape, which is not simply the sum of its parts. Accordingly, in order to solve the problem of the nature of landscape as a whole, the researcher has to reconstruct the essence of the landscape in his mind, deducing his conclusions from his systematic investigations of landscape phenomena by using different levels of observation.

Phenomenology and its existentialistic world view, as offered by Merleau-Ponty (1964), has apparently also influenced the concerned Canadian environmentalist Neil Evernden (1984), who expressed a fierce opposition to economic and utilitarian reasoning in our relations with nature. He is very critical of environmental management based on an "objective" science of ecology. He claims that ecology can help us manage natural resources, namely the process of changing a particular landscape into a resource, but it does not challenge this concept of resourcism. It is based on the prevailing tendency of "reification" – the conversion of a person, place or idea into a thing.

Similar biocentric ideas have also been expressed by the followers of the "deep ecology" movement (Devall and Sessions 1985) and by their intellectual leader, the Norwegian environmental philosopher Arne Naess (Naess 1986, Naess and Rothenberg 1989). Rejecting the anthropocentric piecemeal "reformistic" environmental improvement, they demand much more far-reaching environmental policies based on a normative total cultural view of "ecosophy" – or wisdom of the household – in which ethical fundamental positions should be combined with practical arguments.

As claimed also by Everden (1985), there is, indeed, an urgent need for a new, post-industrial reconciliation, and even symbiosis between human society and nature (Naveh and Lieberman 1994). In this – in the terms of the great Jewish cultural philosopher and teacher Martin Buber (1970) – our present estranged I-IT relation with nature (and therefore also with our biosphere landscapes) should be transformed into a new reciprocal I-THOU relation. An essential part of these relations is the spiritual experience gained from the self-transcendent openness not only of natural, but also of cultural landscapes and their in intrinsic, amenity and aesthetic values. *These are far too rich to be reduced to dollars in markets and are more than complementary to the aesthetic values attached to human art objects.* Therefore all biosphere natural and cultural landscapes should not be viewed merely as a source for our materialistic satisfaction, but also as a source of enlightenment and enjoyment, and – as shown by Roszak (1992) in a lucid exploration of eco-psychology – also as a vital source of mental health.

Stephen Kellert (1996), a social scientists at the Yale School of Forestry and Environmental Studies has recently illuminated the cultural roots of "biophilia" namely the evolutionary deeply ingrained need for affiliating with wildlife and nature by comparing the attitudes and people's feeling and beliefs towards 9 basic values in the United States with those of Germany, Japan and several developing countries. These values include an aesthetic attraction for animals and nature, a dominstic interest in exercising mastery and control over wildlife, an ecologistic and scientific inclination to understand the biological function of organisms and their habitat, a humanistic affection and emotional bonding with animals, a moralistic concern for ethical relations with the natural world, a naturalistic interest in experiencing direct contact with wildlife and the outdoors, a symbolic use of animals and nature for communication and thought, utilitarian interest in pragmatically exploiting wildlife and nature, a negativistic avoidance of animals and natural environments for reasons of fear, dislike, or indifference. He reached rather optimistic conclusions, stating that the cultural traditions of Western, Eastern, industrial, developing, and hunter-gatherer societies posses powerful elements of appreciation and concern for natural and living diversity. His practical applications for the conservation of biological diversity have much in common with those suggested here in a broader context for ecodiversity and landscape conservation. He maintains that managing for biological diversity may require new concepts of human wellbeing stretching beyond the limits of immediate experience, including the view of the human species as both the consequence and creators of evolution. He also makes a strong point for the need of environmental education and ethics.

THE DYNAMIC CONSERVATION OF BIOSPHERE LANDSCAPES AND THEIR ECODIVERSITY

The Need for a Balanced Ecocentric Approach and Better Communication Tools

A major conclusion from this discussion is, that for this purpose, a more balanced and down-to-earth approach than that presented by Evernden (1985) and by some "deep ecologists" is required. This approach should be guided neither by arrogant and short-sighted instrumental anthropocentric views nor by a one-sided and utopian egalitarian biocentric view, *but by a balanced ecocentric and transdisciplinary natural and cultural view.* In order to ensure the sustainability of healthy, viable and attractive natural and cultural biosphere landscapes with maximum overall landscape ecodiversity. There is need for an *optimization of soft, intangible values with non-economic richness and hard monetary landscape values, providing combined, biological, ecological, cultural and socioeconomic services.*

This can be realized only if the narrow and short-sighted monetary evaluations in environmental cost/benefit calculations and land-use decisions would be replaced by much broader ecological and socioeconomic functional evaluations of the soft and hard values provided by these landscapes. A valuable practical tool for such a functional evaluation has been provided recently by De Groot (1992). Of great significance, in this respect, is the emergence of a new synthetic field of ecological economy, aiming at the achievement of ecologically and economically sustainable development (Costanza 1996). It has therefore, like landscape ecology, a transdiciplinary goal.

Both conceptions of healthy and attractive landscapes and sustainable economic development have been combined and successfully implemented in the first case study of the Working Group for Future Regional Models at the Department of Applied Landscape Ecology of the Environmental Research Institute, Leipzig-Halle. This inter-disciplinary team under the leadership of the well-known landscape ecologist and systems scientist has succeeded to initiate the ecological, cultural and socio-economic revitalization of a small town in North Germany near the Lüneberger Heide, plagued by unemployment. The conceptual basis for this project – as for all others, carried out presently – is an integrated transdisciplinary "Total Human Ecosystem" framework. Its aim is to foster the synergies between people, economy and ecology. In this case these were the citizen of the township, their economy (providing new jobs and cyberskills) and the restoration of the degraded forest, wetland and river landscapes (making the city more attractive for recreation and tourism). This was achieved by combining advanced and dynamic holistic systems and simulation models with close "grass root" work with the citizens and their political decision makers. The revitalization of both the economy and the biosphere landscapes surrounding the city was met with enthusiastic participation of the residents and the mayor, and it enjoyed the support of local economic enterprises (Grossmann et al. 1997).

As mentioned above, a further important methodological contribution to these goals could be the application of fuzzy logic and its advanced computerized models, described above. However the, ultimate success of such a balanced, holistic landscape-planning-and-management policy will depend on the creation of awareness of the value of these landscapes and of their ecodiversity, concern about their future, and motivation for active involvement in their sustainable utilization, conservation and restoration. Such a cultural shift from the prevailing one-sided instrumental and exploitative perceptions and attitudes will be an important step toward this urgently needed environmental post-industrial revolution. It requires, above all, an educational process which should reach all those who care for these landscapes, those who live from them, and those who deal with them at all levels of decision-making inland-uses. For this purpose much more efficient communication tools are required to bridge the gap between academicians and professionals, the conservation- and restoration-minded ecologists and production-minded foresters, agronomists and economists, and between all these specialists and the public at large.

These communication tools should be able to transform "semantic" scientific information (expressed in words and figures), into "pragmatic", usable information that leads to actual conservation action. This information, combining confirmation with novelty, becomes meaningful through its feedback on the receiver when it leads to actions. In other words, these tools should provide factual updated information on the present status of the landscapes and their ecodiversity. These should be collected by integrated field surveys and remote-sensing, dynamic Geographic Information Systems utilizing artificial intelligence of fuzzy systems and expert systems, simulation models and other advanced landscape ecological methods, which have been applied in the above-mentioned study by Grossmann et al. (1997). These will provide information as "confirmation", and at the same time they should serve as prescribing and anticipating tools regarding the fate of these landscapes and thereby introducing new information as "novelty". Such pragmatic information is vital for the implementation of innovative landscape conservation and development strategies ensuring the evolution and thereby the sustainable future of natural, and cultural biosphere landscapes in the post-industrial global information society.

Such an innovative tool for the provision of pragmatic information has been suggested by Naveh (1993a) and by Naveh and Lieberman (1994) as "Red Books" (later on called "Green Books") for the conservation and restoration of endangered highly valuable landscapes. They should present in clear non-technical terms with ample maps and illustrations not only recent, adverse changes endangering both natural and cultural assets and scenic and economic values but also suggest alternative, sustainable land-use strategies based on holistic landscape planning and dynamic conservation management. In this way they could help to change the attitudes of politicians and decision-makers and provide practical guidance for holistic, sustainable and multi-beneficial land use planning and management. The first such Green Book case study was initiated by the above-mentioned IUCN-CESP-Working Group for Landscape Ecology and Conservation and was carried out in West Crete with the support of the European Union. Some of its results will be reported below, after discussing briefly the dynamic non-equilibrium nature of cultural biosphere landscapes.

The Homeorhetic Flow Equilibrium of Perturbation-Dependent Semi-Natural Landscapes

As described in detail by Birks et al. (1988), the so-called "natural" landscapes of Europe are in reality relics of earlier types of land-use. They represent therefore different gradients of human-modified, cultural landscapes. These were maintained by extensive methods which became uneconomical and were abandoned and regenerated in response to other uses and non-uses. In most cases, a much less diverse "potential natural vegetation canopy" took over, which was given the unfortunate term of "climax", with its misleading connotation of a homeostatic stable-state equilibrium.

This fiction of "undisturbed and virgin" and, therefore, "natural" ecosystems, which should be protected completely from human interference, and which could assumably be restored entirely, is true also for all other continents. Today there remain very few larger stretches of land which have not been influenced to lesser or greater degrees by human perturbations and their cultural inputs of energy/matter and/or information, even in the Arctic and Antarctic. These landscapes are therefore either "close-to-natural or "semi-natural". In the latter, their natural – that means spontaneously occurring and reproducing organisms – have been regularly influenced by human uses to such degrees that they have even become dependent on these anthropogenic perturbations, together with the naturally occurring perturbations. In the Mediterranean, these periodic anthropogenic and chiefly cyclical perturbations have been introduced by regular, centuries-lasting, rotational grazing, browsing, burning, cutting and coppicing regimes, together with cultivation and other human land-uses. These defoliation pressures and other perturbations were incorporated at different spatio-temporal scales in the landscape together with natural perturbations of climatic annual and seasonal fluctuations. They resulted in the establishment of a *man-maintained dynamic flow equilibrium or "homeorhesis"* (from the Greek, meaning "preserving the flow") (sensu Waddington 1975) between trees, shrubs, herbs, grass layers, and the cultural artifacts of forests, woodlands, shrublands, grasslands at different spatio-temporal scales.

In contrast to the homeostatic systems, these homeorhetic systems are not returned to a stationary state. They are going on to move along the same trajectory of change, as long as these cyclic perturbations are continued with similar intensities and time intervals. These human-perturbation systems have thereby acquired longterm adaptive resilience and metastability.

In the Mediterranean Basin, the long-term maintenance of this homeorhetic flow equilibrium in combination with the great macro- and microsite heterogeneity of the rocky and rough terrain and the fine-grained agro-silvo-pastoral land-use patterns induced the above-mentioned unique ecological and cultural landscape diversity – or in short – the ecodiversity of the Mediterranean uplands (Naveh 1991).

The close interdependence between this human-perturbation dependent ecodiversity and biodiversity has been shown in several studies, in Israel and elsewhere in the Mediterranean and has far reaching implications for the management of cultural biosphere landscapes (Naveh 1993b). Thus, in all our studies in Israel and southern France and California, highest plant and animal species diversity was obtained where above-mentioned moderate defoliation pressures have been continued. However, where these were either too heavy or were released completely, structural, floristic and faunistic diversity was significantly reduced (Naveh and Whittaker 1979 also published as Chapter 2 in this anthology). These findings were confirmed also on a much larger scale in *all* nature reserves of Northern Israel (Kaplan 1992, Noy Meir and Kaplan 1991).

Also in all other Mediterranean countries, as in many other countries, the disruption of this homeorhetic flow equilibrium by cessation of human agro-silvo-pastoral activities and defoliation processes on one hand and the rapid and careless shift from diversified and stable traditional agriculture to intensive large scale, agro-industrial farming in combination with land abandonment and indiscriminate planting of pine and eucalyptus monocultures has created already large stretches of monotonous and highly inflammable vegetation. This has resulted in lower structural and floristic diversity of herbaceous plants – including many endemics and ornamental geophytes- and at the same time in the loss of overall landscape heterogeneity. In general, it is combined also with increased vulnerability to fire. These issues have been discussed in two landscape-ecological conferences (Baudry and Bunce 1991, Farina and Naveh 1993).

A typical example of such a homeorhetic flow equilibrium and the results of its present distortion are the extensive oak savannas of the "Montados" in southern Portugal and the "Dehesas" in South-West Spain, presented in Fig. 4. As described in more detail by Pinto-Correia (1993), in the complementary multi-beneficial utilization of the open evergreen woodlands of cork oaks (*Quercus suber*), olives and chestnut trees for cork, fruit, wood and livestock production were rotated every 6 years with cultivation of cereals. This agro-silvo-pastoral system was well adapted to the low edaphic and climatic potentials of these uplands. Managed extensively for many decades with minimum human inputs, it created a fine-grained heterogeneous and attractive landscape with relatively high biodiversity. At the same time it also furnished the necessary "hard" economic outputs for the dispersed rural population. However, presently by intensification of mechanized crop production, the soil is eroding and the oak trees are severely damaged, and as crop yield are too low, the

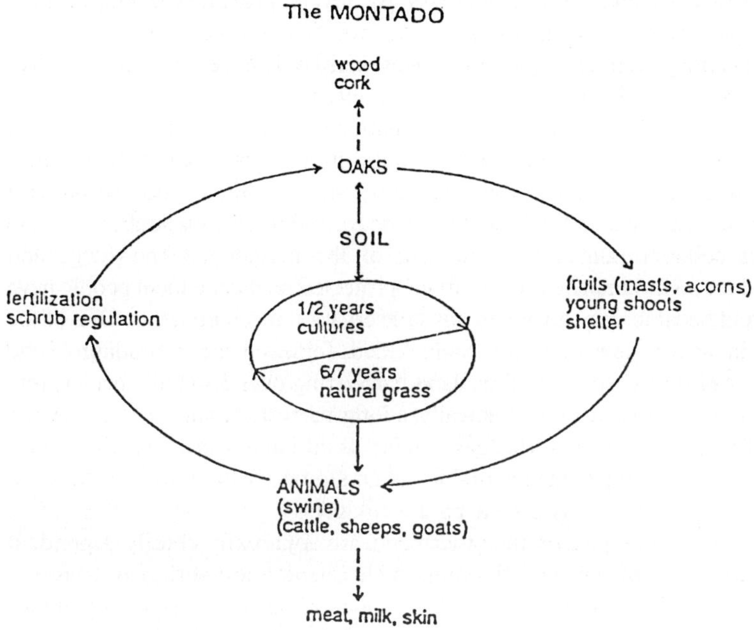

Fig. 4. Homeorhetic land-use cycles in montado landscapes of southern Portugal (Pinto-Correia 1993).

land is finally abandoned, or may be used for Eucalyptus tree afforestation. In other locations, because of the end of pig raising and the "modernized" economy, the land is progressively used more extensively by cessation of cultivation and grazing. The resulting brush encroachment prevents the efficient utilization of the trees and the land is abandoned. In this way, both intensification and extensification are endangering not only the biological and economic productivity, but also the ecodiversity and stability of these unique cultural landscapes.

It is therefore not sufficient to counter further modern landscape degradation and despoliation merely by protective measures. Landscape ecodiversity and all vital biological, ecological, cultural and socioeconomic functions and values can be ensured only by the maintenance of a dynamic homeorhetic flow equilibrium on micro- and macroscales. This requires the well-controlled continuation, simulation and restoration of all natural and cultural landscape patterns and processes, which have shaped these landscapes.

The need for the conservation of ecological processes (Ricklefs et al. 1984), is certainly also true for all other semi-natural, perturbation-dependent landscapes, such as forests, grasslands and wetlands which have been utilized for many centuries by local tribes and villagers. Thus, for instance, in the East African savannas this flow equilibrium is maintained – like in the mediterranean-type shrublands and woodlands – by a mixture of natural and anthropogenic factors, namely by climatically-conditioned rotational cycles of grazing and browsing by wild ungulates and livestock, and by periodical fires, set by pastoralists. In Tanzania Maasailand – like in the Sahel – the distortion of these grazing and browsing cycles by additional watering places and boreholes, has enhanced desertification and induced increasingly severe drought-starvation cycles both of livestock and people (Naveh 1989).

In India, the banning of grazing and other traditional landuses, such as food collection in protected forests and national parks has not only interrupted their homeorhetic flow equilibrium, maintained by these vital defoliation pressures and perturbations. It has caused fierce and sometimes even violent opposition of the local people whose livelihood is depending on these resources. Thus, as, reported by Ashish Kothari (1996), a senior staff member of the Indian Institute for Public Administration, instead of seeking their collaboration and making use of the indigenous knowledge and culture, essential for sound management of many protected areas, the local people have been alienated and became very hostile to any conservation measurements. One of the worst conflicts in which seven villagers were killed, followed the upgrading of the famous Bharatpur bird reserve, a small wetland harbouring over 350 bird species, into the Keoladeo National Park, in 1980. Ironically, a long-term study showed that the ban has adversely affected the wetland: Buffalo grazing as an integral part of this ecosystem had helped to counter the tendency of the wetland to turn into grassland. Now the Park authorities had to allow grass-cutting by the villagers.

The South American Pampas, on the other hand, are apparently chiefly dependent on natural perturbations of periodic flooding cycles, which are seriously distorted by drainage and hydrological "regulations" (C. Ghersa, personal communication). The huge Yellowstone National Park fire, resulting from a long period of disruption of the natural fire cycles and those experienced with increasing frequency and destructive powers in the Californian Chaparral, provide ample proof of the disastrous results of distorting the natural and/or anthropogenic flow equilibrium of fire cycles, either by too heavy "modern" human impacts, or by their complete cessation.

Some Lessons for Ecodiversity Conservation From the West Crete Green Book Study

The outcome of the above-mentioned West Crete Green Book case study initiated by above-mentioned IUCN Working Group, has corroborated most of these findings. It has, above all, provided strong evidence for the interdependence between biological diversity and cultural diversity and for the urgent need of dynamic landscape management to conserve both. Its practical recommendations could serve also as an illustration for our balanced ecocentric and very down-to-earth landscape-ecological approach.

With the support by the European Union, this study has been carried out by two multidisciplinary teams composed of landscape ecologists, physical and cultural geographers, and specialists for pasture, forestry and remote sensing from the universities of Thessaloniki and Cambridge. On the basis of their research report (Grove et al. 1993), also an abridged and well illustrated Greek Green Book version has been produced by the leader of the Greek team, Vasilios Papanastasis This has been distributed among all relevant official and public bodies. The study area covered a broad transect of about 350 km across West Crete from the Aegean to the Libyan Sea, including the White Mountains and the famous Samaria Gorge National Park. This area is one of the most beautiful, fine grained and harmonious example of a montainous Mediterranean landscape, rich in cultural features resulting from the long interplay of a rugged topography with human activities and wild plants and animals. Not only vernacular buildings and terrace walls, but also huge ancient cypress trees and the endemic Cretan spike elm – *Zelkova cretica* – are relics from Roman, Byzantinic, Venecian and Turkish times. Some of its many ornamental, and endemic plants are growing not only in the wild rocky gorges, but also on and around remote fortresses and lofty chapels, vernacular houses and terraces and individual trees. Thus, for instance, the arrestingly beautiful, endemic *Petromarula pinnata* with its tall blue spikes can be found on Venetian masonry and country chapels, *Verbascum arcturus* is hanging over cave chapels, and in the Omalos Plains, leading to the Samaria Gorge National park, the rare, Zelkova trees and the famous tulip – *Tulipa bakeri* can easily be seen by the public. The same is true also for the unique Cretan wild goat – the Agrimi, which has increased considerably since its protection after the establishment of the park.

However, presently there are serious threats to all these assets not only by uncontrolled touristic and recreational pressures, but also by intensive agricultural developments, supported by funds from the European Union. The latter have encouraged large-scale, indiscriminate bulldozing of the steep mountain slopes covered by rich maquis vegetation for economically highly doubtful olive plantations. These are fertilized and sprayed by herbicides and pesticides and are irrigated by pumping up water hundreds of meters and carrying it across the land in ugly black plastic pipes from the already overused aquifers in the plains. They demonstrate clearly the unfortunate results of replacing sustainable traditional land uses and there cultural biosphere ecotopes by "modern" intensive and unstable agriculture. There are also ambitious plans to impound the scenically outstanding and biologically and culturally rich Omalos Plains by ecologically disastrous and economically unjustified irrigation schemes. These serve as the main buffer zone for the Samaria Park, and according to the Green Book recommendations should be included in it, or be protected as a biosphere reserve. In Italy similar careless plans for impounding the scenic biologically and culturally outstanding Grosseto Plains in Tuscany have been prevented, thanks to the recommendation of a team of Dutch and Italian landscape ecologists and experts on land development, and the foresight of the Tuscan Regional Government (Pedroli et al. 1988).

The Samaria Gorge Park itself has been established according to the misconception of the above-mentioned undisturbed natural wilderness paradigm in which minimum or none management should be carried out. For the same reason and therefore all previous traditional human impacts have been removed. However, presently, close to a quarter of million visitors are passing through it each year on a narrow foot path, putting heavy strains to its access roads and surroundings, but without the opportunity to observe its unique archeology and history. These have been neglected, and at the same time the park is encroached by highly inflammable pine trees, which will lead sooner or later to disastrous conflagrations. The fine but deserted vernacular buildings in the old village of Aya Roumeli outside the park are failing into decay, but the new village has only ugly intrusive buildings. Altogether the unique opportunity to maintain and present an example of the traditional Cretan community, its crafts and landscapes has been lost.

In the Green Book recommendation (Grove et al. 1993) a strong point has been made that present rapid and uncontrolled changes in land use are threatening to erode the meaning and unique character of the Cretan island, to "decretinize" it by robbing its biological and cultural-historical heritage. This can be only rectified by integrative and comprehensive masterplans for environmental management and dynamic landscape conservation, for sustainable development and highest quality of life.

Unfortunately, most of the population of Crete is completely indifferent to these issues and does not realize that these irreversible damages inflicted to their cherished Cretan way of life, will persist long after the economic benefits have been forgotten. Their implementation will therefore require great education and persuasion efforts at all public, professional and political levels of Crete. The first step in this direction has been made by the preparation and distribution of the Greek Green Book.

DISCUSSION AND CONCLUSIONS

The conceptualization of the uniqueness of those landscapes that contain both natural and cultural values indicates clearly that they should not be studied and managed with conventional, mechanistic and formal ecological methods. These cultural landscapes should be considered as a special class of biosphere landscapes to which innovative, transdisciplinary landscape-ecological and other holistic methods should be applied. Their structural and functional dynamics as perturbation-dependent, non-equilibrium, but metastable systems has resulted in a rich total landscape ecodiversity. This has been maintained in many cases in a homeorhetic flow equilibrium by the cyclic interplay of natural and cultural processes which cannot be preserved by classical conservation strategies, applied in nature reserves and protected areas. For this purpose innovative quantitative methods such as fuzzy logic, fractal dimensions, functional landscape evaluation and comprehensive tools such as Green books for the conservation and restoration of endangered highly valuable landscape, as well as also others, not mentioned here, could be useful.

The conservation strategies for such cultural biosphere landscapes should be guided by the recognition that *ecodiversity is the tangible expression of the dynamic interplay between ecology and culture, as determined by biodiversity, ecological micro-and macro-site heterogeneity and human land uses and their cultural artifacts, either enriching or impoverishing both biodiversity and overall landscape heterogeneity.*

These are coupled by positive, mutually amplifying and self-augmenting feedbacks: The greater the ecological landscape heterogeneity, the greater the chance for biological diversity and at the same time floristic, faunistic and structural vegetation diversity enhances ecological heterogeneity. But also the opposite is true, and therefore landscape homogenization, fragmentation, and despoliation diminishes biodiversity which in turn reduces further landscape heterogeneity. This is demonstrated in Fig. 5.

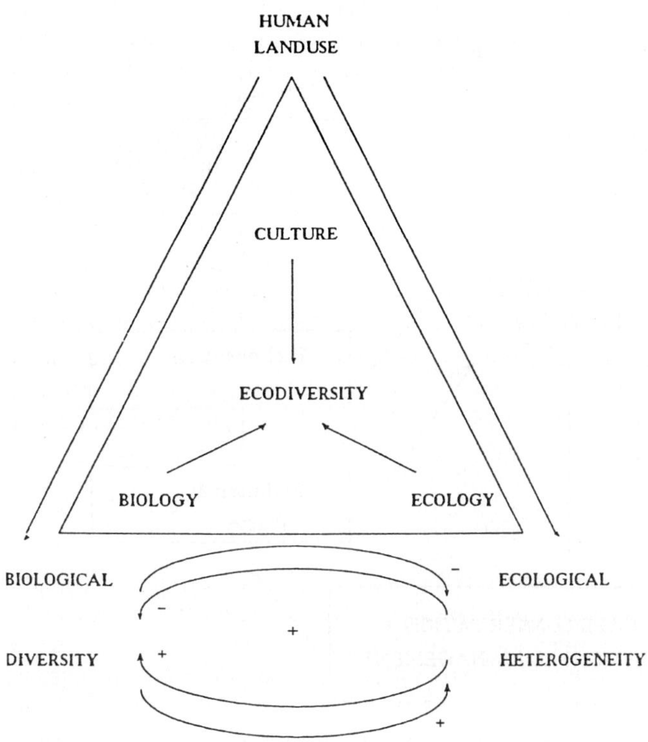

Fig. 5. The landscape shaping forces of the biology-ecology-culture triangle determining ecodiversity.

However, it has to be realized that presently the conservation of biodiversity, ecodiversity and ecological stability by biosphere landscape conservation and sustainable environmental planning and management, has to face many strong demographic, political and socioeconomic forces. These are functioning as a vicious circle of destabilizing feedback couplings. They all drive in the direction of the accumulative loss of biodiversity, ecodiversity and stability, and thereby are enhancing landscape degradation. But they can be removed only by sound landscape conservation and environmental management. This is illustrated in Fig. 6.

Because of these strong forces, our environmental crisis is chiefly a cultural crisis which cannot be resolved only by scientific, technological, political and economical means. Its remedies must include the spheres of cultural, spiritual and ethical values as expressed in our appreciation of life, nature and its intrinsic values. This will be difficult to achieve without a radical shift from consumption to conservation and from unrestrained quantitative growth to lasting qualitative improvement and development. This is, essentially, the post-industrial environmental revolution, required as a vital part of our choice of evolution and not extinction during the Grand Transition towards the age of global informatization, as demanded by Laszlo (1994).

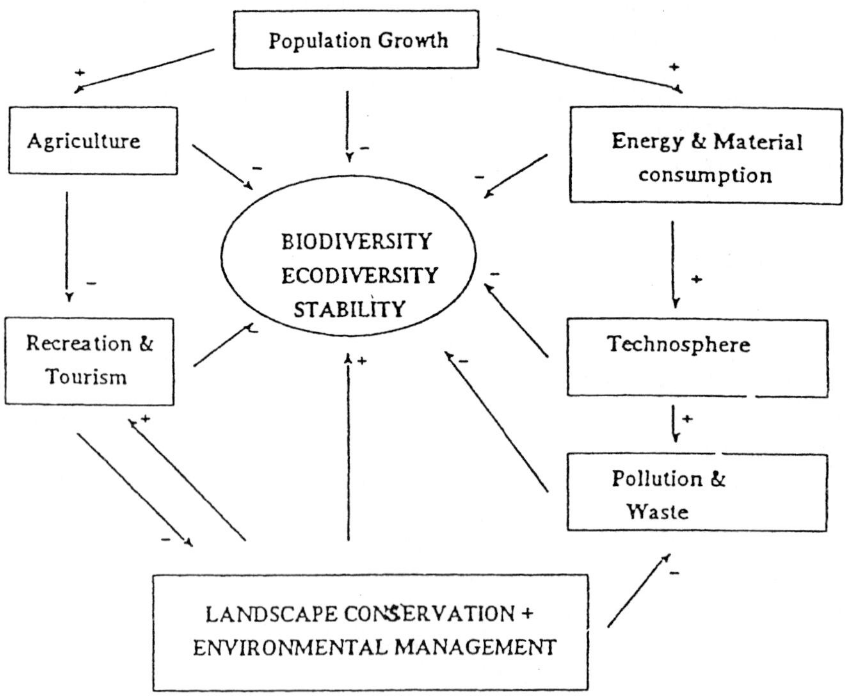

Fig. 6. Landscape conservation and environmental management as major restraining forces of the biological and cultural impoverishment and destabilization of biosphere landscapes.

In this diagnosis of our present environmental debacle, it would be a dangerous cultural misconception to put all the blame on modern technology and to indulge in a romantic nature fundamentalism, which preaches an utopistic return to primitive, untouched nature which does not exist and cannot be restored. We cannot deny the evolution of our neocortex and the noospheric cultural evolution, achieved by our self-reflexive mentation. These endowed us not only with our scientific knowledge and technological power, but also with our ecological wisdom and consciousness to control both. These have given us the present superior status in the ecological hierarchy of our Total Human Ecosystem. We cannot return to the landscapes of the Garden of Eden, nor to the primeval symbiotic status of man with nature. Nor can we save all of the living organisms abounding still on Earth. But, at the same time, we cannot continue to expand our technosphere landscapes, with an arrogant, ignorant and exploitative, anthropocentric attitude.

There are already many encouraging signs for the urgently needed cultural evolution process of the reconciliation of human society with nature. This is indicated in the evolution of new social norms and ethics arising in a strong movement dealing with these intrinsic and "ṡoft" values of nature, as reflected in the actual decision making process in land use planning and management. Recognition of the urgent need for this global environmental revolution and the feasibility of its realization, if humanity will accept this challenge in time, is no longer regarded as merely the utopian dream of some "radical" environmentalists and "deep" ecologists. It is now shared by some of the most prominent economists, managers, political decision makers, and systems analysts and modelers (Gore 1991, King and Schneider. Club of Rome 1991, Meadows et al. 1992, Tolba 1992). This environmental revolution will not be achieved by a careless technological fix, but it can also not by achieved *without innovative, non-polluting, energy saving and regenerative technologies and their careful and efficient applications.*

According to a recent historical-sociological analysis (Schmidt 1996), there is a growing environmental consciousness trickling down from groups of higher status which have access to and are receptive to scientific knowledge in the economically and technologically most advanced countries to the lower strata. Both play an important role in the rise, expansion and refinement of behavioural standards. The researchers suggest that by attending to the role of status motives in the process of ecologisation, environmental policy makers could use it in further expanding their policy. This seems to happen already in the promotion of "environment-friendly" products by many industrial and commercial enterprises. These are becoming more and more sensitive to their ecological reputation and make attempts to realize green policies in their production processes for the local and international market. A most recent example which is symptomatic for this encouraging trends can be witnessed in the restoration of the formerly world-famous Leipziger Messe on a huge area, easily accessible from the airport, and served by shuttle buses and electric trains: Here, even before the building construction were completed an attractive green belt with trees, lawns and a lake were established, and even inside the huge messe hall green islands of trees were planted. A great deal of its advertising is devoted to the realization of this ecological message.

There are already encouraging signs that the efforts of ecologists to create pragmatic information, aimed at saving the tropical rain forests and its biodiversity through a reconciliation between so-called "primitive" tribal cultures and the modern consumption culture of the market economy, are bearing fruits. Some examples are presented in the May–June 1996 issue of "Earthwatch", titled "Culture, Nature & Sustainability". This journal is published for members of the worldwide and very laudable organization, recruiting each year hundreds of volunteers for its "Earth Corps", to help in scientists in their field work. The mission of this research is to "improve human understanding of the planet, the diversity of its inhabitants, and the processes that affect the quality of life on earth". Thus, for instance, as reported by Bloch (1996), the entomologist Larry Orsak learned from his "Earth Watch" – supported research in Papua New Guinea, that in order to save some of the most attractive species of butterflies, he has first to save the forest they lived in. Which meant giving local people a reason to save the forest. This was achieved by helping to set up a commercial butterfly farming operation for vines that attract wild butterflies to lay their eggs. Some of the hatched butterflies are sold to foreign collectors, and some are released to the wild to breed again. Thereby the farmers now equate intact rain forest with substantial income and save not only these extraordinary insects species, but also the rain forest as a whole.

Within the broader context of the protection of the natural resources, there is a steadily growing recognition among conservationists that unless we work with local people and take into account their indigeneous culture and tradion and their socioeconomic needs and expectations, these resources are doomed. As described in the August 1996 issue of the IUCN "World Conservation 11 journal, for this purpose collaborative conservation and restoration management strategies are already widely applied, especially in developing countries and are further developed by its Commission for Environmental Strategies and Planning. Among several recent publications on this issue, the volume on "Expanding Partnership in Conservation (McNeely 1995) is of special significance. But this development is not restricted only to parks and nature reserve: Thus, for instance, in eastern India a broad-based grassroots movement to restore denuded forests, involving 12 000–15 000 villages have mobilized to protect the natural forest regeneration of 1–2 million hectares. In India's central tribal belt about 10 000 groups of the country's Forest Protection Committees are operating.

Finally, also the Indian Forest Service – one of the oldest and largest in the world, with more than 150 000 foresters was forced to follow suit and is gradually departing from its highly controversial and fiercely opposed monoculture plantation of exotic, fast growing species, by recognizing the Joint Forest Management Programmes in which the protection responsibilities and rights of villages over small tracts of public forest lands are secured.

As mentioned above, there are also more and more promising attempts for the creation of sustainable, agricultural landscapes, based on organic farming. Some of these have been described lucidly by Lyle (1994). He showed, that these could replace our present "modern" high input and one-way throughput flow agricultural linear technosphere systems by regenerative systems with cyclical flows, providing continuous replacement through their own functional process. Such an innovative regenerative system, combining interdisciplinary education, research and demonstration has been designed by Lyle and was established at the Center for Regenerative Studies of the California State Polytechnic University at Pomona. In the functional integration of such "neotechnic", and productive agricultural landscapes with other biosphere landscapes lies the promise for the creation of a new dynamic "balance of nature" on a higher level of the Total Human Ecosystem with emerging qualities of ecological diversity, metastability, productivity, beauty and utility or in one word – health – in our new post-industrial cultural landscapes.

In addition to these and many other encouraging initiatives, the conservation and restoration of our highly valuable and most cherished cultural landscapes can be regarded as one important step toward this cultural and environmental revolution. In order to ensure its success, scientists will have to integrate scientific knowledge with ecological wisdom and ecological ethics. This could help us learn from past experience, comprehending the present and envisaging the future of our cultural landscapes, and translating these insights into sound, large-scale conservation and restoration activities in the field.

REFERENCES

Allen TFH, Hoekstra TW (1992) Toward a unified ecology. Columbia University Press, New York, 384 pages.

Allesch CG (1990) The space of landscape and the space of geography. In: Svobodova H (ed.) Cultural aspects of landscape. Pudoc, Wageningen pp 17–23.

Bastian O, Schreiber K-F (1994) Analyse und ökologische Bewertung der Landschaft. Gustav Fischer Verlag, Jena and Struttgart, 502 pages.

Baudry J, Bunce RGH (ed.) (1989) Land abandonment and its role in conservation. Options Mediterraneenes, Ser A 15:1–190.

Bloch A (1996) Bullish on butterflies. Earthwatch XV (3):11–13.

Birks H, Kaland PE, Dagfilm M (eds) (1988) The cultural landscape, past, present and future. Cambridge University Press, Cambridge, 520 pages.

Bohm D (1980) Wholeness and the implicate order. Routiedge and Kegan, London, 284 pages.

Bohm D, Peat FD (1987) Science, order, and creativity. Bantam Books, Toronto, 280 pages.

Buber M (1970) I and Thou. Scribner's, New York, 140 pages.

Burrough PA (1981) Fractal dimensions of landscapes and other environmental data. Nature 294: 243.

Burrough PA, MacMillan RA, van Deursen W (1992) Fuzzy classification methods for determining land suitability from soil profile observations and topography. Journal of Soil Science 43:193–210.

Costanza R (1996) Ecological Economics: Reintegrating the study of humans and nature. Ecological Applications 6:978–990.

Dansereau P (1975) Inscape and Landscape. Columbia University Press, New York. 118 pages.

De Groot RS (1992) Functions of Nature. Wolters-Noordhoff, Amsterdam. 315 pages.

Devall B, Sessions G (1985) Deep Ecology: Living as if Nature Mattered. Peregrine Smith Books, Salt Lake City, USA. 295 pages.

Di Castri F (1981) Mediterranean-type shrublands of the world. In: di Castri F, Goodall GW, Specht RL (eds) Ecosystems of the World II. Mediterranean-type Shrublands. Elsevier, Amsterdam, pp 1–52.

Egler FE (1964) Pesticides in our ecosystem. American Scientist 52:110–136.

Evernden N (1985) The Natural Alien. Humankind and Environment. University of Toronto Press, Toronto. 160 pages.

Farina A (1997) Principles and Methods in Landscape Ecology. Chapman and Hall, London. (in press).

Farina A, Naveh Z (eds) (1993) Landscape Approach to Regional Planning: The Future of Mediterranean Landscapes. Landscape & Urban Planning 24. Special Issue, July 1993.
Forman RTT, Godron M (1986) Landscape Ecology. John Wiley, New York. 619 pages.
Forman RTT (1995) Land Mosaics. The Ecology of Landscapes and Regions. Cambridge University Press, Cambridge. 632 pages.
Frankl VE (1969) Reductionism and nihilism. In: Koestler A, Smithies JR (eds) Beyond Reductionism. New Perspectives in the Life Sciences. Hutchinson, London, pp 396–408.
Golley FB (1990) Love of the land. Editor's comment. Landscape Ecology 4: 81–82.
Gore A (1992) Earth in Balance - Forging a New Common Purpose. Earthwatch Publisher, London, 380 pages.
Green B (1985) Countryside Conservation. The Protection and Management of Amenity Ecosystems. 2nd edition. Unwin Hyman, London. 251 pages.
Grinker RR (1976) In memory of Ludwig von Bertalanffy's contribution to psychiatry. Behavioral Science 21:207–217.
Grossman WD, Meiss K-M, Fraenzle S, Multhaup T, Roesch A in collaboration with Costello DF, Simon F, Sorkin M (1997) Alternative Landschaftsplan für eine kleine attraktive Stadt in der Informationsgesellschaft- Beispiel Visselhoevede. UFZ-Umwelt-forschungszentrum Leipzig-Halle, Arbeitsgruppe Regionale Zukunftsmodelle, Leipzig. 127 pages.
Grove AT, Ispikoudies J, Kazaklis AM, Papanastasis V, Rackham O (1993) Threatened Mediterranean Landscapes: West Crete. Research Report to European Union. Department of Geography, Cambridge University, Cambridge, U.K.
Haber W (1990) Basic concepts of landscape ecology and their application in land management. In: Kawanabe H, Ohgushi T, Higaschi M (eds) Ecology for Tomorrow. Physiology and Ecology, Japan 27 (Special Number), pp 131–146.
IUCN/UNEP/WWF (1991) Caring for the Earth. A Strategy for Sustainable Living. IUCN, Gland, Switzerland. 310 pages.
Jantsch E (1980) The Self-Organizing Universe. Scientific and Human Implications of the Emerging Paradigm of Evolution. Pergamon Press, Oxford. 320 pages.
Kaplan D (1992) Responses of Mediterranean grassland plants to gazelle grazing. In: Thanos, C.A. (ed.) Proceedings 6th International Conference on Mediterranean Climate Ecosystems, September 1991, University of Athens, Athens, Greece, pp 75–79.
Kellert SR (1996) The Value of Life. Biological Diversity and Human Society. Osiand Press/Shearwater Books, Washington, D.C. 480 pages.
King A, Schneider B (Club of Rome) (1991) The First Global Revolution. From the Problematique to the Resolutique, "Blueprint for the Twenty-first Century" by the Council of the Club of Rome. Pantheon Books, New York.
Kosko B (1993) Fuzzy Thinking: the New Science of Fuzzy Logic. Hyperion, New york 318 pages.
Kothari A (1996) India's protected areas: The journey to joint management. World Conservation 2/96:8–9.
Kroeber AL, Kluckholm C (1949) Culture. A Critical Review of Concepts and Definitions. Harvard University Press, Cambridge, MA. 430 pages.
Laszlo E (1972) Introduction to Systems Philosophy: Toward a New Paradigm of Contemporary Thought. Harper Torchbooks, New York. 328 pages.
Laszlo E (1987) Evolution – The Grand Synthesis. Shambhala, New Science Library, Boston. 211 pages.
Laszlo E (1994) The Choice: Evolution or Extinction. A Thinking Person's Guide to Global Issues. C. P. Putnam & Sons, New York. 215 pages.
Leser H (1991) Landschaftsökologie. Eugen Ulmer, Stuttgart. 647 pages.
Li BL (1993) Facing the uncertainty challenge in ecology. Ecological Modelling.
Lucas PHC (1992) Protected Landscapes. A Guide for Policy- Makers and Planners. Chapman & Hall, London. 297 pages.
Lyle JT (1994) Regenerative Design for Sustainable Development. John Wiley, New York. 338 pages.
Mandelbrot BB (1982) The Fractal Geometry of Nature. Freeman & Co., New York. 325 pages.
Mansvelt JD, Mulder JA (1993) European features for sustainable development: a contribution to the dialogue. Landscape and Urban Planning 27:67–90.
McNeely A (ed.) (1995) Expanding Partnership in Conservation. Island Press, Washington, D.C.

Meadows DH, Meadows DL, Randers J (1992) Beyond the Limits. Global Collapse or a Sustainable Future. Earthscan Publications, London. 300 pages.
Merleau-Ponty M (1962) Phenomenology of Perception. Routledge & Kegan Paul, London. 393 pages.
Milne BT (1991) The utility of fractal geometry in landscape design. Landscape & Urban Planning 21:81–90.
Naess A (1986) Intrinsic value: Will the defenders of nature please rise? In: Soulé M (ed.), Conservation Binc. Publishers, Sunderland, Massachusetts, USA, pp 504–515.
Naess A, Rothenberg D (1989) Ecology, Community and Cambridge University Press, Cambridge. 223 pages.
Nassauer JI (1992) The appearance of ecological systems as a matter of policy. Landscape Ecology 6:239–250.
Naveh Z (1982) Landscape ecology as an emerging branch of human ecosystem science. Advances in Ecological Research 12:189–237.
Naveh Z (1984) The vegetation of the Carmel and Nahal Sefunim and the evolution of the cultural landscape. In: Ronen A (ed.) Sefunim Prehistoric Sites in Mount Carmel, Israel. BAR International Series 230. Oxford, pp 23–63.
Naveh Z (1987) Biocybernetic and thermodynamic perspectives of landscape functions and land use patterns. Landscape Ecology 1:75–85.
Naveh Z (1989) The challenges of desert landscape ecology as a transdisciplinary problem-solving oriented science. Journal of Arid Environments 17:245–253.
Naveh Z (1990a) Landscape ecology as a bridge between bio-ecology and human ecology. In: Svobodova H (ed.) Cultural Aspects of Landscape. Pudoc, Wageningen, pp 1–73.
Naveh Z (1990b) Fire in the Mediterranean – a landscape ecological perspective. In: Goldammer JG, Jenkins MJ (eds) Fire in Ecosystem Dynamics. SPB Academic Publishing, The Hague, pp 1–20 (See also Chapter 3).
Naveh Z (1991) Mediterranean uplands as anthropogenic perturbation dependent systems and their dynbamic conservation management. In: Ravera O (ed.) Terrestrial and Aquatic Ecosystems – Perturbation and Revovery. Ellis Horwood, New York, pp 545–557.
Naveh Z (1993a) Red Books for threatened Mediterranean landscapes as an innovative tool for holistic landscape conservation. Introduction to the western Crete Red Book case study. Landscape & Urban Planning 24:241–249.
Naveh Z (1993b) Biodiversity and landscape management. In: Kim KC, Weaver RD (ed.). Biodiversity and Landscapes: A Paradox of Humanity. Cambridge University Press, New York, pp 175–195.
Naveh Z (1995a) Transdisciplinary landscape-ecology education and the future of post-industrial landscapes. In: Farina A, Naveh Z, (eds) Symposium on Educating Landscape Ecologists for the 21st Century: The Role of Landscape Ecology in Scientific and Professional Training. IALE World Meeting 1995, Toulouse. Bolletin dei Museo di Storia Naturale de la Lunigiana, Vol. 9. Supplemento, Aulla, pp 13–26.
Naveh Z (1995b) Interactions of landscapes and cultures. Landscape and Urban Planning 32:43–54.
Naveh Z, Lieberman AS (1994) Landscape Ecology Theory and Applications. 2nd Edition. Springer-Veriag, New York. 360 pages.
Naveh Z, Whittaker RH (1979) Structural and floristic diversity of shrublands and woodlands in Northern Israel and other Mediterranean areas. Vegetatio 44:171–190.
Negoita CV (1985) Expert Systems and Fuzzy Systems. Benjamin/Cummings Publishing Company, Menlo Park, California. 160 pages.
Noy-Meir E, Kaplan D (1991) The Effect of Grazing on the Herbaceous Mediterranean Vegetation and Its Implications on the Management of Nature Reserves. Interim Report to the Nature Conservation Authorities, Jerusalem (in Hebrew). 120 pages.
O'Neill RV, DeAngelis DL, Waide JB, Allen TFH (1987) A Hierarchical Concept of Ecosystems. Princeton University Press, Princeton, NJ. 384 pages.
Pedroli B (1989) The Nature of Landscapes. CIP-Gegevens Konnninkiije Bibliotheek, Den Hague. 146 pages.
Pedroli GMB, Vos WD, Rossi R (1988) The Farma River Barrage Effect Study. Giunta Regionale Toscana. Marsilio Editori, Firerize, Italy. 370 pages.

Pignatti S (1994) Ecologia del Paesaggio. Unione Tipografica Editrice Torinese (UTET), Torino, Italy. 228 pages.
Pinto Correia T (1993) Threatened landscapes in Alentejo, Portugal: the 'montado' and other agro-silvo-pastoral systems. Landscape and Urban Planning 24:43–48.
Rappaport RA (1979) Ecology, Meaning, and Religion. North Atlantic Books, Richmond, California, pp 97–144.
Ricklefs RR, Naveh Z, Turner RE (1984) Conservation of Ecological Processes. IUCN Commission on Ecology Paper no. 8. IUCN, Gland, Switzerland.
Roszak T (1992) The Voice of the Earth. An Exploration of Ecopsychology. Simon and Schuster, New York. 367 pages.
Schmidt K (1996) Status and the environment. Change 32:10–12.
Steiner GH (1978) Fontana/Collins, London. 420 pages.
Steinhardt U (1996) Applying fuzzy set theory for medium and small scale landscape assessment. (in Press).
Straus E (1963) The Primary World of Senses. Fontana/Collins, London 320 pages.
Svobodova A (ed.) (1990) Cultural Aspects of Landscape. Pudoc, Wageningen. 179 pages.
Syrbe R-U (1996) Fuzzy -Bewertungsmethoden für Landschaftsökologie und Landschaftsplanung. Archiv für Natur und Landschaft 34:181–206.
Tolba M (1982) Saving Our Planet- Challenges and Hope. Chapman and Hall, London. 369 pages.
Troll C (1971) Landscape ecology (geo-ecology) and bio-ceonology – a terminology study. Geoforum 8:43–46.
Tuan Yi-Fu (1974) Topophilia. A study of Environmental Perception, Attitudes and Values. Prentice Hall, Englewood Cliffs, New Jersey. 260 pages.
Waddington CH (1975) A catastrophe theory of evolution. The Evolution of an Evolutionist. Cornell University Press, Ithaca, N.Y. pp 253–266.
Zadeh LA (1973) Fuzzy sets and systems. In: Fox J (ed.) System Theory. Microwave Research Institute Symposia Series IV. Polytechnic Press, New York, pp 29–37.
Zonneveld IS (1995) Landscape Ecology. SPB Academic Publishing, Amsterdam. 199 pages.

CHAPTER 11

NAVEH Z. (1998)

ECOLOGICAL AND CULTURAL LANDSCAPE RESTORATION AND THE CULTURAL EVOLUTION TOWARDS A POST-INDUSTRIAL SYMBIOSIS BETWEEN HUMAN SOCIETY AND NATURE. RESTORATION ENERGY 6: 135–143.

Natural regeneration of *Q. ithaburensis* after cessation of uncontrolled goat and cattle grazing in the Menashe Hills, Israel.

Abstract I discuss ecological and cultural restoration within the broader context of the critical transition period from the fossil fuel age to the post-industrial global information age. In this cultural evolutionary process, the restoration of natural and cultural landscapes should play a vital role. For this purpose, it has to be guided by a holistic and transdisciplinary systems approach, aiming not only at the organismic but also at the functional and structural restoration of ecological and cultural diversity as total landscape ecodiversity. For the development of suitable restoration strategies, a clear distinction has to be made between different functional classes of natural and cultural solar-powered biosphere and fossil-powered technosphere landscapes, according to their inputs and throughputs of energy and materials, their organisms, their control by natural or human information, their internal self-organization and their regenerative capacities. Not only technosphere landscapes but also intensive agro-industrial landscapes have lost these capacities and are heavily subsidized by fossil energy and chemicals, to the detriment of the environment and human health. They therefore have to be rehabilitated by more sustainable but not less productive agricultural systems based on organic farming. But their natural regenerative capacities can be restored only by regenerative systems, with the help of cultural "neotechnic" information. The promise for an urgently required evolutionary symbiosis between human society and nature in a sustainable post-industrial total human ecosystem lies in the functional integration of such innovative regenerative systems and all natural and cultural biosphere landscapes with healthier and more livable technosphere landscapes. To this goal, ecological and cultural landscape restoration can make an important contribution

INTRODUCTION

Human society is undergoing a critical period of transition from the industrial fossil-fuel age to the post-industrial global information age. The great challenges of these dynamic sociocultural and environmental changes have been presented lucidly by Erwin Laszlo (1994), the internationally renowned systems philosopher and expert on global trends. He points out that this "grand transition" is a crucial cultural evolutionary period of unprecedented global stress. It is a period of great danger but also of great opportunity, in which global survival will be determined by our choice between "extinction or evolution." The fate of our global ecosphere landscapes is closely coupled with these cultural evolutionary trends, and one of the most obvious symptoms of this danger is the accelerating rates of their biological and cultural impoverishment and their ecological destabilization. These are caused by the increasing pressures on the open landscape, resulting from exponential population growth and uncontrolled agro-industrial and urban industrial expansion. Unfortunately, since the Industrial Revolution human society has acted under the illusion that, with the help of science and technology, the human mind can rule and control nature, as envisioned by Bacon. Now, however, if our choice is to prevent gradual extinction by ensuring further evolution, then the human mind has to become nature's conscious partner and not its master, so that a new symbiotic relationship can evolve between human society and nature.

The conservation and restoration of biological and cultural diversity, the richness of our natural and cultural landscapes, and their ecological integrity and health will be among the most relevant expressions of this symbiotic partnership.

They could therefore play a vital role in this cultural evolutionary process. This demands an evolutionary outlook and the broadening of the scope of restoration from ecological to cultural restoration. In order to face this challenge, restoration ecology will have to progress from an interdisciplinary to a transdisciplinary field. This means the merging of relevant fields of the natural sciences with relevant fields from the humanities and arts in close coordination and cooperation towards the common goal of furthering the post-industrial symbiosis between human society and nature by ensuring the evolution of healthier natural and cultural landscapes.

For this purpose, more comprehensive and integrative restoration strategies, based on holistic landscape-ecological principles and methods, must be applied and must be implemented with the help of innovative and more efficient tools for the guidance of land managers and users and for the persuasion of the public and the political decision makers.

I have already discussed some of these issues in this journal (Naveh 1994a). In this paper my intention is to outline briefly the basic holistic and transdisciplinary premises that have to be taken into consideration in ecological and cultural landscape restoration. I will suggest a functional classification of landscapes on which diverse restoration, reclamation, and rehabilitation strategies could be based.

PREMISES FOR ECOLOGICAL AND CULTURAL LANDSCAPE RESTORATION

As discussed in more detail in a recent symposium on ecological and cultural restoration (Allen and Naveh 1996), the goal of ecological landscape restoration is to return as far as possible to the ecological structure and function of a desired landscape, and the goal of cultural landscape restoration is to restore the historical and cultural values of ancient and traditional landscapes. In cultural landscapes this requires the restoration of biological, ecological, and cultural diversity, together with structural and functional landscape integrity and heterogeneity—as *total landscape ecodiversity*, as I have discussed in more detail previously (Naveh 1994a).

By introducing the notion of culture into landscape restoration, we have to broaden our conceptual and methodological scope from the natural sciences to the humanities, from strictly bioecological issues to much more complex human-ecological issues. For this purpose, exclusively discipline-oriented and mostly reductionistic scientific paradigms must be replaced by transdisciplinary concepts and methods, based on a holistic and hierarchical systems view and its innovative approaches to wholeness and complexity. It means, above all, to acknowledge the interconnected, nonlinear, mostly cybernetic and sometimes even chaotic relations between natural systems and human systems. A most relevant development in this direction is the emergence of economic ecology, a transdisciplinary effort to link natural and social sciences broadly, especially ecology and economics (Costanca 1996). Among other things, this led to the economic evaluation of biodiversity, presented by Pierce and Moran (1994) as part of the biodiversity program of The Conservation World Movement of the World Conservation Union.

In our book on landscape ecology (Naveh and Lieberman 1994), we attempted to outline such a transdisciplinary conceptualization for landscape ecology, linking ecological, geographical, historical, human-ecological, and cultural aspects relevant to land and land-use study, planning, and management. Of special significance for our discussion of the evolutionary and human-ecological perspectives of restoration ecology are transdisciplinary and holistic systems studies by Laszlo (1987, 1994), Jantsch (1975, 1980), and Bohm and Peat (1987).

To apply these innovative systems concepts to landscape restoration, it will not be sufficient to regard landscapes merely as scenery or—as is done by many North American landscape ecologists—as "repeated patterns of ecosystems on km wide stretches" (Forman and Godron 1986). For restoration purposes landscapes must be perceived holistically as three-dimensional pieces of earth, water, and atmosphere with all the living organisms inhabiting them—including humans—together with the complex natural and cultural processes sustaining them. As such they have to be well defined in space and time as concrete, ordered wholes and ecological systems in their own right, serving as the spatial matrix for organisms, populations, and communities, and ecosystems on different multi-dimensional scales.

Terrestrial landscape scales range from the smallest mappable landscape cell, or *ecotope,* to the *ecosphere,* the largest, tangible global landscape system of our *total human ecosystem,* which can be viewed as the highest organizational level of the ecological hierarchy, integrating human society and its total natural and physical environment (Naveh and Lieberman 1994). The Total Human Ecosystem is also the highest co-evolutionary ecological entity on earth, and the ecosphere is the tangible spatial entity in which all evolutionary processes take place.

The term ecotope is used chiefly by European landscape ecologists as the basic unit for landscape study, planning, and management (Leser 1991; Zonneveld 1995) and can be regarded also as the actual "site" of an ecosystem (Haber 1990). It is therefore a much more rigorously defined term than the vague "patch," which is generally used by American ecologists.

As shown in Figure 1, the ecosphere is composed of closely interwoven mosaics of sun-powered natural and cultural *biosphere ecotopes* (or bio-ecotopes); human-made, fossil- and nuclear energy–powered, intermediate *agro-industrial ecotopes*; and urban-industrial *technosphere ecotopes* (or techno-ecotopes). *Bio-agro-* and *techno-ecotopes* are spatially integrated in larger, regional landscape units, but they are not structurally and functionally integrated in the ecosphere. Because of the adverse impacts of the latter and the great human pressures on bio-ecotopes, they are even antagonistically related and therefore cannot function together as a coherent, sustainable ecological system. This should serve as the physical and spatial basis for the urgently required cultural evolutionary process, leading to such a synthesis between human society and nature in our total human ecosystem.

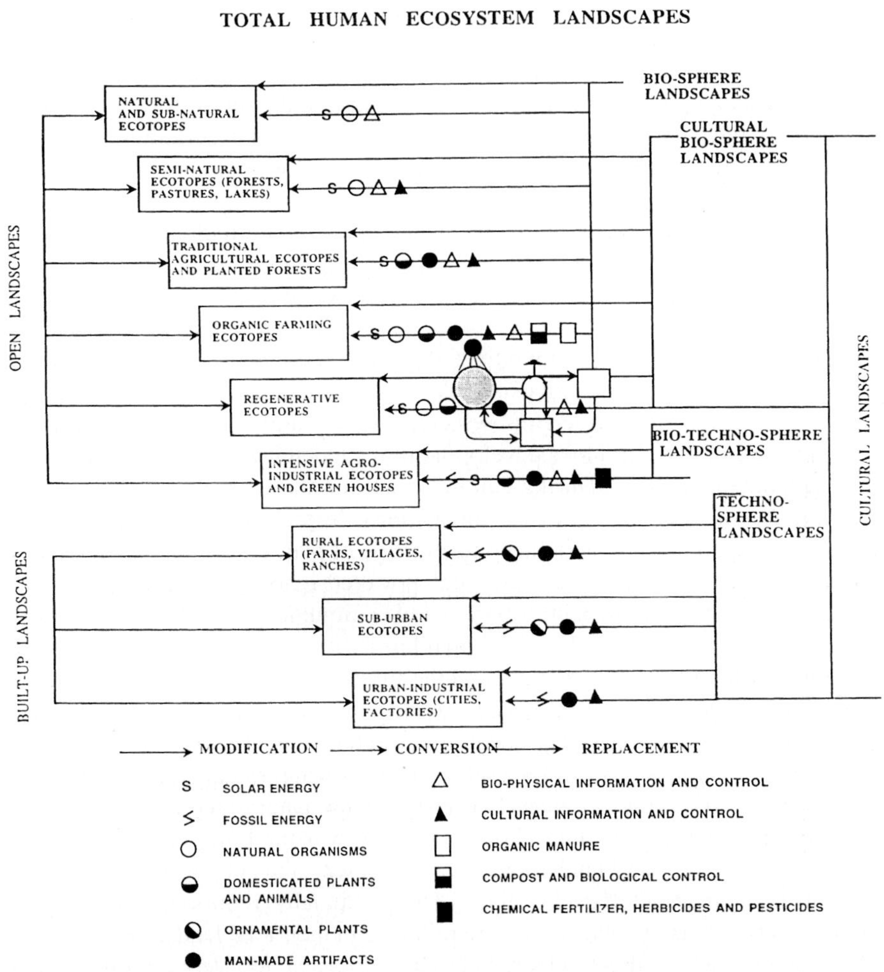

Fig. 1. Functional ordination of landscape ecotopes according to energy, material, and information inputs from the biosphere and technosphere poles. The achievement of a proper balance between these poles should become, with the help of restoration ecology, one of the major goals of landscape ecology.

By restoring, reclaiming, and rehabilitating damaged landscapes; revitalizing wetlands, rivers, lakes and their embankments; creating living corridors and urban biosphere islands in both open and built-up landscapes, restoration ecology, together with other environmental management and improvement measures, is fulfilling an important role in this integration process, and thereby also in this cultural evolution.

THE EVOLUTION OF CULTURAL LANDSCAPES

I have discussed elsewhere in more detail (Naveh 1995) the nature of cultural landscapes and their dynamic interaction with the natural and cultural forces that have shaped them in the past and that are driving them toward their uncertain future. Here I can indicate only some of the most relevant points for ecological and cultural restoration.

In presenting cultural evolution as an integral part of the biological and social evolution within the "grand synthesis" of cosmos and life on earth, Laszlo (1987) viewed culture as inclusive of the characteristic features of all human behavior—not merely as the "high culture" of science, art, and religion but as a basic information pool possessed collectively by all members of society. As such, it contains the sum of rules that code its essential operations and determine to a great extent the stability of society. Throughout history, societies selected this cultural collective information pool in a process of discontinuous development of sudden leaps or "bifurcations" to a higher organizational level. During cultural evolution these bifurcations are driven by the adoption of basic technological innovations. In prehistoric times these were the use of stone tools and fire and the domestication of plants and animals; in the last two centuries they have been the fossil fuel–powered innovations of industrialization. Today, the adoption of new information and communication technologies drives this process towards the post-industrial global level with all the positive and negative implications for human society, nature, and the future of our globe as a whole.

By such bifurcations, human-influenced, modified, and converted open landscapes have been shaped as cultural landscapes through interactions between these cultural evolutionary creative processes of the "self-reflexive mental space of the noosphere" (from the Greek *noos,* "mind"; Jantsch 1980), and the natural and socio-economic environment, producing over time closely interwoven natural and cultural patterns and processes. These cultural landscapes became, thereby, the tangible meeting points between nature and mind (Naveh 1995). These mutually reinforcing, cybernetic interactions between nature and culture are reflected in their total landscape ecodiversity. It is highest where these interactions have operated for long periods over a relatively small and densely settled space, as in the Mediterranean region (Naveh and Kutiel 1990; Naveh 1995, 1997).

Kellert (1996) has made a most valuable contribution to a better comprehension of the role of culture in the appreciation of nature and biological diversity. He generated unique sociological perspectives for the multicultural, common basis of "biophilia," namely, the human affinity for life and the natural world as a whole. In an important lecture on the interaction between biological diversity and cultural diversity, J. A. McNeeley (1995), the chief conservation scientist of the World Conservation Union, has pointed out ways that both cultural and biological diversity can be conserved by approaches adapting to changes, by protecting areas inhabited by indigenous people, and by enabling local people to maintain responsibility for managing their resources, and their natural and cultural assets. For this purpose the application of innovative strategies for ecodiversity restoration and for the rehabilitation of depleted natural and agricultural productivity through organic farming could be of great value. These strategies will be discussed below.

A FUNCTIONAL CLASSIFICATION OF NATURAL AND CULTURAL BIOSPHERE AND TECHNOSPHERE LANDSCAPES AND THEIR MAJOR RESTORATION STRATEGIES

Because of the overwhelming human dominance of the globe, there are now few if any truly natural, undisturbed landscapes left on earth. Therefore, cultural landscapes embrace today a broad global range of open and built-up landscapes with different degrees of modification, conversion, and replacement of their natural elements and processes. In general, the "naturalness" of landscapes is judged by the composition of the flora and fauna, the proportion of spontaneous species, and the diagnostic values of patterns and processes in vegetation and soil in relation to human influences (Van der Maarel 1975). But for the development of suitable strategies for the planning, design, and implementation of ecological and cultural landscape restoration, these distinctions between natural and cultural landscape are not sufficient. They do not take into consideration the basic functional differences between the different sub-classes of cultural biosphere and technosphere landscapes, determined by their differences in energy input—either from solar or from fossil and nuclear sources—by their organisms and natural or converted biotic and abiotic resource materials, and by their regulation by natural or human information, indicated in Figure 1. Another important point of departure that will determine the chances of ecological and evolutionary meaningful restoration is their capacity to organize themselves in a coherent way by maintaining their structural integrity in a process of continuous self-renewal or *autopioesis* (from the Greek, "self-creating"), and thereby also the regenerative capacities owned only by biosphere landscapes. In this respect, in such "hybrid" natural-cultural landscape systems we deal with a mixture of two basic systems of internal self-organizing behavior, defined by Jantsch (1975) as follows: (1) *Adaptive (or organismic) systems* adapt to changes in the environment through changes in their internal structure in accordance with preprogrammed information (genetic templates). This is the domain of biological evolution. (2) *Inventive (or human action) systems* change their structure through internal generation of information (invention) in accordance with their intentions to change the environment. Such information is generated within the system in feedback interactions with the environment. This is the domain of noospheric cultural evolution and is, according to Stebbins (1982), generated by "cultural templates."

One of the major transdisciplinary challenges both for landscape and restoration ecologists will be to accommodate these interacting internal self-organizing systems in their research methods and to define suitable quantitative indices for ecodiversity measurements. As we have pointed out (Naveh and Lieberman 1994), advanced computer models based on knowledge engineering and fuzzy logic (Kosko 1993) may open the way for efficient integration of complex ecological, cultural, and socioeconomic information, qualitative as well as quantitative. These methods could be used also for comparative evaluations of ecodiversity in landscape restoration projects. I have previously offered a further discussion of methods for ecodiversity measurement (Naveh 1994a).

In the hierarchical ordination model of Figure 1, landscape ecotopes have been arranged according to their degrees of modification, conversion, and replacement of natural elements and functions, from natural bio-ecotopes to cultural techno-ecotopes.

(1) Natural and close-to-natural biosphere landscape ecotopes are driven entirely by solar energy and its biological conversion through photosynthesis and assimilation. They contain only natural—meaning spontaneously evolving and reproducing—organisms. As adaptive, self-organizing systems, these landscapes are internally regulated by natural, biological, physical, and chemical information. It is essential to allow the undistorted continuation and/or re-introduction of all those natural ecological processes, ensuring their evolutionary future and thereby also their biological diversity and productivity (Ricklefs et al. 1984). Therefore the aim of restoration strategies should be, wherever necessary, to combine the re-introduction of some of the most important keystone species and those of special ecological, aesthetic, or other cultural value with strictly controlled, dynamic, and continuous multi-factorial restoration and landscape management. This includes the prevention of accumulation of large amounts of highly flammable fuel by controlled burning and the manipulation of the plant–animal complex by the defoliation pressures of browsing, grazing, and cutting (Naveh 1988).

The most recent threats to the integrity of the most precious and attractive nature refuges, in addition to the many other destructive human impacts, is mass tourism, even if it is promoted as "ecotourism." This sad situation, as described by de Palma (1996) for the famous World Heritage Site of the first Canadian park, Banff National Park, occurs all over the world. Here, the closely interwoven assets of bountiful natural beauty and cultural and historical values are "sold off for making money." The multiple threats of heavy tourist pressure, suppression of natural fire, building of roads to accommodate motorized traffic, damming of rivers for a recreational lake, and all other recreational and commercial developments all disrupt the natural ecological and evolutionary processes and the normal cycles of wildlife and are destroying this unique touchstone of Canada's national identity. In this and similar cases, restoration should become an integral part of sound conservation management, helping to prevent and reduce these recreational overdevelopment damages by establishing the most efficient indigenous species and ecotypes for erosion control and as protective vegetation shields and buffers between the protected areas and hotels, parking lots, camping grounds, landing and embankment sites, and roadsides.

All other human-modified and -converted landscapes are controlled to lesser or greater degrees by cultural information. They should be considered therefore as *cultural landscapes*. These have to be subdivided further into three major classes and the following subclasses:

(2) The first major class is that of cultural biosphere landscapes, which, like natural landscapes, are all solar-powered. But they are controlled by different degrees of natural and human information, and they are functioning therefore as mixtures of internal, organismic-adaptive, and human-intervention self-organizing systems. Here we have to distinguish between the following:

(2.1) Human-modified and -used semi-natural ecotopes are those such as forests, woodlands, grasslands, wetlands, and lakes in which biological productivity and diversity are based, as in natural landscapes, on spontaneously reproducing organisms whose productivity is used at least partly for "hard values" as marketable goods for human consumption. As human life-support systems, they fulfill vital food production, regulation, protection and carrier functions. But they also have intrinsic and "soft" spiritual, aesthetic, scientific, and other cultural values. For by far the greatest majority of noncultivated biosphere landscapes, the conservation and restoration of their ecodiversity, which presently is undergoing accelerated biological and cultural impoverishment, is of no lesser importance than that of the biodiversity of natural and close-to-natural bioecotopes. For this purpose, their dynamic, multifunctional conservation and restoration management should include both the natural ecological processes and those ecological and cultural processes that have been introduced by humans throughout their evolutionary history, such as grazing, cutting, prescribed burning, and traditional agricultural practices (Ricklefs et al. 1984; Naveh 1988; Naveh 1991,1994b; Naveh and Lieberman 1994).

In woody vegetation types, the major ecological effect of defoliation pressure is to create favorable light regimes for the herbaceous understory and to prevent the domination of a few shade-tolerant and aggressive native and invasive exotic plants. For this purpose, fire is especially effective. The restoration of natural as well as human-set periodic fire regimes would not only encourage the natural regeneration of fire-adapted species and ecotypes but would also enable the establishment of other desirable reseeded and replanted light-demanding ones, benefiting from the favorable ash seedbeds. This seems to be the case in such different environments as the Mediterranean forests, shrublands, and woodlands in Israel (Naveh 1974, 1990, 1994a); the cold temperatures of Illinois (Lorig 1994); and California (Biswell 1989). Recurring fire may even become the driving force for further speciation and evolution (Anderson 1956). According to Lewis (1982), there is conclusive proof that prescribed fire technologies, well-controlled in time, space, and intensity, were employed by Australian aborigines and North American Indians to control the distribution, diversity, and relative abundance of plant and animal resources. These aboriginal applications of fire apparently differed significantly from natural fires in terms of seasonality, frequency, intensity, and selectivity. In the United States they induced the rich vegetation mosaics along the Californian coast (Lewis 1973) and perpetuated the Midwest prairies, oak forests, and savannas (Curtis 1959; Lorimer 1985). Therefore, wherever appropriate, the controlled simulation of such short- and long-term fire cycles, together with all other natural and cultural landscape-shaping patterns and processes, should become a major aim of restoration strategies. In the Mediterranean, such human-set fires probably opened the dense forests and shrublands as early as prehistoric, biblical, and classical times (Naveh 1974).

In order to implement such strategies, restorationists have to learn as much as possible about the early history of the land to be restored and the fate of its flora and fauna, not only from a purely genetic and ecological point of view, but also from an ethno-historical and anthropological-cultural perspective. They have to become well-versed in fire ecology and, if necessary, learn how safely to apply prescribed burning with different timing and intensities for the different vegetation types and sites as integral parts of their restoration activities.

(2.2) Agro-silvo-pastoral ecotopes include meadows, planted forest groves, and traditionally cultivated fields and orchards. Here, through human agricultural practices, domesticated plants and animals have replaced their natural competitors, and biological production is channeled into economic goods with no or very low inputs of chemical fertilizers and pesticides. Therefore these bio-ecotopes, although human-controlled and maintained, have retained a certain amount of their self-organizing capacities. But for economic, sociocultural, and demographic reasons, the traditional agricultural ecotopes are either abandoned or vanishing rapidly in most industrial countries, together with their biological and cultural diversity. They can be rescued only if local populations and their governments recognize them as valuable cultural heritage landscapes, and therefore are willing to invest sufficient means and manpower to conserve worthwhile examples by restoring and maintaining traditional farming practices and their typical local crop strains and animal breeds, and thereby also their great genetic value. One of the few outstanding examples of an attempt to restore the ecodiversity in such traditional landscapes has been reported by Austad (1996) for the Norwegian wooded hay meadows, including the "pollarding" (cutting leaves and branches for cattle fodder) of their tree layer.

(2.3) Organic farming ecotopes should be regarded as a new, promising subclass of cultural biosphere landscapes. Because of the application of suitable agrotechnological methods, and the build-up of high soil fertility through the use of organic manure and compost, their productivity is much higher than that of traditional farming. It comes now closer and closer to that of the high-input agro-industrial farming systems discussed below, but without the detrimental environmental effects caused by chemical fertilizers, herbicides, and pesticides (Mansvelt and Mulder 1993). The conversion of land cultivated by these modern as well as traditional agricultural practices into organic farming systems can be considered as a desirable way of agricultural landscape rehabilitation (Allen and Naveh 1996), in which the land is being made more productive on a sustainable basis.

(2.4) Regenerative ecotopes are the result of an even more advanced system, not only of agricultural landscape rehabilitation but also of ecological and cultural restoration. Here not only the natural regenerative capacity of cultivated land is restored but also the basic cyclic flows of energy, water, and nutrients of natural biosphere landscapes, based on the inputs of solar radiation. At the same time, their biological production is partly channelled into agricultural production. One of its most outstanding examples is realized at the Center for Regenerative Studies at the California Polytechnical State University at Pomona in a combined inter- and transdisciplinary teaching, research, and demonstration project in regenerative technologies. As described lucidly by its chief designers (Lyle 1994), here, with the help of human cultural "neotechnic" information, the five major bioecosystem functions are restored, (Fig. 2): (1) the conversion of solar energy through photosynthesis; (2) the distribution of energy, water, organisms, and material over the landscape; (3) the filtration of air and water flow by plants and the soil; (4) the assimilation and reassimilation of all material as dead biomass and waste through biological and chemical decomposition as the basic process of revitalizing the earth, which has been severely neglected in modern, intensive agriculture; (5) the storage of energy, water, and material for different periods, ranging from geological times to a few hours and days.

(3) Intensive agro-industrial ecotopes are an intermediate stage between bio- and technosphere landscapes. As in bio-ecotopes, their productivity is dependent on photosynthetic conversion of solar energy but is heavily subsidized by fossil energy. Almost all natural regulation and control mechanisms of bio-ecotopes have been replaced by heavy inputs and throughputs of chemicals, and by agrotechnological information, driven almost solely by a short-sighted, over-consumptive, and socially unjust market economy. They have lost thereby all the internal, self-organizing, and regenerative capacities of adaptive organismic systems. In this respect and in their detrimental effects on the open landscape, its wildlife and biodiversity, the quality of its soil and water, and human health, they come very close to technosphere landscapes. The efforts to minimize these adverse impacts by slow-releasing fertilizers

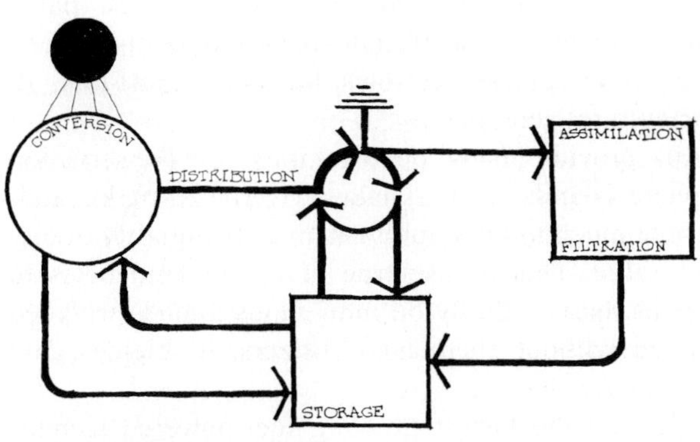

Fig. 2. Major bioecosystem functions of biosphere landscapes. With the help of regenerative technologies such as composting, solar thermal conversion, and photovoltaics, the restoration of these cyclic flows in regenerative landscape replaces the linear throughput flows in agro-industrial and technosphere landscapes and their high outputs of energy, waste, and pollution (After Lyle 1994).

and integrative pest control are only palliative measures that cannot ensure in the long run a sustainable future for our agricultural landscapes and healthy food production. As stated by Lyle (1994), sustainability can be achieved only by replacing these linear technosphere processes of high input and throughput flows with cyclical flows at sources, consumption centers, and sinks, as attempted in regenerative systems.

(4) Technosphere ecotopes and their three subclasses, together with their technological artifacts such as highways, powerlines, quarries, and bridges, which are widely spread over the landscape, are all driven by the technological conversion of high-grade solar energy into low-grade fossil energy, producing great amounts of entropy, pollution, and waste products. As human-made artificial landscapes, they are controlled entirely by human cultural information and completely lack the self-organizing and regenerative capacities of biosphere landscapes. But their subclasses differ in the kind and extent of "biosphere islands" and in their detrimental environmental impacts, as well as in their restoration strategies.

(4.1) Rural ecotopes, such as farms, ranches, and villages, are integral parts of agricultural landscapes and can therefore fulfill an important role in their conservation and restoration. In most industrial countries they are undergoing a rapid process of environmental degradation and urbanization. As Green (1996, pg. 16) has shown in Great Britain and Lucas (1992) elsewhere, this can be prevented by sound countryside planning and management. But future sustainable rural development can be ensured only by the rehabilitation strategies described above with the help of inventive neotechnic information. This can be considered as a most promising model for post-industrial agricultural and rural landscape restoration, and for the creation of diversified and balanced local or regional farming systems, as suggested by Mansvelt and Mulder (1993). At the same time we need to protect, restore, and reclaim the ecodiversity of the remaining noncultivated cultural bio-ecotopes, classified by Agger and Brandt (1988) as "patch and line biotopes" and including natural or artificial lakes and pools, rivers, canals, drainage ditches, hedges, road verges, tree rows, bogs, and marl pits, all surrounded by monotonous agricultural ecotopes.

(4.2) Suburban ecotopes still provide many opportunities for conservation and restoration of larger biosphere islands, such as lakes, riverbeds, parks, and - forest groves. These can contribute much to the improvement of the quality of city life and are therefore of great value. This is also true of recent tendencies to establish low-maintenance gardens, based chiefly on indigenous plants, provided that these gardens are maintained without chemical fertilizers, herbicides, and pesticides.

(4.3) Urban-industrial ecotopes are the fastest growing fuel-powered technosphere landscapes, with the most pronounced adverse environmental impacts. These are indicated by the huge amount of annual energy throughputs, which according to Odum (1993) can be measured in millions of kcal per square meter, compared to only thousands of kcal in solar-powered biosphere landscapes. Here, in general, such restoration options are much more restricted. Therefore, the main efforts have to be devoted to the rehabilitation of derelict urban and industrial complexes and of abandoned mines and quarries, and to the cultural conservation and restoration of the most valuable archeological and historical monuments. That this requires a sound holistic ecological approach has been shown convincingly by the restoration architect Bugod (1996).

CONCLUSIONS

For the future evolution of life on earth and therefore also for the welfare of the post-industrial human society, the ecological and cultural restoration of total land-scape ecodiversity is vital. This requires a holistic, hierarchical systems approach and the broadening of the scope of restoration from the organismic to the evolutionary and functional. For this purpose, restorationists have to recognize landscapes as ecological systems in their own right and not just as ecosystems of kilometer-wide stretches. Only by using the spatial, physical, and biotic context of land- and seascapes as the structural and functional foundation for conservation and restoration can we reconstruct, at least partly, their organismic and functional ecosystems and ensure their further evolution. We have to realize the transdisciplinary nature of cultural landscapes as the tangible meeting point between mind and nature, and to make a clear distinction between biosphere landscapes and technosphere landscapes. In restoration projects we have to overcome the discordance between natural and cultural biosphere landscapes by aiming to restore total landscape ecodiversity through integration of biological and cultural diversity with landscape heterogeneity, health, and integrity. This can be achieved only by preserving and restoring their natural and anthropogenic flow processes, promoting both intrinsic and instrumental values and ensuring their vital life-support functions and further evolution. In cultural technosphere landscapes, ecological restoration, reclamation, and rehabilitation are restricted to only the remaining biosphere islands. This is also true for agro-industrial landscapes. Future sustainable and healthy food production can be ensured only by the rehabilitation of their soil fertility and by the restoration of cyclic flows in neotechnic regenerative systems. These provide for continuous replacement through their own functional processes of the energy, material, and information used in their operation. They realize, thereby, the only possible cybernetic synthesis between the internal self-organizing and self-renewing capacities of adaptive organismic and inventive human active systems in agricultural landscapes.

The promise for the urgently needed symbiosis between human society and nature in a sustainable post-industrial total human ecosystem lies in the functional integration of natural and cultural biosphere landscapes with much healthier and more livable technosphere landscapes. Restoration ecology can play an important role in this process and help thereby in diverting the trajectory of biological and cultural evolution on earth away from creeping extinction and toward further evolution.

REFERENCES

Agger P, Brandt J (1988) Dynamics of small biotopes in Danish agricultural landscapes. Landscape Ecology 3:227–240.

Allen EB, Naveh Z (1996) The limits to ecological and cultural restoration in Mediterranean-type landscapes. In: Steinberger Y (ed.) Preservation of our world in the wake of change. Volume VI B. Proceedings of the Sixth International Conference of the Israeli Society for Ecology and Environmental Quality Sciences (ISEEQS), 30 June–4 July, 1996. ISEEQS, Jerusalem, pp 775–778.

Anderson E (1956) Man as a maker of new plants and new plant communities. In: Thomas WL (ed.) Man's role in changing the face of the earth. University of Chicago Press, Chicago, Illinois, pp 763–777.

Austad I (1996) Effects of abandonment and reintroduction of management on field layer vegetation in wooded hay meadows in western Norway. In: Steinberger Y (ed.) Preservation of our world in the wake of change. Volume VI B. Proceedings of the Sixth International Conference of the Israeli Society for Ecology and Environmental Quality Sciences (ISEEQS), 30 June–4 July, 1996. ISEEQS, Jerusalem, pp 767–770.

Biswell HH (1989) Prescribed burning in California wildlands vegetation management. University of California Press, Berkeley, California.

Bohm D, Peat FD (1987) Science, order, and creativity. Bantam Books, Toronto.

Bugod P (1996) Monuments and their environmental treatment. In: Steinberger Y (ed.) Preservation of our world in the wake of change. Volume VI B. Proceedings of the Sixth International Conference of the Israeli Society for Ecology and Environmental Quality Sciences (ISEEQS), June 30–July 4, 1996. ISEEQS, Jerusalem, pp 764–766.

Costanca R (1996) Ecological economics: reintegrating the study of humans and society. Ecological Applications 6:978–990.

Curtis JT (1959) The vegetation of Wisconsin. University of Wisconsin Press, Madison, Wisconsin.

de Palma A (1996) Popularity brings a huge Canadian park to crisis. The New York Times, Science Supplement, 5 January 1996:3A.

Forman RTT, Godron M (1986) Landscape ecology. Wiley and Sons, New York.

Green B (1992) Countryside conservation: the protection and management of amenity ecosystems. 2nd edition. Unwin Hyman, London.

Green B (1996) Countryside conservation: land ecology, planning and management. 3rd edition. E & FN Spon An Imprint of Thomson & Professional, London.

Haber W (1990) Basic concepts of landscape ecology and their application in land management. In: Kawanabe H, Ohgushi T, Higaschi M (ed.) Ecology for tomorrow. Physiology and Ecology Japan, (Dept. of Zoology, Kyoto University) Vol. 27, special number.

Jantsch E (1975) Design for evolution: self-organization and planning in the life of human systems. George Braziller, New York.

Jantsch E (1980) The self-organizing universe: scientific and human implications of the emerging paradigm of evolution. Pergamon Press, Oxford.

Kellert SR (1996) The value of life: biological diversity and human existence. Island Press, Washington, D.C.

Kosko B (1993) Fuzzy thinking: the new science of fuzzy logic. Hyperion, New York.

Laszlo E (1987) Evolution: the grand synthesis. Shambala, New Library, Boston, Massachusetts.

Laszlo E (1994) The choice: evolution or extinction. A thinking person's guide to global issues. C.P. Putnam & Sons, New York.

Leser H (1991) Landschaftsoekologie. Eugen Ulmer GmbH, Stuttgart.

Lewis HT (1973) Patterns of Indian burning in California: ecology and ethnohistory. Anthropological papers no. 1. Ballena Press, Ramona, California.

Lewis HT (1982) Fire technology and resources management in aboriginal North America and Australia. In: Williams NM, Hunn ES (ed.) Selected symposium 67, American Academic Association of Science. Westview Press, Boulder, Colorado, pp 45–66.

Lorig SM (1994) Turnbull Woods forest preserve: an ecological assessment and restoration management plan 1994. The Chicago Horticultural Society, Glencoe, Illinois.

Lorimer CG (1985) The role of fire in the perpetuation of oak forest. In: Johnson JE (ed.) Challenges in oak management and utilization. Cooperative Extension Service, University of Wisconsin, Madison, Wisconsin, pp 22–34.
Lucas PHC (1992) Protected landscapes: a guide for policymakers and planners. Chapman & Hall, London.
Lyle JT (1994) Regenerative design for sustainable development. John Wiley & Sons, New York.
Mansvelt JD, Mulder JA (1993) European features for sustainable development: a contribution to the dialogue. Landscape and Urban Planning 32:43–54.
McNeeley JA (1995) The interaction between biological diversity and cultural diversity. International Conference on Indigenous Peoples, Environment and Development, Zurich 15–18 May 1995. International Union for the Conservation of Nature, Gland, Switzerland.
Naveh Z (1974) The ecological effects of fire in the Mediterranean region. In: Kozlowski TT, Ahlgren CE (eds) Fire and ecosystems. Academic Press, New York, pp 401–443.
Naveh Z (1988) Multifactorial reconstruction of semi-arid Mediterranean landscapes for multipurpose land uses. In: Allen EB (ed.) Selected symposium 109, American Academic Association of Sciences. Westview Press, Boulder, Colorado, pp 234–256.
Naveh Z (1990) Fire in the Mediterranean: a landscape ecological perspective. In: Goldammer JG, Jenkins MJ (eds) Fire in ecosystem dynamics. ZPB Academic Publishing, The Hague, The Netherlands, pp 1–20.
Naveh Z (1991) Mediterranean uplands as anthropogenic perturbation dependent systems and their dynamic conservation management. In: Ravera O (ed.) Terrestrial and aquatic ecosystems: perturbation and recovery. Ellis Horwood, New York, pp 545–557.
Naveh Z (1994a) From biodiversity to ecodiversity: a landscape ecological approach to conservation and restoration. Restoration Ecology 2:180–189.
Naveh Z (1994b) The role of fire in the conservation of Mediterranean ecosystems and landscapes. In: Moreno JM, Oechel WC (eds) The role of fire in Mediterranean-type ecosystems. Springer-Verlag, Berlin and Heidelberg, pp 163–185.
Naveh Z (1995) Interactions of landscapes and cultures. Landscape and Urban Planning 32:43–54.
Naveh Z (1998) From biodiversity to ecodiversity: holistic conservation of biological and cultural diversity of Mediterranean landscapes. In: Rundel OW, Montenegro G, Jaksic F (eds) Landscape degradation in Mediterranean-type ecosystems. Springer-Verlag, Berlin and Heidelberg. (in press).
Naveh Z, Kutiel P (1990) Changes in vegetation in the Mediterranean basin in response to human habitation and land use. In: Woodwell G (ed.) The earth in transition: patterns and processes of biotic impoverishment. Cambridge University Press, pp 259–300.
Naveh Z, Lieberman AS (1994) Landscape ecology, theory and applications. 2nd edition. Springer Verlag, New York.
Odum EP (1993) Ecology and our endangered life-support systems. 2nd edition. Sinauer Associates, Inc., Sunderland, Massachusetts.
Pierce D, Moran D (1994) The economic value of biodiversity. Earthscan Publications, London.
Ricklefs RR, Naveh Z, Turner RE (1984) Conservation of ecological processes. Paper no. 8, Commission on Ecology. International Union for the Conservation of Nature, Gland, Switzerland.
Stebbins GL (1982) Darwin to DNA: molecules to humanity. W.H. Freeman, San Francisco, California.
Van der Maarel E (1975) Man-made natural ecosystems in environmental planning and management. In: Van Dobbin WH, Loewe-McConnell RH (eds) Unifying concepts in ecology. Dr. W. Junk, The Hague, The Netherlands.
Zonneveld IS (1995) Land ecology. SPB Academic Publishing, Amsterdam.

CHAPTER 12

NAVEH Z. (2001)

TEN MAJOR PREMISES FOR A HOLISTIC CONCEPTION OF MULTIFUNCTIONAL LANDSCAPES.

LANDSCAPE AND URBAN PLANNING 57: 269–284.

Multifunctional agro-pastoral, scenic, and water-rich recreational landscape in the Huleh Valley, Upper Galilee. In the background Mt. Hermon.

Abstract The object of this paper is to present 10 major premises serving as a holistic conception for research on multifunctional landscapes (MFLs). Such a theory should become an integral part of the conceptual foundation of transdisciplinary goal-oriented and mission-driven landscape research. Based on a dynamic systems view, emerging from the recent paradigm shifts and insights from findings on complexity and wholeness, MFLs should be conceived as tangible, mixed natural and cultural interacting systems. They are the concrete, self-transcendent and self-organizing Gestalt systems of our total human ecosystem. Ranging from the smallest mappable ecotope to the global ecosphere landscape, they should be studied, upscaled, managed and evaluated with a biperspectivable systems view. For this purpose MFLs have to be treated simultaneously as products of material, natural biogeophysical systems and as mental, cognitive noospheric systems. This can be achieved with the help of innovative transdisciplinary approaches and research methods, in close cooperation among landscape researchers from natural sciences, social sciences, the humanities and the arts, as well as the professionals involved in all phases of land use decision. By adopting such a transdisciplinary and integrative approach towards the landscape as a whole, landscape ecologists could take part in such joint studies and projects, not only as narrow specialists in their own field of expertise as ecologists or geographers. They could help bridge the gaps between all biological and human ecological aspects, related to land use. Thereby they could play a useful role in ensuring the future of healthy, attractive and stable MFLs as part of the creation of post-industrial symbiotic relations between human society and nature.

Keywords Holistic landscape ecology; Dynamic systems view; Transdisciplinarity; Multifunctional landscapes; Total human ecosystem

1. INTRODUCTION

The object of my paper is to suggest 10 major premises, which should serve as the holistic conception of multifunctional landscapes (MFLs), with a clear vision for the necessity of transdisciplinary goal-oriented landscape research. For this purpose we have to abandon the reductionistic and positivistic assumptions, which are still widely accepted in the natural sciences. Namely that we can achieve complete scientific objectivity and predictability in the transdisciplinary study of MFLs.

The presently occurring rapid changes in most human-influenced landscapes on earth are unpredictable. Therefore, instead of clinging to the classical scientific model of a predictive science, it is essential for landscape research to become an anticipating science, and like medicine, to become also a *prescriptive science.* As pointed out by Holling (1996) and Bright (2000), we have to anticipate environmental surprises and we have to learn to deal also with uncertainties and unpredictability.

We cannot predict the future of our landscapes, but we can help shape their future. We can only attempt to anticipate their fate and the risks involved in their further misuse and degradation and the prospects for their further sustainable development. We can illustrate these anticipations in different scenarios, based on the principle of "if . . . then". We could further help in realizing the most desirable scenarios, both for human society and nature, and prescribe the best remedies for their management, conservation, and restoration. For this purpose a landscape theory cannot be bound down by a rigid, human detached and mechanistic predictive theory, for which classical Newtonian physics has served as a model. It must be guided by a much broader and flexible, future-oriented and holistic view of the world systems and its present deep ecological and cultural crisis.

Humanity has reached a crucial turning point in its relations to nature. The fate of our natural and semi-natural landscapes plays a vital role in these relations. Increasing pressures on their natural resources as well as on their biological and cultural assets, driven by the vicious circle of exponential growth of population and consumption, gravely endanger our sustainable future.

These processes are an integral part of the dramatic transition from the industrial to the post-industrial global information society and its globalized economy. Laszlo (2000) has called these encompassing changes in all spheres of life and systems levels a "macroshift". In these cultural evolutionary trends, our only choice is between further evolution of sustainable life on earth, or its further exponential degradation until its final extinction (Laszlo, 1994).

Transdisciplinary research and education implies an overarching scientific and practical approach, transcending and crossing disciplines and professions, creating a common conceptual base and aiming together towards a common systems goal (Jantsch, 1970). Such a joint, transdisciplinary goal is essential in this context (cf. Naveh, 1995a; Naveh and Lieberman, 1994; Grossmann and Naveh, 2000). It should lead to the creation of new, mutually beneficial symbiotic relations between nature and human society, as realized in healthy, productive, attractive and livable MFLs. Research and action towards this goal have to be based on a sound holistic conception of landscapes and their role in this process (Li, 2000a; Naveh, 2000a).

2. THE MEANING OF TRANSDISCIPLINARITY FOR LANDSCAPE RESEARCH

Such a transdisciplinary challenge does not mean that landscape researchers have to neglect their own unique disciplinary expertise of dealing with the land as a whole. Rather they will have to share it with others, such as economists, anthropologists, environmental psychologists, and sociologists. This requires overcoming the "disease of specialized deafness" for anything outside our own competence. As noted by Allesch (1990, p. 171): "we need specialists that know how special their knowledge is. We need the biologist and the anthropologist, but the biologist who knows that man's nature is more than biology, and the anthropologist who knows that man is part of the ecological circuits".

By opening such a broader integrative "window" for our landscape perception, transcending that of our own discipline, we will not only benefit from knowledge outside our own expertise. We will be able to act both as experts and help to integrate our knowledge with that of the other participants, bridging the gaps between bio-ecology and human ecology (Naveh, 1990). This is essential for achieving our common goal.

From a recent IALE Mission (1998), one can get the impression that the major core theme of landscape ecology is nothing more than studying spatial variations and landscape heterogeneity at different scales on a vaguely defined interdisciplinary level. Instead of retreating to the isolation of a disciplinary academic ivory tower, landscape ecologists have to participate actively at the implementation level. Levin (1999) recently presented such participation as an "integrative science of ecological management". They must become "committed actors", and not only "critical but marginal spectators in this game" (Di Castri, 1997). The Dutch Association for Landscape Ecology (WLO, 1998) expressed forcefully this need to face the challenges of safeguarding and creating sustainable healthy, productive and attractive landscapes for the next millennium. This should be achieved with the help of a broader holistic, conception, applied to transdisciplinary landscape research.

In her final statement at the first IALE conference on cultural aspects of landscape, Nassauer (1990, p. 173) demanded that " . . . we must be courageous in innovating around the conventions of our own disciplines. We must dare to borrow from what is useful in the approaches and knowledge of our colleagues in the arts, social sciences and physical and biological sciences. We cannot afford to be sidetracked into critiques of old, traditional paradigms. Rather, we should move on to invent what works now".

This requires also an appropriate transdisciplinary educational program, to be enriched by a broad evolutionary literacy (Laszlo, 1994) and according to Orr (1992), above all by a new, "life-centered post-modern education" (see also Naveh, 1990, 2000a).

3. TEN MAJOR PREMISES FOR A HOLISTIC CONCEPTION OF MULTIFUNCTIONAL LANDSCAPES (MFLs)

A holistic theory of landscapes cannot be considered in isolation. It has to be based on a hierarchical systems view of the world, rooted in general systems theory (GST) and in its recent holistic and transdisciplinary insights in organized complexity, self-organization and co-evolution in nature and in human society. Presented recently in more details (Naveh, 2000b), these issues will be included in the 10 major premises serving as the core theory for such a holistic theory of MFLs.

3.1. First Premise: MFLs are Part of the Dynamic Synthetic Evolution of Self-Organizing Non-Equilibrium Dissipative Structures

The true meaning of contemporary holistic landscape conception can be fully comprehended only in the broader context of the recent holistic and transdisciplinary "scientific revolution", senso Kuhn (1996). Such a revolution was initiated by a major paradigm shift from parts to wholes and from entirely reductionistic and mechanistic approaches to more holistic and organismic ones. It shifted from breaking down, analyzing and fragmenting wholes into smaller and smaller particles towards new trends for integration, synthesis, and complementarity. It meant the need to replace the reliance on exclusively linear and deterministic processes by non-linear, cybernetic and chaotic processes, based on systems thinking of complexity, networks and hierarchic order. It leads from a belief in the objectivity and certainty of the scientific truth towards the recognition of the limits of human knowledge, the need for a contextual view of reality, and the need for dealing with uncertainties. Presently, this scientific revolution leads beyond multidisciplinarity and interdisciplinarity to transdisciplinarity.

According to Holling et al. (1999) such a holistic paradigm shift is changing already the science and practice of adaptive resources management. It recognizes the limits of our ecological knowledge and the validity of human wisdom and traditional commonsense and its deep cultural values. It has to change also our view of landscapes from a multidisciplinary and interdisciplinary picture of physical, chemical, biological, and other landscape elements and processes into a transdisciplinary view of the landscape and its multifunctional natural and cultural dimensions.

Probably the greatest achievement of the transdisciplinary scientific revolution has been inspired by a great number of new research findings and especially those by Prigogine and his collaborators on the self-organizing properties of non-equilibrium "dissipative structures". These new ordering principles of self-organization occur in non-equilibrium systems with the help of "dissipative structures". These dissipate entropy as part of the continuous energy exchange with their environment. They create "order out of fluctuations" (Prigogine, 1976), and "order out of chaos" (Prigogine and Stengers, 1984) by increasing negentropy within the system. Negentropy is the opposite of entropy and disorder. It is characterized by the increase of effective information and energy efficiency, greater flexibility and creativity, higher structural complexity at each higher organizational systems level.

These should be also the properties of sustainable societies, their economy and landscapes in the emerging post-industrial information society. It constitutes a major transdisciplinary paradigm shift from the neo-Darwinian conception of evolution to an all-embracing conception of synthetic cosmic, geological, biological and cultural co-evolution. This new holistic evolutionary conception has been described in comprehensive ways first by Jantsch (1980) as the "self-organizing universe" and by Laszlo (1987) as the "grand synthesis". Emphasizing cooperation as the creative play of an entire evolving universe, this paradigm is opposed to the Newtonian paradigm of an atomistic world, operating by mechanistic laws of the clockwork-like universe and its more modern view as a bio-chemical and physical machine.

The synthetic evolutionary process should be conceived, as a discontinuous development of sudden leaps by "bifurcations" (from the Greek *furca* means fork) to a higher organizational level. As shown by Laszlo (1994) in Fig. 1, in the cultural evolution these were leaps from the primitive food gathering—hunting to the more and more advanced agricultural and industrial stages. They culminated in societies globally integrated in the emerging information age. Each of these bifurcations is driven mainly by the widespread adoption of basic cultural and technological innovation, such as symbolized presently by the computer. Landscape evolution is an integral part of this cultural evolutionary process.

Systems on a relatively high organization level that can continuously renew themselves and regulate this process in such a way that the integrity of their structure is maintained, are called *autopoietic systems* (from the Greek *autopoiesis*

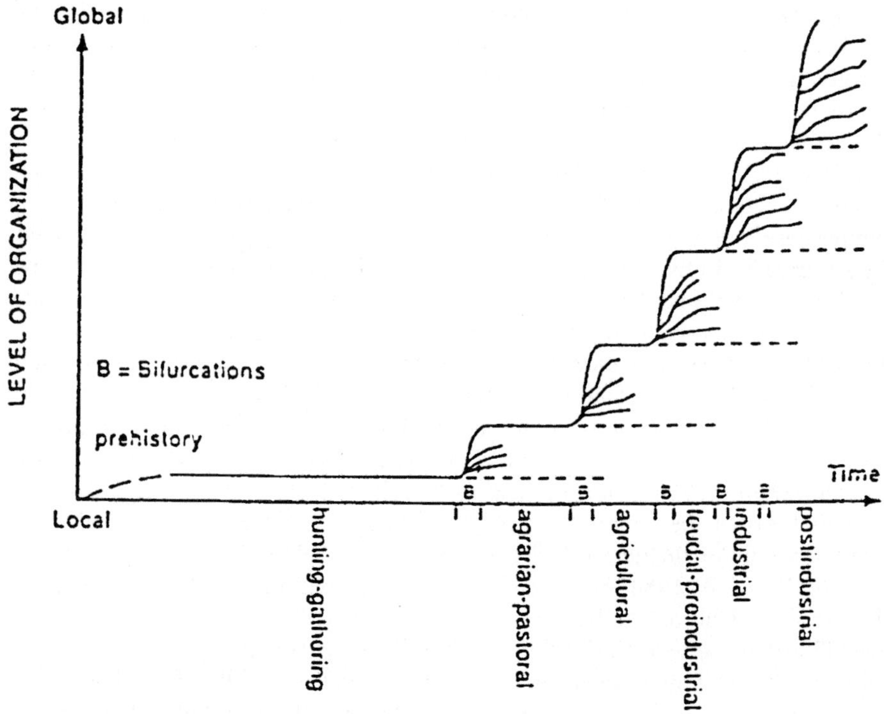

Fig. 1. The convergence of human societies to higher organization levels during cultural evolution by bifurcations (after Laszlo, 1994).

means self-production or self-renewal). They achieve this by *self-organization* as cross-catalytic networks (CNN) of interrelated component-producing processes in a flow of matter and energy and information.

Throughout the span of recorded history, human societies have converged to progressively higher organizational levels. The process began with the hunting-gathering tribes of the Stone Age and currently culminates in the coming of societies globally integrated in the emerging information age. Each bifurcation, driven mainly by the widespread adoption of basic technological innovations, has impelled societies toward more complex, more embracing levels of organization. Today, the widespread adoption of the new information and communication technologies drives the process to the global level.

The integration of the structure-oriented model of self-organization of dissipative structures, rooted in non-linear thermodynamics, and the organization-oriented model of catalytic networks of autopoiesis, has culminated in a coherent theory of living systems, ecosystems and social systems (Jantsch, 1980; Laszlo, 1987; Capra, 1996). It can be adopted also for solar-powered biosphere landscapes, such as natural and semi-natural forests, shrublands, grasslands, wetlands, river-, lake-, and seascapes (Naveh and Lieberman, 1994; Naveh, 2001 also published as Chapter 13 in this anthology). This has far-reaching implications for MFLs, which will be discussed further in Section 3.10.

3.2. Second Premise: MFLs are More Than the Sum of Their Parts, They are Unique Gestalt Systems

This holistic systems paradigm is rooted in GST and should serve also as the major theoretical foundation for a MFL theory. It requires that we regard each landscape on its own right as an open, concrete, space–time defined ecological system. Landscapes serve as the spatial matrix and living space for all organisms, including humans, their populations, and ecosystems. Because of their emergent organizational system properties, landscapes are more than the sum of their measurable components. They become an entirely new entity as an ordered whole of a "Gestalt" system. In these like in an organism (or a melody) all their parts are related to each other by the general state of the whole. Not only the natural but also the cultural components of a regional landscape, its forests, grass-and shrublands, its wetlands and rivers, its agricultural fields, its residential and industrial areas, its roads, traffic- and power-lines, and their history contribute to this truly holistic Gestalt character of the landscape. These elements comprise its various biological- and human-ecological, social, economic, psychological, spiritual, aesthetic and functional aspects. We experience and use these in the landscapes as concrete wholes. Their complex network interactions and their implications for resolving the pressing problems of our present ecological and cultural crisis cannot be comprehended merely by analysis, but only by synthesis within the context of the organization of the whole.

Weiss (1969) demonstrated these essential holistic features as expressed by the invariance of the system, in comparison with the more variant fluctuations of its constituents by a simple mathematical formula. He pointed out that this is the result of systems behavior under the internal degrees of freedom of its components by coordination and control. It is realized by the capacity of cybernetic adaptive self-regulation through negative feedback loops. Thereby the system—and in our case the landscape—becomes more than its parts, not in a quantitative-summative way but in a qualitative-structural way. Li (2000a) has recently illustrated this holistic landscape paradigm in a more formal way with the help of mathematical set theory and non-equilibrium thermodynamics.

3.3. Third Premise: MFLs are Part of the Hierarchical Organization of Nature and of the Global Ecological Holon Hierarchy (or "Holarchy")

Inspired by GST, hierarchy theory has become an important part of the systems approach, and as Simon (1962, 1973), Jantsch (1970) and Weinberg (1975) have shown also a cornerstone for transdisciplinarity. The basic GST paradigm is the view of a hierarchical organization of nature as ordered wholes of multileveled, stratified open natural systems. These range from the lowest, physical levels of space–time fields and quarks up to the astronomic entities of planets, stars, galaxies and their clusters. In this macro-hierarchy of the cosmos, the micro-hierarchy of our planet Earth constitutes the biological levels of organismic complexity and the ecological levels of above-organisms complexities, integrating living systems with their environment (Laszlo, 1972).

In the hierarchical organization of any natural system, each higher level contains the lower one and acquires thereby newly emerging qualities, and is more complex than its lower subsystems. It organizes the level below it and serves as the context of the lower level. At the same time, its lower subsystem gives the function of each system and the purpose is given by its supersystem.

For the representation of a holistic and dynamic systems view of the real world we have to take into consideration its hierarchical structure and to adapt our means of measurement and evaluation to each level. Each higher level displays lower frequency behavior. Classical ecology and ecophysiology have dealt mainly with the fast response loops between the atmosphere and the surface vegetation. Until recently it has been assumed that at the landscape scale we have to deal chiefly with the slower and weaker loops of landscape modifications through anthropogenic activities and their effects on biogeochemical cycling and climate change. But because of the exponential rates of these modifications, and "the acceleration of history" (Brown, 1996), their time scales become faster and faster and their loops become more and more strongly coupled. Therefore the prevention of their adverse effects has become also more and more urgent. This has to dictate our priorities for landscape research and has far-reaching implications on the ways in which the usable "pragmatic" information—senso Naveh and Lieberman (1994)—should be brought to the attention of the stakeholders and decision-makers, to induce their immediate response.

An important development for the recognition of the dichotomic Janus-faced nature of each hierarchy level being at the same time both whole and part, was the introduction of the "holon" concept by Koestler (1969). Holon is a composition of two Greek terms: *holos* means whole and *proton* means part. On each intermediary hierarchic level such holons function as self-contained wholes toward their subordinated subsystems, but at the same time they act also as dependent parts toward their super-systems. In other words, depending on our point of view each holon in the systems hierarchy behaves either as a part or as a whole.

In the universal holon macro-hierarchical organization of natural, multileveled and stratified open systems, MFLs should be considered as part of the global ecological micro-holarchy. Serving as the tangible matrix for all organisms, they form their own holon hierarchy or "holarchy", with *ecotopes* as the smallest structural and functional holons and the *ecosphere* as the largest, global holon (Naveh and Lieberman, 1994).

Accordingly, MFLs should be studied and managed in a complementary way. They should be considered as being at the same time parts of a higher super-level of the space-time and perceptional hierarchy, and as wholes toward their lower sub-level of this landscape holarchy. Thereby we are overcoming the contradiction between entirely one-sided holistic or reductionist perceptions of landscapes. However, we have to realize that their organizational complexity cannot be treated as a rigid one-dimensional spatially structured physical and biological hierarchy. They should be treated only as dynamic, multidimensional space–time, conceptual and perceptual holarchies, from the largest and most complex global ecosphere landscape to the smallest landscape cell or "ecotope". Their upscaling from the lower to the higher holarchy levels is a special challenge for landscape ecologists.

The ecotope can be considered as the smallest, more or less homogeneous and clearly discernible and mappable building block of nature, with all its sub-ordinated landscape elements and fluxes. They have been defined and delineated as concrete systems in much more concise ways than landscape "patches", and are mapped in general on scales of 1:10,000–25,000 (Leser, 1991; Zonneveld, 1995). Their boundaries are determined in a pragmatic way, according to the object and the needs of the study. In this respect such landscape cells differ also from ecosystems. The latter are functional and therefore diffuse in space and more or less intangible (Allen and Hoekstra, 1992; Naveh and Lieberman, 1994).

3.4. Fourth Premise: MFLs are Complex Nature—Culture Interaction Systems

Landscapes, together with living systems and eco-systems are belonging to a special class of "ecological interaction systems" whose elements are coupled with each other by mutual, mostly non-linear and cybernetic relations. Such "middle-numbered" systems, senso Weinberg (1975), are characterized by intermediate numbers of diverse natural biotic and abiotic and (anthropogenic) cultural components with greatly varying dimensions and structural and functional relationships among these components. For the organized complexity neither mechanical nor statistical approaches are satisfactory and innovative approaches and methods are required. This is especially the case with highly fragmented and heterogeneous human modified MFLs, in which natural and cultural patterns and processes are closely interwoven. Whereas their natural elements have evolved and are operating as parts of the geosphere and biosphere, their cultural landscape elements are the creations of the *noosphere* (from the Greek *noos* means mind). This is the sphere of our human mind and consciousness. It should be regarded as an additional natural envelope of life in its totality that *homo sapiens* has acquired throughout the evolution of the human cortex. As the domain of our perceptions, knowledge, feeling, and consciousness, it enabled the development of additional noo-spheric realms of the info-socio- and psycho-spheres that have emerged during our cultural evolution (Jantsch, 1976).

In the following premises, the unique features of middle-numbered autopoietic MFLs will be further elaborated. These will also explain further the major differences between ecosystems and landscapes as middle-numbered systems and the need for more far-reaching innovative methods for their study.

3.5. Fifth Premise: MFLs are the Concrete Gestalt Systems of Our Total Human Ecosystem

Landscapes are to be viewed not only within the ecological/functional and/or geographical spatial context. They have to be treated within a much broader context of the integrated human-nature systems complex, as the larger ecological entity in which we live. I suggested naming this entity the total human ecosystem (THE). It integrates humans and their total environment at the highest co-evolutionary level of the global ecological holarchy. Landscapes are the concrete, space–time defined ordered wholes and unique Gestalt systems of our THE, within different scales of the above described landscape holarchy (Naveh, 1982; Naveh and Lieberman, 1994).

The conventional ecological conceptions consider natural ecosystems as the highest organization level of this ecological hierarchy, above organisms, populations and communities. This is to view humans merely as external factors to natural ecosystems. As such they should create their own social and economic hierarchies outside those of the natural ecological hierarchy.

Human-modified landscapes make by far the majority of the total open landscape area at global scales (Pimentel, 1992). Even the few remaining natural and close to natural landscapes and their terrestrial and aquatic networks are affected directly and indirectly by humans and are shrinking rapidly. Their fate—like that of all other land- and seascapes on earth—depends for good or worse almost solely on the decisions and actions of human society (Vitousek et al., 1997).

If we disregard the close links between natural and social systems, we will not be able to divert the global change trajectory from extinction into future sustainable biological and cultural evolution. The THE concept intends to overcome this disregard at the conceptual level. It perceives humans and their ecological, cultural, social, political and economic dimensions as an integral part of this highest co-evolutionary geo-bio-anthropo level of the ecological hierarchy *above* the ecosystem level. Recognizing the deeply embedded evolutionary connectedness between humans and the rest of the natural world, this holistic THE concept could serve also as the basic cornerstone for an overarching holistic metatheory for landscape science in the broadest sense.

This co-evolutionary connectedness in our THE is closely related to self-transcendent openness (Naveh and Lieberman, 1994). Self-transcendence means reaching beyond the boundaries of one's own existence. According to Jantsch (1980), and his self-organization paradigm, evolution is the result of self-transcendence at all levels.

For a better comprehension of the unique self-transcendent openness of MFLs, we can adopt the example of Frankl (1969), the founder of the psychotherapeutic school of logotherapy. To demonstrate the uniqueness of the human Gestalt system and its intrinsic and self-transcendent openness, he used the projection of three-dimensional models into lower dimensional ones. This is contrary to the still prevailing reductionist tendencies through which human phenomena are reduced to "nothing but" chemical or psychological reactions. Once we have projected humans out of their own dimension into the lower dimensions of biology or psychology this multidimensional self-transcendent openness necessarily disappears.

Thus, as shown in Fig. 2, if we project a drinking cup or an open cylinder out of their three-dimensional space into the two-dimensional plane of the outline of their layout or the side view of their profile, we receive only a circle or a rectangle.

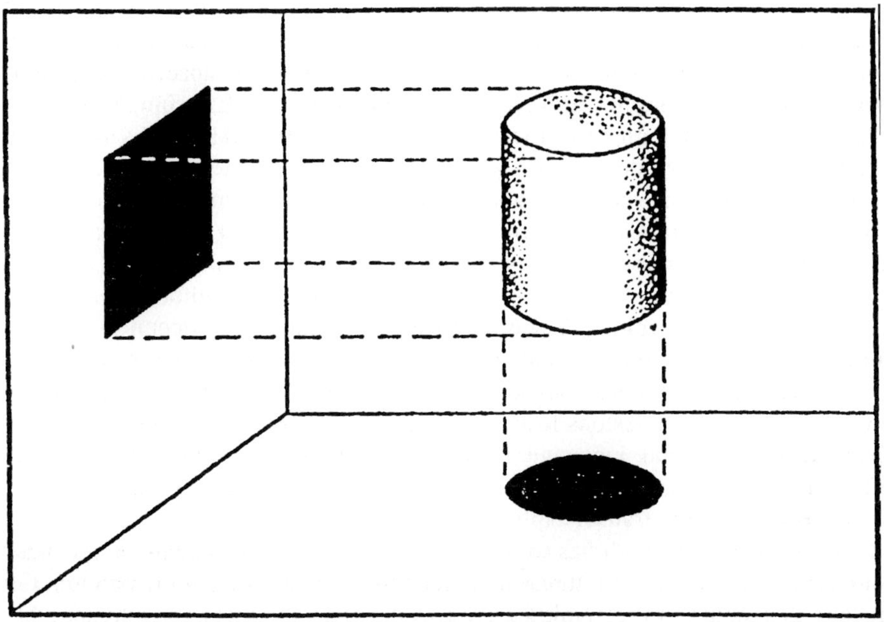

Fig. 2. Projecting of the three-dimensional space of a drinking glass or of an open cylinder into the two-dimensional planes of the layout of its profile (after Frankl, 1969).

In addition, neither of these lower dimensional projections reveals the fact that the drinking cup is an open container and not a closed figure.

The same is happening if we project MFLs out of their unique multidimensional Gestalt wholeness into "nothing but" their lower geophysical, biological, or socio-economical, or mental or artistic dimension. It can be compared to drawing these landscapes only with a pencil, and thereby loosing the unique qualities of the interplay of their colors. These are an intrinsic part of their Gestalt nature, like the rhythm of a melody.

MFLs have not only "formal" openness (Pankow, 1976) to energy/matter and information flow of ecosystems. They have not only the formal structural configurations of the geometrical landscape configurations of patches, corridors, mosaics, and the spatial organization of landscapes. All these features can be measured only with the help of formal languages, such as mathematical equations, models or graphical symbols and maps. Such formal languages cannot represent themselves, but only other objects. To these they are related only by analogy. However, MFLs have also self-transcendent openness. This means they have the capability to represent themselves or to be described adequately by homology with the help of another natural, self-organizing Gestalt system. This is our natural language. It is the organ of consciousness and therefore serves us as our major cultural exchange of personal experience and information. It must therefore be used for a fuller landscape description (Naveh and Lieberman, 1994).

This does not mean that these lower dimensions should not be studied and treated in scientifically sound ways as unidimensional realities, as long as they are recognized as such. The systems view of landscapes can be compared to a perceptional and scientific window through which we are able to look at complex ecological and societal phenomena in their realistic way within the observed context. Such a "contextual window view" is of greatest relevance for transdisciplinary landscape research. Viewed through such a conceptual multidimensional window, these lower dimensions have to be re-interpreted, re-evaluated and represented as parts of the higher whole landscape-ecological Gestalt multidimension. It does also not belittle the great advances achieved in the field of quantitative spatial and functional landscape ecology, based on such formal landscape openness. It only illuminates their perceptional and epistemological limitations for gaining full comprehension of our THE landscapes. If we are aware of this fact, we will never draw inappropriate conclusions from such analytical tools which study landscapes only with these formal language methods, to the same extent that the two-dimensional circle or rectangle are only projections of the open, three-dimensional cylinder. But, as a truly transdisciplinary science, bridging bio-ecology and human ecology, landscape research has to deal also with this self-transcendent openness of our THE landscapes, both in research and education. This is, in my opinion, the only way to account in a comprehensive way for their uniqueness as the tangible meeting point between nature and mind. Some of these unique THE Gestalt features will be elaborated further in the next premises.

3.6. Sixth Premise: In MFLs, a Transdisciplinary Parameter Could Measure Biological Diversity Together with Cultural Diversity and Ecological Heterogeneity, as a Common Index of "Total Landscape Ecodiversity"

As a consequence of the above-described holistic features of MFLs, biodiversity has to be complemented by new, more inclusive and broader transdisciplinary parameters. These should measure biological and cultural diversity together with ecological heterogeneity as "total landscape ecodiversity", as explained further by Naveh (1991b, 1995b, 1998 also published in Chapter 10 in this anthology).

Total landscape ecodiversity can be used as the tangible expression of the dynamic interplay between the biological, ecological and cultural landscape dimensions and their effects on landscape functions. These could be determined by biological diversity, ecological micro- and macro-site heterogeneity and human land uses and their cultural artifacts. These are either enriching or impoverishing both biodiversity and overall landscape heterogeneity at the different scales measured. As shown in Fig. 3, they are coupled by positive, mutually amplifying and self-augmenting feedbacks. The greater the ecological landscape heterogeneity, the greater the chance for biological diversity while at the same time floristic, faunistic and structural vegetation diversity enhances ecological heterogeneity. The opposite is also true. Therefore landscape homogenization, fragmentation, and despoliation diminish biodiversity that in turn reduces further landscape heterogeneity.

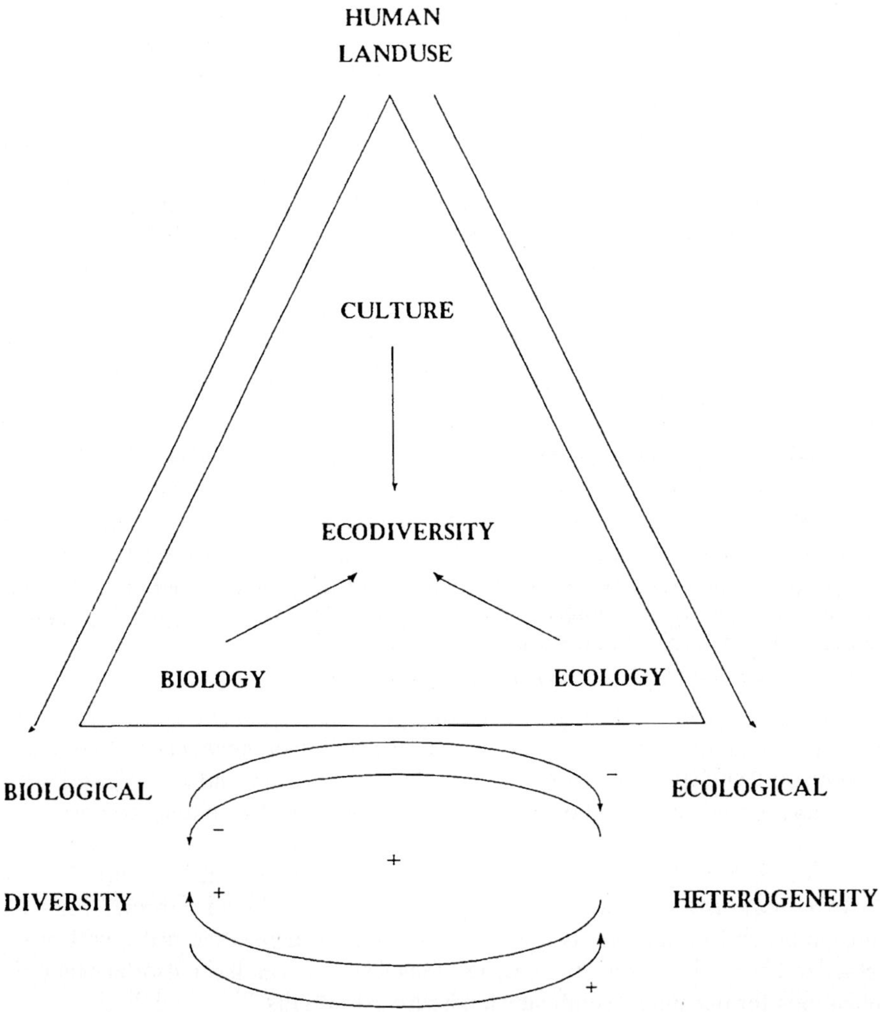

Fig. 3. Total landscape ecodiversity as the result of the interactions between biological, ecological and cultural landscape shaping forces.

3.7. Seventh Premise: Gaining Further Holistic Insights of MFLs Beyond the Archimedian and Cartesian Orders

Bohm (1980), Bohm and Peat (1997) have opened new vistas for holistic and transdisciplinary scientific paradigms. These are also of greatest relevance for the holistic nature of MFLs as autopioetic middle-numbered systems and as unique self-transcendent Gestalt systems. Bohm (1980) has used the advances of the lensless holograph photography for the development of these exciting new holistic ideas and theories, known also as the "Bohm hologram paradigm". In a holograph the light from each part of the object falls onto the entire photographic plate. Therefore, each part of the plate contains information about its interrelated patterns. It is relevant for the perception of the whole by reflecting the whole that, in a sense, becomes enfolded across the holograph. This has served Bohm (1980) as a powerful analogy for a new metatheory of a dynamic holistic whole and undivided order of the universe. To describe the deeper reality, he proposed a "new notion of order" which he named "implicate order" or enfolded order. It is still hidden and lies beneath the regular "explicate order". In this order the fundamental equations are written, using the coordinates of space and time. It gives rise to it in a universal "holomovement". For Bohm, what happens on the plate is simply a momentary frozen version of what is occurring on infinitely vaster scales in each landscape on earth and in each space of the universe. In this "everything is enfolded into everything".

Bohm and Peat (1987) have carried this holistic paradigm even further. They claimed that no one order would fully cover the human experience. As contexts change, orders must be constantly created and modified. This is true also for the Cartesian grid of coordinates, which has dominated the basic order of physical and geographical landscape reality for the last 300 years. They questioned its general appropriateness and arrived at notions of different degrees of order. The flowing river gives a good image of how a simple order of low degree can gradually change to chaotic order of high degree, and eventually to random order. But they show that between the two extremes of simple regular order and chaos there is a rich new field of creativity, as a state of high energy making possible a fresh perception, generally through the mind. Full creativity requires also free play in communication in science.

Bohm and Peat (1987) recognized implicate order as a special case of *generative order*. This order is fundamentally relevant in nature, as well as in human consciousness and in our creative perception and understanding of nature. It is therefore also very important for MFLs. With its help we may be able to reach an entirely new view of consciousness as a generative and implicate order that throws light on nature, mind and society. If such an overall common generative order will bring together science, nature, society and consciousness, this can have also far-reaching implications for our transdisciplinary landscape paradigms.

In a recent comprehensive biography of Bohm and his work, his close collaborator Peat (1997, p. 263) stated that "the implicate order is a door into new ways of thinking and the eventual discovery of new and more appropriate mathematical orders. It is both a philosophical attitude and a method of inquiry".

For landscape research this means that further and deeper insights into the holistic nature of landscapes can be gained if we are ready to free our minds of rigid commitments to familiar notions of order. Only then, we may be able to perceive new hidden orders behind the simple regularity and randomness: "it is possible for categories to become so fixed a part of the intellect that the mind finally becomes engaged in playing false to support them" but clearly, "as context changes so do categories" (Bohm and Peat, 1987, p. 115). Such a change in context occurs when landscapes will no longer be treated as "nothing but" formal, spatial geometric structures and mosaics, describable by Archimedian geometry, and by the Cartesian grid of coordinates. Instead they will be conceived as unique self-transcendent autopoietic THE Gestalt systems. These are imbedded in a hierarchy of subtle, generative, implicate orders, in which human mind, consciousness and creativity play an important role.

The first important steps in this direction beyond the regular Archimedian order are *fractal dimensions,* which occupy already an important role in many landscape studies. Their multiple functions for landscapes have been described recently and very comprehensively by Li (2000b). However, it should be realized that the order of fractals is related to a local order of space, whereas in the implicate and generative order the process of development should be related to the THE landscape as a whole.

For landscape research it will become a major challenge to capture these implicate and generative orders. One possibility could be the combination of the holistic Gestalt interpretation of aerial photography, presented recently by Antrop and Van Eetvelde (2000) as a visual image interpretation of suburban landscapes, with holograph photography. Further new orders may also hopefully emerge through the collaboration of landscape researchers with others in relevant scientific and professional fields. They are essential for the development of practical tools to provide a more integrated appreciation of the aesthetic, ethical and intrinsic functions of nature and their values in the decision making process, beyond narrow economic cost/benefit calculations.

3.8. Eighth Premise: The Dualistic Perception of MFLs can be Overcome by Treating them Simultaneously as Biperspectivable Natural and Cognitive Systems

A major challenge for our transdisciplinary vision is to overcome the great epistemological barriers erected between scientists from the natural and humanistic scientific "cultures" by their contrasting perceptions of landscapes as either entirely physical or entirely mental phenomena. This dichotomy originated from the deeply ingrained Cartesian dualistic view of nature and mind. It has resulted in the positivist and reductionist dualistic interpretations, by which the mental phenomena "do not count", because they cannot be counted, measured and quantified by conventional mathematical and/or biophysical means. However, as shown above, there are many landscape features and functions that cannot be counted by these measures, but they "count" very much in reality. On the other hand, not all those landscape elements which can be counted are worth counting because they do not count at all in reality.

Laszlo (1972) developed an alternative to this dualistic view with the help of a biperspectivable systems view of two major classes of natural systems and cognitive systems. He defined natural systems as "a random accumulation of matter/energy, in a region of physical space–time, organized into co-acting interrelated subsystems or components", and cognitive systems as, "systems constituted by mind events, including perceptions, sensations, volition, feelings, dispositions, thoughts, memories and imaginations, i.e. anything present in the mind".

Both natural and cognitive systems are single, self-consistent mind-events of cognitive systems and natural, physical space–time events of concrete systems. But they are internally and externally observable *simultaneously* as integrated *natural cognitive* and *psychophysical systems.*

As thinking human creatures we are living not only in the physical, and geographical space of these concrete natural systems of the geosphere and the biosphere, which we share with all other organisms. We also live in the conceptual space of the cognitive systems of the human mind, in the noosphere. The MFLs of our THE are the products of both of these internally and externally viewable natural and cognitive systems and their interactions. Therefore they can serve as the "tangible bridge between nature and mind" (Naveh, 1995b). As such they can be perceived by our contextual window view, as well as studied, managed and evaluated simultaneously with such a biperspectivable systems view. Only such a complementary approach allows us to grasp fully the self-transcendent notion of these MFLs. That such a biperspectivable view is essential for any meaningful transdisciplinary research and its practical implementation is shown also in the next premise.

3.9. Ninth Premise: "Hard" and "Soft" Values of MFLs Should be Appraised Jointly by Transdisciplinary Teams

Living systems are driven solely by chemical, physical and biological processes. Thereby they maintain their multifunctionality of life. In ecosystems, the relations between organisms and their biotic and abiotic environment have created additional ecological processes. These lead to a great number of vital regulation, production and protection functions. However, their multifunctionality is only *monodimensional.* It is based solely on natural, material processes of flow of energy/matter and biophysical information. These are investigated by basic and applied disciplines of the natural sciences with the help of formal languages. On the other hand, the multifunctionality of our natural and cultural THE landscapes is *multidimensional* with important reciprocal effects on human society. It is deeply imbedded in their holistic, biperspectivable nature as self-transcendent mixed natural and cultural middle-numbered systems. Their functions are driven both by natural, material–ecological processes of geospheric and biospheric origin, and by cognitive mental processes of noospheric origin. The former processes are transmitted by biophysical information, whereas the latter are transmitted by cultural information chiefly with the help of our natural language.

The evaluation of these multidimensional functions has to include both the *anthropocentric dimensions* of their instrumental values, as measured by their benefits for human society, and the *ecocentric* and *ethical dimension* of the intrinsic existence values of landscape. These are non-instrumental and do not depend on utilitarian values. As explained by Smith (1998, p. 14), "these intrinsic values refer to the natural thing not as a means to an end but as an end in itself". Therefore we have to study landscape functions not only as a mere commodity to be exploited as a resource on which we project our economic interest. They are also a source of value on their own right, even if we cannot put any monetary and social value on their services. It is a grave mistake to assume that we will be able to prevent their extinction by stressing only their socio-economic functions in which the human role is reduced to that of a "homo economus" keystone species (O'Neill and Kahn, 2000). This is true also for the attempts to express nature values in terms of "natural capital". In the conventional economic sense "capital" is accumulated by the production of marketable goods and measured only by monetary values. Such measures can be very misleading, even for the value of the most vital life support functions provided by fertile soil, clean air and water. They are impossible for all these "soft", and intangible values whose importance for the quality of life in the information society will be even greater than in the industrial society (Naveh, 2001 also published as Chapter 13 in this anthology).

Instead, we have to mount a common, transdisciplinary effort with scientists from other natural, social and human disciplines, as well as with artists, planners, architects and with ecopsychologists to study, evaluate and promote landscape multifunctionality based on a much broader holistic view of landscapes and their reciprocal relations with humans.

3.10. The Tenth Premise: The Antagonistic Relations Between Biosphere, Agro-Industrial and Urban-Industrial Technosphere, Threatening Life and Further Evolution can be Overcome Only by a Post-Industrial Symbiosis Between Nature and Human Society

In Section 3.1, I presented cultural evolution as leaps by bifurcations to higher organization levels with the help of mutually reinforcing auto- and cross-catalytic feedback loops. These new insights will be related now to the fate of our THE landscapes.

As the THE expanded throughout human history, along with the growing human population and its increasing consumption and technological power, natural landscapes were converted gradually into human-modified cultural landscapes. Since the industrial fossil fuel revolution a crucial bifurcation has divided these THE landscapes into *biosphere* and *technosphere* landscapes and their ecotopes into *bio-ecotopes* and *techno-ecotopes,* and most recently also into intermediate *agro-industrial ecotopes.*

In contrast to the biosphere autopioetic "regenerative systems", urban-industrial techno-ecotopes are human-made "throughput systems" (Lyle, 1994), driven by fossil and nuclear energy and their technological conversion into low-grade energy. Lacking the self-organizing and regenerative capacities of biosphere landscapes they result in high outputs of entropy, waste and pollution with far-reaching detrimental impacts on the remaining open landscapes and on human health. More recently, high-input agro-industrial ecotopes have replaced almost all low-input cultivated agro-ecotopes in industrial countries and are spreading now also in many developing countries. These agro-industrial ecotopes are much closer to technosphere landscapes than to biosphere landscapes. Although their productivity still depends on photosynthetic conversion of high-grade solar energy, this energy is subsidized to a great extent by low-grade fossil energy. At the same time, their natural control mechanisms have been replaced almost entirely by heavy chemical inputs and throughputs. In this respect, and in their detrimental environmental impacts on the open landscape, its wildlife and biodiversity, and the quality of its natural resources of soil and water, as well as on human health, they come very close to technosphere landscapes. In spite of their high productivity, without heavy financial subsidies, even those agro-industrial systems, that reached highest yields and agro-technological sophistication (e.g. those in Israel and California), are undergoing a deep economic crisis. They have lost thereby not only their ecological sustainability but also their economic sustainability. In this unstable industrial "total landscape" (Sieferle, 1997) all these ecotopes do not function together as a coherent, sustainable ecological system of the global THE ecosphere, like in the pre-industrial biosphere. Their antagonistic relations are a major cause for ecological and economical land use conflicts.

The rapidly growing technosphere and agro-industrial landscapes have caused the destabilization of the geosphere and biosphere by their one-sided, adverse outputs on the biosphere landscapes and their atmosphere, lithosphere and hydrosphere envelope. With the exception of the stabilizing negative feedback couplings, that maintain a dynamic flow equilibrium between the biosphere landscapes and the geosphere, all other interactions are ruled by destabilizing positive feedback loops. This situation not only has far-reaching impacts on the biological and cultural impoverishment of the ecosphere and its landscape ecotopes but also is manifested in threatening global climate changes and in the disruption of the protective ozone layer in the stratosphere. As predicted by Laszlo (1994) this could lead to final extinction, if these threatening trends are not counteracted by cultural regulative controlling and stabilizing ones in all natural and human dimensions of life. This shows clearly that our present environmental crisis has to be recognized and resolved as an all-embracing cultural revolution. Only in this way can we establish a proper balance between productive and attractive biosphere and healthy and livable technosphere landscapes. Their present antagonistic conflicts can be reconciled through the creation of new symbiotic relations between human society and nature. Such an urgently needed post-industrial symbiosis should lead to the structural and functional integration of bio- and technosphere ecotopes into a coherent sustainable ecosphere, in which both biological and cultural evolution can be ensured. In this symbiotic process, the scientific input of landscape research in collaboration with other mission-driven transdisciplinary environmental sciences in restoring, reclaiming, and rehabilitating landscapes as part of comprehensive planning and management for sustainable development towards the information society, could become a driving force.

Thanks to recent insights in self-organization of autopioetic systems and their cross-catalytic networks (CCN), we are now able to express these new symbiotic relations between nature and society in much more robust and even mechanistic terms and translate them into strategies for sustainable development. It would be illusionary to assume that we can restore some of the symbiotic natural feedback loops that existed in the pre-industrial society. However, we are now in a position to create new cultural, information-rich CCN feedback loops, linking natural, ecological, socio-cultural and economic processes of our THE. As a recent multinational and interdisciplinary EU study on modeling sustainable regional development (MOSES, 2000) has shown, such CNN links could be established in the information society between the regional economy and its biosphere landscapes through special taxes devoted to landscape restoration and management. The mathematics of these CNN says that a lasting synergistically acting improvement could arise for humans, nature and economy. This can be achieved in projects with the help of dynamic transdisciplinary systems simulation models and other innovative holistic methods and tools. In these, ecologists, economists and other environmentally concerned scientists collaborate to ensure lasting mutually reinforcing benefit for the people and their physical, mental, spiritual and economic welfare, together with the creation of healthy, productive and attractive landscapes for the emerging information society (Grossmann, 2000; Grossmann and Naveh, 2000).

4. CONCLUSIONS

With the help of a holistic landscape conception, well grounded in systems theory and its recent insights, we will be able to better comprehend and deal with MFLs as an integral part of the natural and socio-cultural processes determining the fate of our THE and overall global survival. The biperspectivable systems view of landscapes, functioning simultaneously as natural and cognitive systems, and therefore as a tangible bridge between nature and mind, opens the way for close cooperation among landscape researchers and scientists from all other relevant disciplines and professions. These have to work together for the joint overarching transdisciplinary vision of a sustainable future of our THE and its landscapes. Equipped with these conceptions, landscape researchers can fulfill an important role by serving in the dual position of experts in their own field and by helping to integrate innovative, future-oriented research and action.

Such research must take into consideration that the recent adoption of new information and communication technologies has caused the rapid development of the infosphere, driving human society through unstable and even chaotic transitional stages towards this global information rich age with all its positive and negative implications, dangers and opportunities. It will depend on the readiness and ability of human society to follow the road toward further evolution and sustainability of our global THE, by choosing the bifurcation, converging towards a higher level of complexity and organization and toward further evolution of life on a higher level of quality. This can be ensured only by the creation of a postindustrial symbiosis between nature and human society, turning the antagonistic relations between the biosphere and the technosphere into mutually beneficial ones, in sustainable, healthy and information-rich biosphere and technosphere landscapes.

This is not an utopian dream that is evident from the many encouraging examples, such as provided in the 1999 State of the World report (Brown et al., 1999), in addition to many others, indicating that we are at the threshold of such a postindustrial environmental sustainability revolution.

Our hope for a sustainable future for the macroshift 2000–2010 lies in the final sentences of Laszlo (2000, p. 114): "endowed with the highest forms of consciousness in our regions of the universe, we are the only species that not only acts, but can also foresee the effects of its actions. As members of a species capable of foresight, we must live up to our responsibility as stewards rather than exploiters of the complex and harmonious web of life on this planet".

ACKNOWLEDGEMENTS

I am grateful for the editors of this special issue Henri Décamps, Anne-Marie d'Hauteserre and Bärbel and Gunther Tress for their useful comments from which my paper has benefited much.

REFERENCES

Allen TF, Hoekstra TW (1992) Toward a Unified Ecology. Columbia University Press, New York.

Allesch CG (1990) The space of landscape and the space of geography. In: Svoboda H (ed.), Cultural Aspects of Landscape. Pudoc, Wageningen, pp 17–23.

Antrop M, Van Eetvelde V (2000) Holistic aspects of suburban landscapes: visual image interpretation and landscape metrics. Landscape Urban Plann. 50, 43–58.

Bohm D (1980) Wholeness and the Implicate Order. Routledge and Kegan, London.

Bohm D, Peat DF (1987) Science, Order, and Creativity. A Dramatic Look at the Roots of Science and Life. Bantam Books, New York.

Bright C (2000) Anticipating environmental surprises. In: Brown LR, Flavin C, French H (eds), State of the World 2000. A Worldwatch Institute Report on Progress Toward a Sustainable Society. Norton & Company, New York, pp 22–38.

Brown LR, 1996. The acceleration of history. In: Brown LR et al. (eds), State of the World 1996. A Worldwatch Institute Report on Progress Toward a Sustainable Society. Norton & Company, New York, pp 9–26.

Brown LR, Flavin C, French H (eds), 1999. State of the World 1999. A Worldwatch Institute Report on Progress Toward a Sustainable Society. Norton & Company, New York.

Capra F (1996) The Web of Life. A New Understanding of Living Systems. Anchor Books Doubleday, New York.

Di Castri F (1997) Editorial: landscape ecology in a changing globalized environment. Landscape Ecol. 12, 3–5.

Frankl VE (1969) Reductionism and nihilism. In: Koestler A, Smithies JR (eds), Beyond Reductionism: New Perspectives in the Life Sciences. Hutchinson of London, London, pp 396–408.

Grossmann WD (2000) Realizing sustainable development with the information society—the holistic double-gain-link approach. Landscape Urban Plann. 50, 179–193.

Grossmann WD, Naveh Z (2000) Transdisciplinary challenges for regional sustainable development toward the post-industrial information society. In: Proceedings of the Third International Conference of the European Society for Ecological Economics, Vienna, 3–6 May, pp 1–3.

Holling J (1996) Surprise for science, resilience for ecosystems and incentives for people. Ecol. Appl. 6, 733–735.

Holling J, Berkes CS, Folke FC (1999) Science, sustainability and resources management. In: Berkes F, Folke C, Colding J (eds), Linking Social and Ecological Systems. Cambridge University Press, Cambridge.

IALE Mission (1998) IALE Mission statement. IALE Bull. 16, 1–3.

Jantsch E (1970) Inter- and transdisciplinarity university: a systems approach to education and innovation. Policy Sci. 1, 403–428.

Jantsch E (1976) Self-transcendence: new light on the evolutionary paradigm. In: Jantsch E Waddington CW (eds), Evolution and Consciousness: Human Systems in Transition. Addison-Wesley, Reading, MA, pp 9–10.

Jantsch E (1980) The Self-Organizing Universe. Scientific and Human Implications of the Emerging Paradigm of Evolution. Pergamon Press, Oxford.

Koestler A (1969) Beyond atomism and holism—the concept of the holon. In: Koestler A, Smithies JR (eds), Beyond Reductionism: New Perspectives in the Life Sciences. Hutchinson of London, London, pp 192–216.

Kuhn TS (1996) The Structure of the Scientific Revolution, 3rd Edition. University of Chicago Press, Chicago.
Laszlo E (1972) Introduction to Systems Philosophy: Toward a New Paradigm of Contemporary Thought. Harper Torchbooks, New York.
Laszlo E (1987) Evolution: The Grand Synthesis. New Science Library, Shambhala, Boston.
Laszlo E (1994) The Choice: Evolution or Extinction. A Thinking Person's Guide to Global Issues. Putnam, New York.
Laszlo E (2000) Macroshift 2001–2010. Creating the Future in the Early 21st Century. San Jose, NY.
Leser H (1991) Landschaftsoekologie. Eugen Ulmer, Stuttgart.
Levin SA (1999). Towards a science of ecological management. Conserv. Ecol. 3, 6–9.
Li B-L (2000a) Why is the holistic approach becoming so important in landscape ecology? Landscape Urban Plann. 50, 27–47.
Li BL (2000b) Fractal geometry applications in description and analysis of patch patterns and patch dynamics. Ecol. Model. 132, 353–361.
Lyle JT (1994) Regenerative Design for Sustainable Development. Wiley, New York.
MOSES (2000) Modeling Sustainable Development in the European Information Society (MOSES). EU Programme Environment and Climate 1994–1998, Area 4. Human Dimensions and Environmental Change, Brussels.
Nassauer JI (1990) In: Svoboda H (ed.), Cultural Aspects of Landscape. Proceedings of the Statements at the Closing Session of the First International Conference on Cultural Aspects of Landscape. Pudoc, Wageningen, p 173.
Naveh Z (1982) Landscape Ecology as an Emerging Branch of Human Ecosystem Science. Academic Press, London. Adv. Ecol. Res. 12, 189–237.
Naveh Z (1990) Landscape ecology as a bridge between bio-ecology and human ecology. In: Svobodova H (ed.), Cultural Aspects of Landscape. Pudoc, Wageningen, pp 45–58.
Naveh Z (1991b) Biodiversity and ecological heterogeneity of Mediterranean uplands. Economica Montana: Linea Ecologica 24, 47–60.
Naveh Z (1995a) Transdisciplinary landscape-ecological education and the future of post-industrial landscapes. Bolletino del Museo dir Store Naturale della Linigiana, 9. Supplemento, Allau, pp 13–26.
Naveh Z (1995b) Interactions of landscapes and cultures. Landscape Urban Plann. 32, 43–54.
Naveh Z (1998) Culture and landscape conservation—a landscape ecological perspective. In: Gopal B, Pathak PS, Saxena KG (eds), Ecology Today: An Anthology of Contemporary Ecological Research. International Scientific Publications, New Delhi, pp 19–48.
Naveh Z (2000a) Transdisciplinary challenges for education towards regional development. In: Braunegg S, Narodoslawsky M (eds), Proceedings of Conference on Higher Education for Sustainable Regional Development. SUSTAIN, Graz, pp. 42–71.
Naveh Z (2000b) What is holistic landscape ecology? A conceptual introduction. Landscape Urban Plann. 50, 7–26.
Naveh Z (2001) Multifunctional, self-organizing biosphere landscapes and the future of our total human ecosystem—a new paradigm for transdisciplinary landscape ecology, World Futures 60: 469–503.
Naveh Z, Lieberman AS (1994) Landscape Ecology Theory and Applications, 2nd Edition. Springer, New York.
O'Neill RV, Kahn JR (2000) Homo economus as a keystone species. BioScience 50, 333–338.
Orr DW (1992) Ecological Literacy. Education and the Transition to a Postmodern World. State University of New York Press, New York.
Pankow W (1976) Openness as self-transcendence. In: Jantsch E, Waddington CW (eds), Evolution and Consciousness: Human Systems in Transition. Addison-Wesley, Reading, MA, pp 16–36.
Peat DF (1997) Infinite Potential. The Life and Times of David Bohm. Addison-Wesley, Reading, MA.
Pimentel D (1992) Conserving biological diversity in agricultural systems. BioScience 42, 354–362.
Prigogine I (1976) Order through fluctuations: self-organization and social systems. In: Jantsch E, Waddington CW (eds), Evolution and Consciousness: Human Systems in Transition. Addison-Wesley, Reading, MA, pp 93–130.

Prigogine I, Stengers I (1984) Order out of Chaos. Man's Dialogue with Nature. New Science Library, Shamabala, Boston.
Sieferle RP (1997) Rückblick auf die Natur. Eine Geschichte des Menschen und seiner Umwelt. Luchterhand Literaturverlag, München.
Simon HA (1962) The architecture of complexity. In: Hierarchy Theory. George Braziller, New York, pp 3–27. Proc. Philos. Soc. 106, 467–482.
Smith MJ (1998) Ecologism Towards Ecological Citizenship. University of Minnesota Press, Minneapolis.
Vitousek PM, Mooney HA, Lubchenco J, Melillo JM (1997) Human domination of Earth's systems. Science 277, 494–499.
Weinberg GM (1975) Introduction to General Systems Thinking. Wiley, New York.
Weiss PA (1969) The living system: determinism stratified. In: Koestler A, Smithies JR (eds), Beyond Reductionism: New Perspectives in the Life Sciences. Hutchinson of London, London, pp 3–55.
WLO (1998) A new identity for landscape ecology in Europe. A research strategy for the next decade: outlines formulated at the European congress. Landscape ecology: things to do, Amsterdam, 6–10 October. The Dutch Association of Landscape Ecology, Wageningen.
Zonneveld I (1995) Land Ecology. An Introduction to Landscape Ecology as a Base for Land Evaluation, Land Management and Conservation. SPB Academic Publishing, Amsterdam.

Zev Naveh (1919), PhD, is Professor Emeritus in landscape ecology at the Technion, Israel Institute of Technology, Haifa. Until 1965 he worked as range and pasture specialist in Israel and in Tanzania. His research at the Technion was devoted to human impacts on Mediterranean landscapes, fire ecology and dynamic conservation management, introduction of drought resistant plants for multibeneficial landscape restoration and beautification. Presently he is involved chiefly in studying theoretical aspects of holistic landscape ecology and sustainable development for the information society. He was the founder and director of the Ecological Garden of the Technion and the cofounder and scientific director of the first Environmental Teaching Project in Israel, and the head of a IUCN working group for landscape conservation. He has been visiting professor and guest lecturer in many universities and keynote speaker in international conferences and is member of the editorial board of Landscape Ecology, Restoration Ecology and Mediterranean Ecology. Professor Naveh has published over 200 scientific publications and is author and co-author of several Hebrew and English books, including the first English monography on landscape ecology.

CHAPTER 13

NAVEH Z. (2004)

MULTIFUNCTIONAL, SELF-ORGANIZING BIOSPHERE LANDSCAPES AND THE FUTURE OF OUR TOTAL HUMAN ECOSYSTEM. WORLD FUTURES 60: 469–503.

Abandoned terraces, covered with Sarcopoterium spinosum.

Address correspondence to Zev Naveh, Faculty of Agricultural Engineering, Technion, Israel Institute of Technology, Haifa 323000, Israel. E-mail: znave@techunix. technion.ac.il

Abstract Solar energy powered autopoietic (self-creating and regenerative) natural and cultural biosphere landscapes fulfill vital multiple functions for the sustainable future of organic life and its biological evolution and for human physical and mental health. At the present crucial Macroshift from the industrial to the post-industrial information age, their future and therefore also that of our Total Human Ecosystem, integrating humans and their total environment, is endangered by the exponential growth and waste products of urban-industrial technosphere landscapes and agro-industrial bio-technosphere landscapes.

This danger can be prevented only by the creation of new symbiotic relations between human society and nature with the help of mutual supportive, restorative cultural and economic cross-catalytic networks in our Total human Ecosystem. This should be part of an all—embracing sustainability revolution, driven by human consciousness and its responsibility to act as stewards rather than exploiters of the complex and harmonious web of life on this planet

Keywords Landscape ecology, evolution, nonequilibrium thermodynamics, autopoiesis, biosphere, technosphere, cross-catalytic networks

INTRODUCTION: THE TRANSITION TO THE GLOBAL INFORMATION AGE—A CRUCIAL PERIOD FOR LIFE ON EARTH

Humankind is undergoing presently closely interwoven changes, embracing all spheres of human life from the biological-ecological to the social-cultural, the economic, technological and political sphere. These global changes are driven by the rapid development of worldwide computer networks of information, allowing the rapid economic built-up expansion and globalization and have been presented by Di Castri (1998) as a multifaceted and interactive picture of globalized gears moved by information flows.

Ervin Laszlo (1994), the renowned systems philosopher and expert on global trends, regarded these changes as a "Grand Transition" from the industrial to the post-industrial global information age, and therefore as an epochal turning point in human socio-cultural evolution. In this, global survival will be determined by the choice of human society between ensuring further evolution of life on Earth, or its final extinction.

Most recently Laszlo (2000) has further described this period of profound and irreversible changes, as a "*Macroshift.*" *This* evolutionary trajectory *is the societal variety of bifurcations in non-human systems, where the consciousness* (and therefore also the behavior) *of the system's members will determine its outcome.*

In his own words (Laszlo, 2000, p. 2);

> It is up to those who live through this Macroshift, whether the era that dawns will be true advance over the era we have been living through, or a lapse into downward spiral: more penury, greater poverty, more violence, and greater degradation of environment and the quality of life.

Some of the most recent findings show in a very convincing way that this dilemma of human society to choose between evolution and extinction, is not a far-fetched doomsday prophecy. On the contrary, as will be shown below, it has become even more urgent since the publication of his book in 1994. These findings show clearly how urgent is a human-induced radical change in the current direction of this Macroshift. However, the values, insights and behavior of all of us, in all parts of the world will decide its outcome.

In the 1999 *State of the World Report* as one of the most important and reliable source of information on critical global environmental issues, published annually by the Worldwatch Institute Brown et al. (1999) summarized this situation as follows:

> Coupled with the rapid population growth and the even faster growing consumption have undermined the ecological foundations of our natural life support systems, on which our civilization depends. Modern industrial society has become thereby a major destabilizing biological, sociological and even geological force and has reached a crucial turning point in its relations to nature.

Thanks to the extraordinary achievements of science and technology we are approaching now this new global, information-rich age. However in the creation of the rapidly expanding technosphere and its urban-industrial and agro-industrial landscapes, the technological power and skill of *Homo Industrialis* and his obsessive believe in the panacea of growth and the increase in material goods, have far exceeded his ecological wisdom, knowledge and ethics. Thus, in spite of the great advances in science and technology, the industrial society has not been able to resolve the deep ecological crisis it has created in the last century. Its leading governments have squandered the historic opportunity to reverse the accelerating speed of Earth's environmental decline during the prosperity of the last years.

In fact, during the last century the world population has increased 4 times, the world economy has increased even 17 times and the living standard of the Western world has improved tremendously. But at the same time, the less industrialized countries have become poorer and hungrier and presently 1, 2 billion people must survive with less than one dollar per day and lack access to clean water and hundreds of millions breathe unhealthy air.

The accelerating speed of the global climate destabilization is illustrated in a most recent report of the Intergovernmental Panel on Climate Change (IPCC). It envisages that the catastrophic physical, ecological and socio-economic consequences of human-induced climatic changes will occur with greater speed as previously predicted. These warnings are supported by solid facts on the accelerating speed of global warming, indicated by the melting of ice sheets in the Arctic and the Antarctic and of mountain glaciers from the Peruvian Andes to the Swiss Alps are proceeding with more speed than expected. The Arctic Ocean ice cap has been reduced by nearly half and the ice shelves of the Antarctic Peninsula are in full retreat and lost until 1997 2,000 square kilometers, but in 1999 alone 3,000 square kilometers. As the ice on land melts and flows to the sea, this is causing not only devastating floods and landslides, like those in Venezuela, worsened by deforestation that killed more than 30,000 people, but also the rise of the sea level. Over the last century, the sea level rose by 20–39 centimeters. However during this century it could rise by as much as 1 meter, and if the Greenland ice sheet will continue to shrink with the present speed, and will melt entirely, then the sea levels could rise even by 7 meters. This will have catastrophic consequences for all Atlantic and Pacific islands and shore lands.

In recent years, the warming atmosphere has spurred already more severe and more frequent weather events, like the December 1999 storms in Central and Western Europe, that caused nearly $10 billion damage. The latest flooding in India have left more than 15 millions inhabitants in terrible distress, without shelter and almost without food. Many of these will join the millions of displaced "environmental refugees" in Asia and South America, created by extreme and disastrous climatic and events.

Natural disasters have cost the world $605 billion over the last decade—as much as in the previous four decades combined. There are also already alarming signals that global climate changes are more rapid than the rate of adaptive changes with which habitats and many of their organisms can cope. Already now 11 percent of all bird species, 25 percent of mammals and 34 percent of fish are in danger of extinction, and 27 percent of the world's coral reefs have been lost already.

It becomes now also more and more obvious that extreme climatic events such as heavy rains, strong winds, drought, and extreme cold or hot periods are becoming more and more frequent and leading to more and more catastrophic outcomes. Their results, in turn, are becoming more and more severe, because of the exponential process of landscape degradation and its destabilizing impacts. This is especially the case in the densely populated Asian countries, in which traditional unsustainable land uses, such as overgrazing, and overcutting of wood for fuel are combined with the "modern" not less detrimental but much larger scale uncontrolled agricultural and industrial intensification. Whereas the former desertification process creates bare soil, the latter creates also large, impenetrable layers of asphalt, using many thousand times more powerful engineering devices and machines for land denudation and destruction.

The problem of desertification is especially severe in the semi-arid northwestern provinces of China. Already suffering from overgrazing and overplowing, but now, the efforts of the Chinese government to compensate for the loss of fertile croplands by plowing out marginal lands is accelerating even more wind and soil erosion.

A recent, well-documented report by Brown (2001) lends special significance to the warning by Laszlo (2000) that "China is an unsustainable catastrophe in the making." This report claims that the dust bowl, resulting from this unsustainable agricultural development is threatening the future of China. The most recent record dust storm is hitting Beijing and most of the other populous cities in North East China, obscuring the sun, reducing visibility along traffic and closing airports. These huge dust plumes from northern China have reached now also the North America, "blanketing areas from Canada to Arizona with layer of dust."

Brown (2001) concludes, that reversing desertification will require a huge effort, but if the dust bowl continues to spread, it will not only undermine the economy, but it will also trigger a massive migration eastward. He states: "The options are clear: Reduce livestock populations to a sustainable level or face heavy livestock loss as grassland turns to desert. Return highly erodible cropland to grassland or lose all of the land's productive capacity as it turns to desert. Construct windbreaks with a combination of trees and, where feasible, wind turbines, to slow the wind or face even more soil losses and dust storms."

Of special relevance for our discussion on the future of multifunctional solar-powered biosphere landscapes are the results of a recent worldwide pilot study. It has been carried out jointly by 175 interdisciplinary scientists under the auspices of the United Nations Development Program and the United Nations Environment Program, the World Bank and the World Resources Institute (World Resources, 2000–2001, available On-Line at www.wli.org/wr2000). These scientists arrived at the conclusion that a critical point has been reached in the earth's capacity to support both nature and human populations. As illustrated by many maps, the researchers studied different landscapes (but used instead only the vaguely defined and delineated ecosystem term), showing that many of these landscapes are reaching now dangerous thresholds.

Some of the major landscape functions of agro-ecosystems, and coastal-, forest-, freshwater- and grassland-ecosystems were evaluated as goods and services and scored for food/fiber production, water quality, biodiversity, carbon storage, recreation, shoreline protection and woodfuel production. The "Bottom Line" was that overall there are numerous signs that the capacity of ecosystems to continue to produce many of these goods and services is decreasing. The erosion of biodiversity is alarming, chiefly because of "loss of habitat area," and their vanishing landscapes. Amongst the most revealing indicators for the loss of crucial landscape functions is the extent of deforestation of watersheds. Almost a third of all watersheds assessed have lost 75 percent of their original forest cover and seventeen have lost more than 90 percent.

Amongst the many other recent findings on the rapidly increasing threats to these natural and semi-natural landscapes and their organic life, is the latest report by the World Wildlife Fund. It warns, that if until the end of this century the atmospheric Carbon dioxide levels will continue to rise and thus be doubled since the beginning of the industrial revolution, then 70 percent of all natural habitats in the northernmost landscapes of Canada, Russia and Scandinavia could be lost, and 50 percent of those in northern European countries altogether.

THE ROLE OF LANDSCAPE ECOLOGY AND THE SUSTAINABILITY REVOLUTION

This acceleration of overwhelming global threats puts additional weight to the claim by Laszlo (1994) that for the choice of further biological and cultural evolution, a far-reaching ecological, social, cultural and political sustainability revolution in all spheres of life is essential. It can be achieved only by a great effort on global dimensions, driven by all those who are concerned with the future of life on earth and the welfare of all its inhabitants and who can provide the scientific and professional leadership to this revolution.

Of special relevance for this purpose are those sciences dealing with the fate of the land and seascapes, their geophysical, bio-ecological and human-ecological, socio-economic and cultural aspects with broad, integrative approaches. Amongst these sciences, landscape ecology occupies a special place. It emerged after in the sixtieths in Central and Eastern Europe as an interdisciplinary ecological-geographical, and problem-solving, holistic discipline of landscape study, planning and management, and has developed in the last twenty years as a dynamic global environmental science. The International Association of Landscape Ecology (IALE) has hundreds of members, academic researchers, professional planners, land managers and users in about 40 industrial and developing countries. A great number of landscape-ecological studies, dealing with highly diverse themes on natural and cultural, rural and urban landscapes are published in "Landscape Ecology," "Landscape and Urban Planning": and in other, related scientific journals. At the same time, the number of books published on these subjects is increasing steadily.

The fate of all these landscapes—and our global ecosphere landscape as a whole—is closely coupled through mutually amplifying feedback relations with these cultural evolutionary trends of human society and its choice between further evolution and extinction. Therefore landscape ecologists have to become involved in this choice and have to become, what Di Castri (1997) in an important editorial in "Landscape Ecology" has called "*committed actors,*" but not remain only "critical but marginal spectators in this game." As pointed out by Naveh (2001), for this purpose, landscape ecology cannot be carried out anymore in the sheltered academic ivory tower of so-called "objective science," detached from people and their values and needs. This has to be reflected both in theory and action in their day-to-day work in education, research and practice. To contribute jointly with scientists from all other relevant disciplines to a sustainable future of healthy, productive and attractive landscapes should become one of the greatest challenges for future-oriented landscape ecologists. It demands first of all an understanding of the transdisciplinary scientific revolution, offering a holistic systems view of the world. It requires the recognition of the far-reaching impacts of these developments on multifunctional landscapes and their management, conservation and restoration as tangible, self-organizing Gestalt systems of our Total Human Ecosystem. This will be discussed further in more detail.

LANDSCAPES AS TANGIBLE MULTIFUNCTIONAL GESTALT SYSTEMS OF OUR TOTAL HUMAN ECOSYSTEM

The true meaning of future-oriented and mission-driven holistic landscape ecology can be fully comprehended only in the broader context of the present holistic and transdisciplinary scientific revolution. According to Kuhn (1996), such a scientific revolution is characterized by major paradigm shifts that are generally accepted in "normal" science. We are dealing here with a far-reaching paradigm shift from conventional reductionistic and mechanistic concepts to holistic and organismic approaches of wholeness, connectedness and ordered complexity. These are grounded in systems theory, offering a unified worldview that seeks to do justice not only to the physical, biological and the socio-economical, but also to the mental, the cultural and the spiritual reality in which we live. It is leading also to a profound postmodern cultural transformation, changing many of the ideas which dominate since the industrial revolution most of science and technology and Western society, its education, economy and culture at large. As explained in more detail elsewhere (Naveh, 2000, 2001), this implies a holistic paradigm shift from perceiving landscapes as nothing but large-scale heterogeneous mosaics of physical, chemical, and biological landscape elements in repeated patterns of ecosystems, into a holistic view of landscapes as *multifunctional Gestalt systems* in their own right. The German term "Gestalt" has been introduced into psychological Gestalt theory, in which humans are perceived as whole persons, fully embedded in the world, and the world is seen more like a living person than like a nonliving mechanism of separate interacting parts. Antrop and Van Eetvelde (2000) have recently pointed out the relevance of some of the laws of psychological Gestalt theory for a holistic perception of complex landscape patterns. The scope of this theory has been broadened by the ecopsychologist Cahalan (1995) to achieve a deepened sense of the full therapeutic function of the natural environment, to which I will refer further below.

Naveh and Lieberman (1994), and more recently Naveh (2000, 2001) have presented detailed discussions on the holistic nature of landscapes and their multidimensional functions. Here it is sufficient to point out that in landscapes all natural and cultural dimensions are intrinsically related to each other by the general state of the whole and its emergent qualities from the smallest, mappable landscape cell or *ecotope,* to the global ecosphere landscape. The different landscape units and types are closely interlaced into a multi-layered, stratified hierarchy of holons, being both parts of their higher level supersystem and wholes towards their lower level subsystems. Therefore, instead of a puzzle of separate particles, forming together a mosaic, we deal with a hierarchically structured interacting network of landscapes at different scales. Together with increasing spatial temporal and perceptional scales, also the complexity of patterns and processes, and their resulting functions are increasing, and a better comprehension of the underlying ecological and historical and cultural dynamics can be reached (Naveh, 1994a also published as Chapter 4 in this anthology).

Further insights in the unique, holistic nature of landscapes can be gained with the help of Bohm's (1980) and Bohm and Peat's (1987) groundbreaking studies on enfolded implicate and generative orders, in which human mind, consciousness and creativity play and important role. These are hidden behind the familiar notions of simple regularity and randomness, describable in landscapes by Archimedian geometry and the Cartesian grid of coordinates which has dominated the basic order of reality for the last three hundred years. Bohm and Peat (1987) showed that between the two extremes of simple regular order and chaos there is a rich new field of creativity as a state of high energy, making possible a fresh perception of nature through the mind. A major challenge for landscape ecologists will be to capture these new orders with the help of innovative transdisciplinary methods in their research, and to develop practical tools for an integrated appreciation of the aesthetic, ethical and intrinsic functions of landscapes as tangible bridges between human mind and nature.

In accordance with this holistic paradigm shift, humans are not apart from nature or even above nature, valuing only its instrumental functions, but they are *integral parts of nature.* They should therefore not be considered as external, "disturbance" factors or be modeled in these landscapes merely as socio-economic factors. Instead, *human aspects and dimensions have to be treated as an intrinsic part of landscape processes and functions.* As such these functions should also be holistically assessed and utilized for the sake of both natural and human systems.

For this purpose, landscapes are to be viewed, studied, and managed not only within the ecological/functional and the geographical/spatial dimensions of the natural sciences. They have to be treated within a much broader context of the of the integrated human-nature systems complex, as the larger ecological geo-bio-anthropo entity in which we live. Following the eminent, first holistic ecologist, Frank Egler (1964), we suggested naming this entity the *Total Human Ecosystem* (THE) *integrating humans and their total environment at the highest co-evolutionary level of the global ecological holarchy.* As the tangible spatial and functional matrix of all organisms, including humans, their populations, communities and ecosystems, landscapes become thereby also the concrete, space/time defined ordered wholes and unique Gestalt systems of our THE along different functional, spatial and perceptional scales and dimensions (Naveh, 1982; Naveh and Lieberman, 1994).

However, as thinking human creatures we are living not only in this physical, ecological and geographical landscape space, which we share with other organisms. We live also in the conceptual space of the human mind of the *noosphere* (from the Greek *noos* = mind).This is an additional natural envelope of life in its totality that *Homo sapiens* acquired throughout the evolution of the human neocortex from the paleo-mammalian brain, as the domains of our perceptions, knowledge, feeling, volition, and consciousness. It enabled the capacity for "*self-reflective mentation*" (Jantsch, 1980) or of "*reflective consciousness*" and self-awareness, namely the ability not only to perceive and feel things, but to know that one perceives and feels them and hence to order them in the light of his purpose (Laszlo, 1994). This led to the development of additional noospheric realms of the info-socio- and psychospheres that have emerged during the cultural evolution of modern man, through which he became a mighty geological agent with both constructive and destructive powers.

The Total Human Ecosystem should be regarded as the overarching conceptional supersystem for both these physical geospheric and mental and spiritual noospheric space spheres. It could serve therefore also as the unifying conceptional foundation for a very much needed holistic paradigm of for the highly fragmented environmental science rooted either in the natural or the social sciences and humanities. Such a complementary systems view enables us to view the evolution of THE landscapes in the light of the new holistic and transdisciplinary insights as a tangible bridge between nature and mind, and as an integral part of the dynamic self-organization and co-evolution in nature and in human societies. It opens the way for a better comprehension of the multifunctionality of cultural landscapes and their natural and cultural multidimensions (Naveh, 2001).

NEW INSIGHTS INTO SELF-ORGANIZATION AND EVOLUTIONARY PROCESSES IN NON-EQUILIBRIUM SYSTEMS

Of great relevance for this conception of THE landscapes are the insights gained on the self-organization of living systems. In these, the spontaneous emergence of new order, creating new structures and new forms of behavior within network patterns is made possible by their self-regulating feedback loops. Such systems on a relatively high organizational levels, which can renew, repair, and replicate themselves as networks of interrelated component-producing processes, in which the network itself is created and recreated in a flow of matter and energy, are called *autopoietic systems* (from the Greek = self-creating or self-renewing). This is true not only for cells and organisms and ecosystems, but also for landscapes, as interacting Total Human Ecosystems of non-human and human living systems. In his last, seminal book, on the "The Self-Organizing Universe," the great transdisciplinary systems thinker and planner Erich Jantsch (1980, p. 10), has defined autopoietic systems as follows:

> An autopoietic system is in the first line not concerned with the production of any output, but with its own self-renewal in the same process structure. Autopoiesis is an expression of the fundamental complementary of structure and function, that flexibility and plasticity due to dynamic relations, through which self-organization becomes possible.

This autopoietic process is made possible by "*autocatalyis*" by which one of the products of the reaction enters a cycle that helps to reproduce itself by creating its own synthesis. In cycles of "*crosscatalyis*" two or more subsystems are linked, so that they can support each other by catalyzing each other's synthesis and thereby mutually increase their growth. This is the case with living cells, which can produce more of themselves and in the same time they preserve themselves in a changing environment. The eminent biologists Eigen and Schuster (1979) have demonstrated that entire chains autocatalytic self-reinforcing and mutually reinforcing cross-catalytic cycles and network relations are the basis of the complex structures which underlie, and make possible, the emergence of life. Their work led to the recognition of "*hypercycles*" of mutually reinforcing chemical and biological processes of systems far from equilibrium, through positive feedback loops, together with the appearance of instabilities, leading to new, and higher forms of organization.

Such hypercycles are shown on the subcellular DNA level in Fig.1. The major auto- and cross-catalytic cycles of the biosphere, driven by solar energy through the trophic food chains and recycling their decay products are shown in Fig. 2. Here Laszlo (1987) presents the biosphere as the global system, driven by solar energy through trophic food chains and recycling their decay products. It should be realized that such energy-material flows and crosscatalytic networks can be completed fully only in natural and semi-natural, solar-energy powered biosphere landscapes. However, as will be described further in more detail below, in the current global Total Human Ecosystem of the industrial society, these cycles are threatened in agro- and urban-industrial landscapes.

Maturana and Varela (1975) have broadened these findings into a comprehensive systems theory of self-organization of biosystems (Fig. 2). Jantsch (1980) has carried

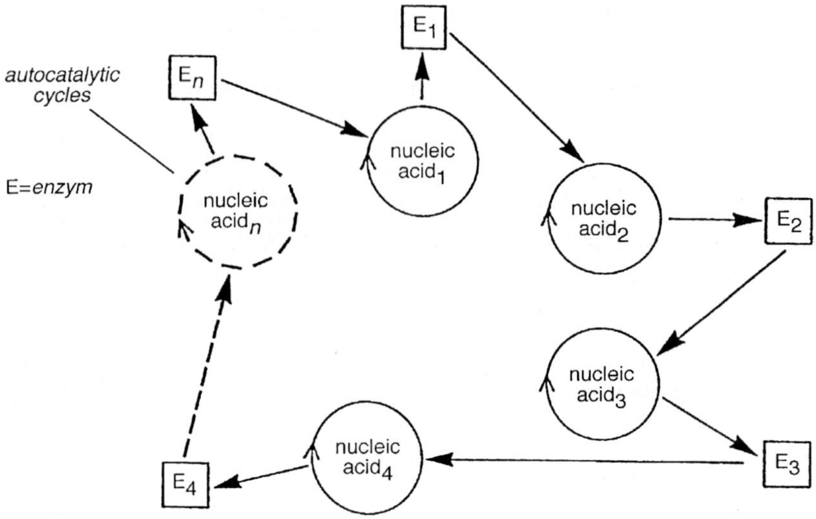

Fig. 1. Autocatalytic cycles in nucleic acid molecules and their enzymes, producing a cross-catalytic cycle with another nucleic acid molecule and forming together a cross-catalytic ("Hyper") cycle, converging thereby to a higher level of cell organization (Laszlo, 1987).

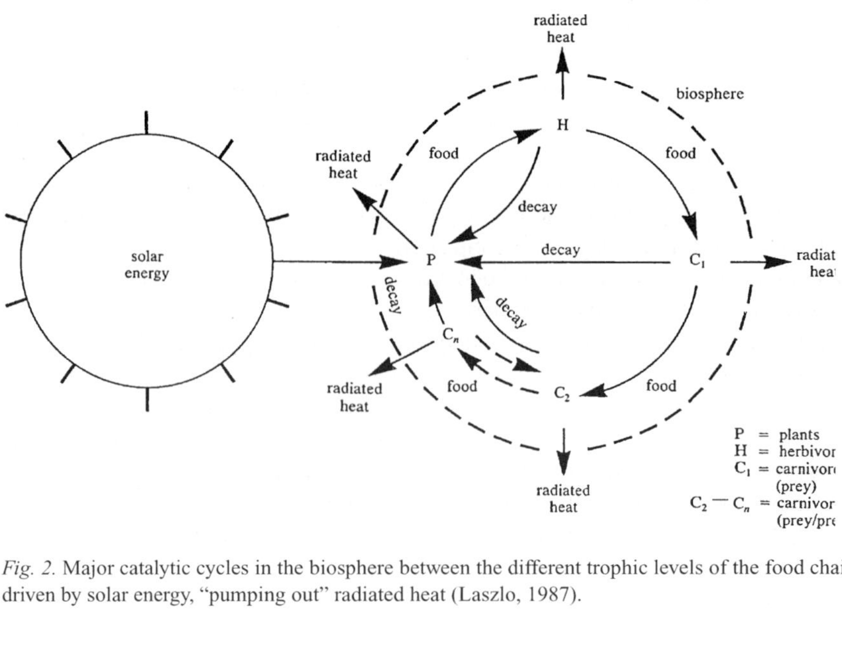

Fig. 2. Major catalytic cycles in the biosphere between the different trophic levels of the food chain, driven by solar energy, "pumping out" radiated heat (Laszlo, 1987).

this new paradigm of dynamic micro-and macro-co-evolution of self-organization in nature even further. He has laid the transdisciplinary foundations for a synthetic view of cosmic, geological, biological, ecological and socio-cultural evolution. It leads to an all-embracing conception co-evolution, emphasizing cooperation as the creative play of an entire evolving universe. This presents a major paradigm shift from the Cartesian and Newtonian view of a mechanistic world and reaches far beyond the still fashionable post-Darwinian and socio-biological interpretations of evolution. Jantsch (1980) achieved this by combining these evolutionary insights with the new ordering principles of self-organization in non-equilibrium systems, developed by the Nobel-Prize winner Prigogine and his Brussels team, to which I will refer further in the context of landscape evolution. Since then, his groundbreaking theories have been corroborated and further developed thanks to recent findings in the sciences of wholeness and the mathematics of complexity. They have been outlined recently by Caprà (1996), the well-known author of the "Tao of Physics," in a remarkable readable and non-formal book "The Web of Life."

In his seminal study on the "Grand Evolutionary Synthesis" Laszlo (1987) further examined the co-evolutionary patterns of change and transformation in the cosmos, organisms and in modern society. In this synthetic evolutionary patterns, systems are not moving in a continuous and linear progress from the simpler to the more complex type of system, and from the lower to the higher level of organization. They leap by the sudden emergence of successive levels of organization from quarks to global socio-cultural systems and to cosmic systems.

These discontinuous developments of sudden leaps from one kind of stable state to another occur as "*bifurcations.*" In systems far from equilibrium subtle "catastrophic" bifurcations can model increasing instability. These may turn chaotic or disappear or lead to a new state of metastability on a higher level of organization.

All these evolutionary steps have been made possible by mutually reinforcing crosscatalytic feedback loops. According to Laszlo (1987), their formation allows dynamic systems to emerge on such successively higher levels of organization on multiple hierarchical levels. On each level the amount of information that can be handled by the cycle is greater than on the lower level, owing to a greater diversity and richness of the components and structures. In Fig. 3, these sudden leaps and their bifurcations are shown on global levels from the lowest organization level of quarks up to the highest level of socio-cultural systems, and according to our conception, up to the Total Human Ecosystem.

Following the findings of Prigogine and his co-workers, these non-linear evolutionary processes can be explained from the thermodynamic viewpoint as new ordering principles that "create order through fluctuations" (Prigogine, 1976) and even "order out of chaos" (Prigogine and Stengers, 1984). They showed that nonlinear thermodynamics of irreversible processes in open systems exchanging energy and material with their environment could lead to the evolution of such new, dynamic globally stable systems. This is opposed to the belief by many physicists that non-equilibrium states are only temporary disturbances of equilibrium, not containing any interesting physical information. Prigogine proved that through the break in time

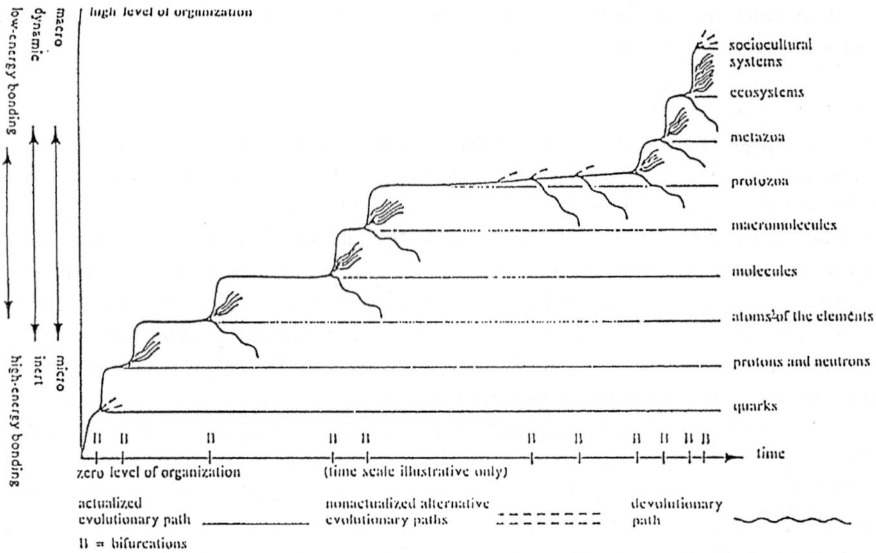

Fig. 3. The sudden evolution of successive higher organization levels from physical to chemical, ecological and cultural systems (Laszlo, 1987).

and space symmetry, non-equilibrium of irreversible processes became sources of order and became a creative evolutionary process. These non-equilibrium systems are called "*dissipative structures*" because they maintain continuous entropy production and dissipate accruing entropy, not accumulating in the system, but being part of the continuous energy exchange with their environment. Dissipative structures constitute the simplest case of spontaneous self-organization in evolution. This has opened the way for realizing that evolution toward increasing complexity and organization is the result of structural fluctuations and innovations that can appear suddenly in a previously stable systems and drive it subsequently to a new regime at a more complex state, as described above.

Jantsch (1980, p. 307) has summarized these findings as follows:

> We stand at the beginning of a great new synthesis. The correspondence of static structures is not its subject, but connectedness of self-organization dynamics—of mind—at many levels. It becomes possible to view evolution as a complex, but holistic dynamic phenomenon of a universal unfolding of order which becomes manifest in many ways, as matter and energy, information and complexity, consciousness and self-reflection.

It is very hard to condense the great amount of information gathered from many studies in these different fields of innovation and presented mostly in highly formal specialist ways. Here I can describe them only in very general and simplified terms. However, fortunately, they have been synthesized so brilliantly by the above-cited transdisciplinary scientists that they are now much more easily accessible. It will assist greatly all those who are concerned about the future of our THE landscapes to get at the roots of these new paradigms and to find thereby new and deeper meaning with many important practical implications for the management of multifunctional landscapes.

THE EVOLUTION AND DYNAMICS OF SELF-ORGANIZING LANDSCAPES, AS DISSIPATIVE STRUCTURES FAR FROM EQUILIBRIUM

The above-described findings enable a much more comprehensive view of landscape dynamics as part of the cultural evolution of our Total Human Ecosystem. In a very simplified way this could be described as proceeding by leaps through bifurcations from the primitive food-gathering hunting stage to the Neolithic agricultural revolution, and from there to the industrial revolution until our present, still chaotic transitional bifurcation stage leading towards information society.

As part of these cultural evolutionary processes the THE has expanded according to the rate of growth of human populations, their consumption and technological power.

This caused the expansion of their "ecological footprints" (Rees, 1995) and "colonization processes" (Fischer-Kowalski and Haberl, 1997), by which natural landscapes were modified and into semi-natural, landscapes and converted into rural, and urban-industrial cultural landscapes. However, during this socio-cultural evolutionary process, and since the industrial fossil fuel revolution with accelerating speed, a crucial symmetry break in time and space occurred. This led to a bifurcation, which has divided these entire natural and cultural Total Human Ecosystem landscapes into 3 major functional landscape classes:

1. *solar-powered, autopoietic and regenerative biosphere landscapes.*
2. *fossil fuel-powered technosphere* landscapes and their ecotopes (or in short *bio-* and *techno-ecotopes*),
3. most recently also into additional intermediate both solar and fossil fuel-powered *agro-industrial ecotopes.*

Because of the vital importance of biosphere landscapes for the development of future-oriented and holistic strategies of sustainable planning, designing, conserving and restoring of natural and cultural landscapes, these bifurcations have far-reaching implications for the safeguarding of life on earth and the sustainable future of our Total Human Ecosystem.

In all biosphere landscapes and their bio-ecotopes, high quality potential and chemical energy (and therefore low entropy producing energy) is derived from solar energy and its conversion through photosynthesis and assimilation into chemical and kinetic energy in the organismic food chain. As has been shown in Fig. 2 part of this energy is dissipated into low quality metabolic heat and respiration as radiated heat. Therefore an increase in structural and spatial heterogeneity, higher species diversity and higher complexity in food chains and webs build up negentropy—as a measure of organizational order and information—in the landscape. Simultaneously entropy production—as a measure of homogeneity, and disorder—is also minimized by the protection and stabilization functions of the "living sponge" of the vegetation cover and its underlying pedosphere. These reduce the rate of kinetic energy and heat flows and their destructive and destabilizing impacts on the landscape. However, according to their human uses, bio-ecotopes have to be divided into natural and close-to natural and cultural bio-ecotopes.

The scientific breakthrough in non-equilibrium thermodynamics and its new ordering principles have also deepened our understanding of the dynamics of landscapes far from equilibrium and their capacity of continuous self-organization from lower to higher holarchical levels (Naveh and Lieberman, 1994).

As has been described elsewhere in more detail (Naveh, 1991, 1998) Mediterranean, as well as most other semi-natural and agro-silvo-pastoral landscapes, behave as such dissipative structures. They are maintained and stabilized only by permanently interchanging energy and entropy with their environment. Driven by positive feedback of environmental and internal fluctuations, they move to new regimes that generate conditions of renewal of higher internal entropy production, while undergoing short- and long-term, and chiefly cyclic fluctuations, far from a homeostatic equilibrium stage. From such an unstable and even chaotic basis, they move to a stage of dynamic stability or "metastability." By "pumping out" entropy as disorder in their autopoietic life-creating process, these landscapes increase their internal negentropy, ensuring more effective information and energy efficiency within the system.

According to Li (2000), these dynamics can be explained now also in more rigorously in thermodynamic and formal terms. With the help of "synergetics," introduced by Haken (1983) Li extended Prigogine's non-equilibrium thermodynamics into more complicated non-linear dynamic situations with which landscapes are confronted. In synergetics, the system consists of a vast number of subsystems and certain conditions of controls are changed even in a very unspecified way. The system can develop new patterns of macroscopic scale of stable states, such as bistability or multistability. The system may also undergo oscillations or random motions and chaos. Li (2000) combined Haken's theories with those of Prigogine and especially his recent study (Prigogine, 1997), for the interpretation of landscape instability or multistability as dissipative systems, in which stochastic fluctuations lead to an intrinsic instability.

In the Mediterranean Basin and elsewhere such cyclical perturbations have been introduced mostly by regular, centuries lasting rotational grazing, browsing, burning, cutting, coppicing regimes, together with cultivation and other human land uses. These human perturbations were superimposed on the seasonal and annual climatic fluctuations, and their resulting defoliation pressures were incorporated in the landscape together with these and other natural perturbations at different spatiotemporal scales. They resulted in the establishment of a human-maintained and dynamic long- and short-term flow. In all these landscapes, the system is not returned to a stationary state of homeostasis, like in traditional climax systems. It is going on to move along the same trajectory of change as it has in the past, as long as these cyclic perturbations are driving these changes and are continued with similar intensities and time intervals. Thereby these human perturbation-dependent systems have acquired long-term adaptive resilience and evolutionary metastability. The long-term maintenance of such homeorhetic flow equilibrium, operating within the great macro- and micro-site heterogeneity of the rocky and rough terrain and producing the fine-grained agro-silvo-pastoral land-use patterns, played apparently a major role inducing the unique combined biological, ecological and cultural landscape "ecodiversity" of these landscapes. Thereby they acquired their most important multifunctional assets.

The disruption of this homeorhetic flow equilibrium is caused not only by too intensive land use pressures, but also by the cessation of all human interference and land abandonment. It is leading to the impoverishment of structural and biological diversity of Mediterranean grasslands, shrublands, woodlands and Maquis, from the Eastern corner to the Western corner of the Mediterranean Basin (Naveh and Whittaker, 1974; Naveh, 1998; Pinto-Correira, 1998). Here, most affected are the light demanding herbaceous species, including many rare endemics and ornamental geophytes (Ruiz de la Tore, 1985), as well as many vertebrates (Warburg et al., 1978), and especially bird species, whose loss is closely coupled with the loss of over-all landscape heterogeneity and diversity (Farina, 1989).

No systematic studies have been carried out up-to-now elsewhere on these aspects. However, we can assume that also in other semi-natural biosphere landscapes outside the Mediterranean Basin a similar homeorhetic flow equilibrium has been maintained either by cyclic natural long or short-tem climatically-induced fire, windrow, flooding and other fluctuations and/or by human-induced fluctuation, such as periodic coppicing, cutting, mowing, grazing and browsing. The disruption of this homeorhetic flow equilibrium by the cessation of traditional, and mostly small-scaled agricultural activities is also one of the major reasons for the loss of some of the most attractive biological and cultural richest rural landscapes and their conversion into dense, homogeneous, secondary forests. This is true not only for Europe, but also for other temperate, as well as subtropical and tropical regions all over the world.

In North American in forests, shrublands and grasslands, such dynamic flow equilibrium has been, most probably maintained both by natural and human set fires. In the African savannas and forests, fire, together with wild and domestic ungulate grazing have fulfilled a similar function. Thus, for instance, in view of the great threats to African elephants, their vital role as a keystone species for diversifying the savannas, the forests, and the swamps is of special importance. This has been demonstrated with many examples by the eminent African wildlife ecologists David Western (1997)* during his devoted, life-long work for the Masai in the Amboseli Park in Kenya and the wildlife of East and Central Africa. He reached also very similar conclusions on the need for innovative holistic and dynamic conservation strategies. These should be based on a holistic shift and "a radical departure from the western view of separateness of Man and Nature" by which nature and society should be intimately linked in our minds.

In the South American Pampas grasslands the cyclic homeorhetic flow equilibrium is apparently maintained by periodic flooding, which ensures the high productivity of these subtropical grasslands. Here it is distorted by draining and irrigation schemes for intensive agriculture. Such periodic flooding seems to be also very important for the maintenance of homeorhetic flow equilibrium in many wetlands and swamps and in maundering riverbeds. In many cases, such as in the Mississippi river, the removal of this flow equilibrium by engineering interventions to "regulate the water flow" has resulted just in the opposite, causing much more severe and damaging floods.

The importance of the re-establishment of this multifactorial homeorhetic flow process by active and dynamic conservation management, furthering the highest attainable multifunctionality of these landscapes, can be supported further by their thermodynamic behavior as dissipative structures.

In Mediterranean shrublands and Maquis, the conditions for the creation of new thermodynamic regimes, leading to their function as dissipative structures are apparently created during their regeneration after periodical perturbations of fire, grazing and cutting. But this will happen only if sufficient time has been allowed for the regeneration phase and thereby also for the import of negentropy through intensified photosynthetic growth and regeneration processes. At the same time, the system can actually use free energy to reorganize itself with increasing structural complexity, biological diversity and productivity. But if these perturbation cycles are too frequent and severe, then the external entropy exchange may become more and more positive and disorder will remain at a high level. The same is true also if these perturbations are stopped altogether, either by total protection and non-interference or by abandonment. In this case negentropy and information rises in the early regeneration phase, but with the lack of further perturbations, the rates of entropy production again increase and disorder becomes more and more positive (Naveh 1990 also published as Chapter 3 in this anthology, 1991, 1994b).

This is expressed by the monotony and low structural, floristic and faunistic diversity of undisturbed Mediterranean shrublands and Maquis thickets, and by their high inflammability. They are therefore not reaching a homeostatic climax stage of rich and stable "mature" ecosystems, because they are aging and becoming more and more stagnant and senescent, more and more inflammable and therefore also more unstable and species-poor in time.

This process can be illustrated with the help of Prigogine's dissipative function, in which entropy (s) and therefore also disorder (D) grows at the rate ds/dt: $D = ds/t$. D may be positive, negative or zero. If it is zero then the system is in a stationary state—as in the homeostatic "climax" state of natural systems. If it is positive ($D > 0$), then it is in a state of progressive disorganization, and conversely the rate of negentropy and information (info) decreases and it looses its capacity for self-stabilization and self-organization, as in the state of too frequent or no perturbations: $D > 0 = dinfo/dt < 0$. But if D is negative ($D > 0$), then the system is in a state of progressive organization and increases its negentropy and information, as in the case of optimum perturbations. Then the homeorhetic metastability and therefore the capacity of constant self-organization and stabilization can be maintained: $D < 0 = dinfo/dt < 0$. Fig. 4 illustrates the results of these 3 different perturbation regimes.

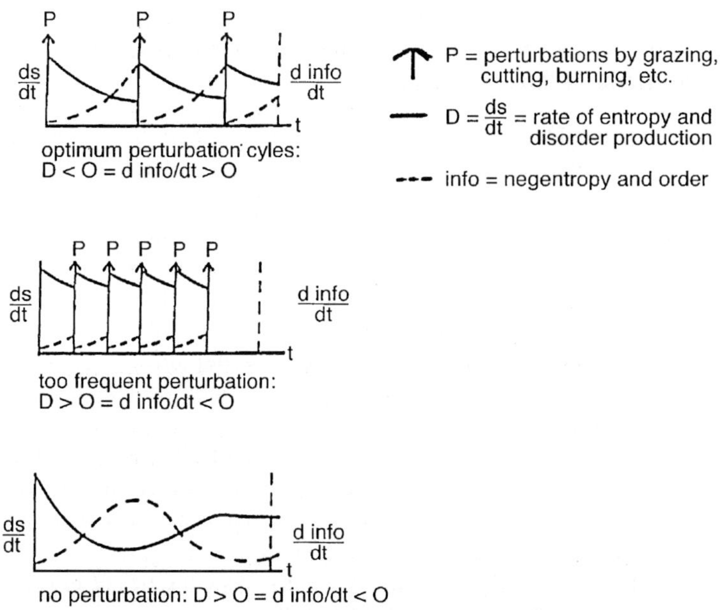

Fig. 4. Mediterranean landscapes as perturbation dependent dissipative structures.

The determination of management strategies for such "optimum" regimes of perturbation and enhancement of biological diversity alone (in nature reserves), or in combination with other land use goals, such as increase of economic production, recreation amenities and scenic values, requires systematic, long-term studies on different landscape scales. Such studies will have to provide the answer if and when controlled burning, both for the reduction of fuel and prevention of destructive wildfires and for dynamic conservation management can be replaced completely by chemical or mechanical means or by domestic or wild animal grazing, and which of these functions are most desired for multiple land uses. These landscape ecological considerations are beyond the now very popular "integrated dynamic ecosystem management."

Many ecologists are already aware of the need for "new paradigm in ecology", replacing the metaphor of a balance of nature maintained without human interference by "the flow of nature" through dynamic change, induced by disturbances including humans and their effects (Pickett et al., 1992). It is regrettable that this new ecological paradigm is not based on the above-described transdisciplinary evolutionary systems concepts and the breakthroughs in non-equilibrium thermodynamics. These are essential for a comprehensive theory of sustainable conservation and restoration management of biosphere landscapes as multifunctional life support systems.

A MULTIFUNCTIONAL CLASSIFICATION OF TOTAL HUMAN ECOSYSTEM LANDSCAPES INTO BIOSPHERE- TECHNOSPHERE- AND AGRO-INDUSTRIAL- LANDSCAPES

For the implementation of the THE landscape paradigm in practice, their classification according to their natural elements and land uses is not sufficient. Thus a new functional classification of natural and cultural biosphere landscapes and technosphere landscapes is urgently required. Its major point of departure should not be based on the degree of so-called "naturalness" of landscapes, but as a clear distinction between these above-mentioned major functional landscape classes and their basic multifunctional differences, according to the following parameters:

1. The different energy and material inputs, throughputs, and outputs
2. The differences in the kind and amount of regulation by natural or human information, as a result of these features
3. The above-described capacity of landscapes to organize themselves in a coherent way by maintaining their structural integrity in a process of continuous self-renewal. The latter determine their evolutionary regenerative capacities, had only by autopoietic natural and seminatural biosphere landscapes.

In addition, we have to take into consideration the two basic system types of internal self-organizing behavior, defined by Jantsch (1975) as follows:

1. *Adaptive (or organismic) systems,* which adapt to changes in the environment through changes in their internal structure in accordance with pre-programmed information (genetic templates). This enables further biological evolution.
2. *Inventive (or human action) systems,* which change their structure through internal generation of information (invention) in accordance with their intentions to change the environment. Such information is generated within the system in feedback interactions with the environment. This enables further noospheric cultural evolution.

In Fig. 5 all THE landscapes and their ecotopes are presented in a hierarchical model, ranging from the natural bio-ecotope pole to the cultural techno-ecotopes pole along a gradient of increasing degrees of modification, conversion, and replacement of natural elements, controls and functions by human-made, artificial ones. These are closely related with increasing throughputs of fossil energy and materials.

1. *Natural and close-to-natural biosphere landscapes* contain only natural (meaning spontaneously evolving and reproducing) organisms. As adaptive self-organizing systems, natural, biological, physical and chemical information internally regulate these landscapes. Contrary to the "non-interference" climax paradigm, it is essential to ensure the undistorted continuation and/or re-introduction of all those natural ecological processes, maintaining their natural homeorhetic flow equilibrium and thereby also their biological diversity and productivity together with all other vital functions (Rickleffs et al., 1984), and hence their evolutionary future. Because of the overwhelming global human dominance, almost no truly undisturbed natural and very few close-to-natural, or sub-natural landscapes are left on earth, in the most inaccessible regions.

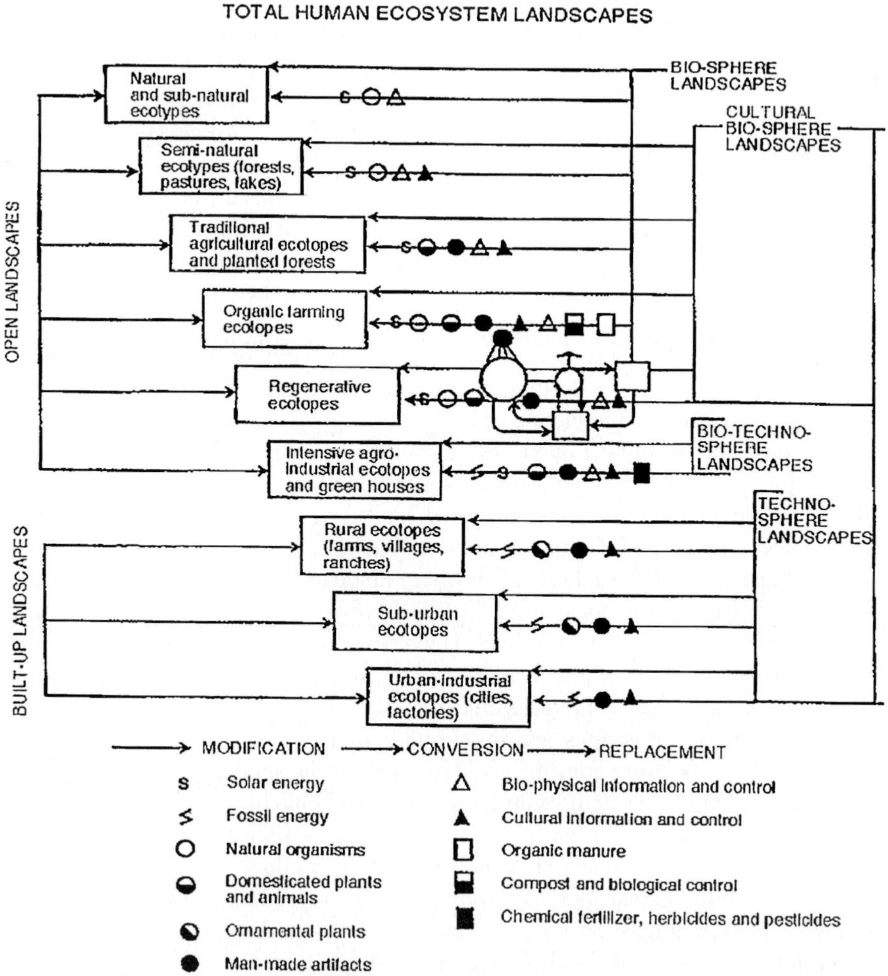

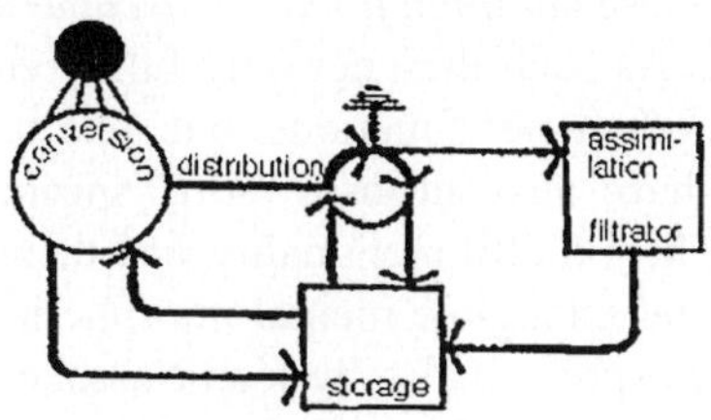

Fig. 5. A functional classification of Biosphere and Technosphere landscapes, according to their energy and material flows, natural of cultural regulation and autopoietic capacities.

However even these are severely threatened both from external pressures, as well as from uncontrolled commercial exploitation of their biodiversity, agricultural encroachment, and poaching. Therefore the last and most precious and attractive protected nature refuges are losing their integrity by mass tourism, even if it is promoted as "eco-tourism." As described by de Palma (1996) for the famous World Heritage Site of the first Canadian park—the Banff National Park—this sad situation is typical for many other cases all over the world. The closely interwoven intrinsic functions and assets of bountiful natural beauty, as well as cultural and historical values are "sold off for making money": The multiple threats caused by the heavy tourist pressures, the suppression of natural fires, the building of roads to accommodate motorized traffic, the damming of rivers for a recreational lake and all other recreational and commercial developments, which disrupt the natural ecological and evolutionary processes and the normal cycles of wildlife, are destroying this unique touchstone of Canada's national identity. The prevention and reduction of these recreational over-development damages and the introduction of sounder and more sustainable management practices are essential for the future of these most valuable biosphere landscapes.

All other human modified and converted landscapes are controlled to lesser or greater degrees by cultural information. they should be considered therefore as *cultural landscapes.* These have to be subdivided further into three major classes and subclasses.

2. The first major class consists of *Cultural biosphere landscapes.* Like natural landscapes these are all *solar-powered.* However, they are controlled by different degrees of both natural and human information, and are functioning therefore as a *mixture of internal organismic-adaptive and human-intervention self-organizing systems.* Also these landscapes are functioning as open, dynamic self-organizing systems, which are far from equilibrium. This enables the spontaneous emergence of new order, creating new structures and new forms of behavior.

2.1. *Human modified and used semi-natural ecotopes,* such as forests, woodlands, grasslands, wetlands, and lakes, in which biological productivity and diversity are based, as in natural landscapes, on spontaneously reproducing organisms. Their productivity is used at least partly for "hard values" as marketable goods for human consumption, such as wood, fiber, forage, and fish production.

But at the same time these *multifunctional landscapes* are important life supporting and improving systems. As such they not only fulfill vital instrumental production, regulation, protection and carrier functions, but, like natural and sub-natural landscapes, their ecotopes have also intrinsic "soft" spiritual, aesthetic, scientific and other cultural values. Their multifunctionality has therefore both tangible physical geospheric-biospheric and intangible mental noospheric dimensions (Naveh, 2001). This has been forcefully expressed also by Joan Nassauer (1997), a leading American landscape ecologist and landscape planner.

More recently ecological economists have suggested that to ensure their sustainable use, their "natural capital," should be presented as a counterpart to the classical "economic capital." In the conventional economic sense "capital" is accumulated by the production of marketable goods, and is evaluated by monetary values. However, there is a danger that if the conservation of this natural capital and its monetary, utilitarian values will become the only justification for conservation management and sustainable development, the other, intangible dimensions and "soft" functions and values will be neglected. Such monetary evaluations are highly doubtful, even for the loss of the most vital "natural capital" namely soil, clean air and fresh water, and they are impossible for such vital life support functions such as the aesthetic and the psycho-hygienic and therapeutic function of these biosphere landscapes. We can even assume that their value in the information society will be of greater importance than in the industrial society. Therefore they should become critical issues in the land use decision-making process.

Of greatest relevance in this context are the findings by the influential American environmental psychologist Stephen Kaplan (1995), on "*the restorative experience of nature*" against the many stresses of modern life. Of special significance in this respect for the information society is the restorative function of these landscapes after "*direct attention fatigue*" caused by continued and intensive mental work, such as performed by High-Tech workers spending many hours in creative work behind the computer.

The emerging transdisciplinary science of ecopsychology is opening new vistas for the recognition of our deeply ingrained and mostly unconscious relations to nature, as presented in all biosphere landscapes. These have been entirely overlooked in modern psychiatry as a typical modern urban-industrial science. But fortunately, now more attention is devoted to these psychotherapeutic biosphere landscape functions, especially by the emergence of the science of ecopsychology.

This has been forcefully expressed by Theodore Roszak (1995, p. 5), the eminent American historian in his introduction chapter to the first anthology on ecopsychology:

> The understanding of human sanity has always stopped at the city limits" . . . "But now are signs that this is beginning to change and a new generation of psychotherapists is seeking ways in which professional psychology can play a role in the environmental crisis of our time.". . . "Unlike other mainstreams schools of psychology that limit themselves to the intrapsychic mechanisms or to a narrow social range, ecopsychology proceeds from the assumption that its deepest level of psyche remains sympathetically bonded to the Earth that mothered us into existence.

For all those committed to the conservation of biosphere landscapes it will be very important to join forces with ecopsychologists, because they can help to reach a better understanding of the deep environmental bond between human beings and the natural environment, during which we evolved and lived for millions of years. In this long period the industrial age was just a split second. But this short span of time was sufficient for the almost total urban alienation from nature and from turning so many of us into compulsive consumers. Ecopsychologists could eventually help society to find ways to overcome this modern syndrome of addiction.

Roszak (1994) has also rightly pointed out to the dangers of "the cult of information," as the most recent expression of "technophilia"—our love affair with machines. In facing the rapidly advancing computer technologies, we should be aware of the danger that, instead of dealing with real three-dimensional, healthy landscapes on which the future of life on Earth depends, the information society will be satisfied with virtual landscapes and the plastic Disney land "nature." In this respect, the comments by Robert Thayer (1994, p. 100) in his important book on technology, nature and the sustainable landscapes, is of great relevance:

> Approaching sustainability will require us to alter radically the various entrenched meanings implicit in our lives, technologies, and landscapes. Just as the need for fundamental change is being felt ever more acutely, an exponential increase is occurring in the ability of information technologies to gloss over real environmental problems and create a landscape fantasy so vivid it threatens to replace reality altogether. Sustainability, I believe, springs not only from guilt associated with technology's physical impact on nature, but also from the growing tendency of technology to replace nature in our minds as well.

By far the greatest majority of all these semi-natural bio-ecotopes are presently undergoing accelerated biological and cultural impoverishment. Therefore the conservation and restoration of their ecodiversity and its resulting biological, ecological and cultural functions, is of no lesser importance than that of the biodiversity of natural and close-to-natural bio-ecotopes. For this purpose, their dynamic multifunctional conservation and restoration management should ensure their homeorhetic flow equilibrium. It can be maintained by those natural as well as cultural ecological processes which have been introduced by human land uses throughout their evolutionary history, such as grazing, cutting, prescribed burning and traditional agricultural practices (Rickleffs et al., 1984; Naveh, 1991, 1994a also published as Chapter 4 in this anthology; Naveh and Lieberman, 1994).

In order to be able to implement such strategies, landscape ecologists, together with all other scientists and professionals involved in this challenge, have to learn as much as possible about the early land-use history not only from a biological-ecological point of view, but also from an ethno-historical and anthropological-cultural perspective. Because of the important role of fire in shaping most semi-natural landscapes, they have to become well versed in fire ecology and its application in different patterns and scales of heterogeneous natural and rural landscapes.

2.2. *Traditional and agro-silvo-pastoral ecotopes* such as meadows, planted forest groves and traditionally cultivated fields and orchards, in which domesticated plants and animals have replaced their natural competitors and their solar-energy driven biological production is channeled into economic goods with no or very low inputs of chemical fertilizers and pesticides. Although human–controlled and maintained, these bio-ecotopes have retained still a certain amount of their self-organizing capacities and they fulfill important functions for the preservation of biological and cultural ecodiversity in the rural landscape. According to Bignal and McCracken (1996), occur more that 50% of the most valued biotopes and in these "low-intensity farmlands" and most efforts should be devoted to the prevention of their loss. However, in the process of urbanization they are undergoing intensification or being abandoned or converted into commercial monospecies conifer or eucalyptus forests. Thereby they are losing both their biological and cultural diversity and scenic attractiveness, that is deeply embedded in their structural and functional heterogeneity as their *total landscape ecodiversity.* This of no lesser importance than that of the biodiversity of natural and close-to-natural and semi-natural bio-ecotopes and should be conserved and restored (Naveh, 1994, 1998 also published as Chapter 6 in this anthology; Farina, 2000).

However, these landscapes can be rescued only if the local populations and their governments are willing to recognize them as valuable cultural heritage landscapes. For this purpose they have to invest sufficient means and manpower to conserve worthwhile examples by restoring and maintaining these traditional farming practices and their typical local crop strains and animal breeds, and thereby also their great genetic value.

To these cultural biosphere landscapes belong also two new and most promising sustainable agricultural systems and their ecotopes, namely:

2.3. *Organic farming ecotopes, cultivated without inputs of chemical fertilizers, herbicides and pesticides.* Thanks to the application of suitable agro-technological methods, higher crop diversity and their rotation, and the building up of high soil fertility through the use of organic manure and compost, their productivity is much higher than that of traditional farming. It comes now close to that of high-input agro-industrial farming systems, but without the detrimental environmental effects caused by chemical fertilizers, herbicides and pesticides (Mansvelt and Mulder, 1993).

On one hand there is a steady proliferation of these chemicals, including those that can interfere with human and animal endocrine systems, but on the other, there are also encouraging signs of a rapidly increasing demand for healthier products, and their sales are growing by 20% a year. Farmers in Europe have doubled the area cultivated with organic methods to 4 million hectares in only three years and in Italy and Austria the share of certified organic products topped 10 percent in 1999 and are now reaching 20 percent and more. The studies by Mansvelt and van Luppe (1999) and by Tress (2000) are fine examples of the important role that landscape ecologists can play in reshaping the European agricultural landscapes in the direction of more sustainable, healthy, and attractive organic farming ecotopes.

2.4. *Regenerative ecotopes* are the result of an even more advanced version of organic farming, in which not only the natural regenerative capacity of cultivated land is restored, but also the basic cyclic flows of energy, water and nutrients of natural biosphere landscapes, based on the inputs of solar radiation. At the same time, their biological production is channeled partly into agricultural production for human uses. Such regenerative farming systems are driven entirely by solar energy and other regenerative non-polluting energy sources. One of the most outstanding examples is realized at the Center for Regenerative Studies at the California Polytechnical State University at Pomona in a combined inter- and transdisciplinary teaching, research and demonstration project in regenerative technologies. As has been described lucidly by its chief designer (Lyle, 1994), here the natural biosphere functions of energy/material conversion, distribution, filtration, assimilation, storage are utilized for intensive agricultural production with the help of human-inventive "neotechnological" information. They provide thereby for continued replacement through their own functional processes of the energy, material and information used in their operation. The symbol for regenerative farming systems in Figure 5 stems from Lyle (1994) and will be shown in more detail in the next figure.

3. *Intensive agro-industrial ecotopes* are an intermediate class between bio- and -technosphere landscapes that have replaced almost all other, low-input cultivated

agro-ecotopes in industrial countries. Although regarded in general as "green landscapes," they should not be confused with all above mentioned, sustainable biosphere landscape. Driven by a shortsighted, over-consumptive and socially unjust market economy, they are spreading now also in many developing countries. Like in bio-ecotopes, their productivity is dependent on photosynthetic conversion of high-grade solar energy. But this energy is subsidized to a great extent by low-grade fossil energy, and the natural biological control mechanisms have been replaced almost entirely by agro-technological information, aiming at maximum production through heavy chemical inputs of fertilizers, pesticides and herbicides. Therefore, these landscapes have lost all self-organizing and regenerative properties. In this respect, and in their detrimental environmental impacts on the open landscape, its wildlife and biodiversity, and on the quality of its natural resources of soil and water, as well as on human health, they come very close to technosphere landscapes, and like these, they are "*Throughput Systems.*"

The efforts to minimize these adverse impacts by slow-releasing fertilizers and integrative pest control are only palliative measures which cannot ensure in the long run a sustainable future for our agricultural landscapes and healthy food production. As stated rightly by Lyle (1994), sustainability can only be achieved by replacing these linear technosphere processes of high input and throughput flows with cyclical flows at sources, consumption centers, and sinks, as attempted in regenerative systems. Even in those countries, in which the highest agricultural production of crops and livestock has been achieved, this type of high-input and throughput of agriculture production is not only ecologically unsustainable, it also cannot be sustained economically without heavy governmental subsidies. Therefore, the time is ripe to realize that the future of sustainable, profitable and healthy agricultural production will depend on the restoration of the fertility and regeneration capacities of cultivated land by organic and especially by regenerative ecotopes. This should be the aim for all future-oriented land and land use research and education.

The major functional brake in cultural landscapes is in their difference in solar or fossil energy inputs:

4. *Technosphere landscapes and their ecotopes* as well as their technological artifacts such as highways, bridges, mines, quarries and power lines are artificial, human-made and are maintained chiefly by fossil and nuclear energy and their technological conversion into low-grade energy and materials. They lack the multifunctionality and self-organizing and regenerative capacities of biosphere landscapes and are regulated solely by cultural information of the internal self-organization systems behavior of human inventive actions. This results in high outputs of entropy, waste and pollution with far-reaching detrimental impacts on the remaining open landscapes and their biological productivity and ecological diversity and stability and human health. Their subclasses differ in these detrimental environmental impacts, but these differences are vanishing rapidly in the process of urbanization.

4.1. *Rural ecotopes,* such as farms, ranches, villages, are closely interwoven with the cultural semi-natural and agricultural landscapes. In Europe and all other industrialized countries they are undergoing a rapid transformation into the other techno-ecotopes. As has been shown by successful examples in Great Britain by Green (1995), and also elsewhere by Lucas (1992) this can be prevented at least partly by sound countryside planning and management in which landscape ecologists are actively involved.

4.2. *Sub-urban ecotopes* can still contain larger biosphere islands, such as lakes, riverbeds, parks and forest groves. These can contribute much to the improvement of the quality of city life and, like rural landscapes, they are therefore of great value. This is also true of recent tendencies to establish low-maintenance gardens, based chiefly on indigenous plants, provided that these gardens are maintained without inputs of chemical fertilizers, herbicides and pesticides.

4.3. *Urban-industrial ecotopes* are the fastest-growing, fuel-powered technosphere landscapes, with the most pronounced adverse environmental impacts. This is indicated by the huge amount of energy that flows annually through western urban-industrialized centers such as New York or Tokyo. They can be measured in millions of kcal per square meters, as compared to only thousands of kcal in solar-powered biosphere landscapes. Fischer-Kowalski and Haberl (1997) have shown that in the transfer from a traditional, chiefly agrarian society into the modern fossil-powered urban industrial society, the social metabolism, as measured by the energy input in GJ/capita/year in Austria, has risen from about 65 to 223 GJ, and that of material input, measured in t/capita/year, from about 4 to 21.5 t. These figures are a very good index for the quantitative measurement of the impact of the technosphere on our THE biosphere landscapes and on the destabilizing of the geosphere and atmosphere. They show clearly the energetic differences between the regenerative natural and cultural biosphere landscapes and the throughput agro- and urban-industrial technosphere landscape systems. As illustrated in Fig. 6, this results in high outputs of entropy, waste and pollution with far-reaching detrimental impacts on the remaining open landscapes and their biological productivity and ecological diversity and stability, as well as on human health.

Rees (1995), applied an innovative index of "*Ecological Footprints*" as the total area of productive land and water required to produce all the resources required by the population of city or a region and to assimilate all its waste produced. He found that in 1991 the population of the city of Vancouver used the productive output of a land area nearly 200 times larger than its political area to maintain its consumer lifestyle.

The improvement of rural and urban life quality is closely coupled with the reduction of these huge throughputs and outputs. This will require far-reaching functional and structural changes in the "metabolism" of the cities and their design, as part of the sustainability revolution, in which landscape ecologists should take an active part. According to UN projections, by 2025 60% of the world's 8.3 billion global citizens will be living in towns and cities and presently half of the world's poorest people, or some 420 million are living in urban settlements and chiefly in

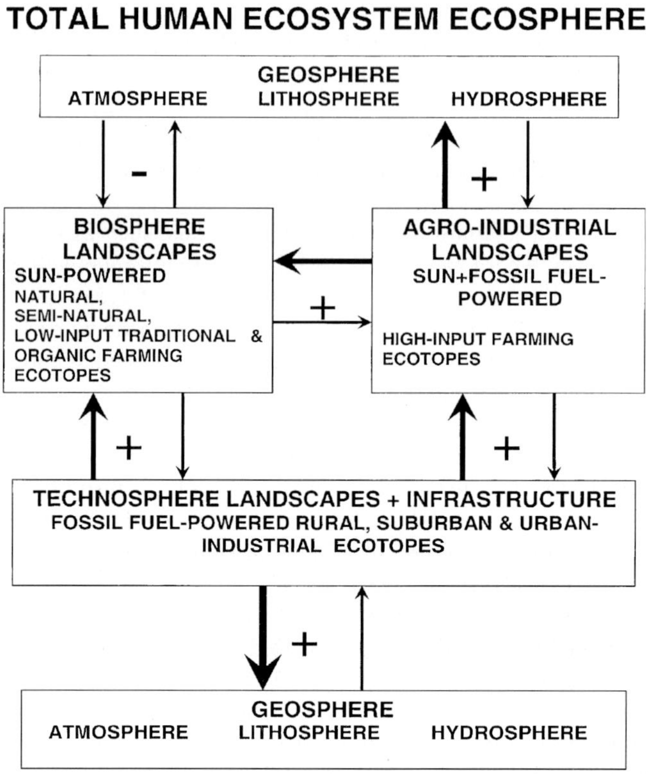

Fig. 6. The disorganized Total Landscape of the industrial Total Human Ecosystem-Ecosphere, and its destabilization by the Technosphere.

"megacities." This "urbanization syndrome" represents clearly one of the most important social, political and ecological challenges for the sustainable future of our Total Human Ecosystem and its landscapes.

The internal self-organizing behavior of technosphere landscapes is dependent entirely on our human inventive actions, driven by our consciousness as a feedback reaction on the information gained from the environment. Therefore there is hope that human society will change these threatening trends before local and global ecological catastrophes will force us to do so. It is up to landscape ecologists and planners to accept this challenge. They have to prove to the decision-makers and the people at large that the creation of biosphere oases in the biological deserts of our cities is by far the cheapest and most efficient way for over-all urban environmental improvement. It does not require fossil energy like all man-made engineering devices and therefore does not add any further burdens of waste, pollution and entropy on the urban landscapes. But is has also much more far-reaching effects: It reduces the soul and body crippling threats of urban stresses and draws the estranged city dwellers closer to a new I-Thou dialogue with nature. It brings thereby a richer dimension to his life, which cannot be measured by dollars or Yen values and material goods, but it will pay high dividend in non-economic richness and urban life-quality.

CONCLUSIONS: THE NEED FOR A POST-INDUSTRIAL SYMBIOSIS BETWEEN HUMAN SOCIETY AND NATURE

Although all these bio-agro- and techno-ecotopes are spatially interlaced into larger regional landscape mosaics, they are antagonistically related because of these overwhelming adverse and destabilizing impacts of the techno- and agro-industrial landscapes and their entropic and wasteful fossil-powered throughput metabolism. They form a disorganized conglomerate that the German landscape historian Sieferle (1997) has called "The Total Landscape." This industrial Total Landscape will be able to function as a coherent, sustainable THE ecosphere landscape only after these antagonistic relations between the biosphere and the technosphere have been reconciled through the creation of an urgently needed post-industrial cultural symbiosis between human society and nature. Such a symbiotic relation should lead, above all, to the structural and functional integration of bio- and techno-sphere ecotopes into a coherent, sustainable ecosphere, in which both biological and cultural evolution can be ensured.

As illustrated in a simplified cybernetic model of the Total Human Ecosystem ecosphere (Figure 6), with exception of the stabilizing negative feedback couplings, maintaining a dynamic flow equilibrium between the biosphere landscapes and the geosphere, all other interactions are ruled by destabilizing positive feedback loops. Because of the rapidly vanishing intact biosphere landscapes and these overwhelming decoupling effects of the technosphere landscapes, the "Gaia hypothesis "(Margulis and Lovelock, 1974), regarding the biosphere together with the atmosphere as a global co-evolutionary self-regulating and self-renewing system, may loose gradually its validity, endangering thereby the future of life on Earth.

A first essential step towards this symbiosis will be the establishment of new, better balanced, complementary relations between healthy, livable and attractive technosphere landscapes and its "hinterland" of vital, attractive and productive biosphere landscapes. This can be achieved through comprehensive landscape planning, conservation, restoration, and design, together with environmental management for sustainable development towards information society.

One of the most important stabilizing feedback couplings is the function of biosphere landscapes as a biological filter and living sponge, absorbing the emissions from the technosphere, including the greenhouse gases and carbon dioxide. At the UN Marrakech meeting on November 7, 2001, some of the major findings of a cluster of eight research EU sponsored projects on the carbon cycle, called "Carbon Europe," were summarized as follows:

1. The European biosphere is a carbon sink that can absorb about 20–30% of the annual European carbon emissions.
2. The European biosphere has additional potential to absorb carbon emissions through forestation projects and improved management methods. This is an

important message, since sinks can be used in part to fulfill the reduction commitments as laid down in the Kyoto protocol.

However, on global scales this symbiosis can be realized only as part of an all-embracing cultural and technological sustainability revolution, initiated by the transition from the "fossil age" to the "solar age" of the new economy, based on the limitless power of the sun as the non-polluting and renewable energy source. In one year the sun provides 15,000 time more energy than that of the total annual consumption of fossil and nuclear energies, and the annual photosynthetic output of the world's vegetation is 10,000 greater than the chemical industry's annual global output. According to reputable studies, such as those conducted by the British Shell concern—which is investing already many millions of dollars in preparation of the emerging shift towards solar energy—by 2050 half of the world's energy needs will be supplied from these renewable sources (Scheer, 2000).

As envisaged by Laszlo (1994), this cultural evolutionary process will guide the bifurcation on the leap towards a higher organizational level of the emerging sustainable information society. But it will be driven not only by the widespread adoption of technological innovations of regenerative and recycling methods and the efficient utilization of solar and other non-polluting and renewable sources of energy: It must to be coupled with more sustainable lifestyles and consumption patterns, caring for nature and even investing in nature.

The former president of the Club of Rome, Ricardo Diez Hochleitner (1999) has expressed these demands in his optimistic vision for the future of the information society as follows: "Humanity's future will only be secured when we are more careful with nature and we shape our interaction with it in a more sustainable way. This will require not only doing everything to increase efficiency of how we use resources; it also implies sustainable lifestyles which will entail giving something up."

An important step to achieve these goals in regional sustainable development will be the replacement of the ruling neo-classical market-economy incentives for quantitative growth, by a more far-reaching and more just integrated socio-ecological approach, based on the Total Human Ecosystem paradigm. This development should be aimed chiefly towards qualitative growth by fostering positive synergies between people, their economy and their open and built-up landscapes.

Thanks to the above-described, recent insights in self-organization of autopoietic systems and their cross-catalytic networks, we are now able to express these new symbiotic relations between nature and society in much more robust and even mechanistic terms and translate them into sustainable development.

In Fig. 7, the major auto- and cross-catalytic cycles and cybernetic feedback loops of our Total Human Ecosystem are presented. In biosphere landscapes, the negative self-stabilizing loops are indicated by a minus sign, and the positive autocatylitic loops symbolizing also autopoiesis and evolution are indicated by a plus sign. The letters CNN in both directions of the arrows indicate the potentials for mutually supporting and amplifying relationships through cross-catalytic networks. However in contrast to the biosphere landscapes, in technosphere landscapes, the plus signs symbolize the "run-away," destabilizing feedbacks of exponential expansion, which can be restrained only by cultural negative feedbacks of sustainable planning and

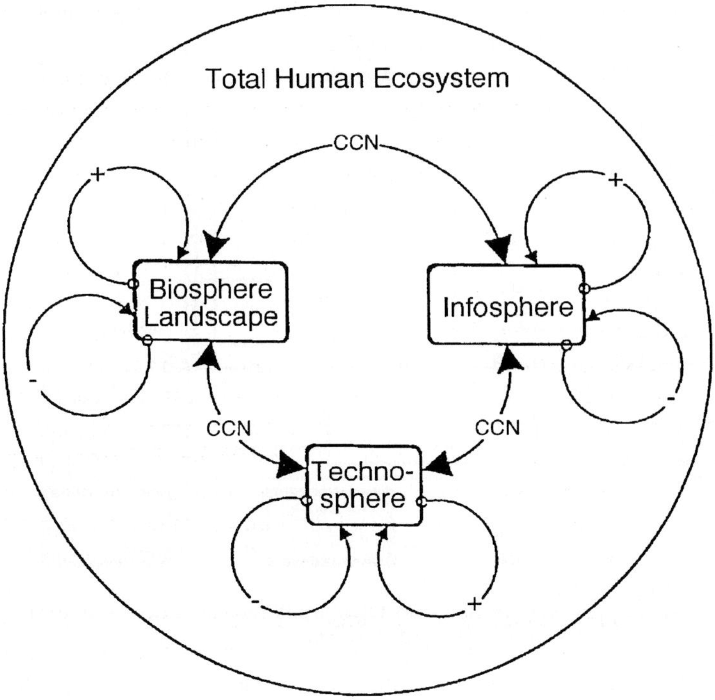

Fig. 7. Major auto- and cross-catalytic cycles and feedback loops in the emerging information society of out Total Human Ecosystem.

environmental management. These are channeled via the information sphere as the noospheric cultural information pool playing a rapidly growing role in the emerging information society.

This conceptual THE model has been applied in a recently completed interdisciplinary and multinational project EU project on "Modeling Sustainable Regional Development in the European Information Society" (EU-Project "MOSES" (2000), carried out under the leadership of Wolf Grossmann in 5 regional case studies. With the help of dynamic systems simulation models, combining a recursive systems-dynamic simulation model with cross–catalytic networks (ISIS) and other innovative methods and tools we could show that we are able to create new cultural, information-rich cross-catalytic and synergistic feedback loops in the emerging information society. In this model we attempted to link natural, ecological, socio-cultural and economic processes of our Total Human Ecosystem, and to create the scientific basis for new cross-catalytic and synergistic feedback loops. These should reduce the destabilizing effects and help to ensure lasting mutually reinforcing—that means synergistic—benefit for the people and their physical, mental, spiritual and economic welfare, together with the creation of healthy, productive and attractive landscapes for the emerging information society.

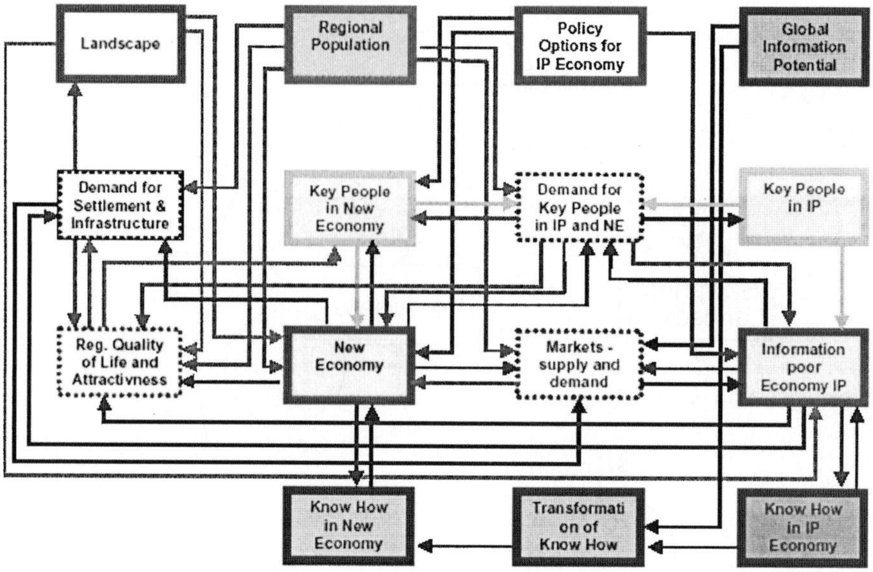

Fig. 8. Overview of the structure of the Information Society Integrated ISIS Model (Prepared by W. D. Grossmann, M. Lintl, and H. Kasperidus, for E-U Project Moses 2000).

The project coordinators W. D. Grossmann, M. Lintl, and H. Kasperidus prepared a very simplified version of the basic structure of this complex holistic model. It is presented in Fig. 8.

This model revealed mutual supportive cross-catalytic network relations in the dynamics of the emerging information society, initiated by the development of younger companies, their innovators and key people, which are comparable to the autopoietic dynamics driving ecological systems and natural and semi-natural biosphere landscapes.

With the help of this model we could further show that the contribution of nature to regional attractiveness is crucial for regional upswing. Thereby the citizens will gain from nature, namely the "soft" intangible and intrinsic values, and the "hard" and marketable values of the regional green" biosphere landscapes.

The model resembles a CCN, because both partners are capable of autocatalytic growth. But because of the unbalanced relationship between the biosphere landscapes, and the technosphere landscapes, only the regional economy gains from nature, increasing its attractiveness. At the same time the destructive feed-back of the regional economy is increasing because of unrestricted growth and unsustainable development. This is shown in Fig. 9.

Our study showed that the new successful economy could create the missing vital symbiotic CCN link of a "double G"—"give and gain" symbiotic relation by "investing in Nature" through paying an adequate fixed share of the tax income dedicated to

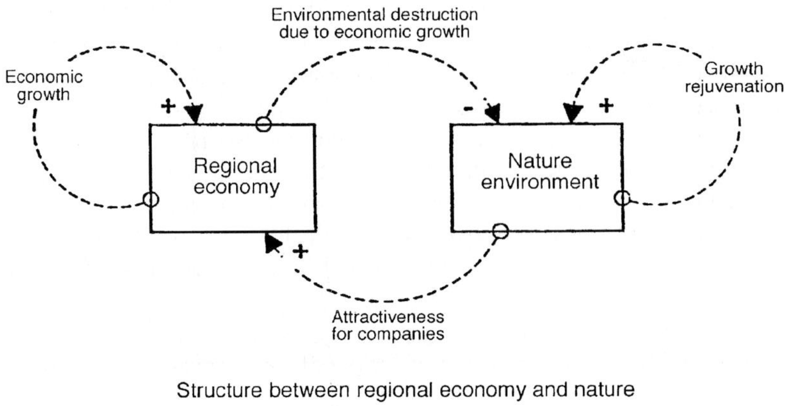

Fig. 9. The unbalanced relationship between the regional economy and nature in the industrial society, leading to its destruction. (Grossmann and Naveh, 2000).

the conservation and restoration and the sustainable design and management of attractive biosphere landscapes. The new link could even allow to purchase and restore more land that becomes available because of the lower demands and negative impacts by human activities of the post-industrial technosphere of the information society (Fig. 10).

If such symbiotic relations could be realized in actual sustainable development projects, they will constitute an important step towards the environmental sustainability revolution.

That this is not an utopian dream can be learned from the encouraging examples provided—in addition to many others—in the 1999 *State of the World report* (Brown et al.,

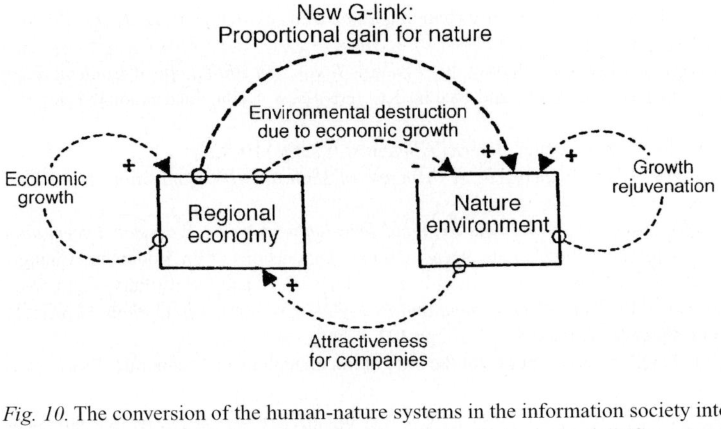

Fig. 10. The conversion of the human-nature systems in the information society into synergistic CCN relations by the creation of symbiotic Double G link of "give and gain" (Grossmann and Naveh, 2000).

1999) indicating the beginning of such an environmental sustainability revolution. To contribute jointly with scientists from other relevant disciplines to these developments should become one of the greatest challenges for future-oriented landscape ecologists. Guided by the holistic worldview and its practical implications they could eventually provide leadership not only as educators, researchers and professional experts, but also as integrators in transdisciplinary teamwork for sustainable development.

In concluding, our hope for a sustainable future for our biosphere landscapes of our post-industrial Total Human Ecosystem as a whole, has been expressed lucidly by Laszlo (2000, p. 114) in the final sentences of his "Macroshift 2000-2010" book:

> Endowed with the highest forms of consciousness in our regions of the universe, we are the only species that not only acts, but can also foresee the effects of its actions. As members of a species capable of foresight, we must live up to our responsibility as stewards rather than exploiters of the complex and harmonious web of life on this planet.

REFERENCES

Antrop M, Van Eetvelde V (2000) Holistic aspects of suburban landscapes: Visual image interpretation and landscape metrics. In: Landscape and Urban Planning, Holistic Landscape Ecology in Action. Palang H, Mander U, Naveh Z (eds) 50, pp 43–58.

Bignal EM, and McCracken DI (1996) Low-intensity farming systems in the conservation of the countryside. *Journal of Applied Ecology,* 33: pp 413–424.

Bohm D (1980) *Wholeness and the Implicate Order.* London: Routledge and Kegan.

Bohm D, and Peat DF (1987) *Science, Order, and Creativity. A Dramatic Look at the Roots of Science and Life.* New York: Bantam Books.

Brown L (2001) Dust bowl threatening Chinas future. *The Earth Policy Alerts.* www.earth-policy.org

Brown LR, Flavin C, and French H (eds) (1999) *State of the World 1999. A World-watch Institute Report on Progress Towards a Sustainable Society.* New York, London: Norton & Company.

Cahalan W (1995) Ecological groundedness in Gestalt Therapy. In: *Ecopsychology. Restoring the Earth healing the Mind.* Roszak T, Gomes ME, and Kanner AD (eds). San Francisco: Sierra Club Books.

Capra F (1996) *The Web of Life. A New Understanding of Living Systems.* Anchor Books, New York: Doubleday.

De Palma A (1996) Popularity brings a huge Canadian park to crisis. *The New York Times, Science Supplement,* 5 January 1996: 3A.

Di Castri F (1997) Editorial: Landscape ecology in a changing globalized environment. *Landscape Ecology* 12: pp 3–5.

Di Castri F (1998) Ecology in a global economy. In: *Ecology Today: An Anthology of Contemporary Ecological Research.* Gopal B, Pathak PS, and Saxena KG (eds) New Delhi: International Scientific Publications, pp. 1–17.

Egler FE (1964) Pesticides in our ecosystems. *American Scientist,* 52: pp 110–136.

Eigen M, and Schuster P (1979) *The Hypercycle: A Principle of Natural Self-organization.* New York: Springer.

EU-Project MOSES (2000) *Modelling Sustainable Regional Development in the European Information Society.* EU Program, Environment and Climate, Area 4: Human Dimensions of Environmental Change.

Farina A (1989) Recent changes of the mosaic pattern in a montane landscape (North Italy) and consequences on vertebrate fauna. In: *Land Abandonment and its Role in Conservation. Options Mediterraneennes.* Baudry J, and Bunce RGH (eds) Ser. A:15, pp 121–134.

Farina A (2000) The cultural landscape as a model for the integration of ecology and economic. *BioScience,* 50: pp 313–320.

Fischer-Kowalski M, and Haberl M (1997) Tons, joules, and money: Modes of production and their sustainability problems. *Society & Natural Resources,* 10: pp 61–85.

Green B (1995) *Countryside Conservation. The Protection and Management of Amenity Ecosystems,* 2nd edn, London: UNWIN HYMAN.

Grossmann WD (2000) Realizing sustainable development with the information society—the holistic Double-Gain-Link approach. In: *Landscape and Urban Planning,* Palang H, Mander U, and Naveh Z (eds) Holistic Landscape Ecology in Action. 50, pp 179–193.

Grossmann WD, and Naveh Z (2000) Transdisciplinary challenges for regional sustainable development toward the post-industrial information society. *Third International Conference of the European Society for Ecological Economics May 3–6 Vienna,* pp 1–3. Vienna.

Haken H (1983) *Synergetics,* 3rd edn. Berlin: Springer Verlag.

Hochleitner RD (1999) Prospects of hope for the future: How we will live tomorrow. *Deutschland,* 6: pp 12–15.

Jantsch E (1975) *Design for Evolution. Self-Organization and Planning in the Life of Human Systems.* New York: George Braziler.

Jantsch E (1980) *The Self-Organizing Universe. Scientific and Human Implications of the Emerging Paradigm of Evolution.* Oxford: Pergamon Press.

Kaplan S (1995) The restorative benefits of nature: Toward an integrative framework. *Environmental Psychology,* 15: pp 169–182.

Kuhn TS (1996) *The Structure of the Scientific Revolution,* 3rd edn. Chicago: University of Chicago Press.

Laszlo E (1987) *Evolution: The Grand Synthesis.* Boston: Shambhala, New Science Library.

Laszlo E (1994) *The Choice: Evolution or Extinction. A Thinking Person's Guide to Global Issues.* New York: C.P. Putnam & Sons.

Laszlo E (2000) *Macroshift 2001–20010. Creating the Future in the Early 21st Century.* San Jose, New York: Excel.

Li B-L (2000) Why is the holistic approach becoming so important in landscape ecology? In: *Landscape and Urban Planning,* Palang H, Mander U, and Naveh Z (eds) Holistic Landscape Ecology in Action. 50, pp 27–47.

Lucas PHC (1992) *Protected Landscapes. A Guide for Policy-Makers and Planners.* London: Chapman & Hall.

Lyle JT (1994) *Regenerative Design for Sustainable Development.* New York: John Wiley.

Maturana H, and Varela F (1980) *Autopoiesis and Cognition.* Dodrecht: D. Reidel.

Mansvelt JD, and Mulder JA (1993) European features for sustainable development: A contribution to the dialogue. *Landscape and Urban Planning,* 32: pp 43–54.

Mansvelt JD, and van der Lubbe MJ (1999) *Checklist for Sustainable Landscape Management.* Final Report to the European Community on "The Landscape and Nature Protection Capacity of Organic/Sustainable Types of Agriculture." Amsterdam: Elsevier.

Margulis L, and Lovelock JE (1974) Biological Modulation of the earth atmosphere. *Icarus,* 21: pp 471–489.

Nassauer JN (1997) Cultural sustainability aligning aesthetics and ecology. In: *Placing Nature. Culture and Landscape Ecology,* Nassauer JI (ed.) Washington, D. C.: Island Press.

Naveh Z (1982) Landscape ecology as an emerging branch of human ecosystem science. *Advances in Ecology Research 12.* London: Academic Press, pp 189–237.

Naveh Z (1990) Fire in the Mediterranean: A landscape ecological perspective. In: *Fire in Ecosystem Dynamics. Mediterranean and Northern Perspectives,* Goldammer JG (ed.) The Hague: SPB Academic Publishing, pp 1–20.

Naveh Z (1991) Mediterranean uplands as anthropogenic perturbation dependent systems and their dynamic conservation management. In: *Terrestrial and Aquatic Ecosystems, Perturbation and Recovery,* Ravera OA (ed.) New York: Ellis Horwood, pp 544–556.

Naveh Z (1994a) From biodiversity to ecodiversity: A landscape ecology approach to conservation and restoration. *Restoration Ecology,* 2: pp 180–189.

Naveh Z (1994b) The role of fire and its management in the conservation of Mediterranean ecosystems and landscapes. In: *The Role of Fire in Mediterranean-Type Ecosystems,* Moreno J and Oechel W. C (eds) New York: Springer, pp 163–185.

Naveh Z (1998) From biodiversity to ecodiversity—holistic conservation of biological and cultural diversity of Mediterranean landscapes. In: *Landscape Disturbance and Biodiversity in Mediterranean-Type Ecosystems,* Montenegro G, Jaksic F, and Rundel PW (eds) Ecological Studies 136. Berlin-Heidelberg: Springer, pp 23–54.

Naveh Z (2000) What is holistic landscape ecology? A conceptual introduction. In: *Landscape and Urban Planning,* Palang H, Mander U, and Naveh Z (eds) Holistic Landscape Ecology in Action. 50, pp 7–26.

Naveh Z (2001) Ten major premises for a holistic conception of mutlifunctional landscapes. *Landscape and Urban Planning,* 57: pp 269–284.

Naveh Z, and Whittaker RH (1979) Structural and floristic diversity of shrublands and woodlands in Northern Israel and other Mediterranean areas. *Vegetatio,* 441: pp 171–190.

Naveh Z, and Lieberman AS (1994) *Landscape Ecology Theory and Applications.* 2nd edn, New York: Springer.

Pickett ST, Parker AVT, and Fiedler PL (1992) The new paradigm in ecology: Implications for conservation biology above the species level. In: Fiedler PL, and Jain SK (eds). *Conservation Biology.* New York: Chapman and Hall, pp. 65–88.

Pinto-Correira T (1993) Threatened landscapes in Alentejo, Portugal: The Montado and other agro-silvo-pastoral systems. *Landscape and Urban Planning,* 24: pp 273–278.

Prigogine I (1976) Order through fluctuations: Self-organization and social systems. In: Jantsch E, and Waddington CW (eds). *Evolution and Consciousness: Human Systems in Transition.* Reading Massachusetts: Addison-Wesley, pp 93–130.

Prigogine L (1997) *The End of Certainty: Time, Chaos, and the New Laws of Nature.* New York: The Free Press.

Prigogine I, and Stengers I (1984) *Order out of Chaos. Man's Dialogue with Nature.* Boston and London: New Science Library Shamabala.

Rees W (1995) Ecological footprints of the future. *People and Planet,* 5: pp 6–8.

Rickleffs RR, Naveh Z, and Turner RE (1984) *Conservation of Ecological Processes.* Commission on Ecology Paper No. 8. IUCN, Gland, Switzerland.

Roszak T (1994) The Cult of Information. *A Neo-Luddite Treatise on High-Tech, Artificial Intelligence, and the True Art of Thinking.* Berkeley, Los Angeles, London: University of California Press.

Roszak T (1995) Where psyche meets Gaia. In: *Ecopsychology. Restoring the Earth Healing the Mind,* Roszak T, Gomes ME, and Kanner AD (eds) San Francisco: Sierra Club Books, pp 1–17.

Ruiz de la Torre JR (1985) Conservation of plants within their native ecosystems. In: *Plant Conservation in the Mediterranean,* Gomez-Campo C (ed.) The Hague: De Junk, pp 197–219.

Scheer H (2000) The second industrial revolution: The solar age. "*Deutschland,*" December 2000: pp 46–49.

Sieferle RP (1997) *Rueckblick auf die Natur. Eine Geschichte des Menschen und seiner Umwelt.* Muenchen: Luchterhand Literatur Verlag.

Thayer RJ (1994) *Gray World, Green Heart. Technology, Nature, and the Sustainable Landscape.* New York: John Wiley & Sons.

Tress B (2000) *Landwirtschafft Landschaft. Umstellungpotential landschaftlicher Konsequenzen der oekologischen Landwirtshaft in Daenemark.* PhD Thesis, Institute for Geography and International Development Studies, Roskilde: University of Roskilde.

Waddington CH (1975) *A Catastrophe Theory of Evolution: The Evolution of and Evolutionist.* Ithaca: Cornell University Press.

Warburg M, Ben Horin RA, and Rankevitch D (1978) Rodent species diversity in mesic and xeric habitats in the Mediterranean region of northern Israel. *Journal Arid Environment,* 1: pp 63–69.

Western D (1997) *In the Dust of Kilimanjaro.* Washington, D.C. Covelo, California: Island Press/Shearwater Books.

World Resources 2000–2001 (2000) *People and Ecosystems The Fraying Web of Life.* Amsterdam: Elsevier Science.

C. RECAPITULATION

CHAPTER 14

NAVEH Z. (2005)

THE TRANSFORMATION OF LANDSCAPE ECOLOGY AND RESTORATION ECOLOGY INTO TRANSDISCIPLINARY SCIENCES OF HOLISTIC LANDSCAPE STUDY, MANAGEMENT AND PLANNING, CONSERVATION AND RESTORATION.

Oak tree growing on rock outcrops in the Ecological Garden of the Technion (Courtesy of Drorah Spitz).

1. INTRODUCTION

In the introduction to this anthology, I mentioned earlier, some of the promising trends towards holistic approaches and concepts in landscape ecology (LE) and in restoration ecology (RE). These trends are the first vital steps for their transformation into transdisciplinary landscape sciences (TLS). In the previous chapters, I presented examples of earlier efforts to meet the theoretical and practical challenges of transdisciplinary landscape research and action. This concluding chapter is a recapitulation of these efforts and a revision of the major concepts in the light of recent findings.

The first part is devoted to the severe ecological, cultural and socio-economic crisis human society is undergoing presently during the transformation from the industrial age to the global information age, which can be resolved only by an urgently needed sustainability revolution in all spheres of life on Earth. This demands from landscape ecologists and landscape restorationists a shift in thinking and action for transforming their sciences into more integrative TLS. In the second part, first, I will discuss one of the major obstacles for such a transformation, rooted in outdated beliefs, attitudes, concepts and reductionistic and mechanistic paradigms on nature–human–culture relations, inherited from classical ecology. Then I will report briefly on the important trends towards transdisciplinarity in systems ecology and integrative "eco-disciplines" like ecological economy. The concluding part is an elaboration on the true meaning of transdisciplinarity and its role for both TLS in the light of new, revolutionary insights gained in the ongoing scientific revolution towards an integral theory of everything, presented most recently by Laszlo.

2. THE ECOLOGICAL, SOCIO-ECONOMIC AND CULTURAL CRISIS DURING THE TRANSITION TO THE GLOBAL INFORMATION AGE

My first point is to show that the need for such a transformation is closely coupled with the imperative to resolve the deep ecological, socio-economic and cultural crisis, accompanying the present transition of humankind from the industrial to the post-industrial global information age. This transition is driven by the rapid development of worldwide computer networks of information, allowing the rapid economic built-up, expansion and globalization. Unfortunately, this drive towards globalization is presently ruled by a cruel market economy which is dominated to a large extent by the demands of profit seeking powerful cooperations. It results in environmental, social and cultural upheavals of global dimensions. The changes induced thereby embrace all closely interwoven spheres of human life from the biological to the cultural, social, economic, technological and political sphere (Capra, 2002).

However, during this current transition, human society has reached a crucial turning point in its relations to nature and its life-supporting and -enhancing functions. These have been distorted by the great technological power and skill of *Homo industrialis* and his obsessive belief in the panacea of growth by an increase in material goods and their wasteful consumption.

Driven by the short-sighted hubris of the ruling economic, technocratic and political decision makers, these did not realize that our scientific and technological power and skill far exceeds our ecological wisdom, scientific knowledge and environmental ethics. Therefore, in spite of the great advances achieved in science and technology, the industrial society has not been able to resolve the deep ecological crisis it has created in the last century. Its leading governments have squandered the historic opportunity to reverse the accelerating speed of Earth's environmental decline during the prosperity of the last part of the twentieth century. Unfortunately the developing countries are following uncritically these mistakes, resulting in the development of similar ecological and economic unsustainable and unhealthy industrial and agro-industrial farming system.

The close links between this ecological crisis and the severe global socio-economic crisis have been shown by Brown (2001a) in a challenging book on the urgently needed "eco-economy". By trusting our fortunes to the working of the free market, the above-described process of globalization is apparently accelerating these dangerous trends and is further widening the income gaps both between and inside nations. These are stressing even further the health of our planet and its landscapes.

At the same time, as many studies have shown, a powerful set of social forces are outweighing the effect which rising incomes have on people's health. Therefore those becoming richer are becoming neither more happy nor healthy. According to the latest FOA report, 842 million people in 2001 and in 2003 10 million more suffered from chronic hunger. Each year 10 million children under the age of five die from starvation. In most countries this is not caused by the lack of food availability, but by the lack of money to buy this food. Although the global economy is causing a rise of those employed in developing countries, about half of the 2.8 billion world's laborers are earning only less than $2 a day and have a hard time to nurse their family. About 1.1 million people have no access to clean drinking water and every 8 seconds one child dies as a result of diseases related to consumption of polluted water. According to UN predictions in 20 years two-thirds of all the world's population will suffer from fresh water deficit because its consumption doubles itself every 20 years, twice as much as the rate of population growth.

It is outside the scope of this chapter to present in detail the full extent of this ecological crisis, but the following is a short review of some of its most alarming effects. We are witnessing now the largest mass extinction of living species since the dinosaurs were wiped out 65 million years ago, but unlike the previous ones, caused by natural phenomena, the present one is almost entirely the result of human economic activities. There is an alarming increase in the "critically endangered" species in those regions, of which we have sufficient information. The current extinction rates are at least 1000 times higher than the background extinction (IUCN 2000). For the first time in the long Earth's history, one species – *H. sapiens*, who has become now *H. industrialis* – has acquired the power to eradicate most of life, together with the life-supporting services of its ecosystems and landscapes. But at the same time he is also endangering his own life. Thus, for instance, in spite of the growing awareness of the dangers for human and nature life of toxic pesticides and herbicides, there are still hundred millions tons of hazardous chemicals produced each year and we are carrying already 500–1,000 times more lead in our body than did our ancestors in the pre-industrial society (Platt and McGinn, 2002).

The sad fact is that the majority of human society and especially its leading political and economic decision makers have not yet accepted the fact that we cannot separate our fate from that of all life on the Earth. Therefore, no sufficient steps have been taken to reduce the threats to the web of life that are increasing even more rapidly than the growth of the world population and its exponential consumption rates with serious consequences for the fate of our global landscapes.

According to the 2005 report of National Accounts prepared by the Global footprint Network [info@footprintnetwork.org], overshoot is continuing to grow: humanity consumed 23% more ecological resources than the Earth produced. This is up from 21% for the prior year, leading to the destruction of ecological assets, such as fisheries and forests, on which our economy depends. Out of 150 nations, tracking their "Footprint trade", China, the United Kingdom and Japan were the largest importers of biocapacity.

Thus, the import of Soya beans from Brazil to China to feed pigs, chicken and fish, has grown since 1995 by 10.7%, reaching in 2004 to 2 billions dollar. This resulted in an alarming increase in the destruction of Amazon's forests, with its 100 000 species, far exceeding the forest cutting for cattle grazing and timber. Since 1995 these forests shrunk by 1.7 millions ha, and in 2004 alone by 600 000 ha. If the present rate of extinction continues, only 20% of all natural forests and grasslands will remain in 2050. China has lost already more than 6 million agricultural lands to the construction of cities, factories, highways, and to deserts, so that the Amazon is paying the price for the twenty-first century Soya "gold rush" of China and the rapid expansion of technosphere landscapes and desertification.

The question of coexistence of nature and humanity has been comprehensively addressed by Rosenzweig (2003). With the help of many examples, he has amply illustrated the nature of this problem and argues rightly that the present conservation and restoration methods, aimed chiefly at protecting nature in "small" reserves alone, are not adequate to meet the enormity of the challenges that humanity faces. He suggests a new kind of "reconciliation ecology", as "the science of inventing, establishing, and maintaining new habitats to conserve species diversity in places where people live, work, or play". This could be also called a "creative landscape science" in which everybody should engage in the conservation and restoration of Earth's biodiversity. This is an attractive proposal but, albeit, will be effective only on smaller scales.

The major cause for species extinctions is the destruction of their habitat by human pressures, but on global scales the biodiversity is also affected adversely by the global climate changes. These have been summarized recently in a detailed and well-illustrated report by Appenzeller and Dimick (2004) as geo-, eco- and time signs from Earth as follows: *Carbon dioxide levels rise. Mercury climbs. Oceans warm. Glaciers melt. Sea level rises. Sea ice thins. Permafrost thaws. Wildfires increase. Lakes shrink. Lake freezes up later. Ice shelves collapse. Drought linger. Precipitation increases. Mountain streams run dry. Winter loses its bite. Spring arrives earlier, autumn comes later. Plants flower sooner. Migration times vary.*

Habitats change. Birds nest earlier. Diseases spread. Coral reefs bleach. Snow packs decline. Coastlines erode. Cloud forests dry. Temperatures spike at high latitudes.

Presently, there is growing evidence for a human generated global disruption of climate and there are already alarming signals that climate changes are progressing more rapidly than predicted. The last ten years have been the warmest since systematic temperature measurements were available around 1850 and most probably in the last thousand years. Enhanced rates of melting of ice and glaciers are observed all over the world on the mountains and in the Arctic and Antarctic Oceans. According to the most recent report by the multinational Arctic Council, this region has warmed in the last 50 years twice as much as the rest of the globe. Arctic Ocean has shrunk as much as 20%, snow cover on land has diminished and the permafrost underlying the tundra has become less stable, endangering already some of the most exposed settlements and their people. In Alaska, East Siberia and West Canada until the end of this century all the ice of the arctic sea could vanish in the summer months. This threatens the seals resting and giving birth on sea ice and the polar bears, using to stalk seals. Loss of tundra would also rob many other animals of their breeding and feeding grounds. As warming imperils wildlife and vegetation, it is also threatening the traditional way of life and the economy of Arctic indigenous people.

In addition to the longer-term temperature trends, recent monthly data also indicate accelerating rise. This comes not as a surprise in view of the rising atmospheric levels of carbon dioxide and other greenhouse gases, making it the most predictable of all environmental parameters. In an Earth Policy Institute news release, Brown (2003) reports that falling water tables and rising temperatures help explain why the world grain harvest has fallen short of consumption in each of the past four years, dropping world grain stocks to their lowest level in a generation. In addition to the serious economic damages chiefly by such reduction of crop yields, there are now also more and more indications of the increasing seriousness of the threats to life on Earth by global climate change, and of the accelerating rates of adaptive changes with which habitats and many of their organisms can cope (see also Chapter 13 for more details).

This is also true for human life. In the 2003 year's warmest August on record in the northern hemisphere (Larsen 2003), many thousands of people perished. As usual, the victims of such environmental catastrophes were the social handicapped – weaker, poorer and homeless – children and especially elders in urban centers. In France alone 14 802 people died from the searing temperatures of 40 °C, remaining unusually high for two weeks. A lack of public recognition of the danger that high temperatures exposures poses, added to the lethality of these heat waves. In Paris many elders died who were left behind in hot apartments and abandoned by their children, who left for their "sacred" August vacation.

The World Meteorological Organization estimates that the number of heat-related fatalities could double in less than 20 years. Already now we are seeing evidence of more frequent heat waves coupled with elevated ground-level ozone concentration exceeding the European Union's health risk threshold. Other findings, supported by computer simulations suggest that the sharp losses of more than 60% from the ozone layer in three out of five winters during unusual severe winter temperatures of the

Arctic's stratosphere could increase the delay of its recovery for two decades or more. Until now, scientists assumed that until 2050 the ozone layer will return to the size of 1980, however at the recent congress of the American Association of geophysicists at San Francisco new evidence was presented that this will happen 15 years later because of the continued use of CFC in old refrigerators and air conditioners. The ozone hole in the Antarctic in September–October 2004 is still extending over 24 million sq km – like the area of North America. It could be much smaller, if not, for the two above reasons. These conditions heighten also the concern regarding possible positive feedback coupling between climate change and stratospheric ozone depletion. At the same time, researchers warn that if nothing is done to stabilize our climate, global sea levels could rise as much 6 meters with dire consequences such as flooding most Pacific Islands and the southern half of Florida.

The 2003 European heat wave and apparently also the 2005 very severe winter in the Balkan, and the widespread flooding in summer 2005 in Europe and especially in Rumania, as well as the severe drought and the destructive wildfires raging in the Iberian Peninsula are indications that with increasing frequencies of extreme climatic events such as strong winds, heavy rainstorms, heavy drought and extreme cold or hot periods, their catastrophic outcomes are becoming more severe. According to the most recent informations, 2005 was the warmest year ever measured on Earth. These will increase further the exponential process of landscape degradation and its destabilizing global impacts. This is especially the case in the densely populated Asian countries. Here, traditional unsustainable land uses – such as overgrazing, overpumping of ground water and overcutting of wood for fuel, are combined with "modern" even more detrimental agricultural and industrial intensification, are occurring on much larger scale. Dust bowl, resulting from this unsustainable agricultural development, is threatening the future of China. Reaching also North America, they are "blanketing areas from Canada to Arizona with layer of dust". Brown (2001b) concludes, that reversing desertification will require a huge effort, but if the dust bowl continues to spread, it will not only undermine the Chinese economy, but will also trigger a massive migration eastward. The rapidly advancing desertification process plagues already up to one-third of the Earth's land area, affecting more than 1 billion people in 110 countries (Earth Policy Institute News, 2003). All these more recent threats are true also for the Mediterranean Region and unfortunately confirm my predictions for global climate changes ten years ago in Chapter 5.

Reducing fossil fuel use is not only essential for reducing pollution of air, water and soil, it is the key to stabilizing climate. The Intergovernmental Panel on Climate Change is projecting a rise of 1.4–5.8 °C during this century if carbon emissions continue to increase. Since the 1960, the world's fossil fuel consumption doubles each ten years and simultaneously global emissions of carbon are rising and have increased by more than 9 per cent and in the United States even by 18 per cent between 1990 and 2000 (Garner 2002). Making out only 6 percent of the world's population, it consumes 20 per cent of all the energy, like India and China. In spite of being the world's greatest polluter, the United States has refused to join the Kyoto agreement in 1998, because President Bush claimed that his country could not afford the expenses of the reduction of greenhouse gases, causing the loss of working places. This served as a good excuse for other much less wealthy, but rapidly developing countries that will become also heavy polluters on global scales, like China and India. However, it is highly doubtful whether the reduction of greenhouse gases until 2012 by 5.2 %, as demanded by the Kyoto agreement from 1998, will make any dramatic impact, in view of the ongoing exponential increase in fossil fuel demands. Therefore, a new Protocol has been signed just now at Montreal UN meeting in December 2005, in which all 157 countries who signed the Kyoto Protocol – and again not including the United States – agreed to discuss an obligatory further reduction of greenhouse gases after 2012. According to the report of this conference in the "The Economist", researchers from the Scripps Institute at San Diego provided proof of close correlations between the increasing intensity of the hurricane and typhoon storms and the greenhouse gases, causing the warming of the oceans in the last 65 years.

These findings are most relevant for the recent events in summer 2005: the Gulf of Mexico that has already been vulnerable to coastal storms, has experienced a number of hurricanes of catastrophic dimensions. These took hundreds of human life and caused tremendous suffering to many more. The material damage, estimated by hundreds of millions of dollars, had far-reaching economic, social and political implications, including a sharp increase of the price of crude oil.

Although hurricanes are regarded as "natural disasters", their recent increasing frequency and destructive potency are considered by most scientists as ominous signs of global warnings. However nobody can doubt that in the case of the most catastrophic Hurricane Katrina hitting the coast with a wind speed of 233 km, short-sighted human land use decisions and careless and greedy treatment of their landscapes could have prevented most damages and especially the flooding of New Orleans and its tragic aftermaths. This has been explained lucidly by Cornelia Dean and Andrew Revkin in the *New York Times* from 19th August, in an article called: "After Centuries of "Controlling" Land, Gulf Residents Learn Who's really the Boss." Their major lesson shows convincingly the vital importance of maintaining the homeorhetic flow equilibrium in perturbation-dependent landscapes and their natural protection and regulation functions. In this case it is the dependence of South Louisiana landscapes, made of loose Mississippi River silt, on regular floods of muddy water replenishing the sediments and keeping the landscape above water. But the natural river flow and its adjacent marshes, wetlands and pasturelands have been replaced by artificial flood control canals and levee, channelling these nourishing sediments to the river delta and out into the deep water of the Gulf of Mexico. As a result, Louisiana's coastline has been sinking and the marshes and barrier islands, protecting New Orleans from hurricanes, are eroding fast. Thus, when the levee along the Inner Navigational Canal was broken by Hurricane Katrina, huge torrents of water were released rushing towards New Orleans, flooding the greatest parts of the city.

However, from the very beginning, already in the early 70's the irrational decision building a port on shifting mud in area regularly ravaged by storms and disease was driven by the founders of New Orleans expecting to make a lot of money, over-rode all objections. In the twentieth century, oil and gas power station were developed in the gulf, followed by petrochemical refineries and factories and protected by an elaborate system of levees, dams, spillways and other installations that could not withstand nature's kinetic energy of a 4–5 category hurricane.

Space does not allow me to mention all the recent wrong decisions made in the last 30 years by the Federal and local government, the Congress and the Corps of Engineers, as well as the citizens of New Orleans. They all were not willing to realize that the potentials for environmental disasters by the increasing destructive forces of coastal storms and hurricanes require also much greater investments and efforts in research and infrastructure to prevent such catastrophes. The most severely suffering are – as usually – the poorest segments of the population. In the case of New Orleans these were almost not only totally left to their own fate at the beginning of the catastrophic event and discriminated during the rescuing operations, but also neglected by the Administration in the assignment of funds to ensure safer and better living conditions.

The best solution to avoid the most serious damages would be truly letting the Mississippi River run again its natural cause and, thereby restoring the natural homeorhetic flow equilibrium, but according to Dean and Revkin this would exact unacceptable heavy costs – costs that would be paid immediately by people in the region and by any politician rash enough to endorse such a plan. Therefore, the search for palliative solutions has to be continued. At the same time, it would be very worthwhile for landscape and restoration ecologist and for all other experts dealing with land use to learn the lesson that the safest and – at the long run – also the cheapest way is *to work with nature and not against nature.* In this case it means to work with periodic floods and not against them.

This is the way Dutch government and its planners are facing the severe threats of global climate change causing the flooding of their low-laying coastline and polders. Instead of relying on levees, the government decided to buy from the farmers their agricultural land that will be cultivated extensively with wheat, but will be reserved as to adsorb the predicted flooding. In addition to this, 12 000 "swimming houses" will be constructed on barges in the most endangered areas.

3. TOWARDS A GLOBAL POST-INDUSTRIAL ECO-ECONOMY AND A SUSTAINABILITY REVOLUTION

Reducing fossil fuel oil consumption is not enough. In their forward-looking essay on the new economy for the twenty-first century, Brown and Flavin (1999) provided further proof that the Western economic model – the fossil-fuel based, automobile-centered, throwaway economy – that so dramatically raised living standards for part of the society, who had access to these technological advances, is not a viable system for the world, or even for the West over the long term. As outlined in detail by Brown (2001a), his visionary book for a new "Eco-Economy, shifts from the one-time depletion of natural resources to one that it based on renewable energy and that continually reuses and recycles materials. *It should be a solar, wind and hydrogen-powered, reuse, recycle economy, one that uses energy, water, land and materials much more efficiently and wisely than we do today. It should be an environmentally sustainable global economy, one where economic and social progress can continue not only in the twenty-first century but many centuries beyond.*

In "Plan B", the follow-up of this *Eco-Economy* book, Brown (2003) has further elaborated what should be done to "inflate the bubble economy before it bursts", and rescue "a Planet under stress and a civilization in trouble". He pointed out to numerous facts that we – that means the industrial countries – are satisfying our excessive demands by consuming the Earth's nature assets in the creation of this global bubble economy. As immediate measures Brown suggested:

Not only do we need to stabilize population, raise water productivity and stabilize climate, but we need to do it at wartime speed. The key to quickly shifting from a carbon-based energy economy to a hydrogen-based one is to incorporate the costs of climate change, including crop damaging temperatures, more destructive storms, and rising sea level, in the prices of fossil fuels. We need to get the market to tell the ecological truth.

It will demand a world wide mobilization – comparable to the US mobilization during the Second World War, but the choice is ours to make the generation *that stabilizes population, eradicates poverty, and stabilizes climate.*

The creation of such an "eco-economy" which is both socially and ecologically sustainable, should be also the final goal of sustainable development. This requires an all-embracing cultural and technological *sustainability revolution,* leading to transition from the "fossil age" to the "solar age" of the new economy, based on the limitless power of the sun as the non-polluting and renewable energy source. As one of the central challenge facing humanity, this can be achieved only by *a true synthesis between nature and human society,* in which full use will be made both of natural and of human scientific and technological potentials for their mutual benefit. This is not a utopian dream. In one year the sun provides 15 000 times more energy than that of the total annual consumption of fossil and nuclear energies. The annual photosynthetic output of the world's vegetation is 10 000 times greater than the chemical industry's annual global output. That there are already many encouraging signs that the technological utilization of this renewable non-polluting natural potential can be learned amongst other many examples from the emergence of eco-economy reported by Brown (2001a) and by Fischlowitz–Roberts (2002). Renewable "green power" and especially wind power purchase in many countries is growing more rapidly than ever predicted. Thus, for instance, since the launch of the renewable energy programme in April 2000 in Germany, the use of such wind power has already prevented the emission of 12 per cent of the total of 150 millions of carbon dioxide in two years (Haas, 2002). On a world-wide basis, the use of wind turbines and photovoltaics is growing now annually over 25 percent and in view of the current high prices of crude oil, these "green fuels" will be soon even more competitive with fossil fuels. This is already the case for the great success with vehicles powered by hybrid engines, producing electric power from internal combustion. Thus, the demand for the Toyota's hybrid "Prius" vehicle is growing faster than its production capacities. Further scientific and technological advances in the use of hydrogen fuel cell will lead to an even greater breakthrough. Among these, are the attempts to produce hydrogen gas from water and sunlight using titanium, dioxide as a catalyst. A great step forward on the "Hydrogen Highway" will be the plan for the construction of scores of hydrogen fuel stations in the State of California, pushed by its Governor Schwsarzenegger. He stated at the inauguration of the first hydrogen fueling station: "We will not just dream about hydrogen cars, we will build them". However, presently the main obstacle for shifting to a solar-powered fuel economy is the fact that gasoline is not priced at its real external costs for the environmental and health costs, which are 10–50 times higher as the present prices.

In his updated Plan B book Brown (2600) will give further impressive examples of great technological advances, showing that almost everything we need to do to move the world economy onto an environmentally sustainable path has been done in one or more countries. On the energy front, for example, an advanced designed wind turbine can produce as much energy as an oil well. Gas-electric hybrid automobiles, getting 55 miles per gallon, are easily twice as efficient as average vehicles on the road. Numerous countries are providing models of the different components which Brown has outlined in his Plan B, both describing where we want to go and offering a roadmap of how to get there.

Another very important economic sector in which much better use could be made of the unlimited natural potentials is the non-polluting production and the consumption of healthier food from organic farming, which has become in 2003 the fastest growing sector in the world agricultural economy.

We have to keep in mind that the forces driving the exponential growth of population and consumption leading to these disastrous processes are culturally deeply ingrained and, therefore our ecological and economic crisis is part of a much deeper cultural crisis. This cannot be resolved merely by piecemeal ecological, technological, political, and economical means. Its solution requires *a far-reaching shift in our worldview*, as a further step in our cultural evolution, including the spheres of cultural, spiritual and ethical values. Guided by our consciousness and beliefs, this has to be expressed in our personal day-by-day behavior and reflected in the professional and political decision-making process. These consequences of ethical environmental behavior have been emphasized also by the influential biologist Paul Ehrlich (2002). He expressed the urgent need to recruit social scientist to collaborate with environmental scientists in seeking solutions to the menacing dilemma of the destruction of human life's support systems. However, he approaches human behavior and ethics chiefly from a micro-evolutionary point of view and does not deal with the broader transdisciplinary macro-evolutionary aspects, driving socio-cultural processes.

The latter has been attempted in an impressive way by Ervin Laszlo (1994), the world-renowned and foremost exponent of systems science and global trends. He surmised that human society is faced now with the choice between extinction of organic life on Earth, or its further sustainable evolution. He insists that to avoid this dilemma, there is need for a far-reaching environmental *and* cultural sustainability revolution. This will require encompassing changes at all spheres of life and systems level with all its dangers and opportunities, which Laszlo (2001) calls *Macroshifts.* These are the human society equivalents of genetic mutations in nature. But while the latter lead to the creation either of new species or their extinction, Macroshifts in human society lead either to *breakthrough* to a new civilization or to *breakdown and chaos*. Like bifurcations in non-human systems, they occur as a sudden and radical change in the evolutionary trajectory of a system persisting far from thermal and chemical equilibrium in all realms of life. They have happened already several times during human cultural evolution, but never in such overwhelming speed and extent, and with such far-reaching, both promising and threatening results. Their outcome will be determined by the evolution of the dominant culture and consciousness – the way our values and views will respond to this change. If these will be resistant to change or change too slowly, the established institutions will be too rigid to change and to allow for timely transformation, then this will lead to breakdown. This can be the doomsday scenario awaiting us by the second decade of the twenty-first century.

Fortunately, there exists an alternative of a *breakthrough phase:* in the mindsets of a critical mass of people, a new way of thinking is evolving with more adapted values and more consciousness which will, hopefully catalyze creativity in society. As envisaged by Laszlo, this Macroshift will guide the bifurcation on the leap towards a higher organizational level of the emerging sustainable information society. It will be driven not only by the widespread adoption of technological innovations of regenerative and recycling methods but also by the efficient utilization of solar and other non-polluting and renewable sources of energy. As a *cultural evolutionary process* it must be coupled with more sustainable lifestyles and consumption patterns, caring for nature and even investing in nature.

Professor Wangari Maathai, the vice minister for environmental quality of Kenya, has received the 2004 Nobel Price for Peace, after gaining already the "Alternative Nobel Price". As the leader of the "Green Belt Movement" to fight forest destruction and desertification, she has initiated the planting of 30 millions of trees in Africa. She is not only the first African woman, but also the first person who has been chosen by the Swedish Academic Nobel Committee for her environmental and social achievements. In her speech she demanded democratic reforms to stop the greediness of the cooperations and called for a shift in thinking so that humankind will stop to be a threat to our planet and to the very ecological system on which his existence depends. She maintained that this can be achieved only if industry and all global institutions will recognize that ensuring economic justice and ecological integrity are more important than to gain additional income. It is very encouraging that finally the most prestigious academic establishment in the world, well known for its conservative attitudes, has recognized the importance of environmental justice as an integral part of social and economic justice. Hopefully, a further step will be to honor not only outstanding ecological scholars but also their achievements as worthwhile for the scientific Nobel Price. This could be interpreted as an important indicator that also the scientific community is entering the breakthrough phase of the Macroshift.

The terrible December 2005 tsunami catastrophe with its enormous human cost lends strong support to these demands for a global sustainability revolution. It must wake us up to learn an additional lecture not only on our global vulnerability but also on the sad fact that the prevailing economic and ecological injustice has not only prevented, but even increased its disastrous results. In this unjustly divided world, these are the poorer countries and their islands and coasts which are most exposed to natural disasters as well as to the human-caused effects of global climate change of the rising Pacific Ocean. But they are lacking the means for the infrastructure to establish early warning systems to avert major catastrophe. In addition, unrestricted tourist, urban, agricultural and aqua-cultural developments, together with pollution and land abuse have further increased the geo-physical vulnerability of shallow seashores through the destruction of most of the protective green walls that nature has provided by mangrove forests and coral reefs. As NGOs from India and Sri Lanka reported, it was in those locations where dense mangrove forests and healthy coral reefs protected the coast from the destructive impact of the waves, saved apparently much human suffering, together with the extinction of many other living organisms and their habitats. There is therefore an urgent need, not only to prevent further destruction, but to restore and to reconstruct as far as possible such green protective belts for the sake of both human and nature life. The industrial countries are learning now the lesson, and will help to establish the infrastructure enabling for early warning systems in this region.

The prestigious "Club of Budapest" under the Presidency of Prof. Ervin Laszlo has made a very timely "Post-Tzunami World Solidarity Declaration: LEARNING FROM THE CATASTROPHE", as a worldwide dialogue on "Steps to build a sustainable civilization". This includes a proposal for a Global Marshal Plan for Achieving the Millenniums Development Plan adopted at the UN Millennium 2000 Summit, and the creation of a Civil Parliament within the United Nation. Let us hope that this initiative will bear fruits very soon.

4. THE NEED FOR A SHIFT IN THINKING AND ACTING BY LANDSCAPE ECOLGISTS AND RESTORATIONISTS

My major thesis is that we – as landscape scientists, together with all those dealing with land planning and management – have to take an active role in this sustainability revolution. However, to be sure, to implement these proposals and to commence without further delay this transition to a sustainable world and its far-reaching implications will not be an easy one. Not only many socio-economic, political and technological obstacles, but also cultural and psychological ones will have to be overcome. For environmental scientists and professionals, and especially for those who have received their academic education and training in the realms of the natural so-called "exact" sciences, a major obstacle is the reluctance of many to abandon outdated beliefs, concepts and paradigms. According to Bohm and Peat (1987), innovative scientific developments are hampered because of this unconscious tendency to cling rigidly to familiar ideas in order to maintain a habitual sense of control and security and not to break the old pattern of thoughts and to block, thereby their mind from engaging in creative play, the ideal process in which experience and knowledge interwove with creative insight is not generally carried out.

This is the case with those paradigms which have created the fragmentation in the academic world between the "two cultures" – senso Snow (1963) – of the naturalistic and the humanistic sciences, and the ensuing unsatisfactory perceptions of the relations between humans and nature. The roots of these paradigms can be found in the Cartesian–Newton mechanistic view of nature, introduced into physics as the model for a complete explanation of reality to be comprehended only by its breaking down into smallest parts. It has lead to the total split between mind and matter and the view of the world as a giant machine to be ruled by humans and their technological power, gained by science, and governed by mathematical rules and linear connections between causes and effects. This mechanistic view had also exerted an overwhelming influence on the development of all other natural sciences, dealing exclusively with quantitative, measurable physical and chemical reducible "facts". By these, humans are regarded like any other animal as a purely biological creature and their mental, spiritual and intellectual capacities and ethical values are disregarded. The humanistic sciences, on the other hand, deal only with qualitative values, such as the cultural, sociological, historical and philosophical issues. At the same time, this mechanistic view of the world, in which the whole existence is considered as made up of unconnected elementary parts, has given strong support to this fragmentary way of thinking. As Bohm (1985, page 24) has pointed out, this has far-reaching implications: *Fragmentation is therefore an attitude of mind which disposes the mind to regard divisions between things absolute and final, rather than as a way of thinking that only a relative and limited range of usefulness and validity leads to the distinction between people, profession, nation, race, religion, ideology and so on, is preventing human being from working together for the common good, or even for survival.*

Unfortunately this attitude and the resulting disciplinary fragmentation between the natural sciences, the social sciences, the humanities and the arts are still dominating in many universities and their faculties; especially if their reputation is based on a long academic tradition, they are afraid to loose by such innovative and far-reaching changes. They are, therefore, reluctant to revise their teaching programmes and structures and in many of these universities there is little, if any, institutional and financial support for innovative transdisciplinary research by which these gaps could be bridged. These handicaps for the implementation of transdisciplinarity will be further discussed below.

To face these challenges, and to help steering this Macroshift towards an all-embracing sustainability revolution, will demand from concerned landscape ecologists and restoration ecologists a shift from narrow disciplinary approaches to much broader integrative systems thinking and acting. In this and many other aspects, LE and RE are closely related. This closeness is symbolized by the fact that Richard Hobbs, who was until 2004 the president of the International Association of Landscape Ecology (IALE), is serving now as the chief editor of "Restoration Ecology". Earlier in 1997, he has rightly emphasized the need for the active involvement of landscape ecologists in determining the future of landscapes by producing a truly integrated science, developing sound landscape design principles and increased interaction with policy, and to take a more active role in their practical solutions such ecological and cultural restoration (Hobbs 1997). Francesco di Castri (1997), one of the most influential ecologists, has also urged landscape ecologists *to become committed actors, and not only marginal and critical spectators in a changing globalized environment.* Fortunately most restorationists are already such "committed actors", carrying out many large-scale rehabilitation and reclamation projects all over the world.

Such a commitment is essential, but it is not sufficient for coping with the wealth of problems we are faced for those who are ready to respond to the demands of the emergent global information society for healthy and attractive landscapes. To cope with the complexity of landscapes as an integrative part of the complex network interactions between nature and modern life, landscape scientists will have to bridge the gaps between the natural sciences, the social sciences, the arts and the humanities, joining forces with concerned, environmentally and ecological-oriented scientists from all these fields in co-active landscape studies. All those engaged in historical, aesthetical or other landscape studies as well as in planning, managing, conserving and restoring landscapes cannot restrict themselves merely to the study of either the cognitive, mental, spiritual and aesthetic aspects or only the geophysical and ecological aspects. They will have to deal with all relevant human-ecological aspects, concerning the people living using, perceiving and shaping these landscapes. In all comprehensive landscape studies, we have to consider not only the material and economic needs of the people, but also their spiritual needs, wants and aspirations of all stakeholders involved. All these demands can be realized only by a far-reaching transformation: *LE has to broaden its interdisciplinary ecological–geographical view into a more inclusive view of the multifunctional natural and cultural landscape dimensions and functions and their closely interwoven bio-ecological and human-ecological pattern and processes. RE has to broaden its scope to include not only strategies for biological–ecological restoration but also human-ecological-oriented cultural restoration in the enhancement of the scientific, aesthetic, historical and traditional landscape values. This requires the transformation of both sciences into mission-driven Transdisciplinary Landscape Sciences.*

5. A CRITICAL APPRAISAL OF NATURE–HUMANS–CULTURE RELATIONS IN CLASSICAL ECOLOGICAL SCIENCES

Probably, the greatest obstacle which landscape ecologists and restorationists will have to overcome is the fact that both disciplines will have to abandon the narrow bio-ecological conceptions, inherited from classical ecology as a branch of the biological sciences and its mechanistic worldview and fragmentation syndrome. As a result, ecologists have shied away from studying the complex human–nature–culture relations, and have restricted their study almost exclusively to the relations between plants and animals and their environment (overlooking also micro-organisms). These specific human dimensions belonged to the "soft" non-experimental and non-quantitative humanities and social sciences, and were therefore considered as inferior and left outside the realm of the conventional natural sciences and ecological discourse. The resulting communication gap between the "hard" sciences and the rejection of these "soft" sciences has seriously distorted the holistic scientific view of closely interwoven nature–human–culture relations, essential for the development of a TLS. Humans were regarded as an external, "disturbing" and chiefly detrimental ecological factor, supposedly distorting the mythos of the "balance of nature" by degrading it from a stable and "mature" climax stage and preventing it to reach it again through a pre-conceived secondary succession. The majority of ecologists have also followed uncritically preconceived deterministic Clementsian relay floristic climax–succession paradigm. They rejected the objections of such far-sighted, holistic ecologists as Frank Egler (1942) and the pleas by another holistic ecologist, Pierre Dansereau (1957) to treat modern man both as biological and as cultural creatures of ecological systems, interacting at a higher hierarchical level of the landscape, and to recognize human ecology as the science studying their influence on these landscapes. (For more details on a critical appraisal of the Clementsian-Braun Blanquet climax–succession theories Mediterranean landscapes, see Naveh 1971; Naveh 1999; Naveh and Lieberman 1994.) Even when humans were considered in more sophisticated landscape studies, they still remain external socio-economic disturbance factors or at the most "*Homo economicus* keystone species" (O'Neill and Kahn 2000).

Unfortunately, geography – the second pillar of the scientific foundations of LE – has lost almost completely its integrative and holistic approach. This was the precious heritage from the "father" of modern geography, Alexander von Humboldt, who 200 years ago coined "landscape" as a scientific term. Regional geography as the classical integrative geographic branch has been neglected or has even been abandoned altogether, and the gap between those branches rooted entirely in the natural sciences, like bio-geography and physical geography, and those which have stronger connections with the social and human sciences, such as cultural geography is widening.

The situation in RE is not much better. According to Higgs (2005), RE as a subdiscipline of ecology is part of the above-described first "culture of the natural sciences", whereas ecological and cultural restoration practice represents in many aspects a much broader second "humanistic culture". However, this conflict could be overcome, if RE, like LE, will be transformed into a broader holistic and transdisciplinary science of restoration (Naveh 2005).

The first comprehensive transdisciplinary anthology on human's role in changing the face of the Earth edited by Thomas (1957) expressed the views of great scholars from relevant disciplines, discussing in depth the history and fate of human land uses in the broadest sense. It contained some clear warnings about the things to come as a result of exploding human populations, consumption and unrestricted technological power. However, only very few ecologists participated in this groundbreaking conference and its fully reported discussions. This important document was, therefore, largely overlooked by mainstream ecologists and had very little impact on the "normal science" of ecology – senso Kuhn (1970) – and its future directions. Only 20 years later, when the deep ecological crisis reached already its peak, the mainstream ecologists could not overlook anymore the destructive role which *H. industrialis* played in nature and became more and more involved in actual environmental issues.

It is symptomatic for this deeply rooted distorted view of human–nature relations that even holistically inclined systems ecologists continued to consider the "natural ecosystem", devoid of humans as their principle object of study. They overlooked completely the broader interpretations of the role of humans in ecosystems by Tansley (1935) in this seminal paper, in which he coined the term "ecosystem". He regarded humans as *a major agent of change and as an integral part of the ecosystem concept,* maintaining that *anthropogenic ecosystems* differ from those developed independently of man and stated that

> *We cannot define ourselves to the so-called "natural" entities and ignore the processes and expressions of vegetation now so abundantly provided us by the activities of man. Such a course is not scientifically sound, because scientific analysis must penetrate beneath forms of the natural entities, and it is not practically useful because ecology must be applied to conditions brought about by human activity* (Tansley, 1935, page 304).

Even Eugene Odum (1971) – who can rightly be called "the father of contemporary holistic ecosystem science" – paid only in the third edition of his very influential "Fundamentals of Ecology" textbook attention to humans as an important ecological factor. But even there, the major part of this volume was devoted almost exclusively to natural ecosystems, their biotic communities and the abiotic environment, supposedly developing towards a "mature" homeostatic climax stage without human interference. Natural ecosystems were conceived as the basic functional unit in nature and as the highest level in the ecological hierarchy. Merely in the last chapter on applications and technology and as applied human ecology did he consider "human dimensions in ecosystems". However much later, in his important book on ecology and the endangered life-support systems, he did recognize humans as part of "urban-industrialized" ecosystems (Odum 1993). He added also landscapes, bioregions and biomes above ecosystems in the ecological hierarchy as "ecosystems along with human artefacts". However, the only organizational criteria for emergent qualities at these higher levels were the *spatial expansion* of ecosystems and not any newly emergent, and chiefly, cultural system qualities of organized landscape complexity (Carmel and Naveh 2002).

The influential German ecologist, H. Ellenberg (1971;1973), who introduced ecosystem research into Germany, was one among the first to realize the dichotomic position of humans, imposed by the ecosystem concept, acting as a "supernatural factor" both inside and outside the ecosystem. His distinction between natural and close-to-natural and artificial urban-industrial ecosystems inspired us to a new functional classification of ecosystems and landscapes, in which we distinguished between natural and seminatural *solar-energy powered biosphere and fossil and nuclear-energy powered technosphere ecosystems and landscapes* (Naveh 1982; Naveh and Lieberman, 1994).

In an extensive review, Alberti et al. (2003) showed recently that in spite of the fact that there is not a single extensive ecosystem left on Earth which has not been influenced and modified to lesser or greater degrees by humans, this "natural ecosystem" paradigm is still very popular. Even advanced ecological systems studies, such as those by Joergensen (1997), are dealing solely with the biological-ecological and chemo-physical dimension of ecosystems. They seem to be treated as if they are all natural ecosystems, devoid of humans and any human-ecological dimensions.

However, as Alberti et al. (2003) has noted rightly, even when humans have been finally recognized as components of ecosystems, the attempts to understand human-dominated systems mostly remain reductionistic and human ecological processes are treated almost exclusively as separate phenomena. The latter is true also with few exceptions in the chapters of the anthology dealing with the "subtle" role of humans in ecosystems (Mc Donnell and Pickett 1993).

Egler (1970) was also one of the first to draw attention to the danger for a mechanistic direction in ecosystem science: *the scientific and ecological concepts may wither and only the dry technology of system analysis is left*. He claimed that the advent of the computer has often encouraged the *trivialization of scholarship and the belief that the things that count are those that can be counted, whereas many things which cannot be counted, count*.

In a critical essay on human ecology as an interdisciplinary concept (one of the few studies on human ecology which were published in a major mainstream ecological publication) also Young (1974) warned that *a general systems approach could return us to a discarded mechanistic focus, ignoring the reality of man*. In a response to this essay, in which Young was not aware at all of the existence of LE, I suggested to consider LE as an emerging branch of human ecosystem science that could bridge the gaps between biological ecology and human ecology (Naveh 1982).

These early warnings by Egler (1970) and Young (1974) were justified, if we accept uncritically the most widely recognized and acclaimed attempt for the integration of humans in ecosystems ecology by H. T. Odum (1971; 1983). Using sophisticated energy flow diagrams and models *as a cognitive basis to see complexity and component human-nature relations at the same time*, these models were promoted widely not only for environmental management but also for environmental education (Odum and Odum, 1980). It should be pointed out that such an oversimplified reductionistic and materialistic interpretation, reducing all human dimension and even culture and religion to nothing more than energy flow in one direction and money flow in the opposite, are not doing any service for the true integration of humans in nature (Naveh and Lieberman 1994).

6. HOLISTIC TRENDS IN THE SCIENCE OF ECOLOGY

In recent years, the very value of the above-described ecosystem paradigms and the distorted nature-human-culture conception is more and more doubted. Thus, for instance, O'Neill (2001), one of the most influential promoters of ecosystem ecology, raised the question "whether it is not time to bury the ecosystem concept altogether?" He stated that the ecosystem concept is based on the "machine analogy" inherent in system analysis, implying an outdated homeostatic balance of nature view. He claimed that *H. sapiens* is not outside the system, but "a keystone species" changing system stability by altering environmental constrains, rate processes and biotic structures.

To avoid this confusion, Naveh and Lieberman (1994), as well as Allen and Hoeckstra (1993), suggested that the ecosystems should be regarded only as functional systems for the flow of energy, material and information. However, as explained and shown in the previous chapters, for a precisely defined "real piece of nature", the tangible, three-dimensional landscape units should serve as the actual unit of study. To ensure a comprehensive appreciation of human's role in such landscape units, these should be studied and managed as parts of both the physical and geographical space and the cognitive space of the human mental and spiritual realms, of our Total Human Ecosystems.

Such a complementary systems view enables us to perceive and treat these landscapes as *the tangible bridge between nature and mind, and also of culture*. It opens the way for a better comprehension of multifunctional landscape complexity and its natural and cultural multidimensions (Naveh, 2001 also published as chapter 11 in their anthology; Carmel and Naveh, 2002). For landscape planners and designers, Makhzuomi and Pungetti (1999) have provided a comprehensive description of 14 different dimensions of landscape, integrating them within the unitary framework of holistic landscape research. These could serve very well also as the foundation for transdisciplinary landscape and restoration research of TLS and TRS.

Most recently, a clear tendency can also be noted among ecologists, who reject the narrow bio-ecological natural ecosystem paradigm altogether. A holistic and transdisciplinary paradigm shift is changing already the science and practice of adaptive resources management and has inspired the emerging of new theories for sustainable futures (Holling 2000). This is also reflected in many studies published in the on-line "Conservation Ecology". An important outcome was the "Resilience Project" (Walker 2000), fostering integrative understanding among economists, ecologists, social scientists, and mathematicians, of complex "social-ecological systems", which have much in common with our "Total Human Ecosystem" concept. It has yielded an important collection of essays attempting to develop an integrative theory of transformations in human and natural systems from the angle of resilience and adaptive cyclic changes (Gunderson and Holling 2002), and a working hypothesis for a participatory approach (Walker et al. 2002). Serious efforts are also made for an integration of humans into urban ecosystems, both in Europe (Breuste, et. al., 1998) and in the United States (Alberti et al., 2003).

Although most of these important integrative efforts were mostly directed to environmental and ecological management and its socio-economic aspects, it is hoped that they will be also influential changing the faulty human-ecosystem syndrome. In "Ecological Conservation", already several important attempts have been made to expand the scope of ecological management, conservation and restoration to the cultural, mental and spiritual human dimensions, replacing the "ecosystem" term by "landscapes" (Hull et al., 2003; Toupal, 2003; Wali, et al., 2003).

"Conservation Ecology" has been now been renamed into "Ecology and Society", with the destination to promote integrative science for resilience and sustainability over short and long term, from local to global, and across scales. Their editors, Folke and Gunderson (2004) stated that learning more and more about these systems, we are realizing that social, political, economical human component, are adding other dimensions of complexity. Therefore they maintain that understanding and governing these complexities are prerequisites for seeking and sustaining future. As will be described below in more detail, for such a full comprehension and treatment **an even broader, explicitly transdisciplinary approach, including also other human-ecological and noospheric components is required**.

Another important attempt has been made recently for innovative and integrative approaches by Muller and Li (2004) to apply a "complex systems approach of systems ecology", broadening the narrow outlooks of classical ecology and to cope with the complex human-environmental relations. They stated that such an approach is missing in "this elaborate group of sub- and inter-disciplines such as ecological economics, ecological engineering, ecotechnology, ecological planning, human ecology, and so on". Their arguments to expand the scope of these interactions in the light of new insights gained from" the science of complexity" have much in common with those presented in this anthology. Thus, for instance, they offer some recent developments in ecological sciences, which we have characterized as indicators for a scientific holistic and transdisciplinary revolution, discussed below. These include holistic synthesis instead of reductionistic analysis; irreversible reactions instead of reductionistic analysis; bifurcations and phase transitions instead of continuity; non-equilibrium instead of equilibrium; non-predictability instead of predictability. Mueller and Li (2004) claimed also, rightly, that the comprehension of ecological units as self-organized systems has provoked a new way of thinking about systems development, disturbance and restoration. As a result these have changed the approach to environmental management towards a new system-based management. Among these changes they mentioned policy mix replacing instrumental orientation; sustainability replacing single-problem view; integrating resorts replacing constrain of resorts; ecology and economy replacing ecology vs. economy, participation replacing top-down regulation. In this broader context of human systems Mueller and Li (2004) recognized those dimensions of quality of life from the realms of social sciences which we have characterized as the realm of the noospheric, cognitive and mental space of our THE. They concluded: *"New methods and approaches in the science of complexity have provided us a new opportunity to comprehensively understand the coupling dynamics of human and ecological systems, and the prospects of sustainable management across a range of spatial and temporal scales. New complex systems approaches also shed a new light on co-evolutionary trajectories and the resilience of coupled social and ecological systems."* (Mueller and Li 2004, page45).

To take full advantage of these developments will require a complete paradigm shift towards a transdisciplinary metatheory integrating humans and nature into a coherent whole with the "Total Human Ecosystem" as the core concept for such a unifying metatheory.

We can note also now the first indications for a shift towards transdisciplinarity in the science of ecology as a whole. The successor of the late Eugene Odum at the University of Athens, Georgia, G. W. Barret (2001) – who is a well recognized American landscape ecologist – called to transfer all ecology from a biological into a transdisciplinary science. Adopting our interpretation of the noosphere (Naveh and Lieberman 1994) as the sphere of the mind which is increasingly affecting the rate of energy flows and the quality of matter and materials functioning within the system, Barrett (2001) suggested that ecology, as an integrative and transdisciplinary science should be based on the "**noosystem**" concept and approach.

An additional suggestion for such a shift has been made by Ward (2003), responding to the recommendation by Hull et al. (2003b) to overcome the dogmatic believe in the "balance of nature", with a "biocultural approach to the ecological sciences". He commented that such a biocultural approach would seem to require the scrapping of that "stultifying" recognition of ecology as a branch of biology. Instead, he suggested defining ecology "as a field of study that tries to understand the mixture of chaotic and orderly interactions within, and between nature and human society. Ecology is not itself a discipline, but draws upon many disciplines and methods".

A simpler definition could be *to* **recognize ecology as a transdisciplinary overarching "meta – science", transcending many disciplines that deal with the interactions within and between all organisms – including humans – and their total environment**. However, this may not be attainable for the whole range of all ecological sub-disciplines. As an alternative, ecological principles knowledge and expertise should be integrated with other disciplines dealing with all those complex environmental problems for which the piecemeal approaches of conventional, fragmentized disciplines are not sufficient. In a proposal for a transdisciplinary educational program for regional sustainable development (Naveh 2002), I suggested applying the term "*ecodisciplinarity*" as a shorthand expression for those innovative efforts to bridge the gaps between ecological knowledge, wisdom and ethics and the narrower disciplinary expertise in the scientific and professional realms.

7. SYNTHETIC ECO-DISCIPLINARY SCIENCES – STEPS TOWARDS TRANSDISCIPLINARITY

Such eco-disciplinary attempts to create "meta-sciences" have been made by *ecological economists*, viewing their science as a transdisciplinary effort to link natural and social sciences, and especially ecology and economics (Costanza and Daly 1992; Costanza 1996). **Eco-psychologists** have started to examine the human conscious and unconscious mind as an integral part of the web of nature, in which human health and well-being is depending on a balanced mutual relation with sustainable habitats, landscapes, and the planet as a whole (Roszak et al.1995). **Social ecologists** are opening new vistas for the study of "social metabolism" of energy and material flow, "the colonization" by human civilizations of the biosphere and their bio-and socio-ecological consequences for a sustainable future (Fischer-Kowalski and Haberl 1998). In the young science of *urban ecology* there is also a clear shift in its focal point from natural sciences to a more transdisciplinary problem-oriented research, dealing with the improvement of the living environment of people in cities (Breuste 1998). The ***"Journal of Ecosystem Health"*** of the Society for Ecosystem Health reflects the steadily growing recognition of the importance of transdisciplinary approaches to health and the close links between the health of nature and people. A very important further development in this direction are the transdisciplinary attempts, deserving close attention both by landscape and restoration ecologists are the attempts for defining and measuring of **ecological integrity**, as an integration of the environment, conservation and health, (Pimentel et al. 2000). This was a part of their "Global Integrity Project", composed of prominent scientists with different disciplinary backgrounds from the natural and human science, such as ecology, conservation, epidemiology, environmental scientists, philosophy, political science. Their project is guided by the two complementary policy imperatives, namely conserving integrity and life's sustainability. Of special relevance for us in this respect is the contribution on maintaining the ecological integrity of landscapes and eco-regions by Reed Noss (2000), a world renowned conservation specialist. Using the term "land ecology", he presented an eco-centric version of ecosystem management as the meeting point between non-human needs, human needs and desires, and technology and economy which comes very close to our approach to both landscape and restoration ecology.

At the same time, it should be emphasized that the term "ecodisciplines" does not fit those established environmentally oriented disciplines, such as "environmental economy" in which our environmental problems are externalized. As Rees (2000 has rightly pointed out, these cannot be supposedly solved merely by legal, policy, technological, economic and engineering means, and without consideration of their much deeper and broader bio- and human ecological implications.

One major conclusion from these critical comments is the recognition that neither general ecology, nor ecosystem ecology can resolve the nature-human-culture syndrome, rooted in the deep communication gaps between natural and human sciences. As pointed out already by Naveh and Lieberman (1994) and further developed below, this will require a complete paradigm shift towards a transdisciplinary metatheory integrating humans and nature into a coherent whole with the "Total Human Ecosystem" as the core concept for such a unifying metatheory.

8. THE ROLE OF TRANSDISCIPLINARITY IN LANDSCAPE ECOLOGY AND RESTORATION ECOLOGY

The first important theoretical contribution to transdisciplinarity in LE was made by Langer (1973) in Germany, claiming that in cultural landscapes the anthropogenic elements and the natural ones form units of a higher level whole of "geosocial" systems which are relevant for the planning process. A further important step towards transdisciplinarity has been made at the first international seminar of IALE at Roskilde in 1984, creating vital bridges between those arriving from different cultural, scientific and professional backgrounds. In this meeting I outlined for the first time the conceptional foundations for a transdisciplinary landscape ecology, planning and management science.

Already two years earlier Isaac Zonneveld (1982), the first president of IALE, raised the flag of a transdisciplinary landscape ecological paradigm at the international conference, organized by the Netherlands Association of Landscape Ecology, in which both European and North American scientists with different approaches and outlooks participated. He called landscape ecology *"both as a science and a holistic state of mind"*.

It was not by chance that the Netherlands Association of Landscape Ecology has taken the lead in innovation and transformation towards inter-and transdisciplinary landscape research and management in Europe. Bridging the conceptual barriers between the bio-ecological and geo- and hydro-ecological approaches Dutch landscape ecologists succeeded to create a truly interdisciplinary and influential science of landscape assessment, planning and conservation management. As reported by Klijn and Vos (2000), this has been expressed forcefully at the 25th anniversary of the Dutch Association of Landscape Ecology. One of its major conclusions was the demand for the application of much broader holistic conception with clearer definitions of the theoretical and practical aims. Without neglecting the natural science disciplines of landscape ecology, they emphasize the need for change and innovation by linking up with other disciplines. Landscape ecology should be extended towards an overarching landscape science integrating it with economics, psychology, sociology, and other principles involved with design planning and management and decision making.

Recognizing the growing need to face these challenges in the future direction of LE, the University of Wageningen and "Alterra Green World Research" have established in Wageningen a joint strategic research program, named DELTA. Its aim is to foster such research by creating bridges between the humanities and the natural sciences and cooperating with stakeholders. As one the first activities, an international seminar has been organized in 2002 on the expectations and practices of interdisciplinary and transdisciplinary landscape studies (Tress et al. 2003), which has produced very useful information for such studies.

Another important attempt for the implementation of transdisciplinary concepts, methods and practices has been made in a comprehensive anthology on the development and perspectives of LE in Germany and Central Europe (Bastian and Steinhardt 2002). It opened new vistas for all those landscape researchers and practicing ecologists who are ready to accept new ideas, enriching and improving their own work. It showed convincingly that:

1. Landscape ecology is a synthetic science in which theoretical and practical aspects are closely interwoven by synergistic, mutually amplifying interactions.
2. Landscape ecology is not a "virtual science" which can be carried out merely by sitting behind the computer in an air conditioned office. It is first of all a **field science**, by which the basic bio-geo-and human ecological data have to be collected. But their evaluation and synthesis has to be carried out with the help of the most advanced and sophisticated tools provide presently by the information technology.
3. Landscape ecology should not be considered merely as a scientific and technical field, rooted in the natural sciences, as implied by "geo-ecology". As a

transdisciplinary science, it has to include also the socio-cultural realms, as integral parts of every landscape, rooted in the social sciences and humanities. In all this respects, as well as in the very rich literature references, this important book is distinguished from most other studies, dealing with theoretical and methodological aspects of LE.

A major advance towards transdisciplinary has also been achieved in the fields of ecological landscape design and planning by Makhazoumi and Pungetti (1999). Although focusing on the Mediterranean Region where their comprehensive research and creative planning projects were carried out, the principles and methods applied have global relevance. In their theoretical and practical work, the authors have succeeded integrating holistic landscape ecological concepts with a broad transdisciplinary outlook, leading to sustainable landscape development. Therefore this study should be considered as an outstanding pioneer example for our transdisciplinary landscape science.

Unfortunately, in the United States, the strikingly holistic and transdisciplinary heritage of landscapes studies by the eminent geographer Carl Sauer was overlooked. He can truly be regarded the pioneer of a post-modern holistic landscape conception, emphasizing its unique humanistic qualities and doubting the possibility to reduce its "quality of understanding at a higher plane" to the formal process of a so-called objective science (Sauer 1925; personal communication 1959). However, under the strong influence of ecosystem and population ecology, most American landscape ecologists followed the well-paved road of formal "normal science" – senso Kuhn (1970). Basing their theory on equations, models, concepts and principles, such as mosaic patterns; process and change, patch-corridor-matrix models; optimal grain size (Forman 1995), and on what could be called "quantitative spatial landscape ecology" (Turner et al. 2001), their research was concerned chiefly with the spatial aspects of landscape heterogeneity, dynamics and change, emphasizing the biophysical landscape elements and processes. Some of these quantitative methods were reviewed critically by Patil et al.(1998). As stated by Gustafson (1998), the quantification of spatial heterogeneity remains problematic and the difficulties associated with predicting the response of ecological entities to spatial patterns has led to few definite tests and applications. The distinction between measurable and mapable heterogeneity and ecologically relevant heterogeneity seems also to be very fuzzy (Hobbs 1997).

More recently, however, there is a clear tendency towards more holistic interpretations of landscape ecology. By these, like in Canada, Europe and elsewhere, American LE is turning increasingly into a problem-solving oriented science of holistic landscape research, planning and management, as discussed in more detail by Naveh (2000). Many joint projects are carried out by applied ecologists, conservationists, restorationists, foresters, wildlife managers and other professional land managers together with landscape ecologists from the Academia. Landscape planners and designers have followed the example of the leading American landscape ecologist and planner, Jane Nassauer, who proclaimed at the first IALE conference on cultural landscape aspects: *"We must be courageous in innovating around the conventions of our own disciplines. We must dare to borrow from what is useful in the approaches and knowledge of our colleagues in the arts, social sciences and physical and biological sciences"* (Nassauer 1990, p.172).

Thanks to the strong involvement of leading landscape ecologists and planners such as Jingle Wu and Laura Musacchio, who organized the 16th US-IALE 2001 Annual Symposium at the Arizona State University in Tempe, transdisciplinarity aspects are getting now also greater attention: The theme for the 19th Annual Symposium 2004 in Las Vegas was "Transdisciplinary Challenges in Landscape Ecology", and a special session was devoted to interdisciplinary and transdisciplinary approaches for creating and maintaining sustainable landscapes in which contribution were presented, dealing with existing and emerging LE concepts and methodologies, furthering the development of a science of sustainable landscapes.

Based on the findings of earlier studies on the conservation management and restoration of Mediterranean uplands, I concluded (Naveh 1988), that these destroyed Mediterranean upland ecosystems cannot be restored any more, to their previous state, but can be rehabilitated by multi-factorial state factor rehabilitation, ensuring the optimization of multi-beneficial production, regulation and carrier function for multiple land uses. This interdisciplinary and multidimensional nature of conservation and reconstruction research and their implementation requires a holistic approach. I stated: *"This integrated landscape approach should be applied to the reconstruction of these deteriorating uplands. For this purpose neither the restoration of a non-existing natural pristine state or of an illusionary climax can serve as models. But they could be reclaimed or rehabilitated according to Bradshaw and Chadwick (1980) by activities which seek to bring them back into beneficial use, by rehabilitation biological potential. In these uplands rehabilitation should be reconciled with the socio-economic advancement of their populations by incorporating multipurpose land uses."* (Naveh 1988, page 235).

In this paper I described in detail the different strategies we used in the last 15 years for the multi-factorial restoration of nature reserves and parks, in highly disturbed ecotopes, and the rehabilitation strategies for highly disturbed, devastated and derelict areas, including limestone quarries and roadsides. I made a strong point of the importance of the restoration of the homeorhetic flow process, (As well in several chapters of this anthology), demonstrating this by the topological models of attractor surfaces with three dimensions as applied in catastrophe theory by Waddington (1975). As described by Naveh and Lieberman (1994), these present multidimensional space, occupied by the innumerable variables simultaneously controlling these dynamic processes. Their surfaces have the shape of valleys with stabilized streams as canalized pathway of changes, or *"chreods"* (from the Greek for "necessary path"). The diversification processes through time are simulated by the branching of the single first valley into several smaller streams, thereby creating the *"epigenetic landscape"* of a river plain.

Mediterranean uplands apparently meandered in a very broad chreod river plain, as long as their developmental dynamics were governed by traditional agro-pastoral land uses. But with the disruptions of this homeorhetic flow equilibrium they are pushed out of the chreods over the watershed by accelerating neotechnological degradation functions. This disruption is also caused by abandonment and total protection from human interference, resulting in the decline of biological diversity, productivity and integrity. The more harsher and/or more fragile the independent state factors of climate, soil parent material, relief, hydrology and biota, the more severe and far reaching are human impacts on dependent functional and structural ecotope variables. Therefore also the farther they are pushed away from their chreods and the sooner total landscape desertification will be reached. (See also chapter 5 in the context of global climate change).

A transdisciplinary approach has been applied in Central Europe and especially in the pioneering work in large-scale whole landscape restoration and rehabilitation has been carried out already for many years in the industrial and despoiled lignite coal mining landscapes, and especially along the Ruhr river district in West Germany under the leadership W. Pietsch (personal communication 1972). Its aim was to restore some of the nature assets and at the same time also to improve the living conditions in this heavily polluted and congested region. It should also prevent the migration of engineers and the best-trained technicians, to other more prospering and attractive regions with higher environmental quality, such as Bavaria. More recently, Pietsch (1999) is leading similar large-scale holistic whole-landscape reclamation projects in the abandoned Lusatian coal mining district in Saxonia.

More recently, also in RE a trend towards holistic and even transdisciplinary restoration ecology and practice, and a growing tendency towards whole landscape restoration is emerging, aiming at the sustainable future of both natural and cultural landscapes for the sake of both nature and human society. As reported in a special session and in a follow-up field workshop at the Sixth International Conference of the Israeli Society for Ecology & Environmental Quality Sciences in Jerusalem 1996, in whole-landscape ecological and cultural restoration in different countries and continents, transdisciplinary approaches are essential. This is especially the case when attempts are made for the restoration of historical and other cultural landscape values of traditional landscapes. (See also chapter 11)

Probably one of the most striking example for the importance of interdisciplinary and even transdisciplinary approaches in the Unites States to the solution of complex environmental problems is the large-scale landscape restoration project of one of the world most ambitious efforts to rehabilitate and manage the Chesapeake Bay, seriously threatened by the heavy urban-industrial and agricultural impacts of a densely populated area at the Eastern Central Atlantic coast of the United States (Hassett et al. 2005). In California transdisciplinary whole-landscape restoration of oak woodlands was introduced by the prominent landscape planner and designer John Lyle (Lyle and Stafford 1996). Lyle (1994) also initiated and designed the Center for Regenerative Studies at the California Polytechnical State University at Pomona as a combined inter-and transdisciplinary teaching, research and demonstration project in regenerative technologies.

The progress towards whole-landscape transdisciplinary restoration and rehabilitation was also very obvious in the contribution of several participants in most recent, important symposium on estuarine restoration, published as a special issue in "Restoration Ecology" in March 2005. The need for such approaches towards restoration was expressed lucidly in this symposium by Eric Higgs (2005), the president of the International Society for Restoration Ecology. He criticized the narrow positivistic and mostly reductionistic concepts still ruling academic education and research and regarded the present science of restoration ecology as part of this narrow "first culture of the natural science". On the other hand, ecological and practical restoration, represent in his opinion a much broader "second humanistic culture". It should however be realized that only by bridging the gaps between both, can the goals of restoration be achieved, as part of a "third alternative culture" by which RE would expand its scope into a transdisciplinary science and practice of restoration, with close links to LE and the transformed landscape science (Naveh 2005). As mentioned above, such close links already exist and could be further enhanced for the mutual benefit of both sciences (Bell et al. 1997). Naveh and Allen (2002) have provided some actual examples of the realization of such synergistic relations in the Mediterranean Basin and the Tropics of Mexico.

One of the greatest challenges in natural and seminatural biosphere landscapes is the conservation and restoration of ecological processes, ensuring the protection, regulation and carrier functions and maximum attainable biodiversity. Therefore the goals of landscape restoration in the broadest sense can be realized only by restoring these functional processes of nature. For transdisciplinary ecological and cultural restoration, these links for both sciences are even of more important, because the new landscape science will have to deal with the study of the functional and structural properties of landscapes and the dynamic change of both their natural and cultural patterns and processes at different scales of time and space, as affected by natural causes and by human-induced cultural impacts on the land. Restoration ecologists will gain from the application of holistic landscape ecological principles and methods by broadening and deepening the scope of their work. For landscape ecologists it provides a great opportunity to be actively involved in one of the most important tools to realize in practice this joint, major goal, for this and future generations. We can therefore conclude this subchapter with the hope that these links will become stronger and a truly win-win relation will evolve, that may, eventually lead to the fusion of both sciences.

9. THE TRUE MEANING OF TRANSDISCIPLINARY FOR BOTH LANDSCAPE SCIENCES

Transdisciplinarity has become in recent year of great significance in almost all spheres of life and many different fields of knowledge. However, the rapidly growing number of publications dealing with transdisciplinarity has not contributed much to a better understanding of its true meaning. In the context of scientific activities this has caused a misunderstanding of the true distinction between interdisciplinarity and transdisciplinarity, regarding the latter merely as a broader range of partnership in research than interdisciplinarity. Such a minimalist distinction has been adopted also by some leading landscape ecologists.

Thus, for instance, Moss (1999) claimed that transdisciplinary integration involves not only the scientific and technological disciplines as in interdisciplinarity, but also the involvement of planners and administrators. However, for future directions of LE, he rejected both inter- and transdisciplinarity as major goals. He argued that for LE to play a distinct role in environmental problem solving, it should become a more discipline–like than an interdisciplinary field, focusing chiefly on the biotic and abiotic landscape elements and developing its own theoretical basis. These arguments would be justified if transdisciplinarity would be misconceived as a vague, non-focused field, requiring only general, superficial knowledge from many diverse fields. Looking beyond the boundaries of their own disciplines and their own familiar way of thinking and discovering common ground with other disciplines through transdisciplinarity does not mean that landscape ecologists have to neglect their own unique disciplinary expertise of "dealing with the land as a whole" (Moss 1999). Working together towards a common overarching transdisciplinary goal, they will have to share this knowledge with scientists, professionals and practitioners with different backgrounds and expertise in relevant fields, not only in the fields of natural sciences but also in those of the social sciences, humanities and arts with human-ecological outlooks (Naveh 2000; Tress and Tress 2002).

If we mean by "land as a whole" those landscapes, serving as the spatial and functional matrix for all organisms, including humans, using and shaping these landscapes for better or worse, then there is not a single problem connected that can be resolved by one or more disciplines from the natural sciences only. In this process not only biotic and abiotic, but many more cultural, socio-economic, historical, political and other factors are involved. Therefore also the demand by Moss (1999) for the creation of a common theoretical basis can be realized only with the involvement of all those, both from the natural sciences as well as from the social sciences, the humanities and arts, sharing the common goal to develop jointly an innovative, problem-solving oriented landscape meta-disciplinary science for dealing in a comprehensive and efficient with this "land as a whole". The same is true for both transdisciplinary landscape sciences of planning and management and of restoration.

Brandt (1998) defined the role of interdisciplinarity in LE *"an expansion of spatially oriented geo-or bio-ecology into adjacent disciplines and applied sciences, planning and management"*, and that of transdisciplinarity *"a broad co-operation of very different mutually inspiring disciplines and practitioners dealing with landscapes"*. In their introduction to a recent important symposium to which I will refer further below in more detail, in a similar vain Tress et al. (2003, page 10) stated that *"by interdisciplinarity we mean projects that involve several unrelated academic disciplines in a way that forces them to cross subject boundaries to solve a common research goal. By transdisciplinarity we mean projects that integrate both academic researchers from different unrelated disciplines and user group participants to reach a common goal"*.

Transdisciplinarity certainly involves a higher level of integration and cooperation, but these distinctions are not sufficient to identify its true conceptual and epistemological meaning and for a full comprehension and realization of its potentials. As the prefix *"trans"* indicates, in contrast to interdisciplinary it *goes* **not between but across and even beyond disciplines** and their related activities, creating *an entirely new type of integrative knowledge, leading to new relationships between researchers and all others involved*. Based on *systems theory and network thinking* a new quality of scientific knowledge should emerge, enabling a better comprehension of the complexities of the real world, which has been fragmentized both by academicians and practitioners into different fields of disciplinary knowledge and interests. The main difference between both concepts lies therefore not only in the broader range of participants in landscape research, but *in the different nature of the mutual relationships of these participants, opening many more options for resolving the complex problems which landscape research is facing*.

The renowned systems philosopher, planner and educator Erich Jantsch (1970) stated in a seminal lecture that transdisciplinarity implies an overarching scientific and practical approach that transcends and crosses disciplines and professions, aiming together towards a common systems goal. It depends on the ability to reach out beyond interdisciplinarity towards an even higher stage of integration and cooperation, towards the fusion of their cross-disciplinary knowledge. From such a fusion a new scientific quality and discipline could emerge for dealing comprehensively with human-nature-culture-landscapes relations.

To illustrate the differences between multi-inter- and transdisciplinarity Jantsch (1970) used a hierarchical systems model for the successive steps of increasing cooperation and coordination: In the one-level structured multidisciplinarity, several disciplines are involved, but with no interaction. In pluridisciplinarity there is uncoordinated interaction between disciplines, and in crossdisciplinarity, interaction and coordination is imposed by a singly discipline. In the two-level structures and multigoal oriented interdisciplinarity. Coordination is determined from the higher level purpose. However, *only in transdisciplinarity, as the most complex interacting multi-level-goal structure, closely interwoven coordination and fullest cooperation towards a higher level common systems goal is achieved.*

For both TLS the **common systems goal should be the contribution to the sustainability revolution, furthering its mutually beneficial symbiotic relations between humans and their landscapes in the post-industrial information society**. An essential condition for advancing towards this goal is an ongoing interactive dialogue, defined by the great holistic physicist and science philosopher, David Bohm (1996) as "a stream of meaning, flowing among, through and between us, out of which may emerge some new understanding". It should lead to a close cooperation between many fields of knowledge, expertise and practice not only for solving "the ills of society", but also further the "possibility for a transformation of the nature of consciousness, both individually and collectively".

To create such an interactive dialogue, one of its major obstacles is the tendency for many specialists to be "deaf" to any knowledge outside their own competence. Unfortunately, this tendency will prevail as long as most meetings are organized around many overlapping sessions in different fields, with only a few frontal "plenary lectures" to be attended by all participants. If these would be followed- up by general discussions, in smaller discussion groups around selected transdisciplinary themes it could create such productive interactive dialogues. However, such a type of conference organization is much more demanding and it requires strong and experienced leadership and a clear definition of its goal and needs sufficient time for reaching joint conclusions. Among the more recent meetings I attended, the conference on "Multifunctional Landscapes, Interdisciplinary Approaches to Landscape Research and Management", at the Centre for Landscape Research, University of Roskilde, Denmark, October 18–21, 2000, organized by Jesper Brandt, Baerbel and Gunther Tress could serve as outstanding example for such successful interactive meetings.

For our purpose, the discussions at the International Transdisciplinary Conference 2000 at Zurich (Scholz et al. 2000), promoting transdisciplinarity for joint problem-solving ventures among science, technology and society have great relevance and deserve therefore special mention. It was organized by the Swiss Federal Institute of Technology, taking a lead in transdisciplinary studies in its Natural and Social Science Interface program. Its object was to foster transdisciplinarity as an effective way for managing complexity to meet the challenges of the 21st century by developing transdisciplinary practice in a wide range of fields, promoting transdisciplinary research and creating favorable institutional structures and powerful incentives for successful transdisciplinarity. In the announcement of the conference, Rudolf Haeberli, its chairman, declared that universities and basic research have an important role to play, but in a global society, they are just one knowledge producer. Therefore, attempts were made to involve also decision makers. During the sessions, a great number of important transdisciplinary initiatives and projects were presented, carried out mainly in Switzerland, including also landscape ecological studies. However, at the same time the great obstacles which have to be overcome for their implementation became very apparent.

To these obstacles a special section was devoted also in an international seminar, dealing with the potential and limitations of transdisciplinary landscape studies (Tress et al. 2003). Here, among others, the epistemological problems with which those carrying out such studies have to struggle were discussed in depth by Winder (2003). He maintained that *"we cannot investigate complex systems without integrating knowledge from more that one knowledge community. The engineer and the sociologist, the anthropologist and the chemist need much help to specify boundary conditions for their own subsystem. This is integrative research. This requires negotiating compromises between researchers that allow them to construct a composite knowledge system that capitalizes on perceived strengths without compromising the logical consistency of the whole."* (Winder 2003, page 84–85)

For this purpose they must ignore some of their own beliefs derived from different knowledge communities by negotiating their temporary suspension. Winder (2003) distinguishes between three types of such beliefs: **Theory-beliefs** which can be set aside quite easily; **creedal beliefs** which can occasionally be suspended for the duration of the project, but are much harder to banish or dismiss, and **cultural beliefs** which are part of our own identity and define the knowledge communities to which we belong and can be temporarily suspended only if the participants are sufficiently well motivated. Such a deeply ingrained cultural belief in Western societies has become the paradigm of "the balance of nature" which we are supposedly not allowed to disturb.

Important information was provided also in a special session, dealing with training. De Noy-van Tol (2003, page 129) from the Wageningen Initiative for Strategic innovation (ISI) of the Wageningen University and Research Center stressed the need for transdisciplinary education approaches from two directions, which are of great importance also for landscape studies.

1. *"Historic insights: real innovation and breakthrough in science and technology only happen on the* ***interfaces*** of specialist discipline
2. *Developments in society: the problems we – society- are faced with when trying to make use of or resources in a sustainable way can only be solved when we look at processes in an integrated manner, a holistic approach, to understand the complexity of what we do".*

It is very obvious that to achieve such genuine collaboration as required in transdisciplinary enterprises is not easy. It requires not only high motivation for cooperation towards their joint systems goal, but also special skills for ***systems network thinking***. For this purpose the new branch of ***systems learning***, resulting from the capacity of combining systems thinking with systems acting is of greatest importance. It seeks to avoid formulating problems according to only one particular perspective to the exclusion of others. In our case both academic restoration researchers and practitioners involved in land-and seascape studies will have to acknowledge the validity and relevance of diverse knowledge and perspectives in encouraging genuine local participation and capacity – building, as well as for building up their own capacities. A very useful source for this purpose has been provided by Senge (1994) in "The Fifth Discipline" on systems thinking and learning for organization and management research. It could inspire also the creation of innovative systems methodologies, essential for cooperative transdisciplinary problem-solving oriented landscapes studies enhancing the communication between researchers, professionals, practitioners and stakeholders.

10. THE HOLISTIC AND TRANSDISCIPLINARY SCIENTIFIC REVOLUTION

Although many of those applying transdisciplinary concepts and methods in their work may not be aware of it, the road towards transdisciplinarity which I have discussed above has been paved actually by what I suggest to call *"the scientific holistic and transdisciplinary revolution"*. It started in the last part of the twentieth century and is still going on, but as will be shown below, has now apparently reached its climax.

According to Kuhn (1970), such a revolution is initiated when new paradigms of conceptional schemes replace gradually those of conventional and well-established paradigms of so-called "normal science". Kuhn explains that for most scientists, major theories or *"paradigms"* are like spectacles, which they put on in order to resolve "puzzles". Every now and then a "paradigm shift" occurs in which these spectacles get smashed and scientists put on new ones and turn everything upside down, sideways or in a different color. Once a paradigm shift takes place in any scientific field, a new generation of scientists is brought up wearing the new glasses and accepting the new vision as natural "true". Kuhn realized that in practice it is impossible for scientists to choose simply the "better" one between two competing theories, as claimed by Popper's positivistic falsification standard of objectivity, because this was a myth. Paradigm shifts are such drastic alterations of perceptions that advocates of theories will not even agree on what constitutes a valid test of such so-called "objective" processes.

The scientific revolution, with which we are concerned here, has lead to a radical shift from reductionistic and mechanistic to holistic and organismic paradigms, based on systems thinking of complexity, networks and hierarchic order. It has replaced the belief in the objectivity and certainty of the scientific truth with the recognition of the limits of human knowledge, leading to a contextual view of reality, and to the need for dealing with uncertainties. And last, but not least, it has replaced mono-and multi-disciplinarity by inter-and transdisciplinarity approaches. According to Holling et al. (1999), such a holistic paradigm shift is changing already the science and practice of adaptive resources management and recognizes human wisdom and traditional commonsense and its deep cultural values.

This scientific revolution has finally lead to the development of transdisciplinary conceptions and research methods, trying to do justice not only to the physical, biological and the socio-economical, but also to the mental, the cultural and the spiritual reality in which we live. This profound postmodern scientific and cultural transformation, has changed already many of the ideas, believes and perceptions, still dominating in Western society. The well-know author of the "Tao of Physic" and the "Web of Life", Capra (2002), has presented some of these changes as "Hidden Connections", integrating the biological, cognitive , and social dimensions of life into a "science of sustainability".

This shift towards holistic paradigm is presently reaching its peak, offering **a unified view of the world**, *"perceived not anymore like a giant mechanism, but as cast organism, in which every part affects every other"*(Laszlo 1994). For this unified view of the world Laszlo (2003; 2004) has formulated a genuine transdisciplinary theory for the *"integration of everything"*. This has been achieved by summarizing, and integrating and interpreting with his remarkable expertise in systems and complexity theory the latest findings in such diverse sciences as quantum physics, physical cosmology, evolutionary-, neuro, – and quantum biology, and in the new field of consciousness studies. Laying the foundations for a post-modern integral science of quantum, cosmos, life and consciousness, both studies should be regarded as the most recent important breakthrough in this scientific revolution. As such they deserve a much more detailed presentation. However, I have to restrict myself chiefly to those aspects which open new vistas for our understanding of the place of humans in nature– and therefore also in their landscapes – as an integral part of on all embracing conception of synthetic cosmic, geological, biological and cultural evolution. These issues are not only of greatest significance for the theoretical foundations of both new transdisciplinary sciences of landscape planning and management, conservation and restoration. They will hopefully also provide for these sciences and for all those concerned with the sustainable future of nature and human life on Earth conceptual, educational and practical tools for attaining the urgently needed post-industrial symbiosis between human society and nature. The following subchapter serves therefore also the concluding one.

11. THE INTEGRAL THEORY OF EVERYTHING AND ITS SIGNIFICANCE FOR THE RECIPROCAL RELATIONS BETWEEN HUMANS AND THEIR LANDSCAPES

In "The Connectivity Hypothesis" Laszlo (2003) has provided a formal synthesis of a coherent global hypothesis for this of connectivity of a unified universe with the help of a general mathematical model. Reviewing this book in the new journal "Ecological Complexity" (Naveh 2004) I have characterized it as follows: "Laszlo's present book should be regarded as the most significant conceptual breakthrough for such a genuine unified theory of the world. Unlike that of Einstein and other attempts, his daring attempt of a global hypothesis of connectivity is not grounded exclusively in physics but embraces quantum, cosmos, life and consciousness and their relevant scientific fields. The formal expression of a coherent unified universe by a general mathematical model has put it without doubt on the cutting edge of the transdisciplinary scientific revolution and its post-modern complexity sciences". (See also Naveh 2004, in a more extensive review of this book).

His more recent book (Laszlo 2004) on "Science and the Akashic Field" is a lucid, nontechnical introduction to this theory of the "informed universe", highlighting the theory's crucial feature, namely *"the revolutionary discovery that at the roots of reality there is an interconnecting, information-conserving and information-conveying cosmic. For thousands of years, mystics and seers, sages and philosophers maintained that there is such a field. In the East they called it the Akashic Field* (Akasha is a Sanskrit word meaning "ether": all pervasive space. The Akashic Record or Chronicle is the enduring record of all that happens, and has ever happened in space and time). *But the majority of Western scientists considered it a myth. Today, at the new horizons opened by the latest scientific discoveries, this field is being rediscovered. The effects of the Akashic Field are not limited to the physical world: the A-field" (as we shall call it) informs all living things – the entire web of life. It also in forms our consciousness* (Laszlo 2004, page 3).

As mentioned above, its is typical for such a scientific revolution, that a growing number of anomalous unexplained phenomena of "puzzles" or "fables", beyond the established theories, explored by cutting-edge researchers are coming to light. However, in our case, this is not happening only in one single discipline, like physics, to which Kuhn (1970) chiefly devoted his attention: In the inquiry of such fables, Laszlo goes across many scientific fields, from cosmology to quantum physics and from biology to psychology, of which many of these are on their own transdisciplinary integrations of different disciplines such as physical cosmology, quantum and evolutionary biology an consciousness research.

The cornerstones for this all-embracing holistic scientific theory that will grow into a genuine theory of everything are two kinds of these anomalies, namely **coherence and correlation**. Coherence is well-known in physics, but the kind of coherence discovered today in a growing number of these sciences is more complex. They indicate *"a quasi-timing together of the elements of a system"*, whether this system is an atom, an organism, an ecological system, or a galaxy. Thereby all these parts of the system of such coherence are so correlated that what happens to these parts, also happens to the other parts. These ecological systems include of course also landscapes and our Total Human Ecosystems, integrating humans and their total environment, of which landscapes are its tangible three-dimensional components.

According to Laszlo (2004) the discovery of this unity of coherence and correlation is at the core of the next paradigm shift in science, offering the best-ever basis for creating *"the long sought but hitherto unachieved integral theory of everything"*. Laszlo (2004) recognized that David Bohm was one among the first surmising that interactions in the domains of nature – as well as of mind – are meditated by a fundamental information field in the quantum vacuum in addition to the energy sea that fills cosmic space.

I have cited often in the previous chapters Bohm's significant contributions for a holistic scientific conception, and therefore will here briefly recapitulate his ideas in the light of Laszlo's findings. In his search for a holistic physics, Bohm – like Laszlo – rejected the reductionistic Newtonian physics, recognizing reality as a collection of separate interacting objects moving in space and time and called this *"the explicate order"*. But he also refused to accept the weird behavior of quantum as a full description of this reality and suggested that the selection of the state of the quantum is not at random, and that the particle behavior is weird and indeterministic only at he surface, but at the deeper level it is determined by the quantum potential. Bohm (1980) called this the subtle *"implicate order"- a holofield where all the states of the quantum are permanently recorded, and enfolded.* Observed reality emerges from the field by constant unfolding it as the explicate order. On this order all present concepts, methods and parameters of LE and RE are still based (See also chapter 11).

Like Laszlo in his present work, Bohm used already then, the lensless holographic photography as a metaphor, comparing it to this implicate, enfolding order. In contrast to the explicate order of normal photography, in which light comes from each part of the object and falls onto the entire photographic plate, in a holograph every tiny part of the plate contains information about **the entire scene**; light in a sense becomes enfolded across the holograph. *What appears to be distant objects will, within the implicate order, lie close together, the one being enfolded in the other.*

The implicate order concept was further developed into an even broader concept of a *generative or creative order by* Bohm and his close associate David Peat (Bohm and Peat 1987), who after Bohm's death published a comprehensive biography on his life and times (Peat 1998) As discussed in chapter 11 the generative order of fractals and contextual scaling in hierarchical levels of contemporary LE and RE is an important attempt for transcending the Euclidian regularities, but they still remain within the categories of natural sciences. Therefore, we claimed that *"one of the greatest challenges for landscape ecology as a transdisciplinary science is the inclusion of further enfolding orders with new categories for intrinsic nature values, health and self-organization, as well as for human consciousness and creativity.*" (Naveh and Lieberman 1994, page S2–14) These enfolding and generative orders have now been transformed by Laszlo into a new category of *the universal information field,* providing the tools for encoding this information with the help of quantum physics.

Exploring the role of information in nature Bohm has already located nature's information field in the *quantum vacuum* that is now recognized, according to Laszlo as a superdense medium filling the universe, that apparently not only transports light, energy, pressure and sound. *"Its vacuum vortices record information on the state of the particles that created them – and their interference pattern records information on the ensemble of the particles of which the vortices have interfered."* (Laszlo 2004, page 52). Like waves that propagate in the sea, producing a connection among vessels, fishes and other objects that generated them, this vacuum carries information on atoms, molecules, organisms, populations and ecologies of organisms, and on the whole state of the universe. However, in contrast to the sea waves, the vacuum is a frictionless medium, and waves and objects could move forever: the "wave memory" of the universe may be eternal.

Laszlo likens these superposed vacuum-interference patterns to nature's "holographs", and offers the "daring but logical" hypothesis that this so-called zero-point field of the quantum vacuum *"generates the holographic field that is the memory of the universe"*. According to Laszlo the ancient **Akashic or A-field**, is taking now its place as a radical innovation, among the other fundamental fields. These fields have been recognized by physicians as the G – gravitational field, the EM electromagnetic field and various nuclear and quantum fields. But as such an universal information field, the A-field has even much more far-reaching ecological social and psychological implications on human's mind, consciousness and on live a whole, and therefore also on our relations with nature and its resulting reciprocal interactions with our landscapes. Laszlo surmises that the A-field conveys the most direct, intense and therefore evident information between things that are closely similar to one another in this isomorphic basic form by superimposed vacuum wave-interference patterns. In these, like in a hologram, everything is meshed or "conjugated". The intensity of this information depends on their isomorphic closeness: Humans are directly informed by fellow humans and less directly by plants and animals, and all of nature. In the words of Laszlo: *"This information through the A-field subtly tunes all things to all other things and accounts for the coherence we find in the cosmos, as well as in living nature."* (Laszlo 2004, page 108).

Laszlo (2004 page 110) further claims that information through the A-field accounts not only for the quasi-instant coherence of all parts of an organism, but also for the subtle but effective correlation between organisms and environments. In this field, the existing linkages between phenome and genome of these organisms can be mechanical, chemical, biochemical, or field-transmitted by electromagnetic and quantum fields. These can apparently trigger not only adaptive mutations in the genome but such quantum fields appear also to link parts of within the whole organism and they may also link it with its external environment. *"The fact is that the organism is amazingly coherent in itself, and is coherently linked with the world around it"*. Laszlo notes that only now the best-conceived fables of the cutting edge of biological research where systems biology meets quantum physics to the fledging discipline of quantum biology, command serious attention.

The hologram of the ecology in which organisms and their populations are embedded correlates all organisms in that ecology, namely the habitat and their niches for plants and animals and the landscapes in which humans live, down to the structure of their genome. This leads Laszlo to the rather revolutionary conclusion that the ongoing variation of the genome is subtly informed, increasing the probabilities that when the environment changes, the genome comes up with mutations that are viable in the new milieu".

Laszlo has cited the cosmologist and mathematical physicist Fred Hoyle (1989) who refuted the Neo-Darwinistic paradigm of random mutations as the major engine of evolution and pointed out that life evolving purely by chance is about as likely as a hurricane blowing through a scrap yard assembling a working airplane. The assumptions of a slow and gradual geological evolution were also contradicted by fossil findings which prompted Gould and Eldredge (1977) to advance the theory of "punctuated equilibrium. In this macro-evolutionary theory, new species arise already in time spans of not more than five to ten thousand years. We can assume that for ecotype interspecific variability this time span could be even much shorter.

These findings corroborate our assumptions on relatively short evolutionary time which was allowed for the far-reaching genetic impacts of the intensive agro-pastoral land use pressures in the Mediterranean and the Southern Levant, discussed in chapter 2. These induced the great adaptive resilience acquired by Mediterranean ruderalic plants as aggressive colonizers, invading successfully California grasslands, lacking the major Neolithic agricultural bifurcation and its ecological aftereffects. (See also chapters 1 and 7).

The realization that the coherence of organisms is operating through the modification of the genome clearly contradicts the neutral and near-neutral theories of molecular evolution. But it leads strong support to the non-random impacts of environmental selective stresses, overriding migration and genetic drift. This has been shown by Nevo and his collaborators in a great number of impressive studies for diverse plant and animal species in Israel and the Near East (Nevo (2001).

The new insights gained for the reciprocal relations between human society and its landscapes are amongst the most important consequences of the integral theory of everything for both TLS. For this purpose we have first of all to revise our holistic and transdisciplinary landscape concepts, discussed in the previous chapters, in the light of Laszlo's Integral Theory of Everything:

The A-field can now be identified as the underlying mechanism for both the physical-geospheric and the cognitive mental and spiritual noospheric space spheres of the Total Human Landscape in which we live. We can now also assume that the far-beyond time and space reaching "non-locality" information, generated by the A-field creates the coherence between all organisms – including humans, their brain and minds and consciousness – and their physical and biotic landscape components at different levels of intensity. This drives the auto-and cross catalytic cycles of mutually supporting feedback correlations between all micro-and macro-quantum systems, ranging from the molecular to the biosphere and our THE enabling their self-organization, described in chapter 13. It is therefore also responsible for the above-mentioned reciprocal relations of the nature-human-culture landscapes triangle.

In Fig. 1 these relations are presented from an evolutionary and historical perspective: Our THE emerged from the Informed Universe through the quantum vacuum of the Akashic Field, accounting for the coherence of all its components. Human systems evolved by auto- and cross-catalytic cycles as the result of their biological and cultural evolution. The strength of the coherence and their underlying correlations between humans and their landscapes is determined by the degrees of intensity of their land use history through cybernetic eco-feedback couplings. These have been modified from "total nature" and its balanced homeostatic state to mixed "nature-culture" and its dynamic homeorhetic states, and partly converted into an

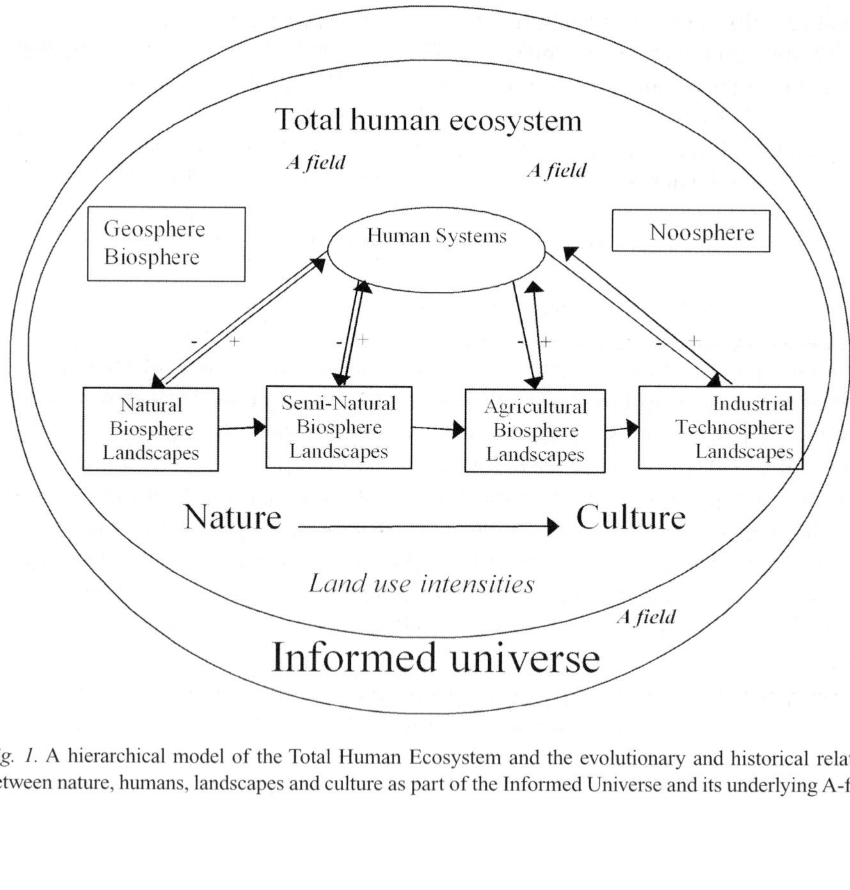

Fig. 1. A hierarchical model of the Total Human Ecosystem and the evolutionary and historical relations between nature, humans, landscapes and culture as part of the Informed Universe and its underlying A-field.

unbalanced "total culture" state, dominated by destabilizing urban-industrial (+) feedback couplings.

12. CONCLUSION: NEW CHALLENGES FOR A TRANSDICIPLINARY LANDSCAPE SCIENCE

A new, post-industrial dynamic balance, reconciling between nature and culture, and ensuring mutually supportive symbiotic relations can be established in these landscapes by introducing stabilizing, negative (-) cultural feedbacks. Both trans-disciplinary landscape sciences can make important contributions to this goal by supplying pragmatic scientific information and knowledge and by their active involvement in the implementation of this knowledge by research and education and by their participation in landscape planning, design, management, conservation and restoration projects, furthering the sustainability revolution.

The German landscape historian Sieferle (1997) has described in a comprehensive way how humans, acting as co-creator of agricultural and cultural landscapes have shaped the great majority of our landscape for good and for worse, and these, in turn have affected human society. He maintains that the "total industrial landscape" is a transformation landscape, in which culture has completely displaced nature. For the prediction of its future we should exclude the possibility that out of this landscape a more stable autonomous culture will emerge for which a technological controlled solar energy system would not be any obstacle. I have emphasized above in a much more affirmative way that such a solar energy system has to be a vital part of this new culture. To enhance this process, the recognition of the role of coherence through the A-field, conveying information for the human information pool of culture and its impacts on the network of connections and correlations of human-landscape relations will have both theoretical but also practical implications on future, more sustainable land uses. It should lead to the realization that our estrangement from nature by the conversion and degradation of biosphere into technosphere landscapes and the resulting density and other urbane stresses, is creating in a relatively short period an entirely different much more aggressive and less healthy and happy subspecies of *Homo sapiens*. This cannot be prevented in spite of the impressive advances in medical and psycho-therapeutic treatments and their heavy economic burden on society, because these changes will be genetically fixed.

According to Roszak et al. (1995), restoration is one of the recent, most promising examples of the positive effects of these reciprocal interactions. It differs from most other human impacts on the landscape by having an **active conservation value**, many times combined also with utilitarian purpose. Now we may be able to provide also more convincing arguments that for those actively involved in restoration it could have also favorable effects on their mental and physical well being.

Such new insights and their positive effect on human behavior and ethics will be greatly enhanced if the surprising technological breakthroughs from this rediscovery of the A-field as the quantum vacuum holofield envisaged by Laszlo (2004) will come true. These may not only enable quantum computation, but also conveying effective "in-formation" without the expenditure of energy and "teleport not just bits of quanta but atoms and molecules, living cells and organs, as well as aspects and elements of consciousness.

The ongoing exponential landscape degradation cannot be prevented by treating landscapes solely as a commodity to be exploited as a resource on which we project our economic interest as measured by monetary parameters and products of the "free market play". We have to recognize also their intrinsic existence values by which they become **not as a means to an end, but as an end itself**. Even the term **"natural capital"** introduced by ecological economists (Costanza and Daly 1992) cannot account fully for the most vital life supporting functions, provided by fertile soil, clean air and water, and not at all for the life fulfilling intangible aesthetic, cultural, spiritual and re-creative values of healthy and attractive biosphere landscapes. The importance of these "soft values" for our quality of life and mental well-being in the emerging information society is greater now than ever. Therefore much greater attention should be paid also to the **psychotherapeutic landscape functions**. These are derived from the restorative experience of nature against the multitude stresses of modern life, and especially against the **"direct attention fatigue"**. According to Kaplan (1995), this occurs after prolonged intensive mental and creative work, such as performed behind the computer by High-Tech workers.

We cannot predict with certainty what will happen to this landscape by extrapolating from what has happened in the past, and we can therefore also not predict in certain terms its future. **But we can help to shape its future**. We can attempt to anticipate their fate and the risks involved in its further misuse and degradation and the prospects for its further sustainable development. We can illustrate these anticipations by modeling different scenarios, realizing the most desirable scenarios, both for human society and nature, and prescribe the best remedies for their management, conservation, and restoration. For this purpose our landscape theory cannot be bound down by a rigid, human detached and mechanistic predictive theory, for which classical Newtonian physics has served as a model. Both landscape sciences have to become "post-normal" sciences as suggested, rightly for environmental management sciences in general by Waltner-Toews et al. (2003). They must strive to become also prognostic and prescriptive sciences and normative ones. They have to be guided by a much broader and flexible, future-oriented and holistic view of the world systems, our global, deep ecological and cultural crisis, and our readiness as scientists to become involved in their solution.

In our present scientific disciplinary fragmentation, the physical geospheric and the mental, cognitive and spiritual noospheric space spheres are imprisoned either in the natural sciences or in the social sciences and humanities. Therefore, instead of leading into a complementary "third culture", they have resulted in the formation of "two cultures" opposing each others (Snow 1963). Fortunately, this holistic core idea that humans are being part of and not apart from nature and its processes and functions, is shared also by ecological economists, anthropologists and psychologists, by social ecologists, and scientists from other eco-disciplines.

New ecological problems are arising, demanding not only strictly ecological, but also ethical, social and economic sound solutions. Thus for instance, already now it has became apparent that even "green technologies", such as wind power farms can cause severe damage to birds, especially if these are established in or near the flyways of migrating birds along routs of bird migration. Further progress in the right direction towards sustainability in the post-industrial information age will be accompanied by other serious ecological problems. Therefore, their importance for successful sustainable development lies not only in the ecological but also in the socio-cultural, ethical and economic realms. This has been also shown by the outcome of the EU MOSES project (2000), in which we attempted to develop a transdisciplinary systems model for sustainable regional development towards the information society, described in chapter 13.

In recent years great advances have been achieved in developing generic or process – oriented simulation models to cope with the complex spatial and temporal landscape dynamics. This was very apparent in the recent special issue of "Ecological Complexity", devoted to this subject (Bollinger et al. 2005; Green and Sadedin 2005). Although, some of these models also consider anthropogenic impacts and population dynamics, they do not attempt to explore the deeper, underlying causes of human behavior, driving these dynamics. Their models stay only within the limits of the biophysical complexity of energy/material flow of ecosystems, not transcending beyond the boundaries of natural sciences into the social sciences and humanities. Therefore they cannot contribute very much to the above-described transdisciplinary goals of our new landscape science and the sustainability revolution.

This is not the case in the ISIS (Information Society Integrated Systems Model) developed in our MOSES-project. In this multinational project we brought together an interdisciplinary team of 15 scientists from economics, regional sciences, consulting, landscape planning, ecology, landscape ecology, systems sciences, and knowledge research with the ultimate aim to develop policies, contributing to influence the present development of the information society and their landscapes in desirable directions. In this systems simulation model we considered four major realms of human factors shaping land use and therefore also landscapes, namely lifestyles, preferences, attitudes, knowledge and the economy. These interacting and rapidly evolving factors were grouped into four "realms": 1) economy 2) culture, attitude, consciousness and human needs and wants, 3) science, know-how and technology, and 4) the natural, semi-natural, agricultural and urban-industrial landscapes as the tangible matrix of the THE. The changes in these realms, driven by their auto-and crosscatalytic interactions, guided us in our model how to encourage the most desirable developments and discourage the undesirable ones from a social and ecological point of view. As explained in more detail by Grossmann and Naveh (2000), this can be achieved by **viable management of viable systems though influence, but not by control**.

A most important tool for this purpose is **information management**, by which the transformation of the knowledge gained from semantic to **pragmatic information**, is changing the reality via its feedback on the receiver. We are not yet able to make meaningful predictions about this transformation. However, we can envisage that using and further developing the Integral Theory of Everything through the A-field effects as the scientific foundation will serve the core concept for a unifying metatheory linking natural, ecological and socio-economic systems in the Total Human Ecosystem. As a major achievement of this scientific transdisciplinary revolution, this will be also a major step towards the creation of a scientifically well-grounded synthetic "third culture" of Snow (1963).

For an even more meaningful scientific breakthrough on the effect of the A-field on these reciprocal relations and their practical application for landscape management and planning, conservation and restoration and for land use decisions in the broadest sense, the close cooperation of open-minded and environmentally oriented quantum physicists and biologists. To gain their participation in this effort will be probably a great challenge for both TLS. They may help us to supply more effective scientific weapons for fighting against the ignorance, short-sightness and greediness which are the main causes for the destruction of nature and climate destabilization and therefore the main obstacle for advancing towards the sustainability revolution.

In conclusion: To fulfill a meaningful role in the urgently needed social and ecological sustainability revolution, landscape ecologists and restorationists have reached a crucial turning point in their science. They cannot go on following the well-paved and secure road of prevailing conventional and chiefly outdated mechanistic and reductionistic scientific paradigms of linear thinking. These have been based on the assumption that the only obligation to society and all their merit as "true scientists" is to provide human-detached so-called "objective" scientific information. They will have to choose now between two alternatives: Either continuing to be satisfied by being one among other disciplinary branches of ecology and geography, filling the scientific journals and books with fine peer-reviewed studies, for describing and measuring landscape heterogeneity, stability, and changes with the most advanced and sophisticated methods and models. However there is little hope that his semantic scientific information will change the reality. Or alternatively, transforming both sciences into goal-oriented and mission-driven "post-normal" transdisciplinary sciences of landscape history and appraisal, planning and management, conservation and restoration, and providing meaningful pragmatic information for helping to change reality.

Accepting this challenge they will have to become "committed actors" and join all those concerned with the future of life on Earth and the welfare of all its inhabitants. Their eco-disciplinary teams should combine professional and scientific expertise and leadership with ecological knowledge, wisdom and ethics, providing pragmatic scientific information towards a sustainability revolution.

A first vital step towards this goal will be the establishment of new, better balanced and complementary relations between healthier, more livable and attractive technosphere landscapes and their "hinterland" of viable and diverse biosphere landscapes and of productive and sustainable agricultural landscapes, on which human well beings depend. Among one of the most urgent tasks will be to ensure the instrumental and non-instrumental live supporting functions of "*keystone biosphere landscapes*" and especially their potentials to act as a biological filter and living sponge, absorbing the emissions from the technosphere. This requires the restoration, reclamation and rehabilitation of damaged landscapes,

the revitalization of wetlands, rivers, lakes and their embankments, the creation of living corridors and viable urban biosphere islands, together with dynamic conservation management of nature parks and reserves. It will be achieved by closest cooperation of both transformed landscape ecology and restoration ecology into transdisciplinary landscape sciences.

REFERENCES

Alberti M, Marzluff JM, Schulenberger E, Bradley G, Ryan C, and Craig Z (2003). Integrating humans into ecology: Opportunities and challenges for studying urban ecosystems. *Bio-Science* **53,** 1169–1179.

Allen TFH, and Hoekstra TW (1992) *Toward a Unified Ecology*. Columbia University Press. New York, USA.

Antrop M (2000) Background concepts for integrated landscape analysis. *Agriculture, Ecosystems and Environment* **77:** 17–28.

Antrop M, and Van Eetveld V (2000) Holistic aspects of suburban landscapes: visual image interpretation and landscape metrics. Special issue, *Landscape Ecology* **50:** 43–58.

Barrett GW (2001) Closing the ecological cycle: the emergence of integrative science. *Ecosystem Health* **7**: 79–84.

Bastian O, and Schreiber K-F (1994) *Analyse und oekologische Bewertung der Landschaft*. Gustav Fischer Verlag Jena, Stuttgart, Germany.

Bastian O, and Steinhardt U (eds) (2002) *Development and Perspectives in Landscape Ecology. Conceptions, Methods, Application*. Kluwer Academic Publishers, Dodrecht, the Netherlands.

Baudry J, Burel F, Thenaile C, and LeCoeur D (2000) A holistic landscape ecological study of the interactions between farming activities and ecological patterns in Brittany, France. Special issue, *Landscape and Urban Planning* **50:** 119–128.

Bohm D (1985) *Wholeness and the Implicate Order*. Ark Paperbacks, London and New York.

Bohm D (1996) *On Dialogue*. Nichol L (ed.) Routledge. New York, USA.

Bohm D, and Peat DF (1987) *Science, Order, and Creativity. A Dramatic Look at the Roots of Science and Life*. Bantam Books. New York, USA.

Bollinger J, Lischke H, and Green DG (2005) Simulating the spatial and temporal dynamics of landscapes using generic and complex models. (Introduction to special Issue) *Ecological Complexity* **2**:107–116.

Brandt J (1998) Key concepts and interdisciplinarity in landscape ecology: A summing-up and outlook. In: Dover JW, and Bunce RGH (eds) *Key Concepts in Landscape Ecology. IALE(UK)*. Garstang, pp 421–434.

Brandt J, and Agger P (1984) *Methodology in Landscape Ecological Research and Planning*. First International Seminar of IALE, Roskilde University Centre, October 1984. Roskilde Universitaetsforlag, Roskilde, Denmark.

Bradshaw AD, and Chadwick MJ (1980). *The Restoration of Land*. Blackwell Scientific Publication, Oxford.

Breuste JH (1998) Preface. In: Breuste J, Feldmann H, and Uhlmann O (eds) 1998. *Urban Ecology*. Springer-Verlag Berlin, Germany, pp V–VII.

Bright C (2000) Anticipating environmental surprises. In: Brown LR, Flavin HF, and French H (eds) *State of the World 2000. A Worldwatch Institute Report on Progress Toward a Sustainable Society*. W.W. Norton & Company, New York, London, U.K, pp 22–38.

Brown LR (2001a) *Eco-Economy: Building an Economy for the Earth*. W.W. Norton & Co, New York, USA.

Brown LR (2001b) *Dust Bowl Threatening Chinas Future*. The Earth Policy Alerts. [online] htpp:// www. earth policy.org

Brown LR (2003) *World Creating Food Bubble Economy Based on Unsustainable Use of Water*. Earth-Economy Update 2003–2.[online] htpp://www.earth-policy.org/Updates22.htm

Brown LR (2003) Plan B: *Rescuing a Planet under Stress and a Civilization in Trouble*. W. W. Norton & Co. New York.
Brown LR (2006) Plan B 2.0: *Rescuing a Planet under Stress and a Civilization in Trouble*. W. W. Norton & Co. New York.
Cahalan W (1995). Ecological groundedness in Gestalt theory. In: Roszak T, Gomes M, and Kanner A (eds) *Ecopsychology Restoring the Earth healing the Mind,* Sierra Club Books, San Francisco, USA, pp 21–223.
Capra F (2002) *The Hidden Connections. Integrating the Biological, Cognitive, and Social Dimensions of Life into a Science of Sustainability*. Doubleday, New York, USA.
Carmel Y, and Naveh Z (2002) The paradigm of landscape and the paradigm of ecosystems – implications for land planning and management in the Mediterranean Region. *Journal of Mediterranean Ecology* **3:** 24–34.
Costanza R (1996) Ecological economics: reintegrating the study of humans and nature. *Ecological Applications* **6:** 987–990.
Costanza R, and Daly HE (1992) Natural capital and sustainable development. *Conservation Biology* **6:**37–46.
Dansereau P (1957) *Biogeography: An Ecological perspective*. Ronald Press, New York.
De Groot RS (1992) *Functions of Nature: Evaluation of Nature in Environmental Planning, Management and Decision Making*. Wolters Noordhoff BV, Groningen, The Netherlands.
De Noy-van Tol, J. 2003. Needs for training of Professionals. In: Tress; B., Tress, G., van der Valk, A. and G. Fry, editors. 2003. *Interdisciplinary and Transdisciplinary landscape studies: Potentials and Limitations*. Delta Series 2, Wageningen 2003, The Netherlands. pp 129–135.
Di Castri F (1997) Editorial: Landscape in a changing globalized environment. *Landscape Ecology* **12:** 3–5.
Egler FE (1942) Vegetation as an object of study. Philosophical Science, **9,** 245–290.
Egler FE (1964) Pesticides in our ecosystem. *American Scientist,* **52:** 110–136.
Egler FE 1970. *The Way of Science. A philosophy for the Layman*. Hafner, New York.
Ehrlich PR (2002) Human natures, nature conservation, and environmental ethics. *BioScience* **52:** 31–43.
Earth Policy News (2003) [online]htpp://www.earth-policy.org/Updates/Update 23.htm
EU-Project MOSES (2000) *Modeling Sustainable Regional Development in the European Information Society*. Final report ENV4-CT97–2600-PL97–0543 for the EU Program, Environment and Climate, 1994–1998 Area 4: Human Dimensions of Environmental Change. UFZ Leipzig-Halle, Germany.
Farina A, and Naveh Z (eds) (1993) A Landscape approach to regional planning: The future of Mediterranean landscapes. Special Issue: *Landscape and Urban Planning,* **24:** 1–4.
Farina A (1998) *Principles and Methods in Landscape Ecology*. Chapman & Hall, London, U.K.
Fischer-Kowalski M, and Haberl H (1998) Sustainable development, long term changes in socio-economic metabolism, and colonization of nature. *International Social Science Journal* **156:** 573–587.
Fischlowitz –Roberts B (2002) Green Power Purchases growing by Leaps and Bounds. *Eco-Economy Update* 2002–4 [online] htp://www.earth-policy.org/Updates/Update9.htm
Flavin C (2001) Rich Planet, Poor Planet. In: Starke L (ed.) *State of the World 2001*. Norton & Company, New York, pp 3–20.
Forman RTT (1995) *Land Mosaics. The Ecology of Landscapes and Regions*. Cambridge University Press, Cambridge, U.K.
Garner G (2002) The challenge for Johannesburg: Creating a more secure world. In: L. Starke (ed.) *State of the World 2002*. Norton & Company, New York, London, U.K., pp 3–23.
Gould SJ, and Eldrege N (1983) Punctuated equilibria: the tempo and mode of evolution reconsidered. *Paleobiology 3*.
Gunderson LH, and Holling CS (eds) (2002) *Panarchy. Understanding Transformations in Human and Natural Systems*. Island Press, Washington DC, Covelo California, USA.
Gustafson EJ (1998) Quantifying landscape spatial patterns: What is the state of the art? *Ecosystems* 1: 143–156.
Green DG, and Sadedin S (2005) Interactions matter-complexity in landscapes and ecosystems. *Ecological Complexity* **2:** 117–130.

Grossmann WD (2000) Realizing sustainable development in the information society. *Landscape and Urban Planning* **50:** 179–194.

Grossmann WD, and Naveh Z (2000) Transdisciplinary challenges for regional sustainable development towards the post–industrial information Society. *Proceedings International Congress European Ecological Economy Society Vienna 2000.*

Green B, and Vos W (2001) Threatened Landscapes Conserving Cultural Environments. Spon Press, London and New York.

Haas L (2002) New energy from the power of the wind. *Deutschland* **6:** 46–49.

Haken, H. 1983. *Synergetics,* 3rd edn, Springer Verlag, Berlin.

Hassett B, Palmer M, Berhardt E, Smith S, Carr J, and Hart D (2005) Restoring watersheds project: trends in Chesapeake Bay tributary restoration. *Frontiers in Ecology and the Environment.* **3:** 259–267.

Higgs E (2005) The two culture problem: Ecological restoration and the integration of knowledge. *Restoration Ecology.*

Hobbs R (1997) Future landscapes and the future of landscape ecology. *Landscape and Urban Planning* **37:** 1–9.

Holling CS (1998) Surprise for science, resilience for ecosystems and incentives for people. *Ecological Applications* **6:** 733–735.

Holling CS (2000) Theories for sustainable futures. Conservation Ecology **4** (2):7 (Editorial)[online]

Holling CS, Berkes F, and Folke C (1999) Science, sustainability and resources management. In: Berkes BF, Folke C, and Colding J (eds) *Linking SocialSystems.* Cambridge University Press, Cambridge.

Hoyle F (1983) *The Intelligent Universe.* Michael Joseph, London.

Hull RB, Robertson DP, Richert D, Seekamp E, and Buhyoff GJ (2003a) Assumptions about ecological scale and nature knowing best hiding in environmental decisions. *Conservation Ecology,* **6:** 12. [online] URL:http://www.consecol.org/vol6/iss2/art12

Hull RB, Robertson DP, and Buhyoff GJ (2003b) Beyond the interventionist-preservationist duality. Conservation Ecology **7,** 1: 4. [online]URL: http://www.consecol.org/vol7/iss1/resp4

Ihse M, and Lindahl C (2000) A holistic model for landscape ecology in practice: the Swedish survey and management of ancient meadows and pastures. *Landscape and Urban Planning* **50:** 3–22.

IUCN (2000) *IUCN* Species Survival Commission Report *Red List of Threatened Species* World Commission- IUCN. Gland, Switzerland.

Jantsch E (1970) Inter-and transdisciplinary university: A systems approach to education and innovation. *Policy Sciences* **1:** 203–428.

Jantsch E (1980) *The Self-organizing Universe. Scientific and Human Implications of the Emerging Paradigm of Evolution.* Pergamon Press, Oxford, U.K.

Kaplan S (1995) The restorative benefits of nature: toward an integrative framework. *Environmental Psychology* **15:** 169–182.

Klijn J, and Vos W (eds) (2000) *From Landscape ecology to Landscape Science.* Kluwer Academic Publishers, Dodrecht, The Netherlands.

Koestler A (1969) Beyond atomism and holism – the concept of the holon. In: Koestler and Smithies JR (eds) *Beyond Reductionism: New Perspectives in the Life Sciences.* Hutchinson of London, London, pp 192–216.

Kroenert R, Steinhard U and Volk M (eds) (2001) *Landscape Balance and Landscape Assessment.* Springer-Verlag, Berlin, Germany.

Kuhn TS (1970) *The Structure of the Scientific Revolution. University of Chicago Press,* Chicago, USA.

Langer H (1970) Die oekologische Gliederung der Landschaft und ihre Bedeutung fuer die Fragestellung der Landschaftspflege. *Landschaft 1 Stadt* **3:** 2–29.

Langer H (1973) Oekologie der geosozialen Umwelt. *Landschaft 1 Stadt* **3:** 2–29.

Larsen J (2003) Record Heat Wave in Europe takes 35, 000 lives. Far Greater Losses May Lie Ahead. Eco-Economy Update **2003–9.** [online] htp://www.earth-policy.org/Updates/Update29.htm

Laszlo E (1972) *Introduction to Systems Philosophy: Toward a New Paradigm of Contemporary Thought.* Harper Torchbooks, New York, USA.

Laszlo E (1987) *Evolution: The Grand Synthesis.* Shambhala New Science Library, Boston, USA.

Laszlo E (1994) *The Choice: Evolution or Extinction. A Thinking Person's Guide to Global Issues*. C. P. Putnam & Sons, New York, USA.
Laszlo E (2001) *MACROSHIFT Navigating the Transformation to a Sustainable World*. BK Berret-Koehler Publishers, Inc. San Francisco, USA.
Laszlo E (2003) *The Connectivity Hypothesis*. State of University of New York Press, Albany.
Laszlo E (2004) Science and the Akashi Field An *Integral Theory of Everything*. Inner Traditions Rochester, Vermont.
Leser H (1991) *Landschaftsoekologie*. Eugen Ulmer Gmb H& CO. Stuttgart.
Levin S (1999) *Fragile Dominion Plexity and the Commons*. Helix Books Perseus Publishing. Cambridge, Massachusetts, USA.
Li B-L (1996) Fuzzy modeling in ecology. *Ecological Modeling* **90:** 109–184.
Li B-L (2000a) Why is the holistic approach becoming so important in landscape ecology. *Landscape and Urban Planning* **50:** 27–47.
Ludwig D (2001) Crisis and transformation. *Conservation Ecology* **5,** 1:11. [online] URL: htpp://www.consecol/org.Vol5/iss1/art11
Luz F (2000) Participatory landscape ecology- A basis for acceptance and implementation. *Landscape and Urban Planning,* **50:** 157–166.
Lyle JT (1994) *Regenerative Design for Sustainable Development*. John Wiley & Sons, New York. USA.
Makhzoumi J, and Pungetti G (1999) Ecological Landscape Design and Planning: The Mediterranean Context. E& FN Spon, London, U.K.
Mansvelt JD, and van der Lubbe MJ (1999) *Checklist for Sustainable Landscape Management*. Final Report to the European Community on the Landscape and Nature Protection Capacity of Organic/Sustainable Types of Agriculture. Elsevier, Amsterdam, The Netherlands.
McDonnell MJ, and Pickett STA (ed.) (1993) *Humans as Components of Ecosystems. The Ecology of Subtle Human Effects and Populated Areas*. Springer New York, Berlin, Heidelberg.
Moss MR (1999) Interdisciplinarity, landscape ecology and the Transformation of Agricultural Landscapes". *Landscape Ecology* **15:** 303–311.
Mueller F, and Bai-Lian Li (2004) Complex systems approach to study human environmental interactions. In: Hong S et al. (eds) Ecological Response and Strategy. Issues in a Changing World State, pp 31–46.
Nassauer JI (1990) Statements at the closing session of the First International Conference on Cultural Aspects of Landscape. In: Svoboda H (ed.) *Cultural Aspects of Landscape*. Pudoc, Wageningen, The Netherlands, p 173.
Naveh Z (1971) Conservation of ecological diversity of Mediterranean ecosystems through ecological management. In Duffy E, and Watt AS (eds) *The Scientific Management of Animal and Plant Communities for Conservation*. Blackwell Scientific Publications, London, pp 605–622.
Naveh Z (1984) Towards a transdisciplinary conceptual framework of landscape ecology. *Proc. First Inter. Seminar on Methodology in Landscape Ecological Research and Planning*. Roskilde University Center, October 1984, pp. 35–45, 1984.
Naveh Z (1988) Multifunctional reconstruction of semi-arid Mediterranean landscapes. In: Allen E.B (ed.) *The Reconstruction of disturbed Arid Lands. An Ecological Approach.* American Association for the Advancement of Sciences. Symposium 109. Westview Press, Inc. Boulder, Colorado, pp 234–256.
Naveh Z (1990) Landscape ecology as a bridge between bio-ecology and human ecology. Svobodova H (ed.) *Cultural Aspects of Landscape.* Pudoc, Wageningen, The Netherlands, p. 173.
Naveh Z (1993) Red Books for threatened Mediterranean landscapes as an innovative tool for holistic landscape conservation. Introduction to the western Crete Red Book case study. Special Issue. *Landscape and Urban Planning,* **24:** 241–248.
Naveh Z (1994) The role of fire and its management in the conservation of Mediterranean ecosystems and landscapes. In: Moreno J, and Oechel WC (eds) *The Role of Fire in Mediterranean-Type Ecosystems.* Springer Verlag, New York, USA, pp 163–185.
Naveh Z (1995) Interactions of landscapes and cultures. *Landscape and Urban Planning* **32:** 43–54.
Naveh Z (1998a) Culture and landscape conservation – a landscape ecological perspective. In: Gopal B, Pathak PS, and Saxena KG (eds) *Ecology Today: An Anthology of Contemporary Ecological Research.* International Scientific Publications. New Delhi, India, 19–48.

Naveh Z (1998b) Ecological and cultural landscape restoration and the cultural evolution towards a post-industrial symbiosis between human society and nature. *Restoration Ecology* **6:** 135–143.
Naveh Z (1998c) From biodiversity to ecodiversity – holistic conservation of biological and cultural diversity of Mediterranean landscapes. In: Montenegro G, Jaksic F, and Rundel PW (eds) *Landscape Disturbance and Biodiversity in Mediterranean-Type Ecosystems*. Ecological Studies 136. Springer Verlag, Berlin-Heidelberg, Germany, pp 23–54.
Naveh Z (2000) What is holistic landscape ecology? A conceptual introduction. Special issue, *Landscape and Urban Planning* **50:** 7–26.
Naveh Z (2001) Ten major premises for a holistic conception of multifunctional landscapes. Special issue, *Landscape and Urban Planning* **57:** 269–284.
Naveh Z (2002) A transdisciplinary education program for regional sustainable development. *International Journal of Ecology and Environmental Sciences* **28:** 167–191.
Naveh Z (2003) The importance of multifunctional, self-organizing biosphere landscapes for the future of our Total Human Ecosystem – a new paradigm for transdisciplinary landscape ecology. In: Brandt, J. sand H, Vejre (eds). Southhampton, U.K. pp 33–63.
Naveh Z (2005) Epilogue: Toward a transdisciplinary science of ecological and cultural landscape restoration. *Restoration Ecology* **13:** 1–7.
Naveh Z, and Lieberman AS (1994) *Landscape Ecology Theory and Applications.* 2nd ed Springer-Verlag New York, USA.
Neef E (1982) Stages in the development of landscape ecology. In: Tjallingi SP and de Veer AA (eds) *Perspectives in Landscape Ecology*. Pudoc, Wageningen, The Netherlands, pp 19–28.
Noss R (2000) Maintaining the ecological integrity of landscapes and ecoregions. In: *Ecological Integrity. Integrating Environment, Conservation, and Health*. Island Press, Washington D. C., Covelo, California, USA, pp 191–208.
Odum EP (1971) *Fundamentals of Ecology.* 3rd edn. W.B. Saunders Company. Philadelphia.
Odum HT (1971). *Environment, Power and Society.* Wiley, New York.
Odum EC, and Odum HT (1980) Energy systems and environmental education. In: T. Bakshi and Z. Naveh editors. *Environmental Education: Principles, Methods and Applications*. Plenum Press, New York and London, pp 213–231.
O'Neill RV (2001). Is it time to bury the ecosystem concept? (with full military honors, of course!) *Ecology* **82,** 3275–3284.
Oreszczyn S (1999) A systems approach to the research of people's relationships with English hedgerows. *Landscape and Urban Planning* **50:** 95–106.
Orr DW (1992) *Ecological Literacy. Education and the Transition to a Postmodern World*. State University of New York Press, New York, USA.
Palang H, Mander U, and Naveh Z (2000) Holistic Landscape Ecology in Action. (Special Issue) *Landscape and Urban Planning.* **50:** 1–6.
Patil GP, Johnson GD, and Myers WL (1998) Statistical ecology of landscape characterization with satellite data for resource assessment and management: an ecological indicator approach. In: Gopal B, Pathak PS, and Saxen KG (eds) *Ecology Today: An Anthology of Contemporary Ecological Research*. International Scientific Publication, New Delhi, pp. 453–466.
Pedroli B (1989) Die Sprache der Landschaft. *Elemente der Naturwissenschaft,* **51:** 25–49.
Pignatti S (1994) *Ecologia del Paesaggio*. Unione Tipografica Editrice Tornewse (UTET), Turino, Italy.
Pimentel D, Westra L, and Noss RF (2000) *Ecological Integrity. Integrating Environment, Conservation, and Health*. Island Press, Washington D. C., Covelo, California, USA.
Platt McGinn A (2002) Our toxic burden. In: L. Starke (ed.) *State of the World Report 2002*. The Worldwatch Institute,. Norton & Company New York, London, U.K., pp 75–100.
Prigogine I (1997) *The End of Certainty: Time, Chaos, and the New Laws of Nature*. The Free Press, New York.
Prigogine I, and Stengers I (1984) *Order out of Chaos. Man's Dialogue with Nature*. New Science Library Shambhala, Boston and London, U.K.
Rees W (2000) Patch disturbance, ecofootprints, and biological integrity: Revisiting the limits to growth (or why industrial society is inherently unsustainable). In: Pimentel D, Westra L and Noss RF (eds)

Ecological Integrity Integrating Environment, Conservation, and Health. Island Press, Washington D.C., Covelo, California, USA, pp 139–156.

Rosenzweig ML (2003) *Win-Win Ecology: How the Earth's Species Can Survive in the Midst of Human Enterprise*. Oxford University Press New York ,USA.

Roszak TR, Gomes ME, and Kanner AD (eds) (1995) *Ecopsychology. Restoring the Earth Healing the Mind*. Sierra Club Books. San Francisco, USA.

Ruzicka M (1998) Contribution of Slovak Landscape Ecology for IALE. *Ecologia* (Bratislava) **17,** Supplement1: 8–13.

Ruzicka M, and Miklos L (1982) Methodology of ecological landscape evaluation for optimal development of territory. In: Tjallingii SP, and de Veer AA (eds) *Perspectives in Landscape Ecology*. Pudoc, Wageningen, The Netherlands, pp 99–108.

Sieferle RP (1997) *Rueckblick auf die Natur. Eine Geschichte des Menschen und Seiner Umwelt*. Luchterhand Literatur Verlag, Muenchen, Germany.

Snow CP (1963) *The Two Cultures: A Second Look*. Cambridge University Press, Cambridge, U.K.

Tansley AG (1935) The use and abuse of vegetational concepts and terms. *Ecology* **43,** 284–307.

Toupal RS (2003) Cultural landscapes as a methodology for understanding natural resource management impacts in he western United States. *Conservation Ecology* **7** [online] URL: http://www.consecol.org/vol7/iss11/art12

Tress B, Tress G, and Decamps H (eds) (2001) Bridging human and natural sciences in landscape research. (Special issue) *Landscape and Urban Planning* **57:** 137–141.

Tress B, and Tress G (2002) Characteristic of disciplinary and meta-disciplinary approaches. In: Bastian O, and Steinhardt U (eds) *Development and Perspectives in Landscape Ecology. Conceptions, Methods, Application*. Kluwer Academic Publishers, Dodrecht, The Netherlands, pp 27–37.

Tress B, Tress G van der Valk A, and Fry G (eds) (2003) *Interdisciplinary and Transdisciplinary landscape Studies: Potentials and Limitations*. Delta Series 2, Wageningen 2003, The Netherlands.

Tuan Yi-Fu (1974) *Topohilia A Study of Environmental Perceptions, Attitudes, and Values*. Prentice-Hall, Inc., Englewood Cliffs, New Jersey, USA.

Turner MG, Gardner R, and O'Neill RVO (2001) *Landscape Ecology in Theory and Practice, Pattern and Process*. Springer-Verlag, New York, USA.

Young GL (1974) Human Ecology as an interdisciplinary concept. A critical inquiry. *Advances in Ecological Research* **8:** 1–105.

Van der Maarel E (1977) *Toward a Global Ecological Model for Physical Planning in the Netherlands*. Ministry of Housing and Physical Planning, The Hague, The Netherlands.

Walker B (2000) *Analyzing integrated social-ecological systems. Executive Summary*. Wallenberg Workshop. September 12–14 2000. Royal Swedish Academy of Sciences Stockholm [online] htpp://www.resiliance.org/reports/Wallenberg-report.dec00.htlm

Wali A, Darlow G, Fialkowski C, Tudor M, del Campo H, and Stotz D (2003) New methodologies for interdisciplinary research and action in an urban ecosystem in Chicago. *Conservation Ecology* **7** [online] URL: http://www.consecol.org/vol7/iss3/art2

Waltner-Toews D, Kay JJ, Neudoerffer C, and Gitau T (2003) Perspective changes everything: managing ecosystems from the inside out. *Frontiers in Ecology and the Environment* **1:** 23–30.

Ward DJ (2003). Redefining ecology. *Conservation Ecology*, **7** [online]URL:http://www.consecol.org/vol7/iss1/resp8art2

Weinberg GM (1975) *Introduction to General Systems Thinking*. John Wiley & Sons. New York, USA.

Weiss PA (1969) The living system: determinism stratified. In: Koestler A, and Smithies J (eds) *Beyond Reductionism: New Perspectives in the Life Sciences*. Hutchinson of London, U.K., pp 2–55.

Weizsaecker E von (1974) *Offene Systeme: Beitraege zur Zeitstruktur von Informationen, Entropie and Evolution*. St.Klett, Stuttgart, Germany.

Winder, N. 2003. Successes and problems when conduting interdisciplinary and transdisciplinary (=intergrative) research. In: Tress, B., Tress, G., van der Valk, A., and G. Fry, editors. 2003. *Interdisciplinary and Transdisciplinary landscape studies: Potentials and Limitations*. Delta Series 2, Wageningen 2003, The Netherlands. pp. 74–90.

World Resources 2000–2001 (2000) *People and Ecosystems. The Fraying Web of Life.* Elsevier Science. Amsterdam, The Netherlands.

Zonneveld IS (1982) Land(scape) ecology, a science or a state of mind. In: S.P. Tjallingii and A.A. de Veer, editors. *Perspectives in Landscape Ecology.* Pudoc, Wageningen, The Netherlands, pp 9–16.

Zonneveld IS (1995) Land Ecology. An Introduction to Landscape Ecology as a Base for Land Evaluation, Land management and Conservation. SPB Academic Publishing, Oxford U.K.

ACKNOWLEDGEMENTS

I acknowledge gratefully the very useful comments by Professor Emeritus Arthur Lieberman and the kind help of Dr. Yohay Carmel for the preparation of this essay.

EPILOGUE: A LIFETIME OF LEADERSHIP IN THE FIELDS OF LANDSCAPE ECOLOGY, RESTORATION ECOLOGY, AND ECOLOGICAL COMPLEXITY – BY EDITH ALLEN, UNIVERSITY OF CALIFORNIA, RIVERSIDE, CALIFORNIA

The foregoing anthology of Zev Naveh's life work shows not only the progression of his creative thought during his lifetime, but also developments in the fields of landscape and restoration ecology that he helped to shape. In addition, because he is a systems ecologist at heart, he participated in founding the new field of ecological complexity well before it became recognized as a field.

Zev became a keen observer of nature during his early years herding domestic animals and reclaiming farmland in the western Galilee. Although he initially concurred with the paradigm of land managers of this time that removing shrubs with no agronomic value would improve productivity, his experience soon showed that the shrubs were "keystone" species responsible for maintaining soil fertility, moisture, and stability. He consequently became an early proponent of shrub restoration in semi-arid rangelands, helping land managers to make the transition to shrub utilization rather than destruction (McKell 1989). Thus he became one of the earliest restoration ecologists, but recognized that humans are part of nature and not apart from nature, as was the emphasis in the early North American restoration movement (Jordan 2003). Restoration goals to recreate a "pre-Columbian" landscape meant an absence or a minimum of human influence. Recent findings in the Americas show this to be a fallacy, and historic land use in the Americas may be analogous to the Mediterranean landscapes Zev frequently illustrates even though the latter are more intensively managed with a longer history of human habitation.

The concepts of restoration and landscape ecology developed in tandem in Zev's writings. People were part of the landscape, they shaped the landscape, including the urban, semi-natural, and even the apparently "wild" components of the landscape. Zev described how the past uses of Mediterranean landscapes created a "dynamic flow equilibrium" that maintained the vegetation composition and ecosystem dynamics that we have come to associate with these landscapes. Components of the landscape may have been impacted by use, but never enough to exceed natural regenerative abilities. The landscape was historically maintained as a shifting mosaic under varying intensities of use. Modern impacts on landscapes have broken this flow equilibrium. Most restoration efforts emphasize reintroducing or managing for certain species or communities, but Zev's writings emphasize that the cultural and natural processes that created the landscape must also be reinstated. He was in fact the first to introduce the term "cultural landscape restoration" (Naveh 1998).

Cultural landscapes occur worldwide; agricultural, silvicultural forests or grazing lands are obvious examples of cultural landscapes, but many lands regarded as "wildlands" are also cultural in origin. These include the historic grazed Mediterranean shrublands that Zev observed, but also many areas of the New World. For instance, we described the human-managed Maya forests during our presentation in the 2001 European IALE Conference in Stockholm (Naveh and Allen 2001). The Maya cultivated land and planted "forest gardens" for more than two millennia. The forests that exist today are legacies of these historic activities, having a high proportion of trees that are considered useful by present-day Maya (Gomez-Pompa et al. 2003). Similarly, the forests of the Amazon Basin have high densities of trees useful to humans, although the human populations have declined greatly since the early Spanish explorations of the mid-1500's. In a remarkable re-analysis, Mann (2005) argues that there were indeed high population densities in the Amazon prior to European contact. Recent archaeological finds, overlooked by previous explorations, support these early high population densities. While these populations practiced some agriculture based on annual crops such as manioc, the de-forested soils are too poor to support large-scale annual crops for the large estimated population. Instead, some 70 species of useful trees were maintained and planted that produced fruit at various seasons of the year. After European diseases and wars decimated indigenous human populations in the mid to late 1500's, their forests remained. New analyses of Amazonian forest species composition need to be done with ethnobotanical uses in mind.

The reanalysis of indigenous vegetation management includes numerous sites in North America, and Mann (2005) cites a additional studies that infer high pre-contact Indian population densities, extensive agriculture, wide-spread burning, and also planting or maintenance of useful tree species. The historian Pryne (1982) stated that "the virgin forest was not encountered in the sixteenth and seventeenth centuries, it was invented in the late eighteenth and nineteenth centuries." Thus far restoration ecologists have considered the virgin forest to be the reference area for restoration, but as the human impact on landscapes is accepted, cultural landscapes will hopefully become part of restoration goals.

These are just a few examples of forests that have been managed by humans, and indicate that even our present-day "wildlands" are in part cultural landscapes. Zev has warned against restoration to a non-existent pristine state (Naveh 1988), but even more than this recognizes the limits to restoration when accelerating human degradation causes changes to landscapes that are far from equilibrium. Even restoration to a semi-natural state may no longer be possible, but these landscapes can be rehabilitated to bring them back to a useful state and improve their biological potential (Naveh 1988).

In developing his ideas on landscape dynamics, Zev also became one of the first theorists in the new discipline of ecological complexity even before this area was founded. The discipline is defined by the editor in the first issue of the new journal "Ecological Complexity" (Li 2004). Zev's concepts for landscape ecology are based on general systems theory, which is the basis for complexity theory as well. Some of the major tenets of complexity theory include non-linear dynamics, as illustrated by the dynamic flow equilibrium of "epigenetic landscapes," or self-organized systems, as illustrated by the biocybernetic systems described in multiple examples in Naveh and Liebermann (1994).

By his remarkable creativity, influential presentations at international meetings, and especially his writings, Zev has become the defining force for three major areas of ecology, and has shown how these areas are interdependent. It has been my great pleasure and honor to be Zev's colleague for the last two decades. We have worked together on an edited book, editing the journal Restoration Ecology, and preparing symposium presentations. Most importantly, though, is the fact that while furthering the theoretical basis for these disciplines, his writings have never lost sight of the fact that landscape and restoration ecology are at their core applied disciplines. In the introduction to this book he writes "The sad fact is that the majority of human society, especially its leading political and economic decision-makers, have not yet accepted the fact that we cannot separate our fate from that of all life on earth . . ." But more than other theoreticians Zev gives specific management recommendations to maintain the connection of culture with nature. And in fact, he explains that nature will not persist, and thus humans, without this essential connection. This is how we must proceed, to keep people as part of nature so they will care for their surroundings and take action.

REFERENCES

Gomez-Pompa A, Allen MF, Fedick SL, and Jimenez-Osornio J (eds) (2003) *The Lowland Maya Area: Three Millennia at the Human-Wildland Interface.* Haworth Press, Binghamton, NY, USA.

Jordan WR (2003) *The Sunflower Forest: Ecological Restoration and the New Communion with Nature.* University of California Press, Berkeley, USA.

Li B-L (2004) Editorial. *Ecological Complexity* 1:1–2.

Mann CC (2005) 1491: *New Revelations of the Americas before Columbus.* Alfred A. Knopf, New York, USA.

McKell CM (ed.)(1989) *The Biology and Utilization of Shrubs.* Academic Press, San Diego, USA.

Naveh Z (1988) Multifunctional reconstruction of semi-arid Mediterranean landscapes. In: Allen EB (ed.) *The Reconstruction of Disturbed Arid Lands. An Ecological Approach.* AAAS Symposium 109. Westview Press, Boulder, CO, USA, pp 234–256.

Naveh Z (1998) Ecological and cultural landscape restoration and the cultural evolution toward post-industrial symbiosis between human society and nature. *Restoration Ecology* 6:135–143.

Naveh Z and Liebermann AS (1994) Landscape Ecology Theory and Applications. 2nd edn. Springer-Verlag, New York, USA.

Naveh Z and Allen EB (2001) Restoration Ecology and Landscape Ecology – A win win relationship. Plenary lecture at the IALE European Conference 2001:Developments of European Landscapes, Stockholm University and University of Tartu. Publicationes Instituti Geographici Universitatis Tartunsis 92, pp 38–41.

Pryne S (1982) *Fire in America: A Cultural History of Wildland and Rural Fire.* Princeton University Press, Princeton, NJ, USA.

AUTHOR INDEX

Agger, P. 177, 292
Alberti, M. 370–372
Allen, E.B. 13, 283, 290, 403, 404
Allen, M.F. 24
Allen, T.F.H. 152, 253, 258, 305, 372, 380
Allen, A.B. 25
Allesch, C.G. 264, 299
Aloni, R. 64
Anderson, E. 42, 193, 289
Antrop, M. 311, 325
Arianoutsou, M. 210
Aronson, J. 192
Arroyo, M.K. 120
Artnano, B. 184
Aschmann, H. 86, 88
Auclair, A.N. 83, 84
Austad, I. 290
Axelrod, D.I. 89, 188, 193, 206, 209

Bahre, C. 86
Bai-Lin, L. 398
Bakshi, T. 30
Bankroft, H. 198
Barbolani, E. 156
Bar-El, C. 157
Bar-Josef, O. 18, 100, 195
Barrett, G.W. 374
Bar-Tuv, A. 138
Bastian, O. 251, 376
Baudry, J. 165, 269
Baumhoff, M.A. 200
Bazzaz, F.A. 84
Beadle, N.C.W. 57
Bean, L.J. 200
Beard, J.S. 65, 88
Beiles, A. 147
Ben Horin, R.A. 93, 134, 185, 352
Bentley, J.R. 38
Berkes, C.S. 16
Berkes, F. 397
Berhardt, E. 397
Bertalanffy, L. Von 238
Bignal, E.M. 340
Birks, H. 267
Biswell, H.H. 6, 10, 11, 52–54, 97, 208, 289
Bloch, A. 276
Blondel, J. 192
Boer, M.M. 137, 139
Bohm, D. 251, 257, 260, 262, 284, 309, 310, 325, 366, 367, 382, 387, 388
Bolin, B. 136
Bollinger, J. 393
Botkin, D.B. 169
Bottema, S. 198
Bradbury, D.E. 92
Bradley, G. 395
Bradshaw, A.D. 13, 378
Branagan, D. 20, 224, 226
Brandt, J. 26, 129, 130, 177, 292, 381, 383
Braun-Blanquet, J. 52
Breuste, J.H. 372, 375
Bright, C. 298
Brown, L.H. 227, 228
Brown, L.R. 304, 315, 322, 349, 357, 359, 360, 362–364
Buber, M. 265
Bugod, P. 293
Buhyoff, G.J. 397
Bunce, R.G.H. 165, 269
Burbidge, N.T. 89
Burcham, L.T. 55
Burel, F. 395
Burrough, P.A. 258
Bussotti, F. 156, 162

Cain, S.A. 36, 73
Capra, F. 302, 329, 356, 385
Carmel, Y. 8, 29, 370, 372
Carr, J, 397
Cenni, E. 183
Chadwick, M.J. 13, 378
Chaim, S. 62
Chapin, F.S. 150
Christensen, N.L. 86
Clark, D. 195
Clauser, F. 183
Clements, F.E. 54, 56
Cody, M.L. 81, 86, 90
Cohen, A. 136
Connell, J.H. 81
Conrad, C.E. 112
Cooper, W.S. 36, 42, 48, 52, 53, 56

Costanza, R. 266, 374, 392
Cowling, R.M. 88, 189
Cozzi, A. 183
Craig, Z. 395
Curtis, J.T. 289

Dafni, F. 190
Dagfilm, M. 277
Daly, H.E. 374, 392
Dan, J. 15, 62, 83, 101, 120–122, 143, 169, 191, 192, 242
Dansereau, P. 263, 369
Darling, F.F. 222
Darlow, G. 400
Davis, F.W. 176
De Groot, R.S. 137, 139, 154, 266
De Noy-van Tol, J. 384
De Palma, A. 288, 338
DeAngelis, D.L. 279
Decamps, H. 400
Del Campo, H. 400
Deleuil, G. 15
Derman, A. 113
Devall, B. 264
Di Castri, F. 7, 149, 168, 188, 191, 300, 320, 324, 368
Diamond, J. 201
Dijkstra, H. 185
Dinoor, A. 45
Doos, B.R. 156
Downes, S. 81

Eetvelde V. van 311, 325, 395
Egler, F.E. 7, 53, 238, 253, 326, 369, 371
Ehrlich, P.R. 364
Eig, A. 36, 37, 43, 47, 51, 53, 64
Eigen, M. 327
Eldrede, N. 389
Ellenberg, H. 12, 238, 370
Elsasser, A.B. 199
Estes, J.E. 183
Evans, F.C. 92
Evenari, M. 45
Evernden, N. 264, 265

Fagg, A.E. 211
Fahima, T. 212
Fallon, L.E. 224
Farina, A. 126, 141, 165, 173, 251, 269, 333, 340
Farmer, G. 138
Fedick, S.L. 405
Ferretti, M. 162
Fialkowski, C. 400
Fiedler, P.L. 352
Fischer-Kowalski, M. 331, 343, 374
Fischlowitz-Roberts, B. 363
Flavin, C. 316, 62
Folke, C. 316, 373, 397
Forman, R.T.T. 257, 284, 377
Fox, J.E.D. 91
Fraenzle, S. 278
Frankl, V.E. 24
French, H. 316, 350
Fresco, L.F.M. 82
Fry, G. 400
Fuentes, E.R. 92

Gacic, M. 157
Gardner, R. 400
Garner, G. 360
Gat, Z. 139
Geiles, A. 133, 185
Gellini, R. 156, 183
Gillman, C. 221
Gitau, T. 400
Glanz, W. 92
Glavac, V. 133, 211
Glenn-Lewin, D.C. 81
Godron, M. 257, 284
Goff, F.G. 83, 84
Golenberg, E.M. 133, 185
Golley, F.B. 263
Gomes, M.E. 350
Gomez-Pompa, A. 404
Gonzales Brnaldes, F. 126
Good, R. 84
Gore, A. 119, 182, 275
Gould, S.J. 389
Graber, M. 136
Grassle, J.F. 90
Green, B. 176, 254, 292, 343
Green, D.G. 393
Grime, J.P. 82
Grinker, R.R. 252
Grisebach, A. 36
Grossmann, W.D. 29, 266, 267, 299, 315, 347–349, 393
Grossoni, P. 183
Grove, A.T. 81, 132, 155, 169, 191, 192, 271, 272
Grubb, P.J. 110, 197
Gunderson, L.H. 372, 373
Gustafson, E.J. 377
Guyot, L. 45
Haas, G. 167

Haas, L. 363
Haber, W. 12, 129, 177, 253, 284
Haberl, M. 331, 343, 374
Hajek, E. 92
Haken, H. 332
Hanawalt, R.B. 92
Hansen, A.J. 149, 151
Hardin, G. 178
Harrs, R. 211
Harlan, H.R. 199, 208
Harper, J.L. 82
Harper, K.T. 88
Hart, D. 397
Harte, J. 137
Hassett, B. 379
Heady, H.F. 231
Heizer, R.F. 199, 200
Henzell, E.F. 230
Higgs, E. 196, 369, 380
Hill, M.O. 67
Hobbs, R. 367, 368, 377
Hochleitner, R.D. 346
Hoekstra, T.W. 53, 258, 305
Holling, C.E. 28
Holling, C.S. 372, 385
Holling, J. 298, 301
Holmes, E. 132
Hoover, S.R. 129, 177
Hopper, S.D. 89
Horowitz, A. 99, 188, 196
Horowitz, L.K. 212
Horton, J.S. 48
Horvat, I. 126, 191
Hoyle, F. 389
Hull, R.B. 372, 374
Hulme, M. 136
Hunt, J.H. 92
Huntingford, G.W.B. 218
Huston, M. 110
Hutchinson, H.G. 228
Huxley, J. 227

Ihse, M. 397
Ikeya, M. 100
Ispikoudies, J. 132, 156, 183, 278
Iversen, J. 203

Jaeger, J. 136
Jaffe, S. 138
Jantsch, E. 28, 20, 240, 252, 259, 284, 286, 287, 299, 301–303, 305, 306, 326, 327, 329, 330, 336, 382
Jeftic, L. 140, 165
Jelinek, A. 100, 196, 202
Jenny, H. 6, 109
Jensen, D. 156
Jepson, W.C. 51
Jimenez-Osornio, J. 405
Johnson, G.D. 399
Jones, B.J. 103
Jones, P.D. 156
Jordan, W.R. 403
Juhren, G. 51
Juhren, M.C. 51

Kaland, P.E. 277
Kametz, H. 220
Kanner, A.D. 350, 352, 400
Kaplan, D. 9, 29, 125, 173, 268
Kaplan, S. 339, 392
Karr, J.R. 66, 81
Katsoulis, B.D. 138
Katznelson, J. 39
Kay, J.J. 400
Kazaklis, A.M. 132, 156, 183, 278
Keck, D.D. 36, 42
Kellert, S.R. 265, 286
Kennan, T.C.D. 231
Kim, K.C. 167, 182
King, A. 119, 182, 275
Kinsky, J. 79
Kislev, M.E. 202
Klijn, J. 376
Kluckholm, C. 259
Knowles, P.F. 55
Koeppen, W. 36
Koestler, A. 108, 304
Korol, A.B. 212
Kosko, B. 258, 287
Kothari, A. 270
Kraebel, H.R. 48
Kroeber, A.L. 259
Kroenert, R. 397
Kruger, F.J. 66
Krummel, J.R. 22
Kuhn, T.S. 300, 324, 369, 377, 384–386
Kupernick, G. 53
Kutiel, P. 97, 103, 120, 138–140, 189, 191, 192, 203, 209, 286
Kutiel, T. 209

LaMarche, V. 92
Lamont, B.B. 88
Lamprey, H.F. 219, 221, 222, 226

Landers, R.G. 45
Langer, H. 375
Larsen, D. 132
Larsen, J. 359
Lascaratos, A. 138
Laszlo, E. 24, 26, 27, 29, 204, 252, 260, 261, 274, 282, 284, 286, 299–304, 311, 314, 315, 320, 3
Lawton, H.W. 200
Law-Yone, H. 163
Le Houerou, H.N. 97, 104, 140, 165
LeCoeur, D. 395
Ledger, H.P. 227, 228
Leser, H. 251–253, 284, 305
Levin, S.A. 300
Lev-Yadun, S. 198
Levyns, M.R. 89
Lewis, H.T. 100, 200, 202, 289
Li, B.-L. 258, 299, 303, 332, 373, 405
Lieberman, A.S. 5, 12, 23, 29, 101, 108, 110, 118, 119, 127, 129, 132, 140, 154, 160, 167, 169, 176, 178, 182, 191, 251, 253, 261, 264, 267, 284, 287, 289, 299, 302, 304–307, 325, 326, 332, 340, 369–372, 374, 375, 378, 388, 405
Lindahl, C. 397
Lischke, H. 395
Lister, A. 212
Litav, M. 53
Loeb, H. 104
Lorig, S.M. 289
Lorimer, C.G. 289
Loucks, O.L. 84
Love, R.M. 57, 103
Lovelock, J.E. 345
Lubchenco, J. 82, 318
Lucas, P.H.C. 131, 152, 175, 176, 250, 254, 292, 343
Ludwig, D. 398
Luz, F. 398
Lyle, J.T. 25, 256, 276, 277, 291, 292, 313, 341, 342, 379

MacArthur, R.H. 66
MacMillan, R.A. 277
Makhzoumi, J.D. 372
Malanson, G.P. 153
Mandel, M. 138
Mandelbrot, B.B. 258
Mander, U. 138
Mankin, J.B. 30
Mann, A. 61, 273
Mann, C.C. 404
Mansvelt, J.D. 254, 290, 292, 341
Mantilla, E. 184
Marcuzzi, G. 161
Margalef, R. 84
Margulis, L. 345
Marzluff, J.M. 395
Maturana, H. 327
May, R.M. 70
McCracken, D.I. 340
McCulloch, J.S.G. 220
McDonnell, M.J. 371
McKell, C.M. 55, 403
McNaughton, S.J. 79, 81
McNeeley, J.A. 286
McNeely, A. 276
McNeill, J.R. 191
McPherson, J.K. 86
Meadows, D.H. 119, 182, 263
Meadows, D.L. 119, 182, 263
Meeus, J. 162
Meiss, K.-M. 278
Melillo, M. 318
Merleau-Ponty, M. 263, 264
Mesarovic M. 236
Miklos, L. 400
Milan, M. 162
Miller, P.C. 88
Milliman, J.D. 158
Milne, B.T. 258
Moldenke, A.R. 86, 88
Moody, J.A. 123, 156, 183
Mooney, H.A. 7, 65, 85, 86, 87, 90, 112
Moral, R. del 81
Moran, D. 284
Morris, M.G. 66
Moss, M.R. 381
Mueller, F. 373
Mulder, J.A. 254, 290, 292, 341
Muller, C.H. 46, 86
Muller, J.G. 238
Multhaup, T. 278
Munz, P.A. 42
Murdoch, W.W. 66
Myers, N. 118, 137
Myers, W.L. 399

Naess, A. 264
Narr, K.J. 203
Nassauer, J.I. 262, 300, 338, 377, 378
Naveh, Z. 5, 8, 10, 12, 14, 15, 17, 18, 21–25, 29, 35, 38, 43, 45, 48, 52–55, 57, 61, 62, 66, 71, 78, 79, 83, 84, 86, 97, 101, 103–106, 108, 110, 113, 117–122, 125, 127, 129,

132, 135, 140, 141, 143, 145, 148, 149, 153, 154, 159, 160, 162, 165–167, 169, 171–176, 181, 182, 187, 189–193, 195, 196, 200, 202, 203, 205–209, 217, 226, 231, 232, 235, 236, 238, 242, 243, 245, 249, 251, 253, 254, 256, 261, 262, 264, 267–270, 281, 283, 284, 286–290, 297, 299, 300, 302, 304–308, 312, 313, 315, 139, 324–326, 332–334, 338, 340, 349, 355, 369–372, 374, 375, 377, 378, 380, 381, 386, 388, 393, 404, 405
Neef, E. 399
Negotia, C.V. 109, 154, 258
Neudorffer, C. 400
Nevo, E. 17, 125, 146–148, 153, 168, 173, 187, 189, 389
Niering, W.A. 66, 81, 82
Norman, M.J.T. 230
Norton, B.G. 118, 120, 167
Noss, R. 375
Noy-Meir, E. 9, 125, 173, 268, 252, 313, 369, 371

O'Neill, J.R. 191
O'Neill, R.V. 30
Odum, E.C. 371
Odum, E.P. 7, 84, 176, 223, 292, 374
Odum, H.T. 242, 371
Oeschger, H. 138
Oguz, T. 157
Olschowy, G. 240
Olsvig-Whittaker, L. 133, 185
Oneko, A. 227
Oreszczyn, S. 399
Orias, E. 81
Orr, D.W. 300
Orshan, G. 53
Owen, M.A. 230

Palang, H. 29
Palmer, M. 397
Pankow, W. 307
Papanastasis, V. 16, 271
Parker, A.J. 129, 177
Parker, A.V.T. 185, 352
Parsons, D.J. 86, 88
Patil, G.P. 377
Payne, W.J.A. 234
Peat, D.F. 24, 257, 262, 284, 309, 310, 325, 366, 387
Pedroli, B. 399
Pedroli, G.M.B. 167, 264, 272
Peet, R.K. 67, 81
Pereira, H.C. 220, 223
Perles, C. 100, 196
Perrier, E.P. 55
Pestel E. 236
Peterson, C.H. 92
Pfeiffer, J.E. 100
Pickett, S.T.A. 84, 110, 169, 335, 371
Pierce, D. 284
Pietrini, R. 183
Pignatti, S. 18, 191, 199, 251
Pimentel, D. 306, 375
Pinto-Correira, T. 170, 171, 269, 333
Platt McGinn, A. 358
Pons, S.A. 101, 191, 203
Poulianos, A.N. 100
Prigogine, I. 11, 28, 111, 141, 169, 301, 329, 332, 334
Pryne, S. 404
Pungetti, G. 372, 377

Quezel, M. 101, 191, 203

Rackham, O. 191, 192
Randers, J. 133, 184, 279
Rankevich, D. 93, 134, 185, 352
Rappaport, R.A. 263
Raunkiaer, C. 36, 73
Raven, P.H. 89
Rees, W. 331, 343, 375
Renfrew, C. 201
Renzoni, A. 236
Richter, G. 184
Rickleffs, R.R. 335, 340
Rindos, D. 197, 201
Robbins, W.W. 54
Robertson, D.P. 96, 397
Roesch, A. 278
Romme, W.H. 54
Ronen, A. 18, 197, 202, 203
Rosenzweig, M.L. 358
Rossi, R. 185, 279
Roszak, T. 182, 265, 339, 374, 391
Roth, R.R. 66, 81
Rothenberg, D. 264
Rotman, A. 157
Rubin, S. 139
Ruebel, E. 36
Ruiz de la Torre, J.R. 126, 173, 333
Rundel, P.W. 16, 103, 159, 188
Ruzicka, M. 400
Ryan, C. 395

Sadedin, S. 393
Salvador, R. 184
Sampson, A.W. 48
Sanders, H.L. 90
Sauer, C. 6, 57, 99, 196, 197, 377
Sbrilli, G. 183
Scepan, J. 183
Scheer, H. 346
Schimper, A.F.W. 36
Schmidt, K. 275
Schneider, B. 119, 275
Schneider, S.H. 136
Schreiber, K.-F. 12, 171, 251
Schulenberger, E. 395
Schulze, E.D. 156
Schuster, P. 327
Sessions, G. 264
Sestini, G. 140
Shavit, I. 98
Shaviv, L. 148
Sieferle, R.P. 134, 345, 391
Simon, H.A. 303
Smith, M.J. 312
Snow, C.P. 367, 392, 394
Solbrig, O.T. 89
Stebbins, G.L. 6, 8, 47, 55, 195, 210, 287
Steinberger, E.H. 139
Steiner, G.H. 264
Steinhardt, U. 258, 376
Stekelis, M. 99, 167, 195
Stengers, I. 169, 301, 329
Stoms, D.M. 183
Stortelder, A.H.F. 126, 162, 164, 166
Stotz, D. 400
Straus, E. 264
Svobodova, A 280.
Swarzboim, I. 207
Sweeney, J.R. 48
Syrbe, R.-U. 258

Talbot, M. 38, 226
Tansley, A.G. 238, 370
Tatskin, A. 196
Taylor, H.C. 66, 84, 88, 89
Tchernov, E. 193, 198, 199
Thayer, R.J. 340
Thenaile, C. 395
Thrower, N.J.W. 92
Tolba, M. 119, 182, 275
Tomaselli, R. 96
Torn, M. 156
Toupal, R.S. 372
Trabaud, L. 97, 898, 103
Tress, B. 26, 341, 376, 381, 383
Tress, G. 26, 376, 381, 383
Troll, C. 253
Trumble, H.C. 38
Tuan, Y.F. 262, 263
Tudor, M. 400
Turner, R.E. 114, 185, 280, 295, 352
Turner, M.G. 377

Valk, M. 397
Van der Lubbe, M.J. 351, 398
Van der Maarel, E. 12, 84, 287
Van der Valk, A. 400
Van Deursen, W. 277
Van Zeist, W. 198
Varela, F. 237
Varney, B.M. 39
Vejre, H. 26
Vernet, J.L. 195, 196
Vester, F. 239, 246
Vita-Finci, D. 211
Vitousek, P.M. 306
Vogl, R.J. 126
Vos, W. 126, 162, 164, 166, 376
Vroom, M.J. 184

Waddington, C.H. 28, 110, 127, 169, 268, 378
Waide, J.B. 279
Wali, A. 372
Walker, B. 372
Walter, H. 97, 112
Waltner-Toews, D. 392
Warburg, M. 9, 81, 124, 172, 333
Ward, D.J. 374
Weaver, R.D. 167, 182
Weinberg, G.M. 303, 305
Weinstein-Evron, M. 190, 194, 196, 198
Weiss, P. 238, 303
Weizsaecker, E. von 400
Went, F. 51
Werger, M.J.A. 88
West, O. 231
Western, D. 22, 333
Westman, W.E. 88, 153
Westra, L. 399
Whittaker, R.H 7, 8, 15, 61, 62, 65, 66, 67, 70, 71, 81, 82, 88, 89, 99, 104, 120, 125, 140, 143, 145, 148, 167, 171–174, 177, 268, 333
Whyte, R.O. 41, 231, 237, 240
Wigley, T.M.L. 138
Wijermans, M.P. 184

Williams, W.A. 57
Williamson, P. 138
Willson, M.F. 66, 81
Wilson, J.G. 220
Winder, N. 383
Woodcock, J. 104
Woodwell, G.M. 29, 137

Yardeni, D. 45
Yiftach-El, O. 163
Yogev, I. 10, 149
Young, G.L. 237, 371

Zadeh, L.A. 258
Zeevalking, H.J. 82
Zinderen Bakker, Sr., E.M. van 89
Zinke, P. 45
Zohary, D. 36, 41, 102
Zohary, M. 48, 61, 63, 64, 83, 188, 191
Zonneveld, I.S. 12, 251, 253, 284, 305, 376

SUBJECT INDEX

Aborigines 101, 197, 200, 289
Acid rain 137
Adaptive resources management 301, 372, 385
Adaptive responses 98, 146, 208, 209
Adaptive systems 287–289, 291, 293, 336
Aesthetic landscape functions 108, 325, 368
Agricultural engineering 2, 10, 14
Agro-industrial ecotopes 254, 284, 291, 313, 331, 341
Agro-industrial landscapes 26, 175, 182, 193, 204, 254, 261, 282, 291, 293, 314, 320, 321, 336, 345
Agroindustrial pollution 150
Agro-silvo-pastoral landscapes 120, 169, 170, 175, 205, 261, 332
Agro-silvo-pastoral system 269, 270
Agrotechnical revolution 202
Allozyme diversity 146, 168
Alpha ecotope ecodiversity 129, 177
Amboseli Park, Kenya 22, 333
Annual grasslands 35, 41, 44, 47, 53, 54, 240
Antarctic Peninsula 321
Anthopogenic erosion 192
Anthracopohytes 104
Anthropogenic ecosystems 370
Anthropogenic polyatomic gases 135
Anti-masiatis 224
Approximate reasoning 258
Archimedean orders 309, 311
Arctic Ocean ice cap 321
Arizona 56, 82, 322, 360
Artificial intelligence 109, 129, 154, 176, 258, 267
Atmopsheric pollutants 162
Atmosphere-biosphere interactions 140, 166
Australia 8, 9, 13, 22, 38, 57, 65, 85, 87–89, 101, 189, 230, 231
Australian aboriginals 197, 289
Australian heath 65, 86, 88, 90
Autopoiesis 254, 301, 320, 327, 346
Autopoietic systems 301, 327, 346
Autosuccession 208

Banff National Park 288, 38
Beta ecotope ecodiversity 129, 177
Bifurcations 18, 24, 195, 198, 201, 203, 204, 260, 261, 286, 301, 302, 313, 315, 320, 329, 331, 346, 365, 373, 389
Biocoenoses 238
Bio-ecological functions 108
Bioecology 176, 299, 308, 381
Bio-ecosystems 238, 239
Bio-landscapes 175, 176, 180
Biophilia 265, 286
Biosphere 12, 23, 25, 27, 107, 118, 140, 166, 177, 205, 237–239, 250, 253–255, 259, 260, 282, 285, 287, 305, 312–315, 320, 327, 328, 336, 345, 370, 374, 390, 391
  ecotopes 23, 253, 272, 284
  landscapes 16, 24–26, 119, 167, 175, 180, 182, 193, 198, 203, 204, 253, 254, 261, 265–268, 272–274, 277, 282, 287–293, 302, 313, 314, 319, 320, 323, 327, 331, 333, 335–339, 341–343, 345, 346, 348–350, 380, 392, 394
  Reserves 131, 151, 175, 272
Biotic diversity 62, 242
Biotic history 35, 41, 46, 55
Biotic landscape diversity 129
Biotopes 15, 129, 130, 238, 292, 340
Bohm hologram paradigm 310
Broad spectrum revolution 202
Byzantine Empire 122

California 6–11, 14, 18, 25, 28, 35–57, 65, 79, 83–90, 98, 100, 103, 113, 125, 153, 162, 167, 173, 176, 193, 199–201, 206, 208, 209, 268, 277, 289, 291, 314, 341, 364, 379, 389, 403
  bunchgrass prairie 54
Canada 322, 323, 359, 360, 377
Cape Province 66, 85, 87–89
Carbon dioxide CO_2 135–137, 153, 323, 345, 358, 359, 363
Carmel forest 25
Carmel valley 8
Cartesian mechanistic orders 24, 257, 259
Cartesian orders 309
Chagga 3, 21

Chaparral 8, 9, 14, 35, 42, 43, 46, 48, 51–53, 57, 65, 85–87, 90, 103, 208, 271
shrubs 43, 56
Chile 7–9, 16, 65, 84–90, 167
Chilean matorral 9, 90
China 322, 358, 360, 361
Classical ecology 169, 304, 356, 368, 373
Climate 7–9, 15, 16, 35–39, 47, 54, 56, 57, 62, 72, 83, 84, 86, 88, 89, 97, 135–143, 151, 152, 188, 198, 199, 206, 209, 219, 223, 225, 304, 314, 321, 322, 358–360, 362, 363, 365, 379, 394
stresss refuges 142, 152
warming 136
Climatic changes 115, 16, 135–137, 140, 151, 199, 304, 314, 322, 358, 359, 360, 362, 363, 366, 379
Climatic trends 73, 146, 166, 198
Climatic variability 15, 143–145, 147
Close-to-natural biosphere landscape ecotopes 150, 252, 253, 268, 288, 289, 306, 332, 336, 340, 370
Club of Rome 236, 252, 275, 346
Communication tools 14, 24, 131, 178, 265, 267
Comparative rainfall distribution 39
Comparative temperature situations 38
Conservation 3, 5, 6, 11, 12, 15–17, 20, 22–24, 27, 28, 62, 84, 96, 97, 104, 112, 113, 117–119, 125, 129–132, 135, 137, 138, 140, 150–152, 154, 155, 159–161, 166, 169, 173–180, 182, 183, 224, 227, 228, 230, 233, 236, 237, 239, 241, 246, 249–251, 259, 265–267, 270–274, 276, 277, 282, 286, 288, 289, 292, 293, 298, 324, 333–335, 338–340, 345, 349, 355, 358, 372, 375, 376, 378, 380, 386, 391, 392, 394, 395
ecology 372, 373
policy 125, 138
Creative landscape science 358
Creedal beliefs 383
Crete 16, 132, 155, 165, 268, 269, 280, 267, 271, 272
Crosscatalysis 327
Cross-catalytic networks (CNN) 210, 302, 314, 320, 346–348
Cult of information 339
Cultural beliefs 383, 384
Cultural biosphere landscapes 16, 26, 175, 180, 182, 193, 204, 253, 254, 256, 261, 266, 267, 268, 273, 282, 287, 289, 290, 293, 320, 336, 338, 341, 343
Cultural diversity 16, 17, 22, 119, 127, 132, 150, 159, 160, 168, 250, 256, 259, 271, 282, 283, 286, 290, 293, 308, 340
Cultural evolution 17, 24, 99, 108, 192, 195, 204, 240, 260–262, 275, 281, 285–287, 293, 301, 302, 305, 306, 313, 314, 320, 323, 326, 329, 331, 336, 345, 364, 365, 385, 390
restoration 404
Cultural landscape 5, 8, 15, 17, 24, 25, 118, 119, 127, 128, 130, 149, 168, 170, 175–177, 182, 187, 192, 199, 203, 238, 250, 251, 253, 257–263, 265, 267, 268, 270, 272, 277, 281–283, 286, 287, 289, 290, 293, 305, 308, 309, 313, 326, 331, 333, 338, 342, 368, 375, 377, 379, 391
restoration 285
Cultural solar-powered biosphere 282
Cultural technosphere landscapes 254, 293
Cybernetic sensitivity model 245, 246

Dehesas 173, 269
Delta land-system ecodiversity 129, 177
Dendrimental environmental impacts 175, 254, 292, 313, 342
Desertification 16, 118, 143, 146, 149, 165, 181, 270, 322, 358, 360, 365, 379
Desiccations 16, 19, 122, 151, 198
Direct attention fatigue 339, 392
Discordant harmonies 169
Disorder 239, 301, 331, 332, 334
Dissipative structures 11, 111–113, 141, 169, 300–302, 330–332, 334, 335
Diversity pattern 73
DNA polymorphism 146, 168
Domestication revolution 202
Drought 4, 9, 10, 14, 16, 18, 20, 22, 36, 51, 55, 73, 75, 83, 90, 98, 110, 137, 138, 140, 147, 148, 152, 167, 173, 188, 193, 199, 200, 206–209, 219, 223, 225, 228, 229, 231, 232, 236, 242, 270, 322, 358, 360
evaders 98, 207
Drought-starvation control 222, 223
Durilignosa communities 36
Dwarf shrubs formations 43, 47, 48, 52, 53, 121, 148, 150, 209, 235, 241
Dynamic conservation 11, 16, 112, 125, 132, 151, 251, 265, 267, 333–335, 395, *see* Conservation
Dynamic systems view 298, 304

Early Holocene 18, 101, 141, 187, 203, 261
Earth's biodiversity 358

East Brabant, the Netherlands 264
Ecodisciplinarity 374
Ecodisciplines 375
Ecodiversity 12, 15, 16, 24, 117–119, 127–132, 159–161, 167, 169, 170, 173, 174, 177, 178, 180–182, 250, 258, 259, 265–268, 270, 272–274, 282, 283, 286–290, 292, 293, 308, 309, 333, 340
  conservation 118, 129, 160, 175–177, 265, 271
Eco-economy 357, 362, 363
Ecological amplitude 35, 42, 47, 57
Ecological debacle 161
Ecological engineering 14, 373
Ecological footprints 331, 343
Ecological heterogeneity 120, 125, 128, 129, 147, 160, 173, 177, 180, 189, 192, 273, 308
Ecological integrity 238, 282, 365, 375
Ecological interaction systems 305
Ecological land appraisal 240
Ecological landscape restoration 283
Ecological management 231, 232, 240, 241, 243, 300, 372
Ecology 3, 5–10, 12, 14, 15, 18, 22–25, 27, 29, 35, 36, 62, 101, 110, 112, 117, 118, 127, 128, 130, 152, 160, 165, 169, 173, 176, 178, 181, 188, 197, 202, 205, 208, 221, 222, 235, 237, 238, 240, 246, 249–251, 253, 255, 256, 258, 264, 266, 273, 283–285, 290, 293, 298, 299, 304, 308, 320, 323, 324, 335, 340, 355, 356, 358, 368–381, 388, 389, 393–395, 403, 405
Economic rationality 263
Ecophysiology 304
Ecotope diversity index 177
Ecotopes 15, 23, 25, 120–123, 129, 150, 177, 178, 201, 238, 253–255, 272, 284, 285, 288–292, 298, 304, 305, 313, 314, 325, 331, 332, 336, 338, 340–343, 345, 378, 379
Eco-tourism 288, 338
Enkudotos 224
Entropy 11, 110, 111, 141, 169, 175, 176, 223, 239, 254, 292, 301, 313, 330–332, 334, 342–345
Environmental Impact Statements 155, 179
Environmental refugees 17, 322
Environmental revolution 13, 24, 117–119, 131, 149, 160, 182, 274, 275, 277
Epigenetic landscapes 378, 405
Epipaleolithic Natufians 18, 188, 200, 201, 205
EU agricultural policy 166
European landscape ecology 28
Eutrophication 140
Evapotransporation 139
Evolution 5, 8, 15, 17–19, 21, 24, 26, 28, 36, 41, 51, 52, 62, 83, 89, 90, 96–99, 101, 107, 108, 122, 135, 147, 150, 167, 169, 181, 187–189, 191–193, 195, 197, 199, 204, 240, 242, 250–252, 257, 259–262, 265, 267, 274, 275, 281–283, 285–287, 289, 293, 299–302, 305, 306, 313–315, 320, 323, 324, 326, 329–331, 336, 345, 346, 364, 385, 389, 390
Expert systems 109, 154, 258, 267
Explicate order 260, 310, 387
Extremadura 161

Fables 386, 389
Facultative root resprouters 97, 98, 206, 207
FAO 21, 165, 221
Fertile crescent 189, 192
Fire 2, 6, 9–12, 17–19, 41, 43, 46, 48, 51–54, 56, 57, 83, 84, 89, 90, 95–113, 126, 129, 141, 142, 153, 167, 171, 187–189, 191–193, 195–203, 205–210, 231, 242, 243, 261, 269–271, 286, 288–290, 333, 334, 340
Fire adaptation 96
Fire ecology studies 6, 10, 18, 101, 188, 197, 202, 205, 208
Fire management 11, 12, 108, 110, 112, 113, 142
First culture of the natural science 380
Flatlands 120
Floods 139, 140, 143, 242, 321, 334, 361, 362
Floristic diversity 8,61, 81, 126, 172, 206, 269
Forest ecosystem 263
Fossil age 346, 363
Fossil fuel 282, 313, 360–363
  age 282
Fossil-fuel powered technosphere landscapes 176, 331
Fossil-powered technosphere landscapes 282
Fractal dimension 258, 273, 311
Fractal geometry 129, 176, 258
Fragmentization 150
France 45, 83, 96, 100, 103, 104, 123, 171, 196, 203, 263, 268, 359
Fuel management 11, 12, 110, 112, 113, 142, 151
Fuzzy linear programming 110, 154
Fuzzy logic 24, 129, 176, 258, 259, 266, 273, 287
Fuzzy set systems 109
Fuzzy set theory 109, 154, 258
Fuzzy sets 109, 258

Gaia hypothesis 345
Galilee 1, 2, 8–11, 28, 35, 37–39, 41, 44–48, 56, 62–64, 75, 100, 101, 104–106, 139, 143, 162, 163, 174, 197, 202, 205, 208, 235, 240, 241, 297, 403
Game ranching 227, 233
Gamma ecotope diversity index 177
Gamma land-unit ecodiversity 129, 177
Garigue 8, 36, 42, 43, 45, 51, 53, 56, 65, 71, 241, 242, 244
General Circulation Models (GCMs) 136
General systems theory (GST) 251, 300, 303, 405
Generative order 24, 257, 258, 310, 311, 235, 387, 388
Genetic diversity 145–147, 168
Geo-ecology 376
Geophytes 67, 68, 70, 72, 77, 86–88, 98, 104, 107, 112, 123, 126, 148, 168, 171–173, 189, 190, 206–208, 269, 333
Gestalt systems 23, 25, 26, 250, 252, 253, 257, 298, 303, 305–307, 309, 311, 324–326
Global climate change 15, 16, 135, 136, 322, 358–360, 362, 368
Global ecological holon hierarchy 303
Gondwanan 9, 89, 90
Grand synthesis 24, 252, 260, 286, 301
Grand Transition 260, 274, 282, 320
Grasslands 8, 14, 19, 35, 38, 41, 44–50, 52–57, 62, 72, 79, 81, 82, 99, 101, 125, 126, 141, 145, 146, 150, 162, 169, 173, 191, 202, 208, 209, 221, 230, 240, 243, 244, 261, 268, 270, 302, 322, 323, 333, 334, 338, 358, 389
Grazing 9–11, 13, 15, 17, 20, 21, 41, 42, 48, 52–55, 57, 62–65, 69, 72–84, 90, 96, 98, 101–104, 109–112, 122, 124–127, 138, 141, 143, 145, 146, 148, 150, 152, 153, 167, 169, 171, 173, 174, 181, 187, 192, 200, 202, 203, 206, 208–210, 219, 221–224, 229–233, 235, 243, 268, 270, 281, 288, 289, 332–335, 340, 358, 404
  pressure 13, 53, 62, 74–79, 83, 102, 124, 126, 145, 146, 148, 173, 174, 206, 222
Greece 16, 83, 100, 138, 168, 189, 196
Green Belt Movement 365
Green book 15, 116, 118, 132, 155, 165, 178–180, 267, 271–273
Green fuels 363
Green revolution 236
Greenhouse effect 137, 138
Hemicryptophytes 70, 74, 98, 112, 172, 206, 207
Herbaceous plants 15, 18, 19, 48, 57, 66, 97, 98, 102–104, 106, 123, 126, 143, 167, 171, 173, 189, 190, 197–199, 205–210, 269
Holistic conception 5, 23, 25, 29, 297–300, 376
Holistic evolutionary conception 301
Holistic landscape
  conception 12, 29, 300, 315, 377
  conservation 15, 130, 180
  ecology 5, 298, 324
  planning 132, 177, 250, 267
  restoration 15
Holocene 18, 101, 141, 187, 192, 195, 203–205, 261
Homeorhesis 110, 127, 141, 169, 268
Homeorhetic flow equilibrium 15
Human ecology 25, 176, 237, 299, 208, 269–371, 373
Human land use 7, 15, 109, 118, 120, 127, 128, 169, 175, 176, 180, 258, 268, 273, 308, 332, 340, 361, 369
Human society 5, 24, 27, 182, 204, 239, 252, 260, 275, 282, 284, 293, 298, 300, 302, 306, 312, 315, 320, 324, 326, 344, 345, 356, 358, 363–365, 374, 379, 389, 391, 392, 405
Human society—nature symbiosis 24, 264, 281–283, 293, 298, 299, 313–315, 320, 345, 386
Human-influenced landscapes 258, 286, 298
Humanity 182, 191, 257, 275, 299, 346, 358, 363
Humankind 320, 356, 365
Human-landscape coevolution 201
Human-landscape interaction 13
Human-modified ecotopes 253, 289, 338
Human-modified landscapes 17, 118, 120, 167, 250, 252, 267, 288, 306, 313, 338
Human-perturbation dependent landscape 16, 127, 170
Hurricanes 137, 361, 362, 389
Hypercycles 195, 327

Iberian Peninsula 16, 162, 360
Ice-core studies 136, 138
Implicate order 257, 260, 310, 311, 387
Incidental domestication 197
Indian Forest Service 276
Indians 18, 54, 100, 188, 199–201, 208, 263, 289
Industrial revolution 119, 203, 261, 266, 282, 323, 325, 331
Inscape-landscape interaction 263
Intelligent Geographic Information Systems (IGIS) 118, 129, 131, 176

Intensive agro-industrial ecotopes 254, 291, 341
Inventive systems 287, 293, 336
Irano-Turanian territory 37, 46, 47, 83
Iron Age 122
ISIS 347, 348, 393
Israel 1, 2, 5–11, 13–15, 13–15, 17–20, 35–57, 61–65, 68–70, 72, 73, 79, 82–87, 89, 90, 98–100, 104, 105, 113, 120, 123, 125, 126, 138–140, 142, 143, 146–149, 151, 159, 162, 163, 165–168, 171–174, 180, 187–189, 191–193, 195, 196, 199, 203, 204, 206, 207, 235, 241, 242, 260, 268, 281, 289, 314, 389
Israel Soil Conversion Authority (ISCA) 2
Israelite tribes 2

Jamal cave 196
Judean Hills of Israel 15, 25, 38, 120, 121

Keoladeo National Park 270
Kilimanjaro 3, 21, 28, 215, 219, 220, 222
Kyoto protocol 346, 360, 361

Landscape
  architecture 14, 237
  concervation 15, 16, 23, 24, 130, 132, 178, 180, 249–251, 253, 265, 267, 272, 274
  contrast index 129, 177
  dynamics 140, 250, 331, 393, 404
  ecology (LE) 3, 5, 6, 12, 22, 27–29, 110, 117, 118, 128, 160, 169, 173, 178, 235, 237, 238, 246, 250, 251, 253, 255, 258, 263, 266, 284, 285, 298, 299, 308, 320, 323, 324, 355, 356, 375–377, 388, 393, 403, 405
    as interdisciplinary human ecosystem science 237
    role in development 235
  eveness index 129, 177
  evolution 24, 195, 204, 250, 259, 301, 329
  holarchy 96, 204, 306
  naturalness of 289
  patchiness index 129, 177
  research 5, 19, 298–300, 304, 308, 310, 311, 314, 356, 372, 376, 377, 382
  restoration 12–15, 24, 25, 28, 118, 281–284, 287, 288, 292, 314, 379, 380, 404
Laszlo's Grand Synthesis 329
Laszlo's Integral Theory of Everything 27–29, 389
Late Acheulian 197
Late Pleistocene 11, 18, 122, 198, 205, 206, 261
Lebanon 37, 42
Leipzigger Messe 275
Life-centered post-modern education 300
Livestock 20, 21, 23, 41, 55, 84, 108, 125, 142, 143, 149, 150, 154, 165, 173, 191, 192, 203, 220, 222–233, 236, 242–244, 246, 270, 322, 342
  production 22, 143, 144, 170, 226, 228, 232, 233, 246, 269
Long rains 219
Lower Galilee 8, 39, 41, 47, 48, 62–64, 75, 143, 174, 235, 240, 241
Lower Paleolithic 195, 196, 205

MAB 250, 337
Macroshift 299, 315, 320, 321, 350, 364, 365, 367
Man made deserts 220
Maquis 35, 36, 41–43, 45–48, 51, 53, 56, 62, 63, 65, 67, 69–71, 75, 81, 83–87, 90, 96–98, 100–105, 107, 110, 111, 113, 121, 123–125, 142, 145, 153, 161, 162, 165, 171–173, 191, 201, 207, 208, 241, 243, 244, 261, 271, 333, 334
Marine storms 140
Masai 20–22, 28, 215, 217–233, 236, 333
Masaitis 224
Maya 404
Mbugas 221, 230
Mediterranean Action Plan (MAP) 166
Mediterranean areas 8, 61, 73, 86, 88, 90
  comparative data 65
Mediterranean biota 108, 146, 195
Mediterranean climate 7–9, 36, 47, 56, 57, 97, 139, 188, 198, 206, 209
  Israel vs. California 37
Mediterranean ecosystems 6, 35, 41, 57, 97, 112, 154, 165, 168, 191
Mediterranean environment 36, 38, 53, 166
Mediterranean landscapes 2, 16, 18, 24, 67, 73, 96, 99, 101, 112, 127, 140, 141, 154, 159, 160, 165–169, 188, 189, 191–193, 203, 204, 209, 243, 271, 335, 369, 403
  current trends 160
Mediterranean oak savanna 35, 47, 48, 56, 143
Mediterranean uplands 10, 11, 14–16, 23, 96, 97, 99, 109, 127, 135, 140, 141, 152, 167, 178, 180, 181, 189, 191, 192, 195, 203, 237, 241, 244, 246, 268, 378
Mediterranean vegetation 36, 39, 41, 43, 57, 62, 97, 153, 191
Mediterranean-climate type ecosystems 7
Mediterranean-type ecosystems 7, 15, 16, 135, 160, 188
  human disturbance in 159

Mentation 259, 261, 275, 326
Meteorological events 136, 138, 139
Middle Paleolithic 99, 195, 197, 205
Middle Pleistocene 83, 99, 141, 192, 195, 196, 204, 205, 260
Middle-numbered systems 305, 309, 312
Milk production 20–22, 228–230
Modified landscapes 3, 84, 118, 120, 258, 306
Moisture gradient 9, 81, 90
Montados 16, 162, 173, 269
MOSES 23, 29, 314, 347, 348, 393
Mt. Carmel 8, 10, 11, 12, 17–19, 33, 61, 63, 65, 71, 85, 100, 103, 104, 124, 135, 147, 148, 171, 172, 187, 188, 192–199, 201–203, 205–209, 260
Mt. Gilboa 9, 63–65, 68, 70, 72, 81, 147, 148, 206
Mt. Kilimanjaro 3, 21, 28, 215
Mt. Meron 63, 65, 67–70, 81
Multifactorial landscape functions 11, 101
Multifunctional Gestalt systems 324, 325
Multifunctional landscapes (MFLs) 6, 25, 26, 154, 297–300, 302–312, 315, 324, 331, 338, 372, 383
Multiple-use benefits 244
Multiplier effect 201, 203
Multi-purpose reafforestation 23, 243

Natufians 18, 100, 188, 195, 198–202, 205
Natural biosphere landscape ecotopes 25, 119, 198, 204, 253, 288, 333, 336, 341, 348, 380
Natural disasters 139, 322, 361, 366
Natural ecotopes 150, 201, 289, 338
Natural landscapes, in Europe 269
Natural solar-powered biosphere 284, 327, 370
Nature islands 117, 119, 129, 143
Nature Protection Authorities 125
Nature-culture interaction systems 305
Negentropy 301, 331, 332, 334
Neolithic revolution 19, 119, 192, 201, 202
Neotechnological landscape degradation 117, 118, 120, 160, 180
Neo-technological landscape despoliation 143, 242
Neoth Kedumim 25
Neural networks 258
Nitratphytes 104
Nomadic Masai ecosystem 222
Nonequilibrium thermodynamics 303, 320, 332, 335
Noogenesis 240
Noosphere 240, 259, 286, 305, 312, 326, 374
Noosystem 374
North American Indians 200, 289

Obligatory root resprouters 97, 98, 206
Obligatory seed regenerators 97, 207
Omalos Plains 271, 272
Open Tabor oak woodland-grassland 9, 15, 72, 80, 208
Optimization 110, 154, 245, 246, 266, 378
Oregon 56
Organic farming ecotopes 290, 341
Organic farming systems 175, 253, 256, 290

Paleolithic layers 196, 198
Pampas 270, 333
Papua New Guinea 276
Pastoralism 225, 226, 228
Perturbation-dependent ecosystems 126, 151
Perturbation-dependent landscapes 110, 141, 170, 191, 267, 270, 361
Peruvian Andes 321
Photochemical oxidants 137, 152, 153, 162
Phytogeographic studies 36
Phytosociologic studies 36
Pleistocene 5, 9, 11, 17–19, 83, 89, 90, 96–99, 122, 141, 167, 187, 188, 191–193, 195–198, 204–106, 210, 260, 261
  fires 96
Post-industrial environmental revolution 118, 274
Post-industrial global information age 5, 282, 320, 356
Post-industrial Information Age 204, 205, 320, 393
Post-industrial neotechnical landscapes 256
Primitive cultures 101, 393
Primitive pastoral 21
Psychohygienic landscape functions 108
Puzzles 325, 384, 386
Pyrophytism 51
Pyrotechnics 200

Rainfall 15, 36, 38–40, 42, 49, 54, 56, 57, 62–65, 72, 73, 81, 86, 102, 138–145, 147, 148, 165, 219–221, 232
Range management 20, 57, 224, 226, 231, 232, 240
Range productivity 229, 233
Recovery processes 143
Red book 267
Red data books 178
Red List 15, 132, 155, 179, 180
Reflective consciousness 326
Regenerative ecotopes 25, 290, 341, 342
Regenerative systems 254, 256, 276, 277, 282, 292, 293, 313, 342
Regional planning 14
Remote sensing 131, 155, 165, 176, 267, 271

Rendzina soil 8, 13, 41, 44, 56, 62–64, 99, 102, 103, 121–123, 196, 199, 208
Research 2, 3, 5, 6, 8–10, 12–15, 19–21, 23, 25–29, 86, 96, 97, 113, 127, 135, 137, 138, 140, 147, 152–155, 166, 168, 200, 204, 221, 226, 231–233, 236, 237, 241, 245, 249, 266, 271, 276, 277, 287, 291, 298–301, 304, 308, 310–312, 314, 315, 324, 325, 341, 342, 345, 356, 362, 367, 370, 372, 375–386, 389, 391, 393
Resilience project 372
Restoration 3, 5, 6, 10, 12–16, 23–25, 27, 28, 110, 113, 117–119, 127–132, 135, 149, 151–155, 177, 180, 204, 250, 266, 267, 270, 273, 275–277, 281–293, 298, 314, 324, 335, 340, 342, 349, 355, 358, 368, 369, 372, 375, 378–381, 386, 391, 392, 394, 403, 404
  ecology (RE) 3, 5, 6, 12, 14, 27, 28, 117, 118, 128, 129, 283–285, 293, 355, 356, 375, 379, 380, 395, 403, 405
  projects 14, 25, 28, 152, 288, 293, 379, 391
Restorative experience of nature 339, 392
Revegetation 14, 23, 243, 246
Rocky Mountains 56
Rugged land 162
Rural ecotopes 292, 343
Russia 323

Samaria Gorge Park 271, 272
Samaria mountains 72, 147, 148
Satellite images 129, 176
Savanna 8, 10, 16, 20, 28, 35, 47, 48, 56, 62, 99, 125, 143, 167, 208, 220, 222, 226, 231, 232, 269, 270, 289, 333
Scandinavia 323
Scenic despoliation 150, 181
Scientific revolution 29, 300, 301, 324, 356, 384–386
Sclerophyll 9, 36, 37, 42, 56, 88, 89, 96, 98–102, 111, 121, 123, 148, 171, 188, 199, 200, 202, 205
SCOPE 237
Second humanistic culture 380
Self-organization 12, 18, 25, 26, 112, 169, 282, 300, 302, 306, 327, 329, 330, 332, 334, 346, 388, 390
Self-organizing biosphere landscape 26, 319
Self-organizing universe 28, 252, 259, 301, 314, 327
Self-reflective mentation 326
Self-transcendence 28, 306
Semi-natural ecotopes 150, 201, 289, 338
Seral bartha dwarfshrub 35
Shannon-Wiener index 67, 68, 77, 126, 177
Shock stimulus 202
Short rains 219
Shrub formation 42, 45, 46, 53, 171
Shrublands 8–10, 16, 61, 62, 65, 67–76, 78, 79, 81, 84–87, 89, 90, 101, 108, 111, 121, 124–127, 141, 145, 150, 169, 171, 173, 174, 188, 243, 244, 261, 268, 270, 289, 290, 302, 303, 333, 334, 404
Site comparisons 67
Social-ecological systems 372
Socio-ecological landscape functions 108, 242
Socio-economic development 240
Socio-economic functions 108, 180, 303
Soil stability 101, 102
Soils 2, 3, 6, 8–11, 13, 14, 16, 20, 35, 38, 41–47, 50, 51, 56, 57, 62–65, 81, 83, 84, 86, 88–90, 98–104, 108, 109, 117, 118, 121–123, 125, 129, 139–143, 145, 148–150, 152, 161, 162, 167, 171, 173, 177, 189, 195–197, 199, 203, 206–209, 217, 220, 221, 224, 227, 230–232, 235, 242, 243, 269, 287, 290, 291, 293, 313, 322, 323, 339, 341, 342, 360, 379, 392, 403, 404
Solar age 346, 363
Solar energy 25, 110, 175, 176, 193, 201, 238, 253, 254, 261, 288, 291, 292, 313, 320, 327, 328, 331, 340–342, 346, 370, 391
Solar-powered biosphere landscapes 26, 175, 282, 289, 292, 302, 323, 331, 338, 343
South Africa 8, 9, 66, 85, 87
South African fynbos 90
Spain 17, 48, 96, 161, 170, 269
Species diversity 9, 62, 68, 82, 84, 89, 90, 108, 110, 120, 132, 145, 147, 173, 189, 243, 268, 331, 358
  index 129, 177
Species richness 9, 66, 70, 74–77, 79, 81, 82, 84, 90, 120, 123–126, 142, 143, 145, 148, 171–174
Steppe 47, 99, 119, 146, 168, 177, 191, 202, 219–221, 227, 230
Steppe-like grasslands 54
Stratal diversity 62, 67
Structural diversity 9, 9, 61, 81, 110, 111, 123, 126, 140, 142, 167, 171–173, 268, 269, 333, 334
Structural vegetation diversity 123, 171, 180, 273, 308

Suburban ecotopes 292, 343
Suburban landscapes 311
Sustainability revolution 315, 320, 323, 343, 346, 349, 350, 356, 362–367, 373, 382, 391, 393, 394
Sustainable development 20, 22, 23, 26, 155, 266, 272, 298, 314, 339, 345, 346, 348–350, 363, 374, 392, 393
Swiss Alps 321
Syndynamics 35, 48, 57
Syngenetics 48
Systems learning 384
Systems network thinking 384

Tabun cave 100, 196, 202
Tanzania 3, 20, 22, 28, 219, 220, 224, 225
  Masailand 20, 21, 217–220, 226, 232, 270
Techno-ecosystems 238, 239, 246, 254, 284, 313, 331, 336, 343, 345
Technophilia 339
Technosphere 23, 26, 27, 118, 205, 238, 239, 254, 255, 285, 292, 313, 315, 320, 321, 342–345, 349, 394
  ecotopes 12, 254, 284, 292, 314, 370
  landscapes 27, 119, 176, 182, 193, 204, 253, 254, 261, 275, 282, 287, 291–293, 313–315, 320, 321, 336, 337, 341–346, 348, 358, 391, 394
Terra Rossa soil 8, 45, 56, 62, 63, 99
Terrestrial landscape 137, 167, 284
Texas 56
THE paradigm *see* Total Human Ecosystem
Theory beliefs 383
Third alternative culture 380
Thornbush deserts 223
Threatened landscapes 155, 179, 180
Throughput systems 313, 342
Tidal surges 140
Topography 8, 41, 56, 84, 129, 169, 177, 271
Topophilia 262
Total Human Ecosystem (THE) 7, 19–20, 23, 25–28, 176, 183, 238–240, 246, 250, 253, 266, 275, 277, 282, 284, 293, 298, 305, 306, 308, 311–315, 319, 320, 324, 326, 327, 329–331, 336, 343–347, 350, 372, 373, 375, 387, 390, 393, 394
  paradigm 19, 20, 22, 28, 29
Total Industrial Landscape 193, 203–205, 391
Total landscape ecodiversity 16, 118, 119, 127, 132, 160, 169, 174, 178, 180, 250, 272, 282, 283, 286, 293, 308, 309, 314, 340, 344
Total non-inference 16
Tourism 17, 20, 22, 108, 141, 160, 161, 227, 233, 242, 266, 288, 338
Transdisciplinarity 23, 27, 298, 299, 301, 303, 356, 367, 374, 375, 378, 380–385
Transdisciplinary landscape sciences (TLS) 3, 19, 28, 29, 356, 368, 369, 372, 375, 377, 381, 391, 394, 395, 382, 389
Tree formation 42
Trypanocarpy 84, 98, 208, 209
Tuscan landscape 164, 166, 168

UNDEP 237
UNEP Mediterranean task force 139
Unique Gestalt systems 303, 306, 326
United Nations Environmental Program (UNEP) 139, 160, 166
United States 82, 83, 88, 119, 237, 265, 289, 360, 361, 372, 377, 379
Urban crisis syndrome 239
Urban ecology 374
Urban-industrial ecotopes 292, 343, 370
Urban-industrial technosphere landscapes 26, 193, 204, 254, 261, 343
Urban-industrialized ecosystems 370
Urbanization 42, 150, 292, 340, 342, 344

Vegetation
  comparison between California and Israel 6, 35–37, 41–43, 48
  disturbance 62
  engineering 13
  samples 62, 90
  types 6, 35–39, 41–43, 46, 48, 56, 57, 73, 101, 126, 127, 129, 147, 188, 222, 289, 290
Venezuela 321
Viable systems management 29, 393

Warriors 217, 226, 233
Water 10, 16, 21, 23, 25, 42, 98, 102, 103, 108, 118, 130, 137, 140, 142, 147–151, 161, 193, 200, 202, 206, 209, 219, 220, 222–225, 227–233, 236, 242, 246, 260, 271, 284, 290, 291, 297, 313, 321, 323, 324, 339, 341–343, 357, 359–364, 392
  stagnation 140
Wildfires 10–12, 16, 18, 19, 57, 96, 97, 99, 101, 103, 112, 113, 126, 142, 150, 161, 165, 173, 193, 197, 200, 205, 207, 208, 262, 335, 358, 360
Wildlands 41, 57, 193, 242, 404

Wildlife 10, 20–23, 112, 126, 151, 206, 209, 221, 222, 226–228, 231–233, 244, 246, 265, 288, 291, 313, 323, 333, 338, 342, 359, 377
Woodland-grassland 47, 72
Woodlands 8, 9, 15, 16, 19, 41, 47, 61, 62, 64, 65, 68–70, 72–88, 90, 99, 101, 104, 111, 124–127, 141, 143, 145, 146, 150, 161, 165, 169, 170, 173, 174, 199, 200, 202, 207, 208, 220, 221, 243, 244, 261, 268–270, 289, 333, 338, 379
Woody plants 8, 10, 11, 51, 56, 73, 74, 77, 88, 103, 104, 142, 143, 147, 148, 168, 188–190, 197, 198, 205, 209
World Heritage Sites 131, 175, 288, 338

Yellowstone National Park fire 129, 270

Zacca series 46
Zoological studies 62, 81, 124, 172

Landscape Series

1. H. Palang and G. Fry (eds.): *Landscape Interfaces.* Cultural Heritage in Changing Landscapes. 2003 ISBN 1-4020-1437-6
2. I. Vanslembrouck and G. Van Huylenbroeck: *Landscape Amenities.* Economic Assessment of Agricultural Landscape. 2005 ISBN 1-4020-3134-3
3. F. Almo: *Principles and Methods in Landscape Ecology.* Towards a Science of Landscape. 2005 ISBN 1-4020-3327-3 Pb: ISBN 1-4020-3328-1
4. D.G. Green, N. Klomp, G. Rimmington and S. Sadedin: *Complexity in Landscape Ecology.* 2006 ISBN 1-4020-4285-X
5. T.S. Terkenli and A.-M. d'Hauteserre (eds.): *Landscapes of a New Cultural Economy of Space.* 2006 ISBN 1-4020-4095-4
6. Z. Naveh (ed.): *Transdisciplinary Challenges for Landscape Ecology and Restoration Ecology.* Anthology of Studies on Mediterranean and Global Issues. 2007 ISBN 1-4020-4420-8
7. H. Palang, H. Sooväli, A. Printsmann (eds.): *Seasonal Landscapes.* 2007 ISBN 1-4020-4982-X
8. F. Kienast, S. Ghosh, O. Wildi (eds.): *A Changing World.* Challenges for Landscape Research. 2007 ISBN 1-4020-4434-8